Encyclopedia of
Genetics
Revised Edition

Encyclopedia of
Genetics
Revised Edition

Volume 2
Hybridomas and Monoclonal Antibodies –
XYY Syndrome
Appendices
Indexes

Editor, Revised Edition
Bryan D. Ness, Ph.D.
Pacific Union College
Department of Biology

Editor, First Edition
Jeffrey A. Knight, Ph.D.
Mount Holyoke College
Department of Biological Sciences

Salem Press, Inc.
Pasadena, California
Hackensack, New Jersey

Editor in Chief: Dawn P. Dawson

Managing Editor: Christina J. Moose   *Acquisitions Editor:* Mark Rehn
Developmental Editor: Tracy Irons-Georges   *Photograph Editor:* Philip Bader
Research Supervisor: Jeffry Jensen   *Page Design:* James Hutson
Assistant Editors: Andrea E. Miller,   *Layout:* William Zimmerman
Desirée Dreeuws

Library of Congress Cataloging-in-Publication Data

Encyclopedia of genetics / editor, revised edition, Bryan D. Ness ;
editor, first edition, Jeffrey A. Knight. — Rev. ed.
    p.   ; cm.
Includes bibliographical references and index.
    ISBN 1-58765-149-1 (set : alk. paper) — ISBN 1-58765-150-5 (v.1 : alk. paper) —
ISBN 1-58765-151-3 (v.2 : alk. paper) —
    1. Genetics—Encyclopedias.
    [DNLM: 1. Genetics—Encyclopedias—English. QH 427 E56 2004] I. Ness, Bryan D.
QH427.E53 2004
576.5′03—dc22

2003026056

First Printing

PRINTED IN THE UNITED STATES OF AMERICA

Contents

Alphabetical List of Contents. xxxi

Hybridomas and Monoclonal
    Antibodies. 441
Hypercholesterolemia 445

Icelandic Genetic Database 447
Immunogenetics. 449
In Vitro Fertilization and Embryo
    Transfer 454
Inborn Errors of Metabolism 458
Inbreeding and Assortative Mating 461
Incomplete Dominance 465
Infertility. 468
Insurance. 471
Intelligence 474

Klinefelter Syndrome 479
Knockout Genetics and Knockout
    Mice 481

Lactose Intolerance 484
Lamarckianism. 485
Lateral Gene Transfer 489
Linkage Maps 491

Mendelian Genetics 494
Metafemales 499
Miscegenation and
    Antimiscegenation Laws 501
Mitochondrial Diseases 503
Mitochondrial Genes 505
Mitosis and Meiosis 509
Model Organism:
    Arabidopsis thaliana 513
Model Organism:
    Caenorhabditis elegans 516
Model Organism:
    Chlamydomonas reinhardtii. 520
Model Organism:
    Drosophila melanogaster 522
Model Organism:
    Escherichia coli 527
Model Organism:
    Mus musculus 533

Model Organism:
    Neurospora crassa. 536
Model Organism:
    Saccharomyces cerevisiae. 539
Model Organism:
    Xenopus laevis 542
Model Organisms 545
Molecular Clock Hypothesis 547
Molecular Genetics 549
Monohybrid Inheritance 555
Multiple Alleles 559
Mutation and Mutagenesis 561

Natural Selection 568
Neural Tube Defects. 572
Noncoding RNA Molecules 575
Nondisjunction and
    Aneuploidy 579

Oncogenes. 583
One Gene-One Enzyme
    Hypothesis. 586
Organ Transplants and HLA
    Genes 588

Parthenogenesis 592
Patents on Life-Forms 594
Paternity Tests 596
Pedigree Analysis 599
Penetrance 602
Phenylketonuria (PKU) 604
Plasmids 606
Polygenic Inheritance 609
Polymerase Chain Reaction 611
Polyploidy 613
Population Genetics 617
Prader-Willi and Angelman
    Syndromes. 623
Prenatal Diagnosis 626
Prion Diseases: Kuru and
    Creutzfeldt-Jakob Syndrome 631
Protein Structure 634
Protein Synthesis. 638
Proteomics 643
Pseudogenes 646

Pseudohermaphrodites 648
Punctuated Equilibrium. 650

Quantitative Inheritance 654

Race 658
Repetitive DNA 664
Restriction Enzymes 667
Reverse Transcriptase 670
RFLP Analysis 672
RNA Isolation 674
RNA Structure and Function 676
RNA Transcription and mRNA
    Processing 681
RNA World 686

Shotgun Cloning. 691
Sickle-Cell Disease 692
Signal Transduction 696
Smallpox 700
Sociobiology 704
Speciation 708
Stem Cells 710
Sterilization Laws 715
Steroid Hormones 717
Swine Flu 720
Synthetic Antibodies. 723
Synthetic Genes 725

Tay-Sachs Disease 727
Telomeres 728

Testicular Feminization Syndrome 731
Thalidomide and Other Teratogens . . . 733
Totipotency 736
Transgenic Organisms. 739
Transposable Elements 742
Tumor-Suppressor Genes 746
Turner Syndrome 748
Twin Studies 750

Viral Genetics 754
Viroids and Virusoids 756

X Chromosome Inactivation 759
Xenotransplants 761
XYY Syndrome 764

Appendices
Biographical Dictionary of
    Important Geneticists. 767
Nobel Prizes for Discoveries in
    Genetics 780
Time Line of Major Developments
    in Genetics 784
Glossary 804
Bibliography 832
Web Sites 859

Indexes
Category Index III
Personages Index. VII
Subject Index XI

Alphabetical List of Contents

Volume 1

Aggression 1
Aging 3
Albinism 9
Alcoholism. 11
Allergies 13
Altruism 16
Alzheimer's Disease 19
Amniocentesis and Chorionic
    Villus Sampling. 23
Ancient DNA 27
Animal Cloning 31
Anthrax 35
Antibodies 38
Antisense RNA 42
Archaea 45
Artificial Selection 48
Autoimmune Disorders. 51

Bacterial Genetics and Cell
    Structure 54
Bacterial Resistance and Super
    Bacteria. 61
Behavior 65
Biochemical Mutations 70
Bioethics. 73
Biofertilizers. 77
Bioinformatics. 79
Biological Clocks 83
Biological Determinism. 86
Biological Weapons 88
Biopesticides 92
Biopharmaceuticals. 96
Blotting: Southern, Northern,
    and Western 98
Breast Cancer 101
Burkitt's Lymphoma. 106

Cancer 109
cDNA Libraries 115
Cell Culture: Animal Cells. 117
Cell Culture: Plant Cells. 120
The Cell Cycle 122
Cell Division 125

Central Dogma of Molecular
    Biology 128
Chemical Mutagens 131
Chloroplast Genes 133
Cholera. 137
Chromatin Packaging 140
Chromosome Mutation 144
Chromosome Structure 147
Chromosome Theory of Heredity 152
Chromosome Walking and
    Jumping 158
Classical Transmission Genetics. 160
Cloning. 166
Cloning: Ethical Issues 170
Cloning Vectors 174
Color Blindness 179
Complementation Testing. 181
Complete Dominance 184
Congenital Defects. 187
Consanguinity and Genetic Disease . . . 191
Criminality 193
Cystic Fibrosis 195
Cytokinesis 198

Developmental Genetics 201
Diabetes 207
Dihybrid Inheritance 210
Diphtheria 214
DNA Fingerprinting 216
DNA Isolation 220
DNA Repair 223
DNA Replication. 227
DNA Sequencing Technology. 233
DNA Structure and Function 237
Down Syndrome 244
Dwarfism 248

Emerging Diseases. 252
Epistasis 255
Eugenics 259
Eugenics: Nazi Germany 264
Evolutionary Biology 267
Extrachromosomal Inheritance. 274

Forensic Genetics 279
Fragile X Syndrome 282

Gel Electrophoresis 285
Gender Identity 287
Gene Families 289
Gene Regulation: Bacteria 291
Gene Regulation: Eukaryotes 295
Gene Regulation: *Lac* Operon 298
Gene Regulation: Viruses 301
Gene Therapy 304
Gene Therapy: Ethical and
    Economic Issues 309
Genetic Code 313
Genetic Code, Cracking of 319
Genetic Counseling 321
Genetic Engineering 326
Genetic Engineering: Agricultural
    Applications 332
Genetic Engineering: Historical
    Development 335
Genetic Engineering: Industrial
    Applications 339
Genetic Engineering: Medical
    Applications 343
Genetic Engineering: Risks 347
Genetic Engineering: Social
    and Ethical Issues 351

Genetic Load 354
Genetic Screening 357
Genetic Testing 360
Genetic Testing: Ethical and
    Economic Issues 364
Genetically Modified (GM)
    Foods 366
Genetics, Historical Development
    of 370
Genetics in Television and Films 376
Genome Size 378
Genomic Libraries 380
Genomics 384

Hardy-Weinberg Law 389
Heart Disease 392
Hemophilia 396
Hereditary Diseases 399
Heredity and Environment 406
Hermaphrodites 411
High-Yield Crops 413
Homeotic Genes 416
Homosexuality 419
Human Genetics 421
Human Genome Project 428
Human Growth Hormone 432
Huntington's Disease 434
Hybridization and Introgression 437

Volume 2

Hybridomas and Monoclonal
    Antibodies 441
Hypercholesterolemia 445

Icelandic Genetic Database 447
Immunogenetics 449
In Vitro Fertilization and Embryo
    Transfer 454
Inborn Errors of Metabolism 458
Inbreeding and Assortative Mating . . . 461
Incomplete Dominance 465
Infertility 468
Insurance 471
Intelligence 474

Klinefelter Syndrome 479
Knockout Genetics and Knockout
    Mice 481

Lactose Intolerance 484
Lamarckianism 485
Lateral Gene Transfer 489
Linkage Maps 491

Mendelian Genetics 494
Metafemales 499
Miscegenation and
    Antimiscegenation Laws 501
Mitochondrial Diseases 503
Mitochondrial Genes 505
Mitosis and Meiosis 509
Model Organism:
    Arabidopsis thaliana 513
Model Organism:
    Caenorhabditis elegans 516
Model Organism:
    Chlamydomonas reinhardtii 520

Model Organism:
   Drosophila melanogaster 522
Model Organism:
   Escherichia coli 527
Model Organism:
   Mus musculus 533
Model Organism:
   Neurospora crassa. 536
Model Organism:
   Saccharomyces cerevisiae. 539
Model Organism:
   Xenopus laevis 542
Model Organisms 545
Molecular Clock Hypothesis 547
Molecular Genetics 549
Monohybrid Inheritance 555
Multiple Alleles 559
Mutation and Mutagenesis 561

Natural Selection 568
Neural Tube Defects. 572
Noncoding RNA Molecules 575
Nondisjunction and
   Aneuploidy 579

Oncogenes. 583
One Gene-One Enzyme
   Hypothesis. 586
Organ Transplants and HLA
   Genes 588

Parthenogenesis 592
Patents on Life-Forms 594
Paternity Tests 596
Pedigree Analysis 599
Penetrance. 602
Phenylketonuria (PKU) 604
Plasmids 606
Polygenic Inheritance 609
Polymerase Chain Reaction 611
Polyploidy 613
Population Genetics. 617
Prader-Willi and Angelman
   Syndromes. 623
Prenatal Diagnosis 626
Prion Diseases: Kuru and
   Creutzfeldt-Jakob Syndrome . . . 631
Protein Structure 634
Protein Synthesis. 638
Proteomics 643

Pseudogenes 646
Pseudohermaphrodites 648
Punctuated Equilibrium. 650

Quantitative Inheritance 654

Race 658
Repetitive DNA 664
Restriction Enzymes 667
Reverse Transcriptase 670
RFLP Analysis 672
RNA Isolation 674
RNA Structure and Function 676
RNA Transcription and mRNA
   Processing 681
RNA World 686

Shotgun Cloning. 691
Sickle-Cell Disease 692
Signal Transduction 696
Smallpox 700
Sociobiology 704
Speciation 708
Stem Cells 710
Sterilization Laws 715
Steroid Hormones 717
Swine Flu 720
Synthetic Antibodies. 723
Synthetic Genes 725

Tay-Sachs Disease 727
Telomeres 728
Testicular Feminization
   Syndrome 731
Thalidomide and Other
   Teratogens 733
Totipotency 736
Transgenic Organisms. 739
Transposable Elements 742
Tumor-Suppressor Genes 746
Turner Syndrome 748
Twin Studies 750

Viral Genetics 754
Viroids and Virusoids 756

X Chromosome Inactivation 759
Xenotransplants 761
XYY Syndrome 764

Appendices

Biographical Dictionary of
  Important Geneticists. 767
Nobel Prizes for Discoveries in
  Genetics 780
Time Line of Major Developments
  in Genetics 784
Glossary 804

Bibliography 832
Web Sites 859

Indexes

Category Index III
Personages Index. VII
Subject Index XI

Encyclopedia of
Genetics
Revised Edition

Hybridomas and Monoclonal Antibodies

Field of study: Immunogenetics

Significance: *In 1975, Georges Köhler and Cesar Milstein reported that fusion of spleen cells from an immunized mouse with a cultured plasmacytoma cell line resulted in the formation of hybrid cells called hybridomas that secreted the antibody molecules that the spleen cells had been stimulated to produce. Clones of hybrid cells producing antibodies with a desired specificity are called monoclonal antibodies and can be used as a reliable and continuous source of that antibody. These well-defined and specific antibody reagents have a wide range of biological uses, including basic research, industrial applications, and medical diagnostics and therapeutics.*

Key terms

ANTIBODY: a protein produced by plasma cells (matured B cells) that binds specifically to an antigen

ANTIGEN: a foreign molecule or microorganism that stimulates an immune response in an animal

ANTISERA: a complex mixture of heterogeneous antibodies that react with various parts of an antigen; each type of antibody protein in the mixture is made by a different type (clone) of plasma cell

PLASMACYTOMA: a plasma cell tumor that can be grown continuously in a culture

A New Way to Make Antibodies

Because of their specificity, antisera have long been used as biological reagents to detect or isolate molecules of interest. They have been useful for biological research, industrial separation applications, clinical assays, and immunotherapy. One disadvantage of conventional antisera is that they are heterogeneous collections of antibodies against a variety of antigenic determinants present on the antigen that has elicited the antibody response. In an animal from which antisera is collected, the mixture of antibodies changes with time so that the types and relative amounts of particular antibodies are different in samples taken at different times. This variation makes standardization of reagents difficult and means that the amount of characterized and standardized antisera is limited to that available from a particular sample.

The publication of a report by Georges Köhler and Cesar Milstein in the journal *Nature* in 1975 describing production of the first monoclonal antibodies provided a method to produce continuous supplies of antibodies against specific antigenic determinants. Milstein's laboratory had been conducting basic research on the synthesis of immunoglobulin chains in plasma cells, mature B cells that produce large amounts of a single type of immunoglobulin. As a model system, they were using rat and mouse plasma cell tumors (plasmacytomas). Prior to 1975, Köhler and Milstein had completed a series of experiments in which they had fused rat and mouse plasmacytomas and determined that the light and heavy chains from the two species associate randomly to form the various possible combinations. In these experiments they used mutant plasmacytoma lines that would not grow in selective culture media, while the hybrid cells complemented each others' deficiencies and multiplied in culture.

After immunizing mice with sheep red blood cells (SRBC), Köhler and Milstein removed the spleen cells from the immunized mice and fused them with a mouse plasmacytoma cell line. Again, the selective media did not allow unfused plasmacytomas to grow, and unfused spleen cells lasted for only a short time in culture so that only hybrids between plasmacytoma cells and spleen cells grew as hybrids. These hybrid plasmacytomas have come to be called hybridomas.

Shortly after the two types of cells are fused by incubation with a fusing agent such as polyethylene glycol, they are plated out into a series of hundreds of small wells so that only a limited number of hybrids grow out together in the same well. Depending on the frequency of hybrids and the number of wells used, it is possible to distribute the cells so that each hybrid cell grows up in a separate cell culture well.

On the basis of the number of spleen cells that would normally be making antibodies against SRBC after mice have been immunized with them, the investigators expected that one well in about 100,000 or more might have a clone of hybrid cells making antibody that reacted against this antigen. The supernatants (liquid overlying settled material) from hundreds of wells were tested, and the large majority were found to react with the immunizing antigen. Further work with other antigens confirmed that a significant fraction of hybrid cells formed with spleen cells of immunized mice produce antibodies reacting with the antigen recently injected into the mouse. The production of homogeneous antibodies from clones of hybrid cells thus became a practical way to obtain reliable supplies of well-defined immunological reagents.

The antibodies can be collected from the media in which the cells are grown, or the hybridomas can be injected into mice so that larger concentrations of monoclonal antibodies can be collected from fluid that collects in the abdominal cavity of the animals.

Specific Antibodies Against Antigen Mixtures

One advantage of separating an animal's antibody response into individual antibody components by hybridization and separation of cells derived from each fusion event is that antibodies that react with individual antigenic components can be isolated even when the mouse is immunized with a complex mixture of antigens. For example, human tumor cells injected into a mouse stimulate the production of many different types of antibodies. A few of these antibodies may react specifically with tumor cells or specific types of human cells, but, in a conventional antisera, these antibodies would be mixed with other antibodies that react with any human cell and would not be easily separated from them. If the tumor cells are injected and hybridomas are made and screened to detect antibodies that react with tumor cells and not with most normal cells, it is possible to isolate antibodies that are useful for detection and characterization of specific types of tumor cells. Similar procedures can also be used to make antibodies against a single protein after the mouse has been immunized with this protein

included in a complex mixture of other biological molecules such as a cell extract.

Following the first report of monoclonal antibodies, biologists began to realize the implications of being able to produce a continuous supply of antibodies with selected and well-defined reactivity patterns. There was discussion of "magic bullets" that would react specifically with and carry specific cytotoxic agents to tumor cells without adverse effects on normal cells. Biologists working in various experimental systems realized how specific and reliable sources of antibody reagent might contribute to their investigations, and entrepreneurs started several biotechnology companies to develop and apply monoclonal antibody methods. This initial enthusiasm was quickly moderated as some of the technical difficulties involved in production and use of these antibodies became apparent; with time, however, many of the projected advantages of these reagents have become a reality.

Monoclonal Reagents

A survey of catalogs of companies selling products used in biological research confirms that many of the conventional antisera commonly used as research reagents have been replaced with monoclonal antibodies. These products are advantageous to the suppliers, being produced in constant supply with standardized protocols from hybrid cells, and the users, who receive well-characterized reagents with known specificities free of other antibodies that could produce extraneous and unexpected reactions when used in some assay conditions. Antibodies are available against a wide range of biomolecules reflecting current trends in research; examples include antibodies against cytoskeletal proteins, protein kinases, and oncogene proteins, gene products involved in the transition of normal cells to cancer cells.

Immunologists were among the first to take advantage of monoclonal antibody technology. They were able to use them to "trap" the spleen cells making antibodies against small, well-defined molecules called haptens and to then characterize the antibodies produced by the hybridomas. This enabled them to define classes of antibodies made against specific antigenic determinants and to derive information about the structure of the antibody-binding sites and how they are related to the determinants they bind. Other investigators produced antibodies that reacted specifically against subsets of lymphocytes playing specific roles in the immune responses of animals and humans. These reagents were then used to study the roles that these subsets of immune cells play in responses to various types of antigens.

Antibodies that react with specific types of immune cells have also been used to modulate the immune response. For example, antibodies that react with lymphocytes that would normally react with a transplanted tissue or organ can be used to deplete these cells from the circulation and thus reduce their response against the transplanted tissue.

Monoclonal Antibodies as Diagnostic Reagents

Monoclonal antibodies have been used as both in vitro and in vivo diagnostic reagents. By the 1980's, many clinical diagnostic tests such as assays for hormone or drug levels relied upon antisera as detecting reagents. Antibodies reacting with specific types of bacteria and viruses have also been used to classify infections so that the most effective treatment can be determined. In the case of production of antibodies for typing microorganisms, it has frequently been easier to make type-specific monoclonal antibodies than it had been to produce antisera that could be used to identify the same microorganisms.

Companies supplying these diagnostic reagents have gradually switched over to the use of monoclonal antibody products, thus facilitating the standardization of the reactions and the protocols used for the clinical tests. The reproducibility of the assays and the reagents has made it possible to introduce some of these tests that depend upon measurement of concentrations of substances in urine as kits that can be used by consumers in their own homes. Kits have been made available for testing glucose levels of diabetics, for pregnancy, and for the presence of certain drugs.

Although the much-hoped-for "magic bullet" that would eradicate cancer has not been found, there are several antibodies in use for tumor detection and for experimental forms of cancer therapy. Monoclonal antibodies that react selectively with cancer cells but not normal cells can be used to deliver cytotoxic molecules to the cancer cells. Monoclonal reagents are also used to deliver isotopes that can be used to detect the presence of small concentrations of cancer cells that would not normally be found until the tumors grew to a larger size.

Human Monoclonal Antibodies

The majority of monoclonal antibodies made against human antigens were mouse antibodies derived from the spleens of immunized mice. When administered to humans in clinical settings, the disadvantage of the animal origin of the antibodies soon became apparent. The human immune system recognized the mouse antibodies as foreign proteins and produced an immune response against them, limiting their usefulness. Even when the initial response to an antibody's administration was positive, the immune reaction against the foreign protein quickly limited its effectiveness. In an attempt to avoid this problem, human monoclonal antibodies have been developed using several methods. The first is the hybridization of human lymphocytes stimulated to produce antibodies against the antigen of interest with mouse plasmacytomas or later with human plasmacytoma cell lines. This method has been used successfully, although it is limited by the ability to obtain human B cells or plasma cells stimulated against specific antigens because it is not possible to give an individual a series of immunizations and then remove stimulated cells from the spleen. Limited success has resulted from the fusion of circulating lymphocytes from immunized individuals or fusion of lymphocytes that have been stimulated by the antigen in cell cultures. Investigators have reported some success in making antitumor monoclonal antibodies by fusing lymph node cells from cancer patients with plasmacytoma cell lines and screening for antibodies that react with the tumor cells.

There has also been some success at "humanizing" mouse antibodies using molecular genetic techniques. In this process, the portion of the genes that make the variable regions of the mouse antibody protein that reacts with a particular antigen is spliced in to replace the variable region of a human antibody molecule being produced by a cultured human cell or human hybridoma. What is produced is a human antibody protein that has the binding specificity of the original mouse monoclonal antibody. When such antibodies are used for human therapy, the reaction against the injected protein is reduced compared to the administration of the whole mouse antibody molecules.

Another application of antibody engineering is the production of bispecific antibodies. This has been accomplished by fusing two hybridomas making antibodies against two different antigens. The result is an antibody that contains two types of binding sites and thus binds and cross-links two antigens, bringing them into close proximity to each other.

Recombinant Antibodies

Advances in molecular genetic techniques and in the characterization of the genes for the variable and constant regions of antibody molecules have made it possible to produce new forms of monoclonal antibodies. The generation of these recombinant antibodies is not dependent upon the immunizing of animals but on the utilization of combinations of antibody genes generated using the in vitro techniques of genetic engineering. Geneticists discovered that genes inserted into the genes for fibers expressed on the surface of bacterial viruses called bacteriophages are expressed and detectable as new protein sequences on the surface of the bacteriophage. Investigators working with antibody genes found that they could produce populations of bacteriophage expressing combinations of antibody-variable genes. Molecular genetic methods have made it possible to generate populations of bacteriophage expressing different combinations of antibody-variable genes with frequencies approaching the number present in an individual mouse or human immune system. The population of bacteriophage can be screened for binding to an antigen of interest, and the bacteriophage expressing combinations of variable regions binding to the antigen can be multiplied and then used to generate recombinant antibody molecules in culture.

Researchers have also experimented with introducing antibody genes into plants, resulting in plants that produce quantities of the specific antibodies. Hybridomas or bacteriophages expressing specific antibodies of interest may be a potential source of the antibody gene sequences introduced into these plant antibody factories.

—*Roger H. Kennett*

See also: Allergies; Antibodies; Autoimmune Disorders; Burkitt's Lymphoma; Cancer; Genetic Engineering; Genetic Engineering: Medical Applications; Immunogenetics; Model Organism: *Mus musculus*; Oncogenes; Organ Transplants and HLA Genes; Synthetic Antibodies.

Further Reading

Gibbs, W. W. "Plantibodies: Human Antibodies Produced by Field Crops Enter Clinical Trials." *Scientific American* 277 (November, 1997). Details experiments in introducing antibody genes into plants.

Hoogenboom, H. R. "Designing and Optimizing Library Selection Strategies for Generating High-Affinity Antibodies." *Trends in Biotechnology* 15 (1997). Contains detailed information about laboratory techniques used to engineer monoclonal antibodies.

Kontermann, Roland, and Stefan Dübel, eds. *Antibody Engineering.* New York: Springer, 2001. A detailed look at basic methods, protocols for analysis, and recent and developing technologies. Illustrations, bibliography, index.

Mayforth, Ruth D. *Designing Antibodies.* San Diego: Academic Press, 1993. Serves as a practical introduction to designing antibodies for use in medicine or science: making monoclonal antibodies, designing them for human therapy, targeting, idiotypes, and catalytic antibodies.

Stigbrand, T., et al. "Twenty Years with Monoclonal Antibodies: State of the Art." *Acta Oncologica* 35 (1996). Provides an overview of the development of monoclonal antibodies.

Van de Winkel, J. G., et al. "Immunotherapeutic Potential of Bispecific Antibodies." *Immunol-*

ogy Today 18 (December, 1997). Looks at the potential uses of bispecific antibodies.

Wang, Henry Y., and Tadayuki Imanaka, eds. *Antibody Expression and Engineering.* Washington, D.C.: American Chemical Society, 1995. Explores monoclonal antibody synthesis and reviews research on the expression of antibody fragments. Illustrated.

Hypercholesterolemia

Field of study: Diseases and syndromes
Significance: *Hypercholesterolemia represents a significant risk factor for coronary artery disease and stroke. Diet as well as genetics influence the development of hypercholesterolemia.*

Key terms

APOLIPOPROTEIN B (APO-B): a protein essential for cholesterol transport

HIGH-DENSITY LIPOPROTEIN (HDL): a small, denser form of cholesterol, popularly known as the "good" cholesterol because it can transport cholesterol from tissues to the liver

LOW-DENSITY LIPOPROTEIN (LDL): the "bad" cholesterol that tends to deposit into the tissues, especially in the vessel walls

Cholesterol's Role in the Body

Cholesterol is a steroid lipid, a type of fat molecule that is essential for life. It is an important component of cell membranes and is used by the body to synthesize various steroid hormones. When cooled, cholesterol is a waxy substance, which cannot dissolve in the bloodstream. It is transported in the bloodstream in complexes of cholesterol and protein called lipoproteins.

There are two different classes of lipoproteins in the bloodstream. Low-density lipoprotein (LDL) cholesterol is the "bad" cholesterol that tends to deposit into the tissues, especially in the vessel walls. High-density lipoprotein (HDL), a smaller, denser molecule, is the "good" cholesterol, because it can transport cholesterol from tissues to the liver.

About one tablespoon of cholesterol circu-

lates in the bloodstream, which is enough to meet the body's needs. Cholesterol is present in animal-derived foods, but is also produced by the liver. The liver manufactures and regulates the amount of lipoproteins in the body. The normal range of total cholesterol is less than 200 milligrams per deciliter (mg/dl) of blood. A total cholesterol level between 200-240 mg/dl is borderline high, and a total cholesterol level above 240 mg/dl is considered high. The normal range of LDL cholesterol is less than 130 mg/dl, and the normal range of HDL cholesterol is greater than 35 mg/dl. Hypercholesterolemia is diagnosed when the total cholesterol level is higher than the normal range, and the term "hypercholesterolemia" is often used to refer to familial cholesterolemia as well.

Causes of Hypercholesterolemia

Hypercholesterolemia itself may be asymptomatic but can still be damaging to the vascular system. Excess amounts of cholesterol in the blood can build up along the walls of the arteries, which results in hardening and narrowing of the arteries, called atherosclerosis. Severe atherosclerosis can lead to a blockage of blood flow. Atherosclerosis in the heart causes cardiovascular disease (such as heart attacks). The result of atherosclerosis in the brain can be a stroke. Atherosclerosis can also occur in the extremities of the body, such as the legs, causing pain and blood clots.

Hypercholesterolemia occurs when the body is unable to use or eliminate excessive amounts of cholesterol. Several diseases can contribute to hypercholesterolemia, such as diabetes, thyroid disorders, and liver diseases. However, the most important cause of hypercholesterolemia is a combination of diet and genetic factors.

Cholesterol naturally exists in animal products, such as meats (particularly fatty meats), eggs, milk, cheese, liver, and egg yolks. Large intakes of these products can certainly increase one's cholesterol level, not only because they have high concentrations of cholesterol itself but, more important, because they contain fats that prompt the body to make cholesterol. The genetic influence on hypercholesterolemia is also significant.

Genetics of Hypercholesterolemia

It is evident that hypercholesterolemia is more common among certain ethnic groups. Cholesterol levels in northern European countries are higher than those in southern Europe. Asians have lower cholesterol levels than Caucasians. A severe form of hereditary hypercholesterolemia called familial hypercholesterolemia typically does not respond to lifestyle changes. Thus, there is no doubt that genes play an important role in the occurrence of hypercholesterolemia.

Hypercholesterolemia is on the increase worldwide. People with hypercholesterolemia often develop coronary heart disease at a younger age than those in a general population as a result of increased LDL cholesterol levels (about two times higher than normal). In cases of extreme hypercholesterolemia (exceeding three or four times normal), high cholesterol levels can be detected in utero or at birth in cord blood. Individuals with extreme hypercholesterolemia usually develop the first cardiovascular event in childhood or adolescence and die by the age of thirty.

Familial hypercholesterolemia is the best understood genetically. It displays autosomal dominant inheritance, which means that either parent with hypercholesterolemia has a high probability of passing it on. This disorder results from defects of the LDL receptor, which ensures the proper movement of LDLs. Thus, dysfunction of this receptor causes increased levels of LDL in the blood. The LDL receptor gene, which is located on the short arm of human chromosome 19, is prone to a variety of mutations that affect LDL metabolism and movement.

Apolipoprotein B (Apo-B) is a protein essential for cholesterol transport. Apo-B can be affected by both diet and genetics. Individuals with one or more specific genotypes (the genetic constitution of an individual) have much greater changes in cholesterol levels in response to diet than do other genotypes.

The other genetic cause is mutations in the gene for the enzyme cholesterol 7-alpha hydroxylase (CYP7A1), which is essential for the normal elimination of cholesterol in the blood. It initiates the primary conversion of cholesterol into bile acids in the liver. Mutations can cause an accumulation of cholesterol in the liver, as the primary route of converting cholesterol to bile acids is blocked. The liver responds to excessive cholesterol by reducing the number of receptors available to take up LDL from the blood, resulting in an accumulation of LDL in the blood.

Implications

Although genetics plays an important role, hypercholesterolemia is often the result of a combination of genetics and lifestyle. Consuming a healthy diet and exercising regularly can help to maintain an optimal cholesterol level and to reduce the risk of cardiovascular disease for people with either a good gene or a bad gene.

—*Kimberly Y. Z. Forrest*

See also: Alzheimer's Disease; Breast Cancer; Cancer; Heart Disease; Hereditary Diseases; Steroid Hormones.

Further Reading

Goldstein, J. L., H. H. Hobbs, and M. S. Brown. "Familial Hypercholesterolemia." In *The Metabolic and Molecular Bases of Inherited Disease*, edited by C. R. Scriver, A. L. Beauder, W. S. Sly, and D. Valle. 7th ed. New York: McGraw-Hill, 1995. Describes the epidemiology and genetic background of familial hypercholesterolemia.

Rantala, M., et al. "Apolipoprotien B Gene Polymorphisms and Serum Lipids: Meta-analysis and the Role of Genetic Variation in Responsiveness to Diet." *American Journal of Clinical Nutrition* 71 (2000): 713-724. Describes genetic variables that can cause individuals to be sensitive to or at greater risk for hypercholesterolemia from a high-fat diet.

Web Site of Interest

American Heart Association. http://www.americanheart.org. Searchable site provides information on familial and hypercholesterolemia.

Icelandic Genetic Database

Fields of study: Bioinformatics; Techniques and methodologies

Significance: *Iceland is the first country to license the rights of an entire population's genetic code to a private company. The potential scientific and health care benefits of the Icelandic Genetic Database are considered significant. However, its creation has led to a worldwide debate concerning genetic research and its role in public health.*

Key terms

GENETIC DATABASE: a set of computerized records of individuals that contain their genetic information and medical histories

GENETIC PROFILE: a description of a person's genes, including any variations within them

INFORMED CONSENT: the right for a potential research subject to be adequately informed of the aims, methods, sources of funding, conflicts of interest, anticipated benefits, potential risks, discomforts involved in a procedure or trial, and the ability to withdraw consent, which should be in a written, signed document

PHARMACOGENOMICS: the study of how variations in the human genome affect responses to medications; can be used to find the most suitable patients for drug therapy trials or to match people with similar genetic profiles to the drugs most likely to work for them

POPULATION DATABASE: a database containing information on the individuals in a population, which can be defined by a variety of criteria, such as location (a state or country) or ethnicity

Why Iceland?

Icelanders have always displayed an intense interest in documenting their genealogical and medical histories. The complete family histories for more than 75 percent of all Icelanders who have ever lived are known. Although standardized recording of extensive and precise medical records became law in 1915, additional records date to the 1600's. These extensive written records of the Icelandic people are of high quality and unique in the world today.

History of the Database

In the mid-1970's, the Icelandic parliament considered collecting these records into a computer database. The idea was abandoned because of a lack of funding, concern over privacy, and inadequate technology. While working on identifying the gene for multiple sclerosis in 1994, Icelander physician and scientist Dr. Kári Steffánsson realized that Iceland's genealogical and medical records would greatly aid in the search for genes involved in complex but common diseases such as heart disease and diabetes. He also believed that since all Icelanders can trace their genetic roots to the same few founders, their genetic backgrounds would be very similar, making it easier and faster to identify the mutations causing diseases. He determined it was financially and technologically feasible to build a computer database integrating genealogical, medical, and genetic profiles for the first time. However, the genetic profiles of the Icelandic population had yet to be determined. Because Iceland has a nationalized health care system, permission of the Icelandic parliament was required.

With private financial backing, Stefánsson founded the company deCODE Genetics in 1996. Two years later, Iceland's parliament enacted the Act on a Health Sector Database for an Icelandic Genetic Database, awarding a twelve-year license exclusively to deCODE. The database immediately became the subject of intense ethical and medical debates. While this controversy continued, deCODE Genetics computerized the Icelandic genealogical records, created the genetic profiles of eight thousand Icelandic volunteers, and uploaded their genetic, medical, and genealogical records.

As of 2003, the Icelandic Genetic Database was not officially operating. Only the data for volunteers could be used until the database passed a government-ordered security test to ensure that the database would be accessible only to those with appropriate permission and the data had been encrypted and privatized.

Current Uses of the Database

From the very beginning, two different but

Kári Steffánsson, founder of deCODE Genetics, speaking before the forty-first annual meeting of the American Society of Hematology in December, 1999. (AP/Wide World Photos)

interrelated objectives for the database were defined: (1) discovering the genes involved in complex diseases and (2) finding new drugs through pharmacogenomics to combat those same diseases once their genes were identified.

Working with volunteers, deCODE launched an initiative to discover genes involved in more than fifty common diseases, such as diabetes and asthma. Genes involved in all fifty diseases have been mapped at least to a chromosome. Three genes have been isolated, including one for schizophrenia. Based upon this early and rapid success, the company has entered into new pharmacogenomic partnerships with additional companies with the aim to discover drugs that can effectively counter the diseases that have been genetically mapped.

Potential Uses

Because the database will contain the information of the entire Icelandic people, it is also considered a population genetic database. Its data could be used not only to determine an in-

dividual's predisposition to a particular disease but also to predict diseases within the entire population of Iceland before they actually occur. This new form of medical intervention could be used to plan public health policies for groups of people. Predicting diseases is a significant departure from current public health practice, which develops treatment regimens only after a disease appears, not before. What began as a single country's genetic database has now grown into the recognition of the potential role of genetics in worldwide public health policy and planning.

Ethical Concerns

The Act on a Health Sector Database is silent on what data were to be used, how they would be used, informed consent issues, and the right to privacy. Heavily encrypting all the information in the database, removing all personal information that could identify patients individually, and security testing the database were a result of these privacy concerns.

Informed consent issues have created the most serious problems. The act presumes informed consent unless an individual "opts out," which many feel violates the intent of consent. Icelandic physicians have filed a lawsuit to clarify this issue, since Icelandic law requires that physicians guarantee full informed consent.

A second major concern is the licensing of Iceland's complete genetic profile to a company. Because Iceland has a nationalized health plan, medical records have always been considered a national resource. Many feel that Icelandic genetic records are also a national resource and should remain with the people. Related to this issue is concern that granting the rights to only a single company will prevent scientific research both in Iceland and elsewhere on any genes deCODE may identify.

Although controversial, the database continues to provide guidance and lessons for other nations in developing new genetic databases. Ethical, medical, and social issues first raised in Iceland have quickly become issues worldwide as population genetic databases proliferate. This, in turn, has resulted in an active debate on the role of genetic information in worldwide public health and whether it should be permitted to operate in all countries, if at all.

—*Diane C. Rein*

See also: Bioinformatics; Genetic Screening; Genetic Testing: Ethical and Economic Issues; Genomic Libraries; Genomics; Human Genome Project; Linkage Maps; Pedigree Analysis; Population Genetics.

Further Reading

Anna, George J. "Rules for Research on Human Genetic Variation: Lessons from Iceland." *The New England Journal of Medicine* 342, no. 24 (2000): 1830-1833. Deals with the major ethical problems that arose from the creation of the Icelandic Genetic Database and how they could be avoided in the future.

Greely, Henry T. "Iceland's Plan for Genomics Research: Facts and Implications." *Jurimetrics Journal* 40 (2000): 153-191. Covers the history of the database and the ethical and medical issues, presented in a legal context.

Kaiser, Jocelyn. "Population Databases Boom: From Iceland to the U.S." *Science* 298 (1995): 1158-1161. Discusses the development of health and genetic information databases in several countries, including how they are doing it and what new controversies are arising.

Palsson, Bernhard, and Snorri Thorgeirsson. "Decoding Developments in Iceland." *Nature Biotechnology* (1999) 17, no. 5: 406. A short article that covers the early history of the Icelandic Genetic Database. Lists stable URLs for the Icelandic government's Web site on the Health Sector Database Act, as well as a site that contains the full text of most of the articles published about the database.

Wilie, Jean E., and Geraldine P. Mineau. "Biomedical Databases: Protecting Privacy and Promoting Research." *Trends in Biotechnology* 21, no. 3 (2003): 113-116. Addresses the tension that develops between biomedical research with population databases and the need to protect the people whose data resides in the databases.

Web Sites of Interest

Association of Icelanders for Ethics in Science and Medicine. http://www.mannvernd.is. Site of an organization opposed to the Icelandic Genetic Database.

deCode Genetics. http://www.decode.com. Site of the company compiling the Icelandic Genetic Database.

Mapping the Icelandic Genome. http://sunsite.berkeley.edu/biotech/iceland. Site devoted to "the scientific, political, economic, religious, and ethical issues surrounding the deCode Project and its global implications."

Immunogenetics

Field of study: Immunogenetics

Significance: *Immunogenetics is primarily concerned with the major histocompatibility genes that identify self tissues, the genes in B lymphocytes that direct antibody synthesis, and the genes that direct the synthesis of T lymphocyte receptors. This same genetic control that directs immune cell embryonic* *development and activation from an antigenic challenge also explains the basis of organ transplant rejection, autoimmunity, allergies, immunodeficiency, and potential future therapies.*

Key terms

APOPTOSIS: cell death that is programmed as a natural consequence of growth and development through normal cellular pathways or through signals from neighboring cells

CYTOKINES: soluble intercellular molecules produced by cells such as lymphocytes that can influence the immune response

DOWNSTREAM: describes the left-to-right direction of DNA whose nucleotides are arranged in sequence with the 5′ carbon on the left and the 3′ on the right; the direction of RNA transcription of a genetic message with the beginning of a gene on the left and the end on the right

HAPLOTYPE: a sequential set of genes on a single chromosome inherited together from one parent; the other parent provides a matching chromosome with a different set of genes

TRANSPOSON: a sequence of nucleotides flanked by inverted repeats capable of being removed or inserted within a genome

Genes, B Cells, and Antibodies

The fundamental question that led to the development of immunogenetics relates to how scientists are able to make the thousands of specific antibodies that protect people from the thousands of organisms with which they come in contact. Macfarlane Burnet proposed the clonal selection theory, which states that an antigen (that is, anything not self, such as an invading microorganism) selects, from the thousands of different B cells, the receptor on a particular B cell that fits it like a key fitting a lock. That cell is activated to make a clone of plasma cells, producing millions of soluble antibodies with attachment sites identical to the receptor on that B-cell surface. The problem facing scientists who were interested in a genetic explanation for this capability was the need for more genes than the number that was believed to make up the entire human genome.

It was Susumu Tonegawa who first recog-

nized that a number of antibodies produced in the lifetime of a human did not have to have the equivalent number of physical genes on their chromosomes. From his work, it was determined that the genes responsible for antibody synthesis are arranged in tandem segments on specific chromosomes relating to specific parts of antibody structure. The amino acids that form the two light polypeptide chains and the two heavy polypeptide chains making up the IgG class of antibody are programmed by nucleotide sequences of DNA that exist on three different chromosomes. Light-chain genes are found on chromosomes 2 and 22. The specific nucleotide sequences code for light polypeptide chains, with half the chain having a constant amino acid sequence and the other half having a variable sequence. The amino acid sequences of the heavy polypeptide chains are constant over three-quarters of their length, with five basic sequences identifying five classes of human immunoglobulins: IgG, IgM, IgD, IgA, and IgE. The other quarter length has a variable sequence that, together with the variable sequence of the light chain, forms the antigen-binding site. The nucleotide sequence coding for the heavy chain is part of chromosome 14.

The actual light-chain locus is organized into sequences of nucleotides designated V, J, and C segments. The multiple options for the different V and J segments and mixing the different V and J segments cause the formation of many different DNA light-chain nucleotide sequences and the synthesis of different antibodies. The same type of rearrangement occurs between a variety of nucleotide sequences related to the V, D, and J segments of the heavy-chain locus. The recombination of segments appears to be genetically regulated by recombination signal sequences downstream from the variable segments and recombination activating genes that function during B-cell development. Genetic recombination is complete with the immature B cell committed to producing one kind of antibody. The diversity of antibody molecules is explained by the fact that the mRNA transcript coding for either the light polypeptide chain or the heavy polypeptide chain is formed containing exons transcribed from recombined gene segments during B-cell differentiation. The unique antigen receptor-binding site is formed when the variable regions of one heavy and one light chain come together during the formation of the completed antibody in the endoplasmic reticulum of the mature B cell. The B-cell antigen receptor is an attached surface antibody of the IgM class. Binding of the antigen to the specific B cell activates its cell division and the formation of a clone of plasma cells that produce a unique antibody. If this circulating B cell does not contact its specific antigen within a few weeks, it will die by apoptosis. During plasma cell formation, the class of antibody protein produced normally switches from IgM to IgG through the forma-

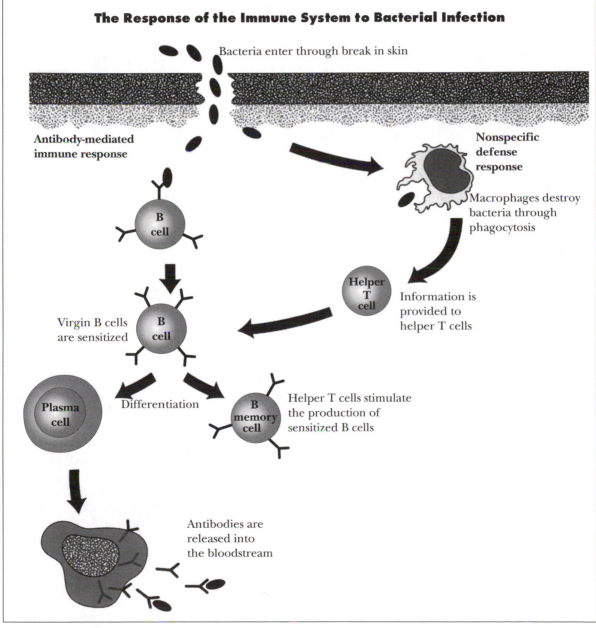

The Response of the Immune System to Bacterial Infection

Bacteria enter through break in skin

Antibody-mediated immune response

Nonspecific defense response

Macrophages destroy bacteria through phagocytosis

B cell

B cell

Virgin B cells are sensitized

Helper T cell

Information is provided to helper T cells

Plasma cell

Differentiation

B memory cell

Helper T cells stimulate the production of sensitized B cells

Antibodies are released into the bloodstream

(Hans & Cassidy, Inc.)

tion of an mRNA transcript containing the exon nucleotide sequence made from IgG heavy-chain C segment rather than the heavy-chain C segment for IgM. The intervening nucleotide sequence of the IgM constant segment is deleted from the chromosome as an excised circle reminiscent of the transposon or plasmid excision process. The result of this switch is the formation of an IgG antibody having the same antigen specificity as the IgM antibody, because the variable regions of the light and heavy polypeptide chains remain the same. Although the activation and development of B cells by some antigens may not need T-cell involvement, it is believed that class switching and most B-cell activity are influenced by T-cell cytokines.

Major Histocompatibility Genes

In humans, the major histocompatibility genes encoding "self antigens" are also called the HLA complex and are located on chromosome 6. The nucleotides that compose this DNA complex encode for two sets of cell surface molecules designated MHC Class I and MHC Class II antigens. The Class I region contains loci *A*, *B*, and *C*, which encode for MHC Class I A, B, and C glycoproteins on every nucleated cell in the body. Because the *A*, *B*, and *C* loci comprise highly variable nucleotide sequences, numerous kinds of A, B, and C glycoproteins characterize humans. All people inherit MHC Class I *A*, *B*, and *C* genes as a haplotype from each of their parents. Children will have tissues with half of their Class I A, B, and C antigens like those of their mother and half like those of their father. Siblings could have tissue antigens identical or totally dissimilar based on their MHC I glycoproteins. Body surveillance by T lymphocytes involves T cells recognizing self glycoproteins. Cellular invasion by a virus or any other parasite results in the processing of antigen and its display in the cleft of the MHC Class I glycoprotein. T cytotoxic lymphocytes with T-cell receptors specific for the antigen-MHC I complex will attach to the antigen and become activated to clonal selection. Infected host cells are killed when activated cytotoxic T cells bind to the surface and release perforins, causing apoptosis.

MHC Class II genes are designated *DP*α and β, *DQ*α and β, and *DR*α and β. These genes encode for glycoprotein molecules that attach to the cell surface in α and β pairs. A child will inherit the six genes as a group or haplotype, three α and β glycoprotein gene pairs from each parent. The child will also have glycoprotein molecules made from combinations of the maternal and paternal α and β pairings during glycoprotein synthesis.

The Class II MHC molecules are found on the membranes of macrophages, B cells, and dendritic cells. These specialized cells capture antigens and attach antigen peptides to the three-dimensional grooves formed by combined α and β glycoprotein pairs. The antigen attached to the Class II groove is presented to the T helper cell, with the receptor recognizing the specific antigen in relation to the self antigen. The specific T helper cell forms a specific clone of effector cells and memory cells.

Genes, T Helper Cells, and T Cytotoxic Cells

The thousands of specific T-cell receptors (TCR) available to any specific antigen one might encounter in a lifetime are formed in the human embryonic thymus from progenitor T cells. The TCR comprises two dissimilar polypeptide chains designated α and β or γ and δ. They are similar in structure to immunoglobulins and MHC molecules, having regions of variable amino acid sequences and constant amino acid sequences arranged in loops called domains. This basic structural configuration places all three types of molecules in a chemically similar grouping designated the immunoglobulin superfamily. The genes of these molecules are believed to be derived from a primordial supergene that encoded the basic domain structure.

The exons encoding the α and γ polypeptides are designated V, J, and C gene segments in sequence and associate with recombination signal sequences similar to the immunoglobulin light-chain gene. The β and δ polypeptide genes are designated VDJ and C exon segments in sequence associating with recombination signal sequences similar to the immunoglobulin heavy-chain genes. Just as there are multiple forms for each of the immunoglobulin variable

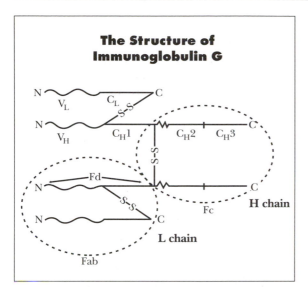

The Structure of Immunoglobulin G

A Y-shaped model of the antibody immunoglobulin G (IgG). V indicates a region of variability that would permit recognition by a wide variety of antigens.
Source: After John J. Cebra's "The 1972 Nobel Prize for Physiology or Medicine," *Science,* 1972.

gene segments, so there are multiple forms for the variable TCR gene segments. Thymocytes, T-cell precursors in the thymus, undergo chance recombinations of gene segments. These genetic recombinations, as well as the chance combination of a completed α polypeptide with a completed β polypeptide, provide thousands of completed specific TCRs ready to be chosen by an invading antigen and to form a clone of either T helper cells or T cytotoxic cells.

Immunogenetic Disease

The HLA genes of the major histocompatibility complex identify every human being as distinct from all other things, including other human beings, because of the MHC Class I and Class II antigens. Surveillance of self involves B- and T-cell antigen recognition because of MHC self-recognition. How well individual human beings recognize self and their response to antigen in an adaptive immune response are determined by MHC haplotypes as well as the genes that make immunoglobulins and T-cell receptors. These same genes can explain a variety of disease states, such as autoimmunity, allergy, and immunodeficiency.

Because immunoglobulin structure and T-cell receptor formation are based on a mechanism of chance, problems involving self-recognition may occur. It is currently believed that thymocytes with completed T-cell receptors are protected from apoptosis when they demonstrate self-MHC molecule recognition. Alternatively, it is believed that thymocytes are also presented with self-antigens processed by specialized macrophages bearing MHC Class I and Class II molecules. Thymocytes reacting with high-affinity receptors to processed self-antigens undergo apoptosis. There also appears to be a negative selection process within the bone marrow that actively eliminates immature B cells with membrane bound auto-antibodies that react with self-antigens. In spite of these selective activities, it is believed that autoreactive T cells and B cells can be part of circulating surveillance, causing autoimmune disease of either single organs or multiple tissues.

It has long been recognized that autoimmune diseases occur in families, and there is growing evidence that an individual with a certain HLA haplotype has a greater risk for developing a particular disease. For example, ankylosing spondylitis develops more often in individuals with *HLA-B27* than in those with another *HLA-B* allele, and rheumatoid arthritis is associated with *DR1* and *DR4* alleles. Myasthenia gravis and multiple sclerosis are two neurological diseases caused by auto-antibodies, and there is evidence that they are related to restricted expression of T-cell variable genes. Genomic studies are providing evidence for the possibility that autoimmune induction occurs because of molecular mimicry between human host proteins and microbial antigens. Among the cross-reacting antigens that have been implicated are papillomavirus E2 and the insulin receptor, and poliovirus VP2 and the acetyl choline receptor.

The genetics of immunity also involves the study of defective genes that cause primary immunodeficiency infectious disease. The deficiency can result in a decrease in an adaptive immune response involving B cells, T cells, or both, as is the case with severe combined immunodeficiency disorder (SCID). There is evidence that SCID can demonstrate either autosomal recessive or X-linked inheritance. One such defect has been located on the short arm of chromosome 11 and involves a mutation of recombination-activating genes that are necessary for the rearrangement of immunoglobulin gene segments and the T-cell receptor gene segments. The inability to recombine the VD and J variable segments prevents the development of active B cell and T cells with the variety of antigen receptors. SCID is essentially incompatible with life and characterized by severe opportunistic infections caused by even normally benign organisms.

Allergies are widely understood to have a genetic component, with the understanding that atopy, an abnormal IgE response, is common to certain families. There is evidence that children have a 30 percent chance of developing an allergic disease if one parent is allergic, while those children with two allergic parents have a 50 percent chance. The genetic control of IgE production can be related to T_{H2} lymphocyte cytokine stimulation of class switching from the constant segment of IgG to the constant segment of IgE on chromosome 14 in an antigen-selected cell undergoing clonal selection.

Immunogenetics and Treatment

Understanding the genetic basis for allergic reactions is resulting in novel approaches to protect against disease. Through genetic engineering, monoclonal mouse/human antibodies can be made that are able to react with serum IgE and down-regulate IgE production. Probably the greatest potential for therapy will parallel the human genome studies that are further elucidating the genetic relationship to immune defense, autoimmunity, and allergy. As science continues to identify those genes that provide protein receptors and messengers, the best drug therapies and molecular manipulation will be discovered.

—*Patrick J. DeLuca*

See also: Allergies; Antibodies; Autoimmune Disorders; Hybridomas and Monoclonal Antibodies; Organ Transplants and HLA Genes; Synthetic Antibodies.

Further Reading

Goldsby, Richard A, Thomas J. Kindt, Barbara A. Osborne, and Janis Kuby. *Immunology.* New York: W. H. Freeman, 2003. A very complete text dealing with the biological basis of immunity, including immunogenetics.

Pines, Maya, ed. *Arousing the Fury of the Immune System.* Chevy Chase, Md.: Howard Hughes Medical Institute, 1998. Informative, well-done report relating different immunological concepts in an entertaining, readable format.

Roitt, Ivan, Jonathan Brostoff, and David Male. *Immunology.* New York: Mosby, 2001. Text and diagrams provide in-depth presentation of immunological concepts, including immunogenetics.

Web Sites of Interest

American Society for Histocompatibility and Immunogenetics. http://www.ashi-hla.org. A nonprofit professional organization for immunologists, geneticists, molecular biologists, transplant surgeons, and pathologists, devoted to advancing the science and exchanging information.

ImMunoGeneTics (IMGT) Database. http://imgt.cines.fr:8104. A database focusing on immunoglobulins, T-cell receptors, and MHC molecules of all vertebrates, including interactive tools.

In Vitro Fertilization and Embryo Transfer

Field of study: Human genetics and social issues

Significance: *The term "in vitro" designates a living process removed from an organism and isolated "in glass" for laboratory study. In vitro fertilization (IVF) is a process in which harvested eggs and sperm can be brought together artificially to form a zygote. The resulting zygote can be grown for a time in vivo, where it can be tested biochemically and genetically, if desired, after which it can be implanted in the uterus of the egg donor or a surrogate.*

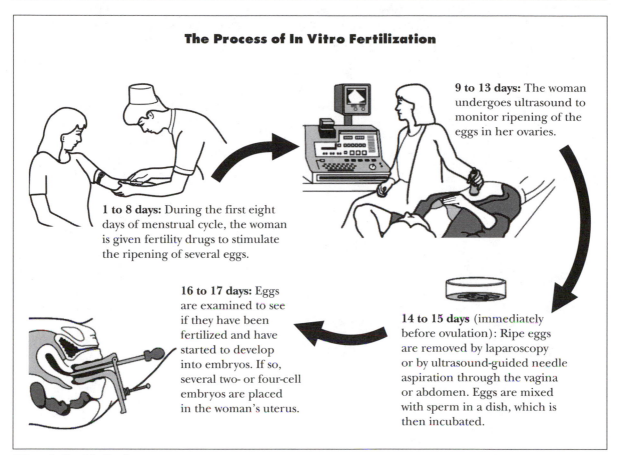

The Process of In Vitro Fertilization

1 to 8 days: During the first eight days of menstrual cycle, the woman is given fertility drugs to stimulate the ripening of several eggs.

9 to 13 days: The woman undergoes ultrasound to monitor ripening of the eggs in her ovaries.

14 to 15 days (immediately before ovulation): Ripe eggs are removed by laparoscopy or by ultrasound-guided needle aspiration through the vagina or abdomen. Eggs are mixed with sperm in a dish, which is then incubated.

16 to 17 days: Eggs are examined to see if they have been fertilized and have started to develop into embryos. If so, several two- or four-cell embryos are placed in the woman's uterus.

(Hans & Cassidy, Inc.)

Key terms

DIPLOID: possessing a full complement of chromosome pairs, as in humans, who have 23 pairs of chromosomes for a total of 46

GAMETE: a germ cell; an egg (ovum or oocyte) or a sperm (spermatozoan)

HAPLOID: possessing a full complement of one of each type of chromosome; mature human gametes are haploid, with 23 chromosomes

SURROGATE: a female that carries an embryo derived from an egg from another female

ZYGOTE: the earliest stage in the development of an organism, just after fertilization

Natural Fertilization

Fertilization, the union of a male gamete (sperm) with a female gamete (ovum), is fundamentally a genetic process. Each of the gametes is haploid, containing half of the genetic information needed for a living organism. Fertilization brings together these two sets, thereby producing a diploid zygote that will develop into an embryo.

Gametes are produced in the gonads (ovaries in females, testes in males) by a special type of cell division called meiosis. Instead of producing diploid daughter cells, as in mitosis, meiosis results in haploid cells. In humans, the natural place for fertilization is in a Fallopian tube of a woman, the channel through which an ovum travels to the uterus. A normal adult woman ovulates each month, releasing a single haploid ovum from one of her two ovaries. Ovulation is under hormonal control.

Sperm from the male's testis are deposited in the woman's vagina during sexual intercourse. Typically, men release hundreds of millions of sperm into the vagina when they ejaculate. From the vagina, these sperm travel through the uterus and into each Fallopian tube in search of an ovum. During this trip, the sperm undergo changes called capacitation. To fuse with the ovum, a sperm must penetrate several surrounding barriers. After fusion of sperm and egg, the nuclear membranes of the two cells break down so that the paternal and maternal chromosomes can congregate in a single nucleus. The resulting zygote divides into two new diploid cells, the first cells of a genetically unique new being.

In Vitro Fertilization and Embryo Transfer

Fertilization can also take place artificially in laboratory culture dishes. Gametes are collected, brought together, and fertilized in a laboratory. After the zygote develops into an embryo, it can then be transferred to a uterus for continued development and eventual birth. This procedure can be done for many species, including humans. The first human conceived by in vitro fertilization (IVF), Louise Brown, was born on July 25, 1978, in England.

Nuclear Transplantation from Donor Eggs

For women who do not produce any viable oocytes because of permanent failure of the ovaries, options for having a child who contains genetic information from the mother are limited. Nuclear transfer into an enucleated donor egg could address this limitation. Since the 1980's, nuclei from relatively undifferentiated mammalian embryonic cells have been successfully transferred to donor eggs. In 1996 researchers at the Roslyn Institute in Scotland advanced nuclear transfer by taking a nucleus from an adult somatic cell and successfully transferring it into an enucleated egg. The result of this work was the birth of the first vertebrate cloned from an adult cell, Dolly the sheep. Since Dolly, nuclear transfer has been successfully performed in cows, pigs, cats, and mice.

Adult somatic cells contain essentially the same genetic information as the single fertilized egg that gave rise to the adult organism. However, unlike the fertilized egg, most adult somatic cells are terminally differentiated and have lost the ability to produce any type of cell in the body, as a fertilized egg can. Nuclear transfer takes a nucleus from an adult somatic cell and places it into an enucleated donor egg. In the environment of the egg, the DNA in the transferred nucleus can "dedifferentiate" and direct the production of a new individual. Because this technique does not involve fertilization, the new individual is considered a clone of the adult organism that contributed the nucleus.

Is the new individual produced really a clone of the adult? The enucleated egg contributes the environment that directs the unfolding of the genetic program that leads to the development of the new individual. Proteins called transcription factors control the expression of individual genes within the DNA. These transcription factors are contributed by the enucleated, donor egg, and they determine what genes will be active, in what cells, and for how long. Proteins contributed by the donor egg will control the early embryonic divisions. The donor egg also contains RNA molecules that serve as templates to create the proteins needed for events in early embryogenesis, essential to the development of the new organism. These molecules will influence how that organism grows and develops and what genes are expressed by its cells.

The nucleus is not the only source of DNA in the animal cell. The donor egg contains organelles called mitochondria that contain their own DNA. Mitochondria reproduce by a process much like bacteria, copying their own DNA and dividing within the cell. All of the mitochondria in an organism produced by nuclear transfer into a donor egg will be derived from the donor egg, not from the cell that donated the nucleus. Mitochondria are responsible for cellular metabolism, and some metabolic diseases can be traced directly to mutations within mitochondrial DNA.

As might be anticipated, this reproductive technique raises ethical questions, as only one parent can contribute a nucleus to the donor egg. Moreover, it involves a great deal of manipulation in vitro, and some suggest that developmental problems can result from such manipulation. Nevertheless, in 2003, as the first "test-tube baby," Louise Brown, celebrated her twenty-fifth birthday, many remarked on how many children had been similarly brought into the world since 1978 and how common the technique had become as an alternative for infertile couples.

—*Michele Arduengo*

In humans IVF is usually used to overcome infertility caused by problems such as blocked Fallopian tubes or low sperm count. IVF is also done in veterinary medicine and for scientific research. IVF also makes genetic diagnoses easier and could eventually lead to more effective gene therapy. Mature sperm for IVF are easily obtained by masturbation. Mature ova are more difficult to obtain. The female is given gonadotropin hormones to stimulate her to superovulate (that is, to produce ten or more mature eggs rather than just one). Ova are later collected by inserting a small suction needle into her pelvic cavity. The ova are inseminated with laboratory-capacitated sperm. Two to four embryos are transferred into the uterus through a catheter. Excess embryos can be saved by a freezing procedure called cryopreservation. These may be thawed for later attempts at implantation should the first attempt fail or a second pregnancy be desired.

Impact and Applications

Technology such as the polymerase chain reaction (PCR) permits assessment of genetic information in the nucleus of a single cell, whether diploid or haploid. IVF gives physicians access to sperm, ova, and very early embryos. One or two cells can be removed from an eight-cell embryo without damaging the ability of the remaining cells to develop normally following embryo transfer. Thus IVF permits genetic diagnosis at the earliest stages of human development and even allows the possibility of gene therapy.

Preimplantation genetic diagnosis (PGD) is used clinically to help people with significant genetic risks to avoid giving birth to an abnormal child that might die in infancy or early childhood. If tests show that the embryo is free of genetic defects, it can be transferred to the uterus for implantation; if found defective it can be destroyed. PGD is successful in avoiding pregnancies with embryos that will develop cystic fibrosis, Huntington's disease, Lesch-Nyhan disease, Tay-Sachs disease, and other genetic abnormalities. Prior to the development of PGD, detection of genetic defects was possible only by prenatal diagnosis during pregnancy. If a defect is detected, termination of the pregnancy through elective abortion becomes an option. Not only does abortion represent a higher risk to the mother, it is an unacceptable choice for many people because of ethical and moral concerns.

Access to gametes prior to fertilization and to embryos prior to implantation also opens the possibility of gene therapy. Gene therapy in human embryos presents insurmountable ethical issues, at present, and has been banned pending more study. Genetic modification of the embryos of other species, especially those of commercial interest, carries no such ethical concerns and is routinely practiced.

IVF also opens the possibility of genetic cloning. Cloning is the process of creating multiple individuals with identical genetic characteristics. This can be accomplished by dividing an early embryo, allowing each group of cells to develop into a separate embryo. A few of these embryos can then be implanted, saving the others for future attempts, or all can be implanted, using several different females as surrogate mothers. Through the use of cryopreservation, these pregnancies could occur years apart. It is even possible to remove the nucleus from an isolated cell and replace it with a nucleus taken from an adult. The cell with the transplanted nucleus is able, using special procedures, to develop into an embryo that can be implanted. The offspring will be genetically identical to the adult source of the transplanted nucleus. Most people recognize cloning technology as inappropriate in human medicine, but it has acceptable applications in agriculture and veterinary medicine.

—Armand M. Karow, updated by Bryan Ness

See also: Amniocentesis and Chorionic Villus Sampling; Cloning; Genetic Counseling; Genetic Screening; Genetic Testing; Genetic Testing: Ethical and Economic Issues; Hereditary Diseases; Infertility; Prenatal Diagnosis; Stem Cells; Totipotency; Turner Syndrome.

Further Reading

Bonnicksen, Andrea L. *In Vitro Fertilization: Building Policy from Laboratories to Legislature.* Reprint. New York: Columbia University Press, 1991. Examines two facets of IVF: the public's political, legal, and ethical concerns surrounding the technique, and the personal, pragmatic world of the individual patients who seek a cure for infertility.

Brinsden, Peter R., ed. *A Textbook of In Vitro Fertilization and Assisted Reproduction: The Bourn Hall Guide to Clinical and Laboratory Practice.* 2d ed. New York: Parthenon, 1999. Details the clinical and laboratory protocols used in assisted reproductive technology and covers such topics as therapeutic options for infertile men, superovulation strategies, the new gonadotropins, polycystic ovaries, oocyte recovery and embryo transfer techniques for fertilization, ectopic pregnancy, oocyte and embryo donation, surrogacy, and ethical aspects.

Elder, Kay, and Brian Dale. *In Vitro Fertilization.* 2d ed. New York: Cambridge University Press, 2000. Surveys advances and protocols of IVF technology. Illustrated.

Grobstein, Clifford. *From Chance to Purpose: An Appraisal of External Human Fertilization.* Reading, Mass.: Addison-Wesley, 1981. A world-renowned embryologist presents a view of IVF before the advent of PGD.

Seibel, Machelle M., and Susan L. Crockin, eds. *Family Building Through Egg and Sperm Donation.* Boston: Jones and Bartlett, 1996. The editors are, respectively, a physician and a lawyer, and they examine the issue of assisted reproduction from medical, legal, and ethical perspectives.

Trounson, Alan O., and David K. Gardner, eds. *Handbook of In Vitro Fertilization.* 2d ed. Boca Raton, Fla.: CRC Press, 2000. Provides a theoretical and practical guide to techniques used in assisted reproduction, with each chapter containing detailed background information and technical accounts of procedures employed. Illustrated.

Web Sites of Interest

American Society for Reproductive Medicine. http://www.asrm.org. Site includes information on infertility and reproduction.

International Council on Infertility Information Dissemination. http://www.inciid.org. Site provides "fact sheets" on in vitro fertilization.

Inborn Errors of Metabolism

Field of study: Diseases and syndromes
Significance: *Inborn errors of metabolism are hereditary genetic defects found in varying frequencies in human populations. Diagnosis and cure of these genetic diseases is a continuing focus of medical research.*

Key terms

METABOLIC PATHWAY: enzyme-mediated reactions that are connected in a series
METABOLISM: the collection of biochemical reactions occurring in an organism

Early Observations

In 1902, Sir Archibald Garrod, a British physician, presented a classic paper in which he summarized his observations and analyses of a rather benign condition known as alkaptonuria. The condition is easily diagnosed because the major symptom is dark urine caused by the excretion of homogentisic acid. Since homogentisic acid is not normally found in urine and is a by-product of certain amino acids with particular ring structures, Garrod reasoned that individuals with alkaptonuria had a defect in the utilization of these amino acids. Garrod also noted that the condition is often found in two or more siblings and postulated that the occurrence of this condition may be explained by the mechanism of inheritance.

In 1908, in "Inborn Errors of Metabolism," Garrod extended his observations on alkaptonuria to other diseases such as albinism and cystinuria. In each case, he argued that the abnormal or disease condition was caused by a defect in metabolism that resulted in a block of an important metabolic pathway. He speculated that when such a pathway is blocked, there would be an accumulation of products that are not seen in normal individuals, or important substances would be missing or abnormal. Some of these abnormal metabolic events might be harmless, such as in alkaptonuria, but others could lead to serious disease. He traced the inheritance of these conditions and discovered that they could be passed on from one generation to the next. He was the first to use the term "inborn errors of metabolism" to describe these conditions. Other investigators have studied more than three thousand additional diseases that can be included in this category. A few of these conditions occur at relatively high frequency in humans. In the U.S. Caucasian population, cystic fibrosis occurs in about 1 in 2,000 births. Some conditions, such as phenylketonuria (PKU), are seen at moderate frequency, about 1 in 10,000. Many of the inborn errors are rare, with frequencies less than 1 in 100,000. A generally accepted definition of an inborn error of metabolism is any condition with actual or potential health consequences that can be inherited in the fashion described by Gregor Mendel in the nineteenth century.

Malfunctioning Proteins and Enzymes

The biochemical causes of the inborn errors of metabolism were discovered many years af-ter Garrod presented his ideas. In 1952, Von Gierke's disease was found to be caused by the defective enzyme glucose-6 phosphatase. After this discovery, many inborn errors of metabolism were traced to defects in other enzymes. Enzymes are proteins that catalyze biochemical reactions. They are responsible for increasing the rates of reactions that occur in all cells. These reactions are important steps in metabolic pathways that are responsible for processes such as utilization of nutrients, generation of energy, cell division, and biosynthesis of substances that are needed by organisms. There are many metabolic pathways that can be affected if one of the enzymes in the pathway is missing or malfunctions. In addition to enzymes, defective proteins with other functions may also be considered as candidates for inborn errors of metabolism. For example, there are many types of defective hemoglobin, the protein responsible for oxygen transport.

Nine-year-old Andy Burgy in 2003. He suffers from an incurable inborn metabolic error known as epidermolysis bullosa, which makes his skin blister at the touch. (AP/Wide World Photos)

These defective hemoglobins are the causes of diseases such as sickle-cell disease and thalassemia.

Genetic Basis of Inborn Errors

The cause of these defects in enzymes and proteins has been traced to mutations in the genes that code for them. Alterations in the structure or nucleotide composition of DNA can have various consequences for the structure of the protein coded for by the DNA. Some of the genetic alterations affecting metabolism simply represent normal variation within the population and are asymptomatic. An example of such a genetic alteration is the ability of some individuals to experience a bitter taste after exposure to chemical derivatives of thiourea. Some asymptomatic variations may lead to complications after environmental conditions are changed. There are a few "inborn errors" that can be induced by certain drugs. Another class of alterations may be minor, with the resulting protein having some degree of function. Individuals with such alterations may live long lives but will occasionally experience a range of problems associated with their conditions. Depending on the exact nature of the mutation, some of the alterations in the resulting protein structure can lead to a completely nonfunctional protein or enzyme. Consequences of this type of mutation can be quite severe and may result in death.

Many of the inborn errors of metabolism are inherited as autosomal recessive traits. Individuals are born with two copies of the gene. If one copy is defective and the second copy is normal, enough functioning protein or enzyme can be made to prevent the individual from exhibiting any symptoms of the disease. Such individuals will be classified as carriers for the defect since they can pass on the defective gene to their offspring. About one in twenty Caucasians in the U.S. is a carrier for the cystic fibrosis gene, and about one in thirty individuals of Eastern Jewish descent carries the gene for the lethal Tay-Sachs disease. When an individual inherits two defective copies of the gene, the manifestations of the disease can be much more severe.

Some inborn errors of metabolism such as Huntington's disease are manifested as dominant genetic traits. Only one copy of the defective gene is necessary for manifestations of the abnormal condition. There are some inborn errors of metabolism that are sex-linked. Diseases that involve mutations carried on the X chromosome may be severe in males because they have only one X chromosome but less severe or nonexistent in females because females carry two X chromosomes.

Diagnosis and Treatment

Significant progress has been made in the diagnosis of inborn errors of metabolism. Prior to 1980, much of the diagnosis for metabolic defects relied on symptoms detected during clinical examination. Biochemical tests are used to detect various substances that accumulate or are missing when an enzymatic defect is present. The commonly used screening for phenylketonuria (PKU) relies on detection of phenylketones in the blood of newborns. For cases in which the genetic defect is known, DNA can often be used for the purpose of genetic testing. Genetic counselors will help parents determine their chances of having a child with a severe defect when parents are identified as carriers. Small samples of cells can be used as a source of DNA, and such cells may even be obtained from amniotic fluid by amniocentesis. This allows diagnosis to be made prenatally. Some parents choose abortion when their fetus is diagnosed with a lethal or debilitating defect.

Although strides have been made in diagnosis, the problem of treatment still remains. For some inborn errors of metabolism such as phenylketonuria, dietary modification will often prevent the serious symptoms of the disease condition. Individuals with phenylketonuria must limit their intake of the amino acid phenylalanine during the critical stages of brain development, generally the first eight years of life. Treatment of other inborn errors may involve avoidance of certain environmental conditions. For example, individuals suffering from albinism, a lack of pigment production, must avoid the sun. For other inborn errors of metabolism, there are no simple cures on the horizon. Since the early 1990's, some medical pioneers have been involved in clinical trials of gene therapy, an attempt to replace a defective gene by insertion of a normal, functioning version. Although theoretically promising, gene therapy has not met with significant success. In addition, there are many ethical issues raised when gene therapy trials are proposed before potential hazards have been completely eliminated. Nevertheless, scientists are looking more and more toward genetic cures to genetic problems such as those manifested as inborn errors of metabolism.

—*Barbara Brennessel*

See also: Amniocentesis and Chorionic Villus Sampling; Biochemical Mutations; Complementation Testing; Cystic Fibrosis; Genetic Screening; Genetic Testing; Genetics, Historical Development of; Hereditary Diseases; Huntington's Disease; Phenylketonuria (PKU); Tay-Sachs Disease.

Further Reading

Econs, Michael J., ed. *The Genetics of Osteoporosis and Metabolic Bone Disease.* Totowa, N.J.: Humana Press, 2000. International experts discuss the genetic and molecular dimensions of their own research into various aspects of the clinical features and pathophysiology of metabolic bone disease.

Lee, Thomas F. *The Human Genome Project: Cracking the Genetic Code of Life.* New York: Plenum Press, 1991. The diagnosis of inborn errors of metabolism, development of molecular methods for diagnosis of these genetic defects, and prospects for treatment of these conditions by gene therapy are highlighted within the context of the Human Genome Project.

O'Rahilly, S., and D. B. Dunger, eds. *Genetic Insights in Paediatric Endocrinology and Metabolism.* Bristol, England: BioScientifica, 1999. Examines endocrine and metabolic diseases among infants, children, and adolescents. Illustrated.

Pacifici, O. G. M., Julio Collado-Vides, and Ralf Hofestadt, eds. *Gene Regulation and Metabolism: Postgenomic Computational Approaches.* Cambridge: MIT Press, 2002. Explores current computational approaches to understanding the complex networks of metabolic and gene regulatory capabilities of the cell.

Scriver, Charles, et al., eds. *The Metabolic and Molecular Bases of Inherited Disease.* 8th ed. 4 vols. New York: McGraw-Hill, 2001. These authoritative volumes on genetic inheritance, by some of the biggest names in the field, survey all aspects of genetic disease, including metabolic disorders. The eighth edition has been thoroughly updated; more than half of the contents are new.

Web Sites of Interest

Children Living with Inherited Metabolic Diseases (CLIMB). http://www.climb.org.uk. A national British organization supporting families and research on a host of inherited metabolic disorders; includes information and links to sites on specific disorders.

Society for Inherited Metabolic Disorders. http://www.simd.org. A nonprofit professional organization promoting worldwide advancement of research and medical treatment of inherited disorders of metabolism. Includes a searchable database of detailed descriptions and diagnoses for specific inborn errors.

Inbreeding and Assortative Mating

Field of study: Population genetics
Significance: *Most population genetic models assume that individuals mate at random. One common violation of this assumption is inbreeding, in which individuals are more likely to mate with relatives, resulting in inbreeding depression, a reduction in fitness. Another violation of random mating is assortative mating, or mating based on phenotype. Many traits of organisms, including pollination systems in plants and dispersal in animals, can be understood as mechanisms that reduce the frequency of inbreeding and the cost of inbreeding depression.*

Key terms

ALLELE: any of a number of possible genetic variants of a particular gene locus
ASSORTATIVE MATING: mating that occurs when individuals make specific mate choices

based on the phenotype or appearance of others

HETEROZYGOTE: a diploid genotype that consists of two different alleles

HOMOZYGOTE: a diploid genotype that consists of two identical alleles

INBREEDING: mating between genetically related individuals

INBREEDING DEPRESSION: a reduction in the health and vigor of inbred offspring, a common and widespread phenomenon

RANDOM MATING: a mating system in which each male gamete (sperm) is equally likely to combine with any female gamete (egg)

Random Mating and the Hardy-Weinberg Law

Soon after the rediscovery of Gregor Mendel's rules of inheritance in 1900, British mathematician Godfrey Hardy and German physician Wilhelm Weinberg published a simple mathematical treatment of the effect of sexual reproduction on the distribution of genetic variation. Both men published their ideas in 1908 and showed that there was a simple relationship between allele frequencies and genotypic frequencies in populations. An allele is simply a genetic variant of a particular gene; for example, blood type in humans is controlled by a single gene with three alleles (A, B, and O). Every individual inherits one allele for each gene from both their mother and father and has a two-allele genotype. In the simplest case with only two alleles (for example, A and a), there are three different genotypes (AA, Aa, aa). The Hardy-Weinberg predictions specify the frequencies of genotypes (combinations of two alleles) in the population: how many will have two copies of the same allele (homozygotes such as AA and aa) or copies of two different alleles (heterozygotes such as Aa).

One important assumption that underlies the Hardy-Weinberg predictions is that gametes (sperm and egg cells) unite at random to form individuals or that individuals pair randomly to produce offspring. An example of the first case is marine organisms such as oysters that release sperm and eggs into the water; zygotes (fertilized eggs) are formed when a single sperm finds a single egg. Exactly which sperm cell and which egg cell combine is expected to be unrelated to the specific allele each gamete is carrying, so the union is said to be random. In cases in which males and females form pairs and produce offspring, it is assumed that individuals find mates without reference to the particular gene under examination. In humans, people do not choose potential mates at random, but they do mate at random with respect to most genetic variation. For instance, since few people know (or care) about the blood type of potential partners, people mate at random with respect to blood-type alleles.

Inbreeding and assortative mating are violations of this basic Hardy-Weinberg assumption. For inbreeding, individuals are more likely to mate with relatives than with a randomly drawn individual (for outbreeding, the reverse is true). Assortative mating occurs when individuals make specific mate choices based on the phenotype or appearance of others. Each has somewhat different genetic consequences. When either occurs, the Hardy-Weinberg predictions are not met, and the relative proportions of homozygotes and heterozygotes are different from what is expected.

The Genetic Effects of Inbreeding

When relatives mate to produce offspring, the offspring may inherit an identical allele from each parent, because related parents share many of the same alleles, inherited from their common ancestors. The closer the genetic relationship, the more alleles two individuals will share. Inbreeding increases the number of homozygotes for a particular gene in a population because the offspring are more likely to inherit identical alleles from both parents. Inbreeding also increases the number of different genes in an individual that are homozygous. In either case, the degree of inbreeding can be measured by the level of homozygosity (the percentage or proportion of homozygotes relative to all individuals).

Inbreeding is exploited by researchers who want genetically uniform (completely homozygous) individuals for experiments: Fruit flies or mice can be made completely homozygous by repeated brother-sister matings. The increase in the frequency of homozygotes can be calcu-

Two children in the Indian state of Bihar in July, 2000. Many children in the area suffer from deformities. Activists blame uranium mining in the area, whereas government officials blame inbreeding, malnutrition, and unsanitary conditions. (AP/Wide World Photos)

lated for different degrees of inbreeding. Self-fertilization is the most extreme case of inbreeding, followed by sibling mating, and so forth. Sewall Wright pioneered computational methods to estimate the degree of inbreeding in many different circumstances. For self-fertilization, the degree of homozygosity increases by 50 percent each generation. For repeated generation of brother-sister matings, the homozygosity increases by about 20 percent each generation.

Inbreeding Depression

Inbreeding commonly produces inbreeding depression. This is characterized by poor health, lower growth rates, reduced fertility, and increased incidence of genetic diseases. Although there are several theoretical reasons why inbreeding depression might occur, the major effects are produced by uncommon and deleterious recessive alleles. These alleles produce negative consequences for the individual when homozygous, but when they occur in a heterozygote, their negative effects are masked by the presence of the other allele. Because inbreeding increases the relative proportion of homozygotes in the population, many of these alleles are expressed, yielding reduced health and vigor. In some cases, the effects can be quite severe. For example, when researchers wish to create homozygous lines of the fruit fly *Drosophila melanogaster* by repeated brother-sister matings, 90 percent or more of the lines fail because of widespread genetic problems.

Assortative Mating

In assortative mating, the probability of particular pairings is affected by the phenotype of the individuals. In positive assortative matings, individuals are more likely to mate with others of the same phenotype, while in negative assortative mating, individuals are more likely to mate with others that are dissimilar. In both cases, the primary effect is to alter the expected genotypic frequencies in the population from those expected under the Hardy-Weinberg law. Positive assortative mating has much the same effect as inbreeding and increases the relative frequency of homozygotes. Negative assortative mating, as expected, has the opposite effect and increases the relative proportion of heterozygotes. Positive assortative mating has been demonstrated for a variety of traits in humans, including height and hair color.

Impact and Applications

The widespread, detrimental consequences of inbreeding are believed to shape many aspects of the natural history of organisms. Many plant species have mechanisms developed through natural selection to increase outbreeding and avoid inbreeding. The pollen (male gamete) may be released before the ovules (female gametes) are receptive, or there may be a genetically determined self-incompatibility to prevent self-fertilization. In most animals, self-fertilization is not possible, and there are often behavioral traits that further reduce the probability of inbreeding. In birds, males often breed near where they were born, while females disperse to new areas. In mammals, the reverse is generally true, and males disperse more widely. Humans appear to be an exception among the mammals, with a majority of cultures showing greater movement by females. These sex-biased dispersal patterns are best understood as mechanisms to prevent inbreeding.

In humans, individuals are unlikely to marry others with whom they were raised. This prevents the potentially detrimental consequences of inbreeding in matings with close relatives. This has also been demonstrated in some birds. Domestic animals and plants may become inbred if careful breeding programs are not followed. Many breeds of dogs exhibit a variety of genetic-based problems (for example, hip problems, skull and jaw deformities, and nervous temperament) that are likely caused by inbreeding. Conservation biologists who manage endangered or threatened populations must often consider inbreeding depression. In very small populations such as species maintained in captivity (zoos) or in isolated natural populations, inbreeding may be hard to avoid. Inbreeding has been blamed for a variety of health defects in cheetahs and Florida panthers.

—*Paul R. Cabe*

See also: Consanguinity and Genetic Disease; Genetic Load; Hardy-Weinberg Law; Heredity and Environment; Hybridization and Introgression; Lateral Gene Transfer; Mendelian Genetics; Natural Selection; Polyploidy; Population Genetics; Punctuated Equilibrium; Quantitative Inheritance; Sociobiology; Speciation.

Further Reading

Avise, John, and James Hamrick, eds. *Conservation Genetics: Case Histories from Nature.* New York: Chapman and Hall, 1996. Examines case studies of germ plasm resources and population genetics, focusing in one chapter on inbreeding in cheetahs and panthers.

Hartl, Daniel. *A Primer of Population Genetics.* Rev. ed. Sunderland, Mass.: Sinauer Associates, 2000. Covers genetic variation, the causes of evolution, molecular population genetics, and the genetic architecture of complex traits.

Hedrick, Philip. *Genetics of Populations.* 2d ed. Boston: Jones and Bartlett, 2000. For those with quantitative experience in the field, this text integrates empirical and experimental approaches with theory, describing methods for estimating population genetics parameters as well as other statistical tools used for population genetics.

Krebs, J., and N. Davies. *An Introduction to Behavioral Ecology.* Malden, Mass.: Blackwell, 1991. Discusses inbreeding avoidance and kin recognition.

Laikre, Linda. *Genetic Processes in Small Populations: Conservation and Management Considerations with Particular Focus on Inbreeding and Its Effects.* Stockholm: Division of Population Genetics, Stockholm University, 1996. Aimed at conservation biologists and addresses the management of inbreeding in small populations. Illustrated.

Soulé, Michael, ed. *Conservation Biology: The Science of Scarcity and Diversity.* Sunderland, Mass.: Sinauer Associates, 1986. Good discussions of inbreeding in birds and mammals, the effects of inbreeding depression in plants and animals, and issues related to the conservation of natural heritage.

Thornhill, Nancy Wilmsen, ed. *The Natural History of Inbreeding and Outbreeding: Theoretical and Empirical Perspectives.* Chicago: University of Chicago Press, 1993. Researchers from several disciplines provide a comprehensive review of ideas and observations on natural inbreeding and outbreeding, among both wild and captive populations. Illustrated.

Incomplete Dominance

Field of study: Classical transmission genetics

Significance: *In most allele pairs, one allele is dominant and the other recessive; however, other relationships can occur. In incomplete dominance, one allele can only partly dominate or mask the other. Some very important human genes, such as the genes for pigmentation and height, show incomplete dominance of alleles.*

Key terms

ALLELE: one of the alternative forms of a gene

CODOMINANCE: the simultaneous expression of two different (heterozygous) alleles for a trait

COMPLETE DOMINANCE: expression of an allele for a trait in an individual that is heterozygous for that trait, determining the phenotype of the individual

HETEROZYGOUS: having two different alleles at a gene locus, often symbolized Aa or a^+a

HOMOZYGOUS: having two of the same alleles at a gene locus, often symbolized AA, aa, or a^+a^+

PHENOTYPE: the expression of a genotype, as observed in the outward appearance or biochemical characteristics of an organism

RECESSIVE TRAIT: a genetically determined trait that is expressed only if an organism receives the gene for the trait from both parents

Incomplete vs. Complete Dominance

Diploid organisms have two copies of each gene locus and thus two alleles at each locus. Each locus can have either a homozygous genotype (two of the same alleles, such as AA, aa, or a^+a^+) or a heterozygous genotype (two different alleles, such as Aa or a^+a). The phenotype of an organism that is homozygous for a particular gene is usually easy to predict. If a pea plant has two tall alleles of the height locus, the plant is tall; if a plant has two dwarf alleles of the height locus, it is small. The phenotype of a heterozygous individual may be harder to predict. In most circumstances, one of the alleles (the dominant) is able to mask or cover the other (the recessive). The phenotype is determined by the dominant allele, so a heterozygous pea plant, with one tall and one dwarf allele, will be tall. When Gregor Mendel delivered the results of his pea-plant experiments before the Natural Sciences Society in 1865 and published them in 1866, he reported one dominant and one recessive allele for each gene he had studied. Later researchers, starting with Carl Correns in the early 1900's, discovered alleles that did not follow this pattern.

When a red snapdragon or four-o'clock plant is crossed with a white snapdragon or four-o'clock, the offspring are neither red nor white. Instead, the progeny of this cross are pink. Similarly, when a chinchilla (gray) rabbit is crossed with an albino rabbit, the progeny are neither chinchilla nor albino but an intermediate shade called light chinchilla. This phenomenon is known as incomplete dominance, partial dominance, or semidominance.

If the flower-color locus of peas is compared with the flower-color locus of snapdragons, the differences and similarities can be seen. The two alleles in peas can be designated W for the purple allele and w for the white allele. Peas that are WW are purple, and peas that are ww are white. Heterozygous peas are Ww and appear purple. In other words, as long as one dominant allele is present, enough purple pigment is made to make the plant's flower color phenotype purple. In snapdragons, R is the red allele and r is the white allele. Homozygous RR plants have red flowers and rr plants have white flowers. The heterozygous Rr plants have the same kind of red pigment as the *RR* plants but not enough to make the color red. Instead, the less pigmented red flower is designated as pink. Because neither allele shows complete dominance, other symbols are sometimes used. The red allele might be called c^R or C_1, while the white allele might be called c^W or C_2.

The Enzymatic Mechanism of Incomplete Dominance

To understand why incomplete dominance occurs, metabolic pathways and the role of enzymes must be understood. Enzymes are proteins that are able to increase the rate of chemical reactions in cells without the enzymes themselves being altered. Thus an enzyme can be used over and over again to speed up a particular reaction. Each different chemical reaction in a cell needs its own enzyme. Each enzyme is composed of one or more polypeptides, each of which is coded by a gene. Looking again at flower color in peas, the *W* allele codes for an enzyme in the biochemical pathway for production of purple pigment. Whenever a *W* allele is present, this enzyme is also present. The *w* allele has been changed (mutated) in some way so that it no longer codes for a functional enzyme. Thus *ww* plants have no functional enzyme and cannot produce any purple pigment. Since many biochemicals such as fibrous polysaccharides and proteins found in plants are opaque white, the color of a *ww* flower is white by default. In a *Ww* plant, there is only one copy of the allele for a functional enzyme. Since enzymes can be used over and over again, one copy of the functional allele produces sufficient enzyme to make enough pigment for the flower to appear purple. In snapdragons the *R* allele, like the *W* allele, codes for a functional enzyme, while the *r* allele does not. The difference is in the enzyme coded by the *R* allele. The snapdragon enzyme is not very efficient, which leads to a deficiency in the amount of red pigment. Flowers with the reduced amount of red pigment appear pink.

Phenotypic Ratios

Phenotypic ratios in the progeny from controlled crosses are also different than for simple Mendelian traits. For Mendelian traits, crossing two heterozygous individuals will produce the following results: $Ww \times Ww \rightarrow \frac{1}{4}WW + \frac{1}{2}Ww + \frac{1}{4}ww$. Since both *WW* and *Ww* look the same, the $\frac{1}{4}WW$ and the $\frac{1}{2}Ww$ can be added together to give $\frac{3}{4}$ purple. In other words, when two heterozygotes are crossed, the most common result is to have $\frac{3}{4}$ of the progeny look like the dominant and $\frac{1}{4}$ look like the recessive—the standard 3:1 ratio. With incomplete dominance, each genotype has its own phenotype, so when two heterozygotes are crossed (for example, $Rr \times Rr$), $\frac{1}{4}$ of the progeny will be *RR* and look like the dominant (in this case red), $\frac{1}{4}$ will be *rr* and look like the recessive (in this case white), but $\frac{1}{2}$ will be *Rr* and have an intermediate appearance (in this case pink)—a 1:2:1 ratio.

Codominance

One type of inheritance that can be confused with incomplete dominance is codominance. In codominance, both alleles in a heterozygote are expressed simultaneously. Good examples are the *A* and *B* alleles of the human ABO blood system. ABO refers to chemicals, in this case short chains of sugars called antigens, that can be found on the surfaces of cells. Blood classified as A has *A* antigens on the surface, B blood has *B* antigens, and AB blood has both *A* and *B* antigens. (O blood has neither *A* nor *B* antigens on the surface.)

Genetically, individuals that are homozygous for the A allele, $I^A I^A$, have *A* antigens on their cells and are classified as type A. Those homozygous for the *B* allele, $I^B I^B$, have *B* antigens and are classified as type B. Heterozygotes for these alleles, $I^A I^B$, have both *A* and *B* antigens and are classified as type AB. This is called codominance because both alleles are able to produce enzymes that function. When both enzymes are present, as in the heterozygous $I^A I^B$ individual, both antigens will be formed. The progeny ratios are the same for codominance and incomplete dominance, because each genotype has its own phenotype.

Whether an allele is called completely dominant, incompletely dominant, or codominant often depends on how the observer looks at the phenotype. Consider two alleles of the hemoglobin gene: H^A (which codes for normal hemoglobin) and H^S (which codes for sickle-cell hemoglobin). To the casual observer, both $H^A H^A$ homozygotes and $H^A H^S$ heterozygotes have normal-appearing blood. Only the $H^S H^S$ homozygote shows the sickling of blood cells that is characteristic of the disease. Thus H^A is dominant to H^S. Another observer, however, may note that under conditions of oxygen deprivation, the blood of heterozygotes does sickle. This looks like incomplete dominance. The phenotype is intermediate between never sickling, as seen in the normal homozygote, and frequently sickling, as seen in the $H^S H^S$ homozygote. A third way of observing, however, would be to look at the hemoglobin itself. In normal homozygotes, all hemoglobin is normal. In $H^S H^S$ homozygotes, all hemoglobin is abnormal. In the heterozygote, both normal and abnormal hemoglobin is present; thus, the alleles are codominant.

Incomplete Dominance and Polygenes

In humans and many other organisms, single characteristics are often under the genetic control of several genes. Many times these genes function in an additive manner so that a characteristic such as height is not determined by a single height gene with just two possible alternatives, as in tall and dwarf peas. There can be any number of these genes that determine the expression of a single characteristic, and very often the alleles of these genes show incomplete dominance.

Suppose one gene with an incompletely dominant allele determined height. Three genotypes of height could exist: *HH*, which codes for the maximum height possible (100 percent above the minimum height), *Hh*, which codes for 50 percent above the minimum height, and *hh*, which codes for the minimum height. If two height genes existed, there would be five possible heights: *AABB* (maximum height); *AaBB* or *AABb* (75 percent above minimum); *AAbb*, *AaBb*, or *aaBB* (50 percent above minimum); *Aabb* or *aaBb* (25 percent above minimum); and *aabb* (minimum). If there were five genes involved in height, there would be *aabbccddee* individuals with minimum height; *Aabbccddee*, *aaBbccddee*, and other individuals having genotypes with only one of the incompletely dominant alleles at 10 percent above the minimum;

AAbbccddee, *aaBbccDdee*, and other individuals with two incompletely dominant alleles at 20 percent above the minimum; all the way up to *AABBCCDDEE* individuals that show the maximum (100 percent above the minimum) height. The greater the number of genes with incompletely dominant alleles that affect a phenotype, the more the distribution of phenotypes begins to look like a continuous distribution. Human skin, hair, and eye pigmentation phenotypes are also determined by the additive effects of several genes with incompletely dominant alleles.

Incomplete Dominance and Sex Linkage

In many organisms, sex is determined by the presence of a particular combination of sex chromosomes. Human females, for example, have two of the same kind of sex chromosomes, called X chromosomes, so that all normal human females have the XX genotype. Human males have two different sex chromosomes; thus, all normal human males have the XY genotype. The same situation is also seen in the fruit fly *Drosophila melanogaster*. When genes with incompletely dominant alleles are located on the X chromosome, only the female with her two X chromosomes can show incomplete dominance. The apricot (w^a) and white (w) alleles of the eye color gene in *D. melanogaster* are on the X chromosome, and w^a is incompletely dominant to w. Male flies can have either of two genotypes, $w^a Y$ or wY, and appear apricot or white, respectively. Females have three possible genotypes: $w^a w^a$, $w^a w$, and ww. The first is apricot and the third is white, but the second genotype, $w^a w$, is an intermediate shade often called light apricot.

In birds and other organisms in which the male has two of the same kind of sex chromosomes and the female has the two different sex chromosomes, only the male can show incomplete dominance. A type of codominance can also be seen in genes that are sex linked. In domestic cats, an orange gene exists on the X chromosome. The alleles are orange (X^O) and not orange (X^+). Male cats can be either black (or any color other than orange, depending on other genes that influence coat color) when they are X^+Y, or they can be orange (or light orange) when they are X^OY. Females show those same colors when they are homozygous (X^+X^+ or X^OX^O) but show a tortoiseshell (or calico) pattern of both orange and not-orange hairs when they are X^+X^O.

—*Richard W. Cheney, Jr., updated by Bryan Ness*

See also: Biochemical Mutations; Complete Dominance; Dihybrid Inheritance; Epistasis; Mendelian Genetics; Monohybrid Inheritance; Multiple Alleles.

Further Reading

Grant, V. *Genetics of Flowering Plants.* New York: Columbia University Press, 1975. Thoroughly reviews heredity in plants and covers incomplete dominance.

Lewis, Ricki. *Human Genetics: Concepts and Applications.* 5th ed. New York: McGraw-Hill, 2003. An introductory text for undergraduates with sections on fundamentals, transmission genetics, DNA and chromosomes, population genetics, immunity and cancer, and the latest genetic technology.

Nolte, D. J. "The Eye-Pigmentary System of *Drosophila.*" *Heredity* 13 (1959). Covers *Drosophila* eye pigments.

Searle, A. G. *Comparative Genetics of Coat Color in Mammals* (New York: Academic Press, 1968). Addresses mammalian coat colors.

Yoshida, A. "Biochemical Genetics of the Human Blood Group ABO System." *American Journal of Genetics* 34 (1982). Covers the genetics of the ABO system.

Infertility

Field of study: Diseases and syndromes
Significance: *Infertility is a disease of the reproductive system that impairs the conception of children. About one in six couples in the United States is infertile. The risk that a couple's infertility may be caused by genetic problems such as abnormal sex chromosomes is approximately one in ten.*

Key terms

IN VITRO FERTILIZATION (IVF): a process in which harvested eggs and sperm are brought together artificially to form a zygote

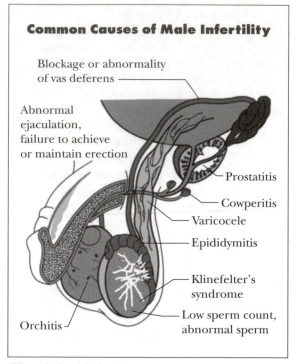

Common Causes of Male Infertility

Blockage or abnormality of vas deferens

Abnormal ejaculation, failure to achieve or maintain erection

Prostatitis

Cowperitis

Varicocele

Epididymitis

Klinefelter's syndrome

Orchitis

Low sperm count, abnormal sperm

(Hans & Cassidy, Inc.)

SEX CHROMOSOMES: the chromosomes that determine the sex of an individual; females have two X chromosomes, while males have one X and one Y chromosome

A Reproductive Disease

Infertility is a disease of the reproductive system that impairs a couple's ability to have children. Sometimes infertility has a genetic cause. The conception of children is a complex process that depends upon many factors, including the production of healthy sperm by the man and healthy eggs by the woman, unblocked Fallopian tubes that allow the sperm to reach the egg, the sperm's ability to fertilize the egg when they meet, the ability of the fertilized egg (embryo) to become implanted in the woman's uterus, and sufficient embryo quality. If the pregnancy is to continue to full term, the embryo must be healthy, and the woman's hormonal environment must be adequate for its development. Infertility can result when one of these factors is impaired. Physicians define infertility as the inability to conceive a child after one year of trying.

Genetic Causes of Infertility

The most common male infertility factors include conditions in which few or no sperm cells are produced. Sometimes sperm cells are malformed or die before they can reach the egg. A genetic disease such as a sex chromosome abnormality can also cause infertility in men. A genetic disorder may be caused by an incorrect number of chromosomes (having more or fewer than the normal forty-six chromosomes). Having a wrong arrangement of the chromosomes may also cause infertility. This situation occurs when part of the genetic material is lost or damaged. One such genetic disease is Klinefelter syndrome, which is caused by an extra X chromosome in males. The loss of a tiny piece of the male sex chromosome (the Y chromosome) may cause the most severe form of male infertility: the complete inability to produce sperm. This form of infertility can arise from a deletion in one or more genes in the Y chromosome. Fertility problems can pass from father to son, especially in cases in which physicians use a single sperm from an infertile man to inseminate a woman's egg.

Female infertility may be caused by an irregular menstrual cycle, blocked Fallopian tubes, or birth defects in the reproductive system. One genetic cause of infertility in females is Turner syndrome. Most females with Turner syndrome lack all or part of one of their X chromosomes. The disorder may result from an error that occurs during division of the parent's sex cells. Infertility and short stature are associated with Turner syndrome. Other genetic disorders in females include trisomy X, tetrasomy

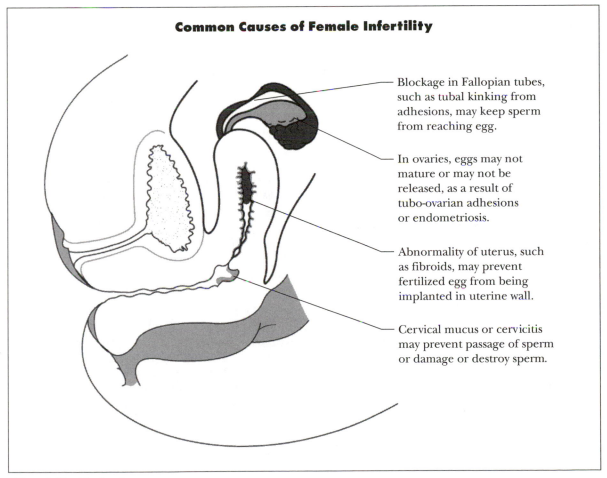

Common Causes of Female Infertility

Blockage in Fallopian tubes, such as tubal kinking from adhesions, may keep sperm from reaching egg.

In ovaries, eggs may not mature or may not be released, as a result of tubo-ovarian adhesions or endometriosis.

Abnormality of uterus, such as fibroids, may prevent fertilized egg from being implanted in uterine wall.

Cervical mucus or cervicitis may prevent passage of sperm or damage or destroy sperm.

(Hans & Cassidy, Inc.)

X, and pentasomy. These syndromes are the female counterparts of Klinefelter syndrome and can be associated with mental retardation.

At least 60 percent of miscarriages or pregnancy losses are caused by chromosomal abnormalities. Most babies with these abnormalities would not survive even if they were born. Chromosomal problems are more common if the mother is older and has a history of requiring longer than a year to conceive. Men who are older or who have a history of being subfertile can also contribute to genetic abnormalities. After the age of thirty-five, the structure within a woman's eggs is more likely to become damaged. Men over the age of forty-five have an increased risk of damage to the structure of the chromosomes in their sperm.

Scientists believe that as their understanding of the genetic basis of infertility problems increases, new therapies will be developed to treat them. Most infertility cases are treated with drugs or surgery to repair the reproductive organs. No treatment is available to correct sex chromosomal abnormalities such as Turner syndrome. However, some women with Turner syndrome can have children. For women who cannot conceive, possible procedures include in vitro fertilization (fertilizing a woman's egg with sperm outside the body) and embryo transfer (moving the fertilized egg into a woman's uterus). Adoption is another option for infertile men and women.

—Fred Buchstein

See also: Amniocentesis and Chorionic Villus Sampling; Cloning; Genetic Counseling; Genetic Screening; Genetic Testing; Genetic Testing: Ethical and Economic Issues; Hereditary Diseases; In Vitro Fertilization and Embryo Transfer; Prenatal Diagnosis; Stem Cells; Sterilization Laws; Totipotency; Turner Syndrome; X Chromosome Inactivation; XYY Syndrome.

Further Reading

Bentley, Gillian R., and C. G. Nicholas Mascie-Taylor. *Infertility in the Modern World: Present and Future Prospects.* New York: Cambridge University Press, 2000. Discusses changes in human reproduction brought on by the intersection of biology, the environment, and culture.

Jansen, Robert, and D. Mortimer, eds. *Towards Reproductive Certainty: Fertility and Genetics Beyond 1999.* Boca Raton, Fla.: CRC Press, 1999. Surveys the status of conception in controlled circumstances outside the body, including ethical, medical, and psychological considerations.

Lewis, Ricki. *Human Genetics: Concepts and Applications.* 5th ed. New York: McGraw-Hill, 2003. An introductory text for undergraduates with sections on fundamentals, transmission genetics, DNA and chromosomes, population genetics, immunity and cancer, and the latest genetic technology.

McElreavey, Ken, ed. *The Genetic Basis of Male Infertility.* New York: Springer, 2000. Explores medical progress in understanding the genetics of spermatogenesis and male infertility. Illustrated.

Marrs, Richard, et al. *Dr. Richard Marrs' Fertility Book.* New York: Dell, 1997. Covers advances in reproductive technology, how emotions can delay or stop ovulation, male sperm count that is borderline or subnormal, which fertility drugs work best and the associated side effects, chances of multiple births, and when to change doctors or see a specialist.

Rosenthal, M. Sara. *The Fertility Sourcebook: Everything You Need to Know.* 2d ed. Los Angeles: Lowell House, 1998. Addresses advances in fertility treatments, including issues for same-sex partners, ethical considerations, and basic information about treatment options.

Web Sites of Interest

American Association of Reproductive Medicine. http://www.asrm.org. Site includes information on infertility and reproduction.

International Council on Infertility Information Dissemination. http://www.inciid.org. Site provides "fact sheets" on in vitro fertilization.

National Institutes of Health, National Libary of Medicine. http://www.nlm.nih.gov/medlineplus/infertility.html. Provides information on all aspects of infertility.

Insurance

Field of study: Bioethics; Human genetics and social issues

Significance: *Many social policy analysts and public health advocates worry that as genetic screening becomes more widely available, individuals considered high-risk may be denied health and life insurance coverage. On the other hand, if such information is withheld from insurance companies, individuals might purchase extra life insurance, causing insurance companies to unknowingly carry unacceptably high risks. Some kind of balance between appropriate disclosure and privacy rights will need to be established.*

Key terms

ALZHEIMER'S DISEASE: a degenerative brain disorder usually found among the elderly; sufferers gradually lose cognitive function and become unable to function independently

CHRONIC ILLNESS: an ongoing condition such as diabetes or hypertension

HIGH RISK: characterized by being likely to someday suffer from a particular disease or disabling condition

PREEXISTING CONDITION: a disease or disorder that is diagnosed prior to a person's application for insurance coverage

High-Risk Individuals and Preexisting Conditions

As tests for a steadily increasing number of genetic defects are perfected, concern has grown among both health experts and the general public that negative results could lead to the denial of health insurance coverage to these high-risk individuals. The insurance industry has always been reluctant to insure people identified as being at high risk or who suffer from preexisting conditions, a reluctance that has intensified as health care costs have increased. For example, people with a family medical history of coronary artery disease have long been considered a higher risk than members of the general population. As a consequence, based on information provided through disclosures of family histories, these people occasionally have been denied health insurance coverage or have been required to pay higher premiums.

Similarly, people who suffer from conditions such as diabetes or hypertension and who change jobs or insurance carriers occasionally discover that their new medical insurance will not pay for any treatment for medical conditions that had been diagnosed prior to obtaining the new insurance. Such "preexisting" conditions are considered ineligible for payment of benefits. While some insurance companies will put a time limit on the restrictions for coverage of preexisting conditions of a few months or a year, providing there are no active occurrences of the disorder, other insurers may exclude making any payments related to a preexisting condition for an indefinite period of time. A person with a chronic condition such as diabetes may discover that while a new insurer will pay for conditions unrelated to the diabetes, such as a broken leg, the individual will be solely responsible for any diabetes-related expenses for the remainder of his or her life. Alternately, the sufferer of a chronic condition may discover that health insurance is available, but only at a much higher premium.

Insurance and Genetic Screening

Insurance companies are just beginning to confront the problems of genetic tests for genetic predisposition to disease. In one court case, *Katskee v. Blue Cross Blue Shield of Nebraska* (1994), the plaintiff had been diagnosed with a 50 percent chance of developing breast and/or ovarian cancer. Consequently, she was seeking payment from her insurance company to cover the costs of prophylactic removal of her ovaries. Initially, the insurance company approved the surgery, but later it reversed that decision, saying that the plaintiff was not covered because her condition was not a "disease" or "bodily disorder." The suit occurred because the plaintiff proceeded with the surgery anyway and then looked to the courts to help her collect from her insurer. The first ruling was in favor of the insurance company but was reversed on appeal, the higher court considering a 50 percent predisposition as meeting the definition of a disease.

Two responses from the insurance industry

are possible as cases like these become more common. They could choose to cover prophylactic treatments as a way to cut long-term costs associated with development of the genetic disease, or they could choose to exclude such conditions under a "preexisting condition" argument. So far the trend has been toward trying to exclude treatment, by the latter approach, including considering the later development of the full-blown disease as a preexisting condition. This is a disturbing trend, as it would tend to discourage genetic screening, ruining the opportunity for preventive measures. If the courts decide to require that insurers must fund prophylactic or preventive treatments, then another quandary occurs: At what percentage predisposition will insurers be required to cover the costs? A predisposition of 50 percent seems like a reasonable number, but what about 45 percent, also high? Covering any level of predisposition would be unreasonable, as it would bankrupt the system, so a line must be drawn, but where? Much more information will be needed before such lines can be drawn without being arbitrary.

In the case of degenerative disorders such as Alzheimer's disease or Huntington's disease, for which there is no prophylactic treatment available, patients may live for many years following the initial diagnosis of the disease while they become progressively more helpless and eventually require extended hospitalization or custodial care. An insurance company that wrote plans to cover nursing-home care could decide to exclude people identified as carrying a gene putting them at risk of developing Alzheimer's disease. The insurance company's reasoning would be that because Alzheimer's sufferers may require many more years of custodial care than the average nursing-home resident, it would be unprofitable to insure known future Alzheimer's sufferers. Such people would be seen as simply being too high-risk.

A number of geneticists and other analysts have suggested that another inherent difficulty with genetic screening is that it opens the door for possible restriction of access to health insurance while not holding out any hope of a treatment or cure for the patient. It is now possible to detect the genetic markers for many condi-

tions for which no effective preventive treatment exists. Alzheimer's disease provides a particularly poignant example. As of the late 1990's, the connection between genes identified as appearing in some early-onset Alzheimer's disease patients and the disease itself was still unclear. People who underwent genetic screening to discover if they carried that particular genetic marker could spend many decades worrying needlessly about their own risk of developing Alzheimer's disease while knowing that there was no way to prevent it. At the same time, the identification of the genetic marker would have identified the patient as a high risk for medical insurance. Huntington's disease represents an even more serious case, in which the test is nearly 100 percent predictive. A positive test is essentially a guaranteed prediction of early death. Tests like these may provide no benefit to affected individuals, and may even cause harm if the information is freely available to insurers.

On the other hand, in some cases the benefits of genetic screening may outweigh its potential costs. For example, certain cancers have long been recognized as running in some families. Doctors routinely counsel women with a family history of breast cancer to have annual mammograms and even, in cases where the risk seems particularly high, to undergo prophylactic mastectomy or lumpectomy. The discovery of genetic markers for breast cancer suggests that women who are concerned that they are at higher-than-average risk for the disease can allay their fears through genetic screening rather than subjecting themselves to disfiguring surgery. Still, the very act of screening could become a double-edged sword. A positive test not only would confirm a woman's worst fears but also could result in her being denied high insurance coverage. Many patients with a high-risk family profile fear that even if the screening turns out negative, simply requesting the test will serve as a flag to health insurers, and they, too, will be assessed higher premiums or denied coverage based on their family histories.

In a climate of rising medical costs and efforts by both traditional insurance providers and health maintenance organizations to reduce expenses, many people feel there is good reason to fear that genetic screening will serve primarily as a tool to restrict access to health insurance. In response to these concerns, a number of government studies have been undertaken to assess potential remedies. Some possible solutions include making the results of all genetic tests confidential, available to insurers and employers only with permission from the individual; passing laws that prohibit discrimination of individuals with genetic predispositions by insurers and employers; and universal health coverage with clearly defined guidelines based on extensive research. A number of states have already enacted laws prohibiting insurance companies from denying coverage to individuals with genetic defects. It is also possible that the Americans with Disabilities Act could be cited against genetic discrimination.

Impact of Medical Genomics

With the mapping of the human genome completed in 2003, it suddenly became clear that nearly all human disease—from complex chronic conditions such as cancer, Alzheimer's, and diabetes to the predisposition for infectious disease and even trauma—has some genetic basis. Although genome sequences are essentially the same among all individuals, what variation there is accounts for many of the differences in disease susceptibility and other health-related differences. All of this has made the drive to study human genomics as it affects human health a burgeoning new field, medical genomics, that promises to affect every medical field. The basis for this discipline will be data gleaned from large, well-designed and controlled clinical studies that are being developed and implemented in several nations to provide information on how genes influence a wide range of traits, from disease states to behavior.

Given the dangers outlined above, it would seem that such studies pose an increased concern regarding issues of privacy, discrepancies in access to health care, and even threats to individuals' jobs, as more and more employers have been forced to "self-insure"—essentially becoming insurers themselves and thus being forced to consider employees' health in decisions of hiring and continued employment. However, in an op-ed piece for *The Wall Street Journal* (December 20, 2002), William R. Brody, president of The Johns Hopkins University, sounded an ominous yet potentially positive note. In view of the inevitable discovery that nearly all disease conditions have some genetic basis, he predicted that, given the difficulty private insurers will soon face in discriminating among conditions, they might also be facing their own demise as medical insurers:

> If legislatures pass laws banning insurers from using genetic screening data, those companies will protect themselves by continually raising premiums to consumers. Some may even go bankrupt because purchasers of insurance will be the more knowledgeable in the transaction. Yet if we allow insurers to use genetic data, many more individuals will be left without coverage because they will be deemed too high-risk to warrant insurance at affordable prices. Given this conundrum, there is only one solution that can preserve the concept of health insurance: universal coverage.

Based on the concept of "community rating," such coverage would spread risk across a large group of individuals (now confined to smaller groups), and hence cost would spread across a national pool, allowing individual traits and hence risks to be diluted. Brody predicts that "that day is coming sooner than many people imagine."

—*Nancy Farm Männikkö and Bryan Ness*

See also: Aging; Alzheimer's Disease; Bioethics; Bioinformatics; Breast Cancer; Congenital Defects; Eugenics; Eugenics: Nazi Germany; Forensic Genetics; Gene Therapy; Gene Therapy: Ethical and Economic Issues; Genetic Counseling; Genetic Screening; Genetic Testing; Genetic Testing: Ethical and Economic Issues; Genomic Libraries; Genomics; Hereditary Diseases; Human Genetics; Icelandic Genetic Database; Prenatal Diagnosis; Race; Sickle-Cell Disease; Sterilization Laws.

Further Reading

Brody, William R. "A Brave New Insurance." *The Wall Street Journal*, December 20, 2002. In this op-ed article, Brody addresses the impact that swift progress and refinement of genetic screening and testing will have on the insurance industry in the United States.

Hubbard, Ruth, and Elijah Wald. *Exploding the Gene Myth: How Genetic Information Is Produced and Manipulated by Scientists, Physicians, Employers, Insurance Companies, Educators, and Law Enforcers.* Boston: Beacon Press, 1999. Argues against genetic determinism and biotechnology and attacks scientists who cite DNA sequences as the presumed basis for a genetic tendency to cancer, high blood pressure, alcoholism and criminal behavior.

Orin, Rhonda D. *Making Them Pay: How to Get the Most from Health Insurance and Managed Care.* New York: St. Martin's Press, 2001. A consumer guide to health insurance and managed care programs that explains how to read and understand a health plan and how to work with insurance companies to get the benefits to which one is entitled.

Rifkin, Jeremy. *The Biotech Century: Harnessing the Gene and Remaking the World.* New York: Jeremy P. Tarcher/Putnam, 1998. Discusses a variety of concerns regarding biotechnology and shows how genetic screening fits into a much wider area of debate in modern science.

U.S. Congress. Senate. Committee on Health, Education, Labor, and Pensions. *Fulfilling the Promise of Genetics Research: Ensuring Nondiscrimination in Health Insurance and Employment.* Washington, D.C.: Government Printing Office, 2001. Committee formed to explore possible connections between genetics research and health insurance and job discrimination, and to ensure against discrimination in these areas.

_____. *Protecting Against Genetic Discrimination: The Limits of Existing Laws.* Washington, D.C.: U.S. Government Printing Office, 2002. Examines existing laws and proposed legislation to prevent genetic discrimination in the form of health insurance loss or denial or losing one's job.

Zallen, Doris Teichler. *Does It Run in the Family? A Consumer's Guide to DNA Testing for Genetic Disorders.* New Brunswick, N.J.: Rutgers University Press, 1997. Provides readers with the knowledge they need to make decisions regarding genetic testing and does so in an easy-to-understand way.

Web Site of Interest

National Human Genome Research Institute, Health Insurance in the Age of Genetics. http://www.nhgri.nih.gov/news/insurance. Discusses the need for health insurance regulation at the federal level to prevent discrimination against individuals because of their genetic makeup.

Intelligence

Field of study: Human genetics and social issues

Significance: *The study of the genetic basis of intelligence is one of the most controversial areas in human genetics. Researchers generally agree that mental abilities are genetically transmitted to some extent, but there is disagreement over the relative roles of genes and environment in the development of mental abilities. There is also disagreement over whether different mental abilities are products of a single ability known as intelligence and disagreement over how to measure intelligence.*

Key terms

DIZYGOTIC ORGANISM: an organism developed from two separate ova; fraternal twins are dizygotic

INTELLIGENCE QUOTIENT (IQ): the most common measure of intelligence; it is based on the view that there is a single capacity for complex mental work and that this capacity can be measured by testing

MONOZYGOTIC ORGANISM: developed from a single ovum (egg); identical twins are monozygotic because they originate in the womb from a single fertilized ovum that splits in two

PSYCHOMETRICIAN: one who measures intellectual abilities or other psychological traits

Evidence for Genetic Links to Intelligence

Much of the research into the connection between genes and intelligence has focused on attempting to determine the relative roles of biological inheritance and social influence in de-

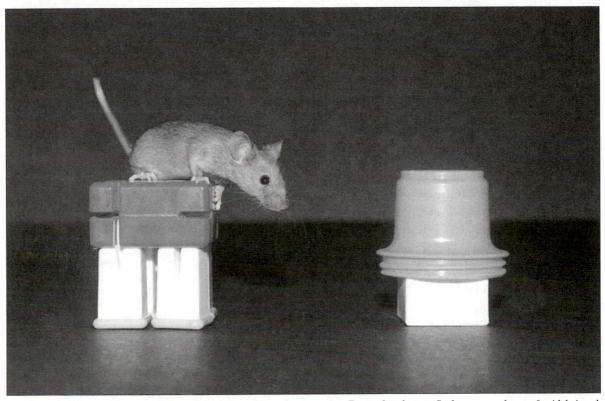

A genetically engineered smart mouse performs a learning and memory test. Researchers hope to find causes and cures for Alzheimer's disease, and possibly ways to increase human intelligence. (AP/Wide World Photos)

veloping intelligence. Such attempts have usually involved a combination of four methods: associations of parental intelligence with the intelligence of offspring, associations of the intelligence of siblings (brothers and sisters), comparisons of dizygotic (fraternal) twins and monozygotic (identical) twins, and adoption studies.

To the extent that mental qualities are inherited, one should expect blood relatives to share these qualities with each other more than with nonrelatives. In an article published in 1981 in the journal *Science*, T. J. Bouchard, Jr., and Matt McGue examined studies that looked at statistical relationships of intellectual abilities among family members. These studies did reveal strong associations between mental capacities of parents and children and strong associations among the mental capacities of siblings. Further, if genes are involved in establishing mental abilities, one should expect that the more genes related people share, the more similar they will be in intelligence. Studies have indicated that fraternal twins are only slightly more similar to each other than are nontwin siblings. Identical twins, developing from a single egg with identical genetic material, have even more in common. Bouchard and McGue found that there was an overlap of about 74 percent in the intellectual abilities of identical twins and an overlap of about 36 percent in the intellectual abilities of fraternal twins.

Family members may be similar because they live in similar circumstances, and identical twins may be similar because they receive nearly identical treatment. However, studies of adopted children show that the intellectual abilities of these children were more closely related to those of their biological parents than to those of their adoptive parents. Studies of identical twins who were adopted and raised apart from each other indicate that these twins have about 62 percent of their intellectual abilities in common.

Twin studies, in particular, have helped to establish that heredity is involved in a number of intellectual traits. Memory, number ability, perceptual skills, psychomotor skills, fluency in language use, and proficiency in spelling are only a few of the traits in which people from common genetic backgrounds tend to be similar to each other. However, psychometricians have not reached agreement on the extent to which mental abilities are products of genes rather than of environmental factors such as upbringing and opportunity. Some researchers estimate that only 40 percent of intellectual ability is genetic; others set the estimate as high as 80 percent.

It is important to keep in mind that even if most differences in mental abilities among human beings were caused by genetics, members of families would still show varied abilities. If, for example, there is a gene for high mathematical ability (gene A) and a gene for low mathematical ability (gene a), it is quite possible that a woman who has inherited each gene (Aa) from her parents will marry a man who has inherited each gene (Aa) from his parents. In this case, there is a 1 in 4 probability that they will have a child who is mathematically gifted (AA) and a 1 in 4 probability that they will have a child who is mathematically slow (aa). This example, although grossly simplified, gives an idea of the effect of variation in the genes inherited.

The Problem of Defining and Measuring Intelligence

Debates over genetic links to intelligence are complicated by the problem of precisely defining and accurately measuring intelligence. It may be that abilities to build houses, draw, play music, or understand complex mathematical procedures are inherited as well as learned. Which of these abilities, however, constitute intelligence? Because of this debate, some people, such as Harvard psychologist Howard Gardner, have argued that there is no single quality of intelligence but rather multiple forms of intelligence.

If there is no single ability that can be labeled "intelligence," this means that one cannot measure intelligence or determine the extent to which general intellectual ability may be genetic in character. Intelligence quotient (IQ), the measure of intelligence most commonly used to study genetic links to intellectual ability, is based on the view that there is a great deal of overlap among various mental traits. Although a given individual may be skilled at music or writing and poor at mathematics, on the average, people who are proficient in one area also tend to be talented in other areas. Proponents of IQ measures argue that this overlap exists because there is a single, underlying, general intelligence that affects how people score on tests of various kinds of mental abilities. The opponents of IQ measures counter that even if one can speak of intelligence rather than "intelligences," it is too complex to be reduced to one number.

Impact and Applications

The passing of mental abilities from parents to children by genetic inheritance is a politically controversial issue because genetic theories of intelligence may be used to justify existing social inequalities. Social and economic inequalities among racial groups, for example, have been explained as differences among groups in inherited intelligence levels. During the nineteenth century, defenders of slavery claimed that black slaves were by nature less intelligent than the white people who held them in slavery. After World War I, the Princeton University psychologist C. C. Brigham concluded from results of army IQ tests that southern European immigrants had lower levels of inherited intelligence than native-born Americans and that blacks had even more limited intelligence. White supremacists and segregationists used Brigham's results to justify limiting the access of blacks to higher education and other opportunities for advancement. In 1969, Berkeley psychologist Arthur R. Jensen touched off a storm of debate when he published an article that suggested that differences between black and white children in educational success were caused in part by genetic variations in mental ability.

Wealth and poverty, even within racial and ethnic groups, have been explained as consequences of inherited intelligence. Harvard psychologist Richard Herrnstein and social critic Charles Murray have argued that American society has become a competitive, information-based society in which intellectual ability is the primary basis of upward mobility. They have maintained, furthermore, that much of intellectual ability is genetic in character and that people tend to marry and reproduce within their own social classes. Therefore, in their view, social classes also tend to be intellectual classes: a cognitive elite at the top of the American social system and a genetically limited lower class at the bottom.

Scientific truth cannot be established by accusing theories of being inconvenient for social policies of equal opportunity. Nevertheless, it is not clear that genetic differences in intelligence are necessarily connected to social status. Even those who believe that inherited intelligence affects social position generally recognize that social status is affected by many other factors such as parental wealth, ed-

---

Genetics and IQ

The genetics of intelligence continues to prompt controversy and often emotional debate centering on the relative roles of genetics and environment in shaping intelligence and multiple intelligence. The dictionary defines intelligence as the capacity to acquire knowledge, process information by reasoning, and make rational decisions. It follows that some individuals may have a greater facility for acquiring and analyzing information than others. Even the concept of multiple intelligence implies that some individuals are more intelligent than others. This is not to say, however, that individuals with a higher level of intelligence will always be more successful while individuals with lower levels of intelligence will always be failures. That is, the abilities conferred by higher levels of inherited intelligence are probabilistic rather than determinate and are shaped by many factors other than genetics alone.

Out of the enormous amount of debate certain facts have emerged about the inheritance of intelligence in humans. First, geneticists, behavioral geneticists, and neurobiologists have consistently demonstrated that there is a significant genetic contribution to intelligence, although the exact genes that code for intelligence have not yet been discovered. Embedded within this inheritance pattern, however, is the undeniable and at times substantial contribution of the social environment in development of intelligence in individuals.

The role of environment in shaping intelligence has also been consistently demonstrated to be a vital factor in shaping intelligence. This is most clearly shown where environmental factors adversely influence growth and development of the central nervous system. Low birth rate, anoxia, malnutrition, childhood trauma, income, occupation, parent sep-aration, and divorce have all been shown to influence the development of intelligence by as much as 40 percent, leading some environmentalists and sociologists to claim that culture is the major factor in intelligence. It is precisely because of the influence of such environmental factors in shaping intelligence that performance gains can be increased somewhat.

Support for the genetic contribution to intelligence comes primarily from studies of identical and fraternal twins, siblings, and family groupings. For example, the correlation of intelligence between identical twins (monozygotic twins) reared together is consistently well over 0.8 (1.0 being the highest correlation), with highest scores measured at 0.86. Scores of fraternal (dizygotic) twins and siblings are lower but still higher than less closely related kin such as cousins and uncles. Furthermore, intelligence measures of adopted siblings show lower correlations compared to intelligence correlations between or among natural siblings. Thousands of such data have led most authorities to suggest that between 40 and 80 percent of an individual's intelligence is shaped by genetics.

The basic genetic mechanisms underlying these observations are not, as yet, well understood. Geneticists, behavioral geneticists, and neurobiologists argue that genes code for brain size, number of brain cells, and number of connections, all of which probably play roles in determining intellectual ability, though the relative contributions of each remain unclear. Further evidence for the role of genetics comes from cases of chromosomal deletions, trisomy, and other genetic abnormalities.

—*Dwight G. Smith*

ucational opportunity, and cultural attitudes.

It seems evident that there are genetic links to mental ability. At the same time, however, the extent to which genes shape intellectual capacities, whether these capacities should be combined into one dimension called intelligence, and the validity of measures of intelligence remain matters of debate. The scientific debate, moreover, is difficult to separate from social and political debates.

—Carl L. Bankston III

See also: Aging; Biological Determinism; Chromosome Mutation; Congenital Defects; Criminality; Developmental Genetics; Down Syndrome; Eugenics; Eugenics: Nazi Germany; Fragile X Syndrome; Genetic Counseling; Genetic Screening; Genetic Testing; Genetic Testing: Ethical and Economic Issues; Hereditary Diseases; Heredity and Environment; Human Genetics; Human Growth Hormone; Klinefelter Syndrome; Nondisjunction and Aneuploidy; Phenylketonuria (PKU); Prader-Willi and Angelman Syndromes; Prenatal Diagnosis; Race; Twin Studies; X Chromosome Inactivation; XYY Syndrome.

Further Reading

Bock, Gregory R., Jamie A. Goode, and Kate Webb, eds. *The Nature of Intelligence.* New York: John Wiley & Sons, 2001. Presents the debate between evolutionary psychologists, who argue against general intelligence and for an intelligence that develops and evolves based on particular, extraspecies domains, and behavior geneticists, who believe general intelligence is fundamental and who focus their work on intraspecies differences. Illustrations, bibliography, index.

Cooper, Colin. *Intelligence and Abilities.* New York: Routledge, 1999. Discusses the origins of intelligence and the question of genes versus environment in determining intellectual abilities. Illustrations, bibliography, index.

Devlin, Bernie, et al. *Intelligence, Genes, and Success: Scientists Respond to "The Bell Curve."* New York: Springer, 1997. Presents a scientific and statistical reinterpretation of *The Bell Curve*'s claims about the heritability of intelligence and about IQ and social success. Bibliography, index.

Fish, Jefferson M., ed. *Race and Intelligence: Separating Science from Myth.* Mahwah, N.J.: Lawrence Erlbaum, 2002. An interdisciplinary collection disputing race as a biological category and arguing that there is no general or single intelligence and that cognitive ability is shaped through education. Bibliography, index.

Fraser, Steven, ed. *The "Bell Curve" Wars: Race, Intelligence, and the Future of America.* New York: Basic Books, 1995. Brief, critical response to the book by Herrnstein and Murray by scholars from a variety of disciplines and backgrounds. Bibliography.

Gardner, Howard. *Frames of Mind: The Theory of Multiple Intelligences.* 10th anniversary ed. New York: Basic Books, 1993. Argues that there is no single mental ability to be inherited. New introduction, bibliography, index.

Gould, Stephen Jay. *The Mismeasure of Man.* New York: Norton, 1996. An influential criticism of IQ as a measure of intelligence and of the idea that intellectual abilities are inherited. Bibliography, index.

Herrnstein, Richard J., and Charles Murray. *The Bell Curve: Intelligence and Class Structure in America.* New York: Free Press, 1994. The authors maintain that IQ is a valid measure of intelligence, that intelligence is largely a product of genetic background, and that differences in intelligence among social classes play a major part in shaping American society. Illustrations, bibliography, index.

Heschl, Adolf. *The Intelligent Genome: On the Origin of the Human Mind by Mutation and Selection.* Drawings by Herbert Loserl. New York: Springer, 2002. Chapters include "Learning: Appearances are Deceptive," "The 'Wonder' of Language," "How to Explain Consciousness," and "The Cultural Struggle of Genes."

Scientific American 9, no. 4 (Winter, 1998). A special issue on "Exploring Intelligence."

Web Site of Interest

Human Genome Project Information, Behavioral Genetics. http://www.ornl.gov/tech resources/human_genome/elsi/behavior .html. Behavioral genetics information that includes the study of the genetic basis of intelligence.

Klinefelter Syndrome

Field of study: Diseases and syndromes

Significance: *Klinefelter syndrome is a sex chromosome disorder in which males have an extra X chromosome. It accounts for ten out of every one thousand institutionalized mentally retarded adults in industrialized nations and is one of the more common chromosomal aberrations.*

Key terms

AZOOSPERMIA: the absence of spermatozoa from the semen

CYTOGENETICS: the study of chromosome number and structure, including identification of abnormalities

GYNECOMASTIA: a condition characterized by abnormally large mammary glands in the male that sometimes secrete milk

HYPOGONADISM: a condition resulting in smaller than normal testicles in males

KARYOTYPE: a pictorial or verbal description of the chromosomes of a single cell

MOSAICISM: a condition in which an individual has two or more cell populations derived from the same fertilized ovum, or zygote, as in sex chromosome mosaics in which some cells contain the usual XY chromosome pattern and others contain extra X chromosomes

Definition and Diagnosis

Klinefelter syndrome is a relatively common genetic abnormality named after Harry Klinefelter, Jr., an American physician. The fundamental chromosomal defect associated with the syndrome is the presence of one or more extra X chromosomes. The normal human male karyotype (array of chromosomes) consists of twenty-two pairs of chromosomes, called autosomes, plus the XY pair, called sex chromosomes. The female also has twenty-two autosome pairs but with an XX pair in place of the XY pair for the sex chromosomes. Klinefelter syndrome affects 1 in every 500 to 600 men. The incidence is relatively high in the mentally retarded population.

Because individuals with Klinefelter syndrome have a Y chromosome, they are always male. Sometimes Klinefelter syndrome is the result of mosaicism, with males having both normal (XY) karyotypes in some cells and abnormal karyotypes (usually with an extra X chromosome) in others. Individuals with sex chromosome complements of XXYY, XXXY, or XX can also be diagnosed with Klinefelter syndrome. Individuals with Klinefelter syndrome that have a sex chromosome complement of XX are male because although an entire Y chromosome is not present, a portion of a Y chromosome is often attached to another chromosome. This condition can sometimes be diagnosed by a careful karyotype analysis.

Signs and Symptoms

The classic type of Klinefelter syndrome usually becomes apparent at puberty, when the secondary sex characteristics develop. The testes fail to mature, causing primary hypogonadism. In this classic type, degenerative testicular changes begin that eventually result in irreversible infertility. Gynecomastia is often present, and it is usually associated with learning disabilities, mental retardation, and violent, antisocial behavior. Other common symptoms include abnormal body proportions (disproportionate height relative to arm span), chronic pulmonary disease, varicosities of the legs, and diabetes mellitus (which occurs in 8 percent of those afflicted with Klinefelter's). Another 18 percent exhibit impaired glucose tolerance. Most people affected also have azoospermia (no spermatozoa in the semen) and low testosterone levels. However, men with the mosaic form of Klinefelter syndrome may be fertile.

Congenital hypogonadism appears as delayed puberty. Men with hypogonadism experience decreased libido, erection dysfunction, hot sweats, and depression. Genetic testing and careful physical examination may reveal Klinefelter syndrome to be the reason for the primary complaint of infertility. Mental retardation is a frequent symptom of congenital chromosomal aberrations such as Klinefelter syndrome because of probable coincidental defective development of the central nervous system. Early spontaneous abortion is a common occurrence.

Treatment and Psychosocial Implications

Depending on the severity of the syndrome, treatment may include mastectomy to correct gynecomastia. Supplementation with testosterone may be necessary to induce the secondary sexual characteristics of puberty, although the testicular changes that lead to infertility cannot be prevented. Any mental retardation present is irreversible. Psychotherapy with sexual counseling is appropriate when sexual dysfunction causes emotional problems. In people with the mosaic form of the syndrome who are fertile, genetic counseling is vital because they may pass on this chromosomal abnormality. Therapists should encourage discussion of feelings of confusion and rejection that commonly accompany this disorder, and they should attempt to reinforce the victim's male identity. Hormonal therapy can provide some benefits, but both benefits and side effects of hormonal therapy should be made clear. Some men with Klinefelter syndrome are sociopathic; for this population, careful monitoring by probation officers or jail personnel can assist in identifying potential violent offenders, who can be offered psychological counseling.

—*Lisa Levin Sobczak, updated by Bryan Ness*

See also: Hereditary Diseases; Infertility; Intelligence; Mutation and Mutagenesis; Nondisjunction and Aneuploidy; X Chromosome Inactivation; XYY Syndrome.

Further Reading

Bock, Robert. *Understanding Klinefelter Syndrome: A Guide for XXY Males and Their Families.* Bethesda, Md.: Department of Health and Human Services, Public Health Service, National Institutes of Health, National Institute of Child Health and Human Development, 1997. Discusses a range of issues, including defining the syndrome, causes, communicating with family and friends, language, education, legal concerns, teaching tips, treatment, sexuality, and more.

Manning, M. A., and H. E. Hoyme. "Diagnosis and Management of the Adolescent Boy with Klinefelter Syndrome." *Adolescent Medicine* 13 (June, 2002). Discusses treatment options and guiding the child through the transition to puberty and adolescence.

Parker, James N., and Phillip M. Parker, eds. *The Official Parent's Sourcebook on Klinefelter Syndrome: A Revised and Updated Directory for the Internet Age.* San Diego: ICON Health, 2002. Discusses topics including the essentials, seeking guidance, the treatment process, and learning more about the syndrome using the Internet. Includes appendices, glossaries, and an index.

Probasco, Terri, and Gretchen A. Gibbs. *Klinefelter Syndrome: Personal and Professional Guide.* Richmond, Ind.: Prinit Press, 1999. Covers diagnosis, characteristics, education, and emotional concerns, and provides information on community resources.

Web Sites of Interest

American Association for Klinefelter Syndrome Information and Support (AAKSIS). http://www.aaksis.org/index.cfm. The national support organization, with links to information, publications, support, and other resources.

Intersex Society of North America. http://www.isna.org. The society is "a public awareness, education, and advocacy organization which works to create a world free of shame, secrecy, and unwanted surgery for intersex people (individuals born with anatomy or physiology which differs from cultural ideals of male and female)." Includes links to information on such conditions as clitoromegaly, micropenis, hypospadias, ambiguous genitals, early genital surgery, adrenal hyperplasia, Klinefelter syndrome, androgen insensitivity, and testicular feminization.

National Institute of Child Health and Human Development. http://nichd.nih.gov. Site includes a link to "Understanding Klinefelter Syndrome: A Guide for XXY Males and Their Families," which includes an introduction to the syndrome and more.

Knockout Genetics and Knockout Mice

Field of study: Genetic engineering and biotechnology

Significance: *In knockout methodology, a specific gene of an organism is inactivated, or "knocked out," allowing the consequences of its absence to be observed and its function to be deduced. The technique, first and mostly applied to mice, permits the creation of animal models for inherited diseases and a better understanding of the molecular basis of physiology, immunology, behavior, and development. Knockout genetics is the study of the function and inheritance of genes using this technology.*

Key terms

EMBRYONIC STEM CELL: a cell derived from an early embryo that can replicate indefinitely in vitro and can differentiate into other cells of the developing embryo

GENOME: the total complement of genetic material for an organism

IN VITRO: a biological or biochemical process occurring outside a living organism, as in a test tube

IN VIVO: a biological or biochemical process occurring within a living organism

Knockout Methodology

Before knockout mice, transgenic animals had been generated in which "foreign" DNA was incorporated into their genomes in a largely haphazard fashion; such animals should more properly be referred to as "genetically modified." In contrast, knockout technology targets a particular gene to be altered. Prior to the creation of transgenic animals, any genetic change resulted from spontaneous and largely random mutations. Individual variability and inherited diseases are the results of this natural phenomenon—as are, on a longer time frame, the evolutionary changes responsible for the variety of living species on the earth. Spontaneously generated animal models of human inherited diseases have been helpful in understanding mutations and developing treatments for them. However, these mutants were essentially gifts of nature, and their discovery was largely serendipitous. In knockout mice, animal models are directly generated, expediting study of the pathology and treatment of inherited diseases.

In a knockout mouse, a single gene is selected to be inactivated in such a way that the nonfunctional gene is reliably passed to its progeny. Developed independently by Mario Capecchi at the University of Utah and Oliver Smithies of the University of North Carolina, the process is formally termed "targeted gene inactivation," and, although simple in concept, it is operationally complex and technically demanding. It involves several steps in vitro: inactivating and tagging the selected gene, substituting the nonfunctional gene for the functional gene in embryonic stem cells, and inserting the modified embryonic stem cells into an early embryo. The process then requires transfer of that embryo to a surrogate mother, which carries the embryo to term, and selection of offspring that are carrying the inactive gene. It may require several generations to verify that the genetic modification is being dependably transmitted.

Usefulness of Knockout Mice

Knockout mice are important because they permit the function of a specific gene to be established, and, since mice and humans share 99 percent of the same genes, the results can often be applied to people. However, knockout mice are not perfect models in that some genes are specific to mice or humans, and similar genes can be expressed at different levels in the two species. Nevertheless, knockout mice are vastly superior to spontaneous mutants because the investigator selects the gene to be modified. Mice are predominantly used in this technology because of their short generation interval and small size; the short generation interval accelerates the breeding program necessary to establish pure strains, and the small size reduces the space and food needed to house and sustain them.

Knockout mice are, first of all, excellent animal models for inherited diseases, the study of which was the initial impetus for their creation. The Lesch-Nyhan syndrome, a neurological disorder, was the focus of much of the early work with the knockout technology. The methodology has permitted the creation of previously unknown animal models for cystic fibrosis, Alzheimer's disease, and sickle-cell disease, which will stimulate research into new therapies for these diseases. Knockout mice have also been developed to study atherosclerosis, cancer susceptibility, and obesity, as well as immunity, memory, learning, behavior, and developmental biology.

Knockout mice are particularly appropriate for studying the immune system because immune-compromised animals can survive if kept isolated from pathogens. More than fifty genes are responsible for the development and operation of B and T lymphocytes, the two main types of cells that protect the body from infection. Knockout technology permits a system-

atic examination of the role played by these genes. It has also proven useful in understanding memory, learning, and behavior, as knockout mice with abnormalities in these areas can also survive if human intervention can compensate for their deficiencies. Knockout mice have been created that cannot learn simple laboratory tests, cannot remember symbols or smells, lack nurturing behavior, or exhibit extreme aggression, which have implications for the fields of education, psychology, and psychiatry.

Developmental biology has also benefited from knockout technology. Animals with minor developmental abnormalities can be studied with relative ease, whereas those with highly deleterious mutations may be maintained in the heterozygous state, with homozygotes generated only as needed for study. The generation of conditional knockouts is facilitating study of the genes responsible for controlling the development of various tissues (lung, heart, skeleton, and muscle) during embryonic development. These genes can be explored methodically with knockout technology.

By 1997, more than one thousand different knockout mice had been created worldwide. A primary repository for such animals is the nonprofit Jackson Laboratory in Bar Harbor, Maine, where more than two hundred so-called induced mutant strains are available to investigators. Other strains are available from the scientists who first derived them or commercial entities licensed to generate and sell them.

Double Knockouts, Conditional Knockouts, and Reverse Knockouts

Redundancy is fairly common in gene function: Often, more than one gene has responsibility for the same or similar activity in vivo. Eliminating one redundant gene may have little consequence because another gene can fulfill its function. This has led to the creation of double knockout mice, in which two specific genes are eliminated. Double knockouts are generated by crossing

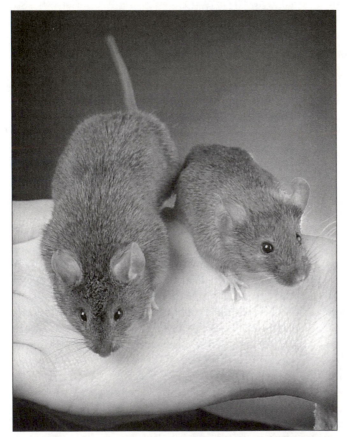

On the left, a "knockout" mouse whose gene for growth/differentiation factor 8 (GDF 8) was removed from its genetic code has grown dramatically larger than the normal mouse (right). (AP/Wide World Photos)

two separate single knockout mice to produce double mutant offspring. Consequences of both mutations can then be examined simultaneously.

Some single knockout mice are deleteriously affected during embryonic development and do not survive to birth. This has led to the generation of conditional knockout mice, in which the gene is functional until a particular stage of life or tissue development triggers its inactivation. The approach is to generate animals with two mutations: The first is the addition of a new gene that causes a marked segment of a gene to be deleted in response to a temporal or tissue signal, and the second is to mark the gene that has been selected to be excised. In these animals, the latter gene remains functional until signaled to be removed.

Knockout methodology involves generation of loss-of-function or null mutations. Its reversal would permit the function of an inoperative gene to be restored. This reversal has been successfully accomplished in mice with the correction of the Lesch-Nyhan defect. Further experimentation may permit it to be applied to humans and other animals. Such targeted restoration of gene function would be the most direct way for gene therapy (the process of introducing a functional gene into an organism's cells) to cure inherited diseases.

—James L. Robinson

See also: Cloning; Developmental Genetics; Genetic Engineering; Genetic Engineering: Medical Applications; Genomics; Model Organism: *Caenorhabditis elegans*; Model Organism: *Mus musculus*; Model Organisms; Transgenic Organisms.

Further Reading

Capecchi, Mario. "Targeted Gene Replacement." *Scientific American* 270 (March, 1994). One of the originators of the technology describes the steps involved and examples of its utility.

Crawley, Jacqueline N. *What's Wrong with My Mouse? Behavioral Phenotyping of Transgenic and Knockout Mice.* New York: Wiley-Liss, 2000. Discusses transgenic technology and the mouse genome. Illustrations, bibliography, index.

Gilbert, Scott F. *Developmental Biology.* 6th ed. Sunderland, Mass.: Sinauer Associates, 2000. Includes a discussion of the knockout methodology. Bibliography.

Mak, Tak W., et al., eds. *The Gene Knockout Factsbook.* 2 vols. San Diego: Academic Press, 1998. Covers six hundred gene knockouts, including their general descriptions, constructs, and phenotypes. Bibliography, index.

Mestel, Rosie. "The Mice Without Qualities." *Discover* 14 (March, 1993). Briefly reports on the creation of knockout mice and their use in understanding the role of the missing gene.

Weaver, Robert F., and Philip W. Hedrick. *Genetics.* 3d ed. New York: McGraw-Hill, 1997. Explains the technique and various applications, including understanding tumor-suppressor genes. Illustrations, bibliography, index.

Web Site of Interest

TBASE: The Transgenic/Targeted Mutation Database, Jackson Laboratory, Bar Harbor, Maine. http://tbase.jax.org. Database of information about transgenic animals generated worldwide, searchable by species, technique, DNA construct, phenotype, laboratory. Features the "Knockout Model of the Month"—a discussion of new animal models—and a glossary.

Lactose Intolerance

Field of study: Diseases and syndromes

Significance: *Lactose intolerance is a common disorder associated with the digestion of milk sugar. It affects a large portion of the human population and creates unpleasant intestinal effects. Its understanding has led to the commercial availability of alternative products that supplement the lack of dairy products in the diet.*

Key terms

CONGENITAL DEFICIENCY: a deficiency that is attributed to inherited genetic causes often apparent at birth

GALACTOSEMIA: a disease attributed to the accumulation of galactose in the blood, caused by a lack of the enzyme that metabolizes galactose; galactosemia is not related to lactose intolerance, which is attributed to the lack of the lactase enzyme

LACTASE: an enzyme that breaks down lactose to the monosaccharides glucose and galactose in the small intestine during the metabolic process; its deficiency is responsible for the ill effects associated with lactose intolerance

LACTOSE: a sugar, also known as milk sugar, that constitutes 2 to 8 percent of milk content and makes up about 40 percent of an infant's diet

The Function of Lactose and Lactase

Milk is the primary source of nutrition for infants. One pint of cow's summer milk provides about 90 percent of the calcium, 30 to 40 percent of the riboflavin, 25 to 30 percent of the protein, and 10 to 20 percent of the calories needed daily. Lactose, also known as milk sugar, exists in the milk of humans, cows, and other mammals. About 7.5 percent of human milk consists of lactose, while cow's milk is about 4.5 percent lactose. This sugar is also one of the few carbohydrates exclusively associated with the animal kingdom; its biosynthesis takes place in the mammary tissue. It is produced commercially from whey, which is obtained as a by-product during the manufacture of cheese. Its so-called alpha form is used as an infant food. Its sweetness is about one-sixth that of sucrose (table sugar).

The metabolism (breaking down) of lactose to glucose and galactose takes place via a specific enzyme called lactase, which is produced by the mucosal cells of the small intestine. Because lactase activity is rate-limiting for lactose absorption, any deficiency in the enzyme is directly reflected in a diminished rate of the sugar absorption. This irregularity should not be confused with intolerance to milk resulting from a sensitivity to milk proteins such as beta-lactoglobulin.

Consequences of Lactase Deficiency

There are three types of lactase deficiency: inherited deficiency, secondary low-lactase activity, and primary low-lactase activity. In inherited lactase deficiency, the symptoms of intolerance develop very soon after birth, as indicated by the presence of lactose in the urine. Patients are recommended a lactose-free diet as well as the consumption of live-culture yogurt, which provides the enzyme beta-galactosidase that attacks the small amounts of lactose that may be in the diet. Beta-galactosidase preparations are also commercially available. Secondary low-lactase activity can be a side effect of peptic ulcer surgery or can occur for a variety of reasons. It may also be present during intestinal diseases such as colitis, gastroenteritis, kwashiorkor, and sprue. Individuals sometimes develop primary low-lactase activity as they get older. A large number of adults, estimated at almost 20 percent, gradually exhibit lactose intolerance, caused by the gradual inability to synthesize an active form of lactase. Susceptible individuals may start developing lactose intolerance as early as four years old.

As a result of lactose intolerance, relatively large quantities of the unhydrolyzed (unbroken) lactose pass into the large intestine, which causes the transfer of water from the interstitial fluid to the lumen by osmosis. At the same time, the intestinal bacteria produce organic acids as well as gases such as carbon dioxide, methane, and hydrogen, which lead to nausea and vomiting. The combined effect also produces cramps and abdominal pains.

Definitive diagnosis of the condition is established by an assay for lactase content in the intestinal mucosa. Such a test requires that the individuals drink 50 grams of lactose in 200 milliliters of water. Blood specimens are then taken after 30, 60, and 120 minutes for glucose analysis. An increase of blood glucose by 30 milligrams per deciliter is considered normal, while an increase of 20 to 30 milligrams per deciliter is borderline. A smaller increase indicates lactase deficiency. This test, however, may still show deficiency results with individuals who have a normal lactase activity.

Lactase deficiency displays remarkable genetic variations. The condition is more prevalent among infants of Middle Eastern, Asian (especially Chinese and Thai), and African descent (such as the Ibo, Yoruba, and other tribes in Nigeria and the Hausa in Sudan). On the other hand, Europeans (especially northern) appear to be statistically less susceptible to the deficiency. Similarly, the Fula tribe in Sudan raises the fulani breed of cattle, and the Eastern African Tussi, who own cattle in Rundi, appear to be rarely affected. It is estimated that 10 to 20 percent of American Caucasians and about 75 percent of African Americans are affected.

The ill effects disappear as long as the diet excludes milk altogether. Often people who exhibit partial lactose intolerance can still consume dairy products, including cheese and yogurt, if the food is processed or partially hydrolyzed. This may be accomplished by heating or partially fermenting milk. Some commercial products, such as Lactaid, are designed for lactose-intolerant people because they include the active form of the lactase enzyme in either liquid or tablet form.

—*Soraya Ghayourmanesh*

See also: Aging; Hereditary Diseases; Inborn Errors of Metabolism.

Further Reading

Auricchio, Salvatore, and G. Semenza, eds. *Common Food Intolerances 2: Milk in Human Nutrition and Adult-Type Hypolactasia.* New York: Karger, 1993. Discusses the health risks associated with not consuming milk. Illustrations, bibliography, index.

Buller, H. A., and R. J. Grant. "Lactose Intolerance." *Annual Reviews of Medicine* 141 (1990). A thorough overview of lactose intolerance.

Hill, John, et al. *Chemistry and Life: An Introduction to General, Organic, and Biological Chemistry.* 6th ed. New York: Prentice Hall, 2000. Includes a section on lactose intolerance.

Ouellette, Robert J. *Organic Chemistry.* 4th ed. New York: Prentice Hall, 1996. Contains a section on lactose metabolism.

Siezen, Roland J., et al., eds. *Lactic Acid Bacteria: Genetics, Metabolism, and Applications.* 7th ed. Boston: Kluwer Academic, 2002. Presents research from a conference held every three years. Illustrations (some color).

Srinivasan, Radhika, and Anil Minocha. "When to Suspect Lactose Intolerance." *Postgraduate Medicine* 104 (September, 1998). Focuses on particular populations in the United States, including Asians, African Americans, and Native Americans. Discusses lactase deficiency, symptoms, and treatment.

Why Does Milk Bother Me? NIH Publication 98-2751 4006187742, DHHS Publication 98-2751 4008061199. Bethesda, Md.: National Digestive Diseases Information Clearinghouse, 1998. An illustrated, twelve-page pamphlet aimed at those suffering from lactose intolerance.

Web Sites of Interest

American Gastroenterological Association. http://www.gastro.org. Site provides a guide to lactose intolerance, including discussion of causes, diagnostics, and treatment, and links to related resources.

National Institute of Diabetes & Digestive & Kidney Diseases. http://www.niddk.nih.gov. This arm of the National Institutes of Health offers resources and links to research on lactose intolerance.

Lamarckianism

Field of study: Evolutionary biology; History of genetics

Significance: *Although some aspects of Lamarckianism have been discredited, the basic premises of nineteenth century French biologist Jean-Baptiste Lamarck's philosophy have become widely accepted tenets of evolutionary theory. Lamarckianism be-*

came intellectually suspect following fraudulent claims by the Soviet scientist Trofim Lysenko that he could manipulate the heredity of plants by changing their environment; by the 1990's, however, scientists had become more willing to acknowledge the influence of Lamarckianism in evolutionary biology.

Key terms

ACQUIRED CHARACTERISTIC: a change in an organism brought about by its interaction with its environment

LYSENKOISM: a theory of transformation that denied the existence of genes

TRANSFORMIST THEORY OF EVOLUTION: a nineteenth century theory that animals gradually changed over time in response to their perceived needs

Lamarckianism Defined

The term "Lamarckianism" has for many years been associated with intellectually disreputable ideas in evolutionary biology. Originally formulated by the early nineteenth century French scientist Jean-Baptiste-Pierre-Antoine de Monet, chevalier de Lamarck (1744-1829), Lamarckianism had two components that were often misinterpreted by scholars and scientists. The first was the transformist theory that animals gradually changed over time in response to their perceived needs. Many critics interpreted this to mean that species could adapt by wanting to change— in other words, that giraffes gradually evolved to have long necks because they wanted to reach the leaves higher in the trees or that pelicans developed pouched beaks because they wanted to carry more fish. Where Lamarck had suggested only that form followed function—for example, that birds that consistently relied on seeds for food gradually transformed to have beaks that worked best for eating seeds—critics saw the suggestion of active intent or desire.

The second component of Lamarckianism, that changes in one generation of a species could be passed on to the next, also led to misinterpretations and abuses of his ideas. In the most egregious cases, researchers in the late nineteenth and early twentieth centuries claimed that deliberate mutilations of animals could cause changes in succeeding generations—for example, they believed that if they cut the tails off a population of mice, succeeding generations would be born without tails. During the twentieth century, the Soviet agronomist Trofim Lysenko claimed to have achieved similar results in plants. Such claims have been thoroughly disproved.

Who Was Lamarck?

Such gross distortions of his natural philosophy would probably have appalled Lamarck. Essentially an eighteenth century intellectual, Lamarck was one of the last scientists who saw himself as a natural philosopher. He was born August 1, 1744, in Picardy, and as the youngest of eleven children was destined originally for the church. The death of his father in 1759 freed Lamarck to leave the seminary and enlist in the military, but an injury forced him to re-

Jean-Baptiste Lamarck (Library of Congress)

Lysenkoism

Although Lamarckian evolutionary theories never enjoyed wide acceptance, a century after Lamarck's death a Russian agronomist, Trofim Denisovich Lysenko (1898-1976), promoted similar theories of heritability of acquired characteristics. Lysenko, born in Ukraine, earned a doctorate in agricultural science from the Kiev Agricultural Institute in 1925.

Lysenko claimed that changing the environment in which plants grew made it possible to alter the fruit they bore, and those alterations would be present in the plants grown from their seed. Unlike Lamarck, who posited gradual change over many generations, Lysenko suggested that dramatic alterations were possible immediately. One of his more outlandish claims was that wheat grown under conditions suited for rye would yield rye seeds, a notion as biologically impossible as the idea that feeding cat food to a dog would result in its giving birth to kittens instead of puppies.

Lysenko's ideas were based on results achieved by an uneducated but successful horticulturalist, Ivan V. Michurin (1855-1935). Michurin developed hundreds of varieties of berries and fruit trees. He credited his achievements to inheritance of acquired characteristics rather than to selective breeding. Lysenko believed similar success was possible with cereal grains, primarily wheat, upon which the Soviet Union relied.

Lysenko used vernalization of wheat as proof that acquired characteristics were heritable. Vernalization involves forcing seeds into responding to the changing of seasons earlier than they would under natural conditions. Bulbs of tulips, for example, when refrigerated for a short time and then placed in a warm environment sprout and bloom and can thereby be forced to blossom midwinter if desired. Lysenko claimed that seeds from vernalized wheat would sprout early without undergoing vernalization themselves. Several ensuing years of good wheat production seemed to validate Lysenko's claims.

Unfortunately for both Soviet science and Soviet agriculture, before it could become evident that Lysenko's seeming successes resulted from good growing conditions rather than from his theories, Lysenko proved more adept at politics than he was at biology. He and his supporters denounced Darwinian evolutionary theories as "bourgeois," contrary to the fundamental principles of Marxism and dialectical materialism as practiced in the Soviet Union. By politicizing science, Lysenko made it impossible for other Russian scientists to pursue research that contradicted Lysenko's pet theories. As director of the Institute of Genetics of the Academy of Sciences from 1940 to 1965, Lysenko wielded tremendous power within the Soviet scientific community. Scientists who challenged his theories not only risked losing their academic positions and research funding but also could be charged with crimes against the state. In the 1940's several of Lysenko's critics were found guilty of anti-Soviet activity, resulting in either their execution or exile to Siberian prison camps.

By the 1950's it was clear that Lysenko's theories did not work. Wheat production consistently failed to achieve promised yields. Agronomists quietly stopped using Lysenko's methods as Lysenko's influence faded, but Lysenko managed to retain his administrative positions for another decade.

—*Nancy Farm Männikkö*

sign his commission in 1768. He sampled a variety of possible vocations before deciding to pursue a career in science.

His early scientific work was in botany. He devised a system of classification of plants and in 1778 published a guide to French flowers. In 1779, at the age of thirty-five, Lamarck was elected to the Académie des Sciences. Renowned naturalist Georges-Louis Leclerc, comte de Buffon, obtained a commission for Lamarck to travel in Europe as a botanist of the king. In 1789, Lamarck obtained a position at the Jardin du Roi as keeper of the herbarium.

When the garden was reorganized as the Museum National d'Histoire Naturelle in 1794, twelve professorships were created; Lamarck became a professor of what would now be called invertebrate zoology.

Lamarck demonstrated through his lectures and published works that he modeled his career on that of his mentor, Buffon. He frequently went beyond the strictly technical aspects of natural science to discuss philosophical issues, and he was not afraid to use empirical data as a basis for hypothesizing. Thus, he often speculated freely on the transformation of species.

Philosophie Zoologique (Zoological Philosophy), now considered his major published work, was issued in two volumes in 1809. In it, Lamarck elaborated upon his theories concerning the evolution of species through adaptation to changes in their environments. An essentially philosophical work, *Zoological Philosophy* is today remembered primarily for Lamarck's two laws:

First Law: In every animal which has not passed the limit of its development, a more frequent and continuous use of any organ gradually strengthens, develops and enlarges that organ and gives it a power proportional to the length of time it has been so used; while the permanent disuse of any organ imperceptibly weakens and deteriorates it, and progressively diminishes its functional capacity, until it finally disappears.

Second Law: All the acquisitions or losses wrought by nature on individuals, through the influence of the environment in which their race has long been placed, and hence through the influence of the predominant use or disuse of any organ; all these are preserved by reproduction to the new individuals which arise, provided that the acquired modifications are common to both sexes, or at least to the individuals which produce the young.

These two tenets constitute the heart of Lamarckianism.

During his lifetime, Lamarck's many books were widely read and discussed, particularly *Zoological Philosophy*. It is true Lamarck's ideas on the progression of life from simple forms to more complex forms in a great chain of being met with opposition, but that opposition was not universal. He was not the only "transformist" active in early nineteenth century science, and his influence extended beyond Paris. Whether or not Lamarck directly influenced Charles Darwin is a matter of debate, but it is known that geologist Charles Lyell read Lamarck, and Lyell in turn influenced Darwin.

Lamarckianism's fall into disrepute following Lamarck's death was prompted by social and political factors as well as scientific criteria. By the 1970's, after a century and a half of denigration, Lamarckianism began creeping back into evolutionary theory and scientific discourse. Researchers in microbiology have described processes that have been openly described as Lamarckian, while other scholars began to recognize that Lamarck's ideas did indeed serve as an important influence in developing theories about the influence of environment on both plants and animals.

—*Nancy Farm Männikkö*

See also: Central Dogma of Molecular Biology; Chromosome Theory of Heredity; Classical Transmission Genetics; DNA Structure and Function; Evolutionary Biology; Genetic Code, Cracking of; Genetic Engineering: Historical Development; Genetics, Historical Development of; Genetics in Television and Films; Genomics; Human Genome Project; Mendelian Genetics.

Further Reading

Burkhardt, Richard W., Jr. *The Spirit of System: Lamarck and Evolutionary Biology, Now with "Lamarck in 1995."* Cambridge, Mass.: Harvard University Press, 1995. Considered by many historians of science to be the most comprehensive examination of Lamarck and his time. Illustrated.

Fine, Paul E. M. "Lamarckian Ironies in Contemporary Biology." *Lancet,* June 2, 1979. Discusses how Lamarckianism has crept into modern evolutionary theory even as some biologists continue to deny any Lamarckian influences.

Lamarck, Jean-Baptiste de Monet de. *Zoological Philosophy: An Exposition with Regard to the Natural History of Animals.* Translated by Hugh Eliott. Chicago: University of Chicago Press, 1984. Lamarck's seminal work, particularly useful to readers curious about the origins of Lamarckianism.

Lanham, Url. *Origins of Modern Biology.* New York: Columbia University Press, 1971. Provides a good general history of biology. Bibliography.

Persell, Stuart Michael. *Neo-Lamarckism and the Evolution Controversy in France, 1870-1920.* Lewiston, N.Y.: Edwin Mellen Press, 1999. Discusses interactions between society, politics, and scientific thought and the rise of anti-Darwinian ideas in late nineteenth and early twentieth century French evolutionary science. Bibliography, index.

Steele, Edward J., Robyn A. Lindley, and Robert V. Blanden. *Lamarck's Signature: How Retrogenes Are Changing Darwin's Natural Selection Paradigm*. Reading, Mass.: Perseus Books, 1998. Argues that some acquired characteristics and immunities (environmental influence), and not just unchanging genetic predispositions, as widely believed, can be passed on from generation to generation. Illustrations, bibliography, index.

Lateral Gene Transfer

Field of study: Population genetics

Significance: *Lateral gene transfer is the movement of genes between organisms. It is also sometimes called horizontal gene transfer. In contrast, vertical gene transfer is the movement of genes between parents and their offspring. Vertical gene transfer is the basis of the study of transmission genetics, while lateral gene transfer is important in the study of evolutionary genetics, as well as having important implications in the fields of medicine and agriculture.*

Key terms

GENE TRANSFER: the movement of fragments of genetic information, whole genes, or groups of genes between organisms

GENETICALLY MODIFIED ORGANISM (GMO): an organism produced by using biotechnology to introduce a new gene or genes, or new regulatory sequences for genes, into it for the purpose of giving the organism a new trait, usually to adapt the organism to a new environment, provide resistance to pest species, or enable the production of new products from the organism

TRANSPOSONS: mobile genetic elements that may be responsible for the movement of genetic material between unrelated organisms

Gene Transfer in Prokaryotes

The fact that genes may move between bacteria has been known since the experiments of Frederick Griffith with pneumonia-causing bacteria in the 1920's. Griffith discovered the process of bacterial transformation, by which the organism acquires genetic material from its environment and expresses the traits contained on the DNA in its own cells. Bacteria may also acquire foreign genetic material by the process of transduction. In transduction a bacteriophage picks up a piece of host DNA from one cell and delivers it to another cell, where it integrates into the genome. This material may then be expressed in the same manner as any of the other of the host's genes. A third mechanism, conjugation, allows two bacteria that are connected by means of a cytoplasmic bridge to exchange genetic information.

With the development of molecular biology, evidence has been accumulated that supports the lateral movement of genes between prokaryotic species. In the case of *Escherichia coli*, one of the most heavily researched bacteria on the planet, there is evidence that as much as 20 percent of the organism's approximately 4,403 genes may have been transferred laterally into the species from other bacteria. This may explain the ability of *E. coli*, and indeed many other prokaryotic species, to adapt to new environments. It may also explain why in a given bacterial genus some members are pathogenic while others are not. Rather than evolving pathogenic traits, bacteria may have acquired genetic sequences from other organisms and then exploited their new abilities.

It is also now possible to screen the genomes of bacteria for similarities in genetic sequences and use this information to reassess previously established phylogenetic relationships. Once again, the majority of this work has been done in prokaryotic organisms, with the primary focus being on the relationship between the domains *Archaea* and *Bacteria*. Several researchers have detected evidence of lateral gene transfer between thermophilic bacteria and *Archaea* prokaryotes. Although the degree of gene transfer between these domains is under contention, there is widespread agreement that the transfer of genes occurred early in their evolutionary history. The fact that there was lateral gene transfer has complicated accurate determinations of divergence time and order.

Gene Transfer in Eukaryotes

Although not as common as in prokaryotes, there is evidence of gene transfer in eukaryotic organisms as well. A mechanism by which gene transfer may be possible is the transposon. Barbara McClintock first proposed the existence of transposons, or mobile genetic elements, in 1948. One of the first examples of a transposon moving laterally between species was discovered in *Drosophila* in the 1950's. A form of transposon called a *P* element was found to have moved from *D. willistoni* to *D. melanogaster.* What is interesting about these studies is that the movement of the *P* element was enabled by a parasitic mite common to the two species. This suggests that parasites may play an important role in lateral gene transfer, especially in higher organisms. Furthermore, since the transposon may move parts of the host genome during transition, it may play a crucial role in gene transfer.

The completion of the Human Genome Project, and the technological advances in genomic processing that it developed, have allowed researchers to compare the human genome with the genomes of other organisms to look for evidence of lateral transfer. It is estimated that between 113 and 223 human genes may not be the result of vertical gene transfer but instead might have been introduced laterally from bacteria.

Implications

While the concept of lateral gene transfer may initially seem to be a concern only for evolutionary geneticists in their construction of phylogenetic trees, in reality the effects of lateral gene transfer pose concerns with regard to both medicine and agriculture, specifically in the case of transgenic plants.

Currently the biggest concern regarding lateral gene transfer is the unintentional movement of genes from genetically modified organisms (GMOs) into other plant species. Such transfer may occur by parasites, as appears to have occurred with *Drosophila* in animals, or by dispersal of pollen grains out of the treated field. This second possibility holds particular significance for corn growers, whose crop is wind-pollinated. Genetically modified corn, containing the microbial insecticide *Bt*, may cross-pollinate with unintentional species, reducing the effectiveness of pest management strategies. In another case, the movement of herbicide resistant genes from a GMO to a weed species may result in the formation of a superweed.

On the beneficial side, lateral gene transfer may also play a part in medicine as part of gene therapy. A number of researchers are examining the possibility of using viruses, transposons, and other systems to move genes, or parts of genes, into target cells in the human body, where they may be therapeutic in treating diseases and disorders.

—*Michael Windelspecht*

See also: Archaea; Bacterial Genetics and Cell Structure; Evolutionary Biology; Gene Regulation: Bacteria; Gene Regulation: Eukaryotes; Gene Regulation: *Lac* Operon; Gene Regulation: Viruses; Hybridization and Introgression; Molecular Genetics; Transposable Elements.

Further Reading

Bushman, Frederick. *Lateral Gene Transfer: Mechanisms and Consequences.* Cold Spring Harbor, N.Y.: Cold Spring Harbor Laboratory Press, 2001. Examines the ability of genes to move between organisms and its implications on the development of antibiotic resistance, cancer, and evolutionary pathways, including those of humans.

Rissler, Jane, and Margaret Mellon. *The Ecological Risks of Engineered Crops.* Cambridge, Mass.: MIT Press, 1996. Introduces the reader to the concept of transgenic crops and then discusses the potential environmental risks of gene flow between genetically modified organisms and nontarget species of plants. Suggests mechanisms of regulation to inhibit environmental risk.

Syvanen, Michael, and Clarence Kado. *Horizontal Gene Transfer.* 2d ed. Burlington, Mass.: Academic Press, 2002. Examines the process of gene transfer from an advanced perspective. Discusses the relationship between gene transfer and phylogenetic analysis, evolutionary theory, and taxonomy.

Linkage Maps

Field of study: Techniques and methodologies

Significance: *Linkage maps can be used to predict the outcome of genetic crosses involving linked genes and, more important, can be used to find the location of genes that are responsible for specific traits or genetic defects.*

Key terms

ALLELES: different forms of the same gene locus; in diploids there are two alleles at each locus

CROSSING OVER: an event early in meiosis in which homologous chromosomes exchange homologous regions

DIHYBRID: an organism that is heterozygous for both of two different gene loci

HOMOLOGOUS CHROMOSOMES: chromosomes that are structurally the same and contain the same loci, although the loci may each have different alleles

LOCUS (*pl.* LOCI): The specific region of a chromosome that contains a specific gene

MEIOSIS: cell division that reduces the chromosome number from two sets (diploid) to one set (haploid), ultimately resulting in the formation of gametes (eggs or sperm) or spores

Linkage and Crossing Over

When Gregor Mendel examined inheritance of two traits at a time, he found that the dihybrid parent (*Aa* or *Bb*) produced offspring with the four possible combinations of these alleles at equal frequencies: ¼*AB*, ¼*Ab*, ¼*aB*, and ¼*ab*. He called this pattern "independent assortment." The discovery of meiosis explained the basis of independent assortment. If the *A* locus and the *B* locus are on nonhomologous chromosomes, then segregation of the alleles of one locus (*A* and *a*) will be independent of the segregation of the alleles of the other (*B* and *b*).

Even simple plants, animals, fungi, and protists have thousands of genes. The number of human genes is unknown, but with the completion of the human genome in 2003 it appeared that the actual number of protein-coding genes was only about 21,000. Human beings have forty-six chromosomes in each cell (twenty-three from the mother and twenty-three from the father): twenty-two pairs of autosomal chromosomes plus two sex chromosomes (two X chromosomes in females and an X and a Y chromosome in males). Since humans have only twenty-four kinds of chromosomes, there must be less than a few thousand genes on the average human chromosome.

If two loci fail to show independent assortment, they are said to be linked and are therefore near one another on the same chromosome. For example, if the alleles *A* and *B* are on one chromosome and *a* and *b* are on the homologue of that chromosome, then the dihybrid (*AB/ab*) would form gametes with the combinations *AB* and *ab* more often than *Ab* and *aB*. How much more often? At one extreme, if there is no crossover between these two loci on the two homologous chromosomes, then ½ of the gametes would be *AB* and ½ would be *ab*. At the other extreme, if the two genes are so far apart on a large chromosome that crossover occurs between the loci almost every time meiosis occurs, they would assort independently, thus behaving like two loci on different nonhomologous chromosomes. When two genes are on the same chromosome but show no linkage, they are said to be "syntenic."

In the first stage of meiosis, homologous chromosomes pair tightly with one another (synapsis). At this stage of meiosis, each homologous chromosome is composed of two chromatids called sister chromatids, so there are four complete DNA molecules (a tetrad) present in the paired homologous chromosomes. A reciprocal exchange of pieces of two paired homologous chromosomes can produce new combinations of alleles between two linked loci if a crossover occurs in the right region. Chromosomes that display a new arrangement of alleles due to crossover are called recombinants. For example, a crossover in a dihybrid with *AB* on one chromosome and *ab* on its homologue could form *Ab* and *aB* recombinants. The average number of crossovers during a meiotic division differs from species to species and sometimes between the sexes of a single species. For example, crossover does not occur in male fruit flies (*Drosophila melanogaster*), and it may occur slightly less often in human males than in females. Nevertheless, within a single sex of a single species, the number of crossovers during a meiotic division is fairly constant and many crossovers typically occur along the length of each pair of chromosomes.

Constructing the Maps

If two loci are very close together on the same chromosome, crossover between them will be rare, and thus recombinant gametes will also be rare. Conversely, crossover will occur more frequently between two loci that are farther apart on the same chromosome. This is true because the location for any particular crossover is random. This fact has been used to construct linkage maps (also called crossover maps or genetic maps) of the chromosomes of many species. The distances between loci on linkage maps are expressed as percent crossover. A crossover of 1 percent is equal to one centi-Morgan (cM). If two loci are 12 cM apart on a linkage map, a dihybrid will form approximately twelve recombinant gametes for every eighty-eight nonrecombinant gametes. Linkage maps are made by combining data from many different controlled crosses or matings. For instance, suppose that a cross between a dihybrid *AB/ab* individual and a homozygous *ab/ab* individual produced 81 *AB/ab* + 83 *ab/ab* progeny (non-crossover types) and 20 *Ab/ab* + 16 *aB/ab* progeny (crossover types). The map distance between these loci would be 100(20 + 16)/(81 + 83 + 20 + 16) = 18 cM.

The table shows the frequency of recombinant gametes from test crosses of three different dihybrids, including the one already described:

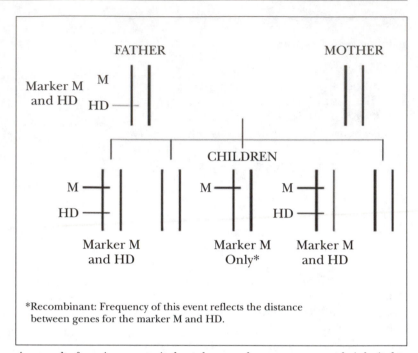

As a result of crossing over, traits located on one chromosome may not be inherited together. Those traits that tend to be inherited together most often also tend to be those located near each other on the chromosome. Those traits that are more distant are more likely to cross over or recombine during the production of gametes (eggs and sperm) and therefore to be absent as a result of crossover. Genetic linkage maps can be constructed based on the frequency of these events. (U.S. Department of Energy Human Genome Program, http://www.ornl.gov/hgmis)

gene pair	cM
a and *b*	18
a and *c*	7

It is clear that the *C* locus must be between the other two loci on the linkage map. The absolute order, *ACB* or *BCA*, is arbitrarily defined by the first person who constructs a linkage map of a species.

```
a            c          b
   ———————————————————————
       7           11
```

In this example, the linkage map is exactly additive. In real experiments, linkage map distances are seldom exactly additive, because the longer the distance between two loci, the greater chance there will be for double crossovers to occur. Double crossovers give the same result as no crossover, and are therefore not detected. Thus, the greater the distance between two loci, the more the distance will be underestimated.

Once a large number of genes on the same chromosome have been mapped, the linkage map is redrawn with map positions rather than map distances. For example, if many other experiments provided more information about linked genes, the following linkage map might emerge:

p	q		a	c		b	r		s
0	6		14	21		32	39		49

The A and C loci are still 7 cM apart ($21 - 14 = 7$), and the other distances on the first map are also still the same.

Very detailed linkage maps have been constructed for some plants, animals, fungi, and protists that are of particular value to medicine, agriculture, industry, or scientific research. Among them are *Zea mays* (maize), *Drosophila melanogaster* (fruit fly), and *Saccharomyces cerevisiae* (baker's yeast). The linkage map of *Homo sapiens* (humans) is not very detailed because it is unethical and socially impossible to arrange all of the desired crosses that would be necessary to construct one. Other techniques have allowed the construction of very detailed physical maps of human chromosomes.

Genetic Linkage Maps and the Structure of Chromosomes

It should be emphasized that the linkage map is not a scale model of the physical chromosome. It is generally true that the relative order of genes on the linkage map and the physical chromosome map are the same. However, the relative distances between genes on the linkage map may not be proportionately the same on the physical map. Consider three loci (A, B, and C) that are arranged in that order on the chromosome. Suppose that the AB distance on the physical map is exactly the same as the BC distance. If the crossover frequency between A and B is higher than between B and C, then the AB linkage map distance will be larger than the BC linkage map distance. It is common to find small discrepancies between linkage maps and physical maps all along the chromosome. Large discrepancies are usually limited to loci close to centromeres. Crossover frequencies are generally very low near centromeres, apparently due to the structural characteristics of centromeres. If two loci are on opposite sides of a centromere, they will appear farther apart on the physical map and much closer on the linkage map.

—James L. Farmer, updated by Bryan Ness

See also: Chromosome Structure; Chromosome Theory of Heredity; Classical Transmission Genetics; Complete Dominance; Dihybrid Inheritance; Gene Families; Genomics; Mendelian Genetics; Mitosis and Meiosis; Model Organism: *Drosophila melanogaster*; Model Organism: *Neurospora crassa*.

Further Reading

Liu, Ben-Hui. *Statistical Genomics: Linkage, Mapping, and QTL Analysis.* Boca Raton, Fla.: CRC Press, 1998. Covers the quantitative and theoretical aspects of genomics, including linkage map construction and merging. Illustrations, glossary, bibliography, index.

Ott, Jurg. *Analysis of Human Genetic Linkage.* 3d ed. Baltimore: Johns Hopkins University Press, 1999. Introductory text that presents basic methods for linkage analysis. Illustrations.

Terwilliger, Joseph Douglas, and Jurg Ott. *Handbook of Human Genetic Linkage.* Baltimore: Johns Hopkins University Press, 1994. Emphasizes computer-based analyses. Illustrations, bibliography, index.

Mendelian Genetics

Field of study: Classical transmission genetics; History of genetics

Significance: *Gregor Mendel was a monk and a science teacher in Moravia when he wrote his famous paper about experimental crosses of pea plants. Little note was taken of it when it was published in 1866, but it provided concepts and methods that catalyzed the growth of modern genetics after 1900 and earned Mendel posthumous renown as the founder of the new science.*

Key terms

GAMETES: reproductive cells that unite during fertilization to form an embryo; in plants, the pollen cells and egg cells are gametes

HYBRID: a plant form resulting from a cross between two distinct varieties

INDEPENDENT ASSORTMENT: the segregation of two or more pairs of genes without any tendency for certain genes to stay together

SEGREGATION: the process of separating a pair of Mendelian hereditary elements (genes), one from each parent, and distributing them at random into the gametes

Early Life

Born Johann Mendel on July 22, 1822, the future teacher, monk, abbot, botanist, and meteorologist grew up in a village in Moravia, a province of the Austrian Empire that later became part of Czechoslovakia (1918) and the Czech Republic (1993). His parents were peasant farmers and belonged to the large, German-speaking minority in this predominantly Czech province. Like most places in Moravia, Mendel's hometown had two names: Hynčice in Czech and Heinzendorf in German.

Johann Mendel was an exceptional pupil, but no local schooling was available for him beyond the age of ten. In 1833, he persuaded his parents to send him to town to continue his education. They were reluctant to let him go because they could ill afford to dispense with his help on the farm or finance his studies. In 1838, Mendel's father was partially disabled in a logging accident, and Johann, then sixteen and still at school, had to support himself. He earned just enough from tutoring to get by. At times, however, the pressure became too much for him. He suffered a breakdown in 1839 and returned home for several months to recuperate. He was to have several more of these stress-related illnesses, but no precise information is available about their causes and symptoms.

In 1840, Mendel completed *Gymnasium*, as the elite secondary schools were called, and entered the University of Olomouc for the two-year program in philosophy that preceded higher university studies. He had trouble supporting himself in Olomouc, perhaps because there was less demand for German-speaking tutors, and his Czech was not good enough for teaching. He suffered another breakdown in 1841 and retreated to Hynčice during spring exams.

That summer, Mendel decided once more against staying and taking over the farm. Since his father could not work, the farm was sold to his elder sister's husband. Johann's share of the proceeds was not enough to see him through

Gregor Mendel. (National Library of Medicine)

the Olomouc program, especially since he had to repeat a year because of the missed exams. However, his twelve-year-old sister sacrificed part of her future dowry so that he could continue. (He repaid her years later by putting her three sons through *Gymnasium* and university.)

Upon finishing at Olomouc in 1843, Mendel decided to enter the clergy. The priesthood filled his need for a secure position and held out possibilities for further learning and teaching, but Mendel did not seem to be called to it. Aided by a professor's recommendation, Mendel was accepted into the Augustinian monastery in Brno, the capital of Moravia, where he took the name Gregor. In 1847, after four years of preparation at the monastery, he was ordained a priest.

Priesthood and Teaching

The Brno monastery was active in the community and provided highly qualified instructors for *Gymnasia* and technical schools throughout Moravia. Several monks, including the abbot, were interested in science, and they had experimental gardens, a herbarium, a mineralogical collection, and an extensive library. Mendel found himself in learned company with opportunities for research in his spare time.

Unfortunately, Mendel's nerves failed him when he had to minister to the sick and dying. Assigned to a local hospital in 1848, he was so upset by it that he was bedridden himself within five months. However, his abbot was sympathetic and let him switch to teaching. A letter survives in which the abbot explains this decision to the bishop: "[Mendel] leads a retiring, modest and virtuous religious life . . . and he devotes himself diligently to scholarly pursuits. For pastoral duties, however, he is less suited, because at the sick-bed or at the sight of the sick or suffering he is seized by an insurmountable dread, from which he has even fallen dangerously ill."

Mendel taught Latin and Greek, German literature, math, and science as a substitute at the *Gymnasium* and was found to be very good at teaching. Therefore, he was sent to Vienna in 1850 to take the licensing examinations so that he could be promoted to a regular position.

These exams were very demanding and normally required more preparation than Mendel's two years at Olomouc. Mendel failed, but one examiner advised the abbot to let him try again after further study. The abbot took this advice and sent Mendel to study in Vienna for two years (1851-1853). There he took courses in biology, physics, and meteorology with some of the best-known scientists of his day, including physicist Christian Doppler and botanist Franz Unger.

For unknown reasons, Mendel returned to Moravia to resume substitute teaching and did not go to Vienna for the exams until 1856. This time he was too nervous to finish. After writing one essay, he fell ill and returned to Brno. Despite this failure, he was allowed to teach regular classes until 1868 even though he was technically only a substitute.

Scientific Work

During his teaching career, Mendel performed his famous experiments on peas in a garden at the monastery. He published the results in an 1866 article, which introduced fundamental concepts and methods of genetics. The first set of experiments involved fourteen varieties of pea plant, each with a single distinguishing trait. These traits made up seven contrasting pairs, such as seeds that were either round or wrinkled in outline or seed colors that were green or yellow. Upon crossing each pair, Mendel obtained hybrids identical to one parent variety. For example, the cross of round with wrinkled peas yielded only round peas; the cross of green with yellow peas yielded only yellow peas. He referred to traits that asserted themselves in the hybrids as "dominant." The others were "recessive" because they receded from view. The effect was the same regardless of whether he fertilized the wrinkled variety with pollen from the round or the round variety with pollen from the wrinkled. This indicated to Mendel that both pollen cells and egg cells contributed equally to heredity; this was a significant finding because the details of plant reproduction were still unclear.

Mendel next allowed the seven hybrids to pollinate themselves, and the recessive traits reappeared in the second generation. For instance, the round peas, which were hybrids of round and wrinkled peas, yielded not only more round peas but also some wrinkled ones. Moreover, the dominant forms outnumbered the recessives three to one. Mendel explained the 3:1 ratio as follows. He used the symbols *A* for the dominant form, *a* for the recessive, and *Aa* for the hybrid. A hybrid, he argued, could produce two types of pollen cell, one containing some sort of hereditary element corresponding to trait *A* and the other an element corresponding to trait *a*. Likewise, it could produce eggs containing either *A* or *a* elements. This process of dividing up the hereditary factors among the gametes became known as segregation.

The gametes from the *Aa* hybrids could come together in any of four combinations: pollen *A* with egg *A*, pollen *A* with egg *a*, pollen *a* with egg *A*, and pollen *a* with egg *a*. The first three of these combinations all grew into plants with the dominant trait *A*; only the fourth produced the recessive *a*. Therefore, if all four combinations were equally common, one could expect an average of three plants exhibiting *A* for every one exhibiting *a*.

Allowing self-pollination to continue, Mendel found that the recessives always bred true. In other words, they only produced more plants with that same recessive trait; no dominant forms reappeared, not even in subsequent generations. Mendel's explanation was that the recessives could only have arisen from

the pollen *a* and egg *a* combination, which excludes the *A* element. For similar reasons, plants with the dominant trait bred true one-third of the time, depending on whether they were the pure forms from the pollen *A* and egg *A* combination or the hybrids from the pollen *A* and egg *a* or pollen *a* and egg *A* combinations.

Mendel's hereditary elements sound like the modern geneticist's genes or alleles, and Mendel usually receives credit for introducing the gene concept. Like genes, Mendel's elements were material entities inherited from both parents and transmitted to the gametes. They also retained their integrity even when recessive in a hybrid. However, it is not clear whether he pictured two copies of each element in every cell, one copy from each parent, and he certainly did not associate them with chromosomes.

In a second set of experiments, Mendel tested combinations of traits to see whether they would segregate freely or tend to be inherited together. For example, he crossed round, yellow peas with wrinkled, green ones. That cross first yielded only round, yellow peas, as could be expected from the dominance relationships. Then, in the second generation, all four possible combinations of traits segregated out: not only the parental round yellow and wrinkled green peas but also new round green and wrinkled yellow ones. Mendel was able to explain the ratios as before, based on equally

---

The Results of Mendel's Pea-Plant Experiments

Parental characteristics	First generation	Second generation	Second generation ratio
Round × wrinkled seeds	All round	5,474 round : 1,850 wrinkled	2.96 : 1
Yellow × green seeds	All yellow	6,022 yellow : 2,001 green	3.01 : 1
Gray × white seedcoats	All gray	705 gray : 224 white	3.15 : 1
Inflated × pinched pods	All inflated	882 inflated : 299 pinched	2.95 : 1
Green × yellow pods	All green	428 green : 152 yellow	2.82 : 1
Axial × terminal flowers	All axial	651 axial : 207 terminal	3.14 : 1
Long × short stems	All long	787 long : 277 short	2.84 : 1

likely combinations of hereditary elements coming together at fertilization. The free regrouping of hereditary traits became known as independent assortment. In the twentieth century, it was found not to occur universally because some genes are linked together on the same chromosome.

Mendel's paper did not reach many readers. As a *Gymnasium* teacher and a monk in Moravia without even a doctoral degree, Mendel could not command the same attention as a university professor in a major city. Also, it was not obvious that the behavior of these seven pea traits illustrated fundamental principles of heredity. Mendel wrote to several leading botanists in Germany and Austria about his findings, but only Carl von Nägeli at the University of Munich is known to have responded, and even he was skeptical of Mendel's conclusions. Mendel published only one more paper on heredity (in 1869) and did little else to follow up his experiments or gain wider attention from scientists.

Mendel pursued other scientific interests as well. He was active in local scientific societies and was an avid meteorologist. He set up a weather station at the monastery and sent reports to the Central Meteorological Institute in Vienna. He also helped organize a network of weather stations in Moravia. He envisioned telegraph connections among the stations and with Vienna that would make weather forecasting feasible. In his later years, Mendel studied sunspots and tested the idea that they affected the weather. He also monitored the water level in the monastery well in order to test a theory that changes in the water table were related to epidemics. A common thread that ran through these diverse research interests was that they all involved counting or measuring, with the goal of discovering scientific laws behind the numerical patterns. His one great success was in explaining the pea data with his concepts of dominance, segregation, and independent assortment.

Mendel felt pleased and honored to be elected abbot in 1868, even though he had to give up teaching and most of his research. He did not have the heart to say good-bye to his pupils. Instead, he asked the school director to announce his departure and give his last month's

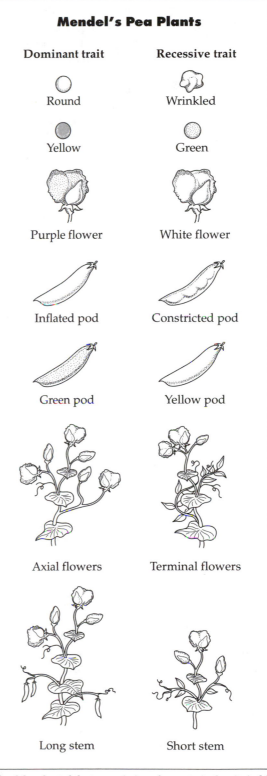

Mendel evaluated the transmission of seven paired traits in his studies of garden peas. (Electronic Illustrators Group)

salary to the three neediest boys in the class. As abbot, Mendel had a reputation for generosity to the poor and to scientific and cultural institutions. He was also an efficient manager of the monastery and its extensive land holdings and a fierce defender of the monastery's interests. From 1874 on, he feuded with imperial authorities over a new tax on the monastery, which he refused to pay as long as he lived. Mendel's health failed gradually in the last years of his life. He had kidney problems and an abnormally fast heartbeat, the latter probably from nerves and nicotine. (A doctor recommended smoking to control his weight, and he developed a twenty-cigar-a-day habit.) He died January 6, 1884, of heart and kidney failure.

Impact and Applications

Years after Mendel's death, a scientific colleague remembered him saying, prophetically, "my time will come." It came in 1900, when papers by three different botanists reported experimental results that were similar to Mendel's and endorsed Mendel's long-overlooked explanations. This event became known as the rediscovery of Mendelism. By 1910, Mendel's theory had given rise to a whole new field of research, which was given the name "genetics." Mendel's hereditary elements were described more precisely as "genes" and were presumed to be located on the chromosomes. By the 1920's, the sex chromosomes were identified, the determination of sex was explained in Mendelian terms, and the arrangements of genes on chromosomes could be mapped.

The study of evolution was also transformed by Mendelian genetics, as Darwinians and anti-Darwinians alike had to take the new information about heredity into account. By 1930, it had been shown that natural selection could cause evolutionary change in a population by shifting the proportions of individuals with different genes. This principle of population genetics became a cornerstone of modern Darwinism.

Investigations of the material basis of heredity led to the discovery of the gene's DNA structure in 1953. This breakthrough marked the beginning of molecular genetics, which studies how genes are copied, how mutations occur,

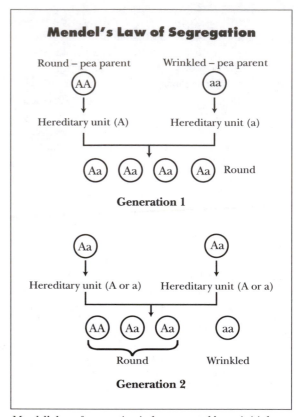

Mendel's law of segregation is demonstrated by an initial cross between true-breeding plants with round peas and plants with wrinkled peas. The round trait is dominant, and the wrinkled trait is recessive. The second generation consists of round-pea plants and wrinkled-pea plants produced in a ratio of 3:1.

and how genes exert their influence on cells. In short, all of modern genetics can trace its heritage back to the ideas and experiments of Gregor Mendel.

—Sander Gliboff

See also: Chloroplast Genes; Chromosome Structure; Chromosome Theory of Heredity; Classical Transmission Genetics; Complete Dominance; Dihybrid Inheritance; Genetic Engineering: Historical Development; Genetics, Historical Development of; Incomplete Dominance; Linkage Maps; Monohybrid Inheritance; Natural Selection; Quantitative Inheritance.

Further Reading

Corcos, A., and F. Monaghan. *Mendel's Experiments on Plant Hybrids: A Guided Study.* New

Brunswick, N.J.: Rutgers University Press, 1993. Covers the seminal work of Gregor Mendel, along with a biography.

Edelson, Edward. *Gregor Mendel and the Roots of Genetics.* New York: Oxford University Press, 1999. Story of Mendel's research into the inheritance of traits in the garden pea. Illustrations (including botanical drawings), bibliography, index.

Henig, Robin Marantz. *The Monk in the Garden: The Lost and Found Genius of Gregor Mendel, the Father of Genetics.* Boston: Houghton Mifflin, 2000. A descriptive look at Mendel's life and work for the general reader. Illustrated.

Iltis, Hugo. *Life of Mendel.* Translated by Eden Paul and Cedar Paul. 1932. Reprint. New York: Hafner, 1966. This first biography of Gregor Mendel is still among the best.

Olby, Robert. *The Origins of Mendelism.* 2d ed. Chicago: University of Chicago Press, 1985. Discusses the history of genetics from the 1700's through the rediscovery of Mendel.

Orel, Vítezslav. *Gregor Mendel: The First Geneticist.* Translated by Stephen Finn. New York: Oxford University Press, 1996. Biography that focuses on how Mendel's work was received by his peers and critics, even after his death. Illustrations, bibliography, index.

Tudge, Colin. *In Mendel's Footnotes: An Introduction to the Science and Technologies of Genes and Genetics from the Nineteenth Century to the Twenty-Second.* London: Jonathan Cape, 2000. Investigates the world of biotechnologies, including cloning, genomics, and genetic engineering. Bibliography, index.

Wood, Roger J., and Vitezslav Orel. *Genetic Prehistory in Selective Breeding: A Prelude to Mendel.* New York: Oxford University Press, 2001. Focuses on the period from 1700 to 1860, before Mendel published the results of his experiments. Illustrated.

Web Site of Interest

MendelWeb. http://www.mendelweb.org. This site, designed for teachers and students, revolves around Mendel's 1865 paper and includes educational activities, images, interactive learning, and other resources.

Metafemales

Field of study: Diseases and syndromes

Significance: *Genetic defects are quite common in humans. The frequency of females born with XXX chromosomes, called multiple X or metafemale syndrome, generally varies between one in one thousand and one in fifteen hundred but may be less in some populations. Although most such females have normal appearance and sexual reproduction, this abnormality needs to be better understood so that the affected individuals' lives are bettered medically and socially.*

Key terms

AUTOSOMES: all chromosomes other than sex chromosomes in a cell nucleus

BARR BODY: named after its discoverer, Murray L. Barr, a dark-stained sex chromatin body in nuclei of females, which represents the inactivated X chromosome; the number of Barr bodies in any cell is generally one less than the number of X chromosomes

LYON HYPOTHESIS: proposed by Mary Lyon in 1962, a hypothesis that during development one of the two X chromosomes in normal mammalian females is inactivated at random; the inactivated X chromosome is a Barr body

MEIOSIS: the process by which gametes (sperm and eggs) are produced in sexually reproducing organisms

NONDISJUNCTION: the failure of homologous chromosomes to disjoin during meiosis I, or the failure of sister chromatids to separate and migrate to opposite poles during meiosis II

SEX CHROMOSOMES: the homologous pair of chromosomes that determines the sex of an individual; in humans, XX is female and XY is male; XX females produce one kind of gamete, X (homogametic sex), and XY males produce two kinds of gametes, X and Y (heterogametic sex)

History and Symptoms

In 1914, Calvin Blackman Bridges discovered nondisjunction of sex chromosomes in the fruit fly, *Drosophila melanogaster.* In 1925, he proposed the genic or sex balance theory, which defined the relationship between sex chromosomes and autosomes (A) for sex determination. According to this theory, the following ratios of sex chromosomes and number of sets of autosomes determine what sex phenotype will emerge in humans. For example, XX + 2 sets of autosomes (2X:2A ratio = 1.0) = normal female; XY + 2 sets of autosomes (1X:2A ratio = 0.5) = normal male; and XXX + 2 sets of autosomes (3X:2A ratio = 1.5) = metafemale, or superfemale.

The term "metafemale" was first applied to the XXX (triple X) condition by Curt Stern around 1959. The frequency of metafemale phenotype in the general human population is approximately one in one thousand to fifteen hundred newborn girls. The XXX females are characterized by the presence of two Barr bodies in their cells. They have a total of 47 chromosomes instead of the normal complement of 46.

Metafemales have variable fertility, ranging from normal to sterile. They may be phenotypically normal but are often slightly taller than average, with longer legs. These individuals may have widely spaced nipples and a webbed neck. Studies have shown that most metafemales lead a normal sexual life and have normal children. In some cases, menstruation may begin at an older age, menstrual cycles may be irregular or temporarily interrupted, and menopause may begin earlier compared to normal XX women.

Genetic Cause

The basic causes of XXX females are best explained through meiosis, the cell division that halves the number of chromosomes in gametes, and nondisjunction. From a single human cell (46 chromosomes) designated for sexual reproduction, meiosis produces four cells, each with 23 chromosomes. Thus, normal human eggs carry one-half (22A + 1X = 23) of the total number of chromosomes (44A + 2X = 46). Occasionally, a mistake occurs during meiosis, called nondisjunction. Nondisjunction during meiosis I or meiosis II can produce eggs with 2X chromosomes (22A + 2X = 24). Usually the nondisjunction that gives rise to XXX females occurs in the female parent during meiosis I.

Fertilization of an egg carrying two X chromosomes by an X-bearing (22A + 1X = 23) sperm results in an individual with 44A + 3X = 47 chromosomes, or a metafemale. The extra X chromosome is not usually transmitted to the children. Thus, metafemales can have normal children. Triple X, triplo-X, trisomy X, and 47 XXX are also the names given to the metafemale phenotype. This genetic condition has also been referred to as "extra X aneuploidy" or "multiple X syndrome."

Social Issues

The IQ of metafemales is usually low normal to normal. In some studies, IQ was found to be lower by 30 points than that of their normal siblings; only a few had an IQ lower than 70. Language learning in XXX children is usually delayed. Emotional maturation may also be delayed. These delays in development are preventable by providing increased psychological, social, and motor stimulation both at home and at school. Tutoring is often needed at some time during development.

The 47 XXX condition can put some affected individuals at risk for speech disorders, learning disabilities, and neuro-motor deficits, which ultimately could lead to decreased psychosocial adaptation, especially during adolescence. One study found young females with 47 XXX to be less well adapted in both their teen and adult years; they described their lives as more stressful. On average, they experienced more work, social, and relationship problems than their siblings. Metafemales may encounter behavioral problems, including mild depression, conduct disorder, immature behavior, and socializing problems. Good parenting and a supportive home may assure a better social and behavioral development.

—*Manjit S. Kang*

See also: Behavior; Biological Clocks; Gender Identity; Hermaphrodites; Homosexuality; Human Genetics; Pseudohermaphrodites; Steroid Hormones; Testicular Feminization Syndrome; X Chromosome Inactivation; XYY Syndrome.

Further Reading

Bender, B., R. Harmon, M. Linden, B. Bucher-Bartelson, and A. Robinson. "Psychological Competence of Unselected Young Adults with Sex Chromosome Abnormalities." *American Journal of Medical Genetics* 88 (1999): 200-206. This article describes research on the social issues of XXX females.

Redei, G. P. *Genetics Manual.* River Edge, N.J.: World Scientific, 1998. Written by an authority with encyclopedic knowledge of genetics, this comprehensive manual provides genetic definitions, terms, and concepts, for the novice and professional.

Rovet, J., C. Netley, J. Bailey, M. Keenan, and D. Stewart. "Intelligence and Achievement in Children with Extra X Aneuploidy." *American Journal of Medical Genetics* 60 (1995): 356-60. This interesting study, conducted in Toronto between 1967 and 1971, tested 72,000 consecutive births. Sixteen females were 47 XXX, of whom 12 participated in the study. They were compared to 16 normal girls, 9 of whom were siblings of the affected girls.

Web Site of Interest

Triplo-X Syndrome. http://www.triplo-x.org. A site that offers social support, a brief introduction to the syndrome, and links to related articles.

Miscegenation and Antimiscegenation Laws

Field of study: Bioethics; History of genetics; Human genetics and social issues

Significance: *Miscegenation is the crossing or hybridization of different races. As knowledge of the nature of human variability has expanded, clearly defining "race" has become increasingly difficult; the study of genetics reveals that the concept of race is primarily a social construct as opposed to a biological reality. Limited understanding of the biological and genetic effects of mating between races, as well as racial prejudice, played a major role in the development of the eugenics movement and the enactment of antimiscegenation laws in the first half of the twentieth century.*

Key terms

EUGENICS: the control of individual reproductive choices to improve the genetic quality of the human population

HYBRIDIZATION: the crossing of two genetically distinct species, races, or types to produce mixed offspring

NEGATIVE EUGENICS: preventing the reproduction of individuals who have undesirable genetic traits, as defined by those in control

POSITIVE EUGENICS: selecting individuals to reproduce who have desirable genetic traits, as seen by those in control

RACE: in the biological sense, a group of people who share certain genetically transmitted physical characteristics

What Is a Race?

Implicit in most biological definitions of race is the concept of shared physical characteristics that have come from a common ancestor. Humans have long recognized and attempted to classify and categorize different kinds of people. The father of modern systematics, Carolus Linnaeus, described, in his system of binomial nomenclature, four races of humans: Africans (black), Asians (dark), Europeans (white), and Native Americans (red). Skin color in humans has been, without doubt, the primary feature used to classify people, although there is no single trait that can be used to do this. Skin color is used because it makes it very easy to tell groups of people apart. However, there are thousands of human traits. What distinguishes races are differences in gene frequencies for a variety of traits. The great majority of genetic traits are found in similar frequencies in people of different skin color. There may not be a single genetic trait that is always associated with people of one skin color while not appearing at all in people of another skin color. It is possible for a person to differ more from another person of the same skin color than from a person of a different skin color.

Many scientists think that the word "race" is not useful in human biology research. Scientific and social organizations, including the American Association of Physical Anthropologists and the American Anthropological Association, have deemed that racial classifications are limited in their scope and utility and do not reflect the evolving concepts of human variability. It is of interest to note that subjects are frequently asked to identify their race in studies and surveys.

It is useful to point out the distinction between an "ethnic group" and a race. An ethnic group is a group of people who share a common social ancestry. Cultural practices may lead to a group's genetic isolation from other groups with a different cultural identity. Since members of different ethnicities may tend to marry only within their group, certain genetic traits may occur at different frequencies in the group than they do in other ethnic or racial groups, or the population at large.

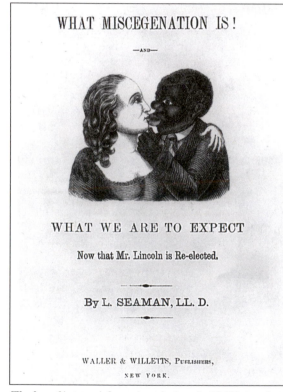

The fear of interracial marriage during the 1860's is only too clear from the title page to this antimiscegenation tract, published after Emancipation near the end of the Civil War. At the time, Charles Darwin had recently published his theory of natural selection, which "social Darwinists" misapplied to justify antiracial social and business policies. Today geneticists can verify that all human beings, despite allelic variations such as skin color, share the same genetic heritage. (Library of Congress)

Miscegenation

Sir Francis Galton, a cousin of Charles Darwin, is often regarded as the father of eugenics. He asserted that humans could be selectively bred for favorable traits. In his 1869 book *Hereditary Genius*, he set out to prove that favorable traits were inborn in people and concluded that

> the average intellectual standard of the Negro race is some two grades below our own. That the average ability of the [ancient] Athenian race is, on the lowest possible estimate, very nearly two grades higher than our own—that is, about as much as our race is above that of the African Negro.

In spite of its scientific inaccuracy by today's standards, the work of Galton was widely accepted by political and scientific leaders of his time. Bertrand Russell even suggested that the United Kingdom should issue color-coded "procreation tickets" issued to individuals based on their status in society: "Those who dared breed with holders of a different colored ticket would face a heavy fine." These "scientific" findings, combined with social and racial stereotypes, led to the eugenics movement and its development in many countries, including England, France, Germany, Sweden, Canada, and the United States.

Laws were passed to restrict the immigration of certain ethnic groups into the United States. Between 1907 and 1940, laws allowing forcible sterilization were passed in more than thirty states. Statutes prohibiting and punishing interracial marriages were passed in many states and, even as late as 1952, more than half the states still had antimiscegenation laws. The landmark decision against antimiscegenation laws occurred in 1967 when the U.S. Supreme Court declared the Virginia law unconstitutional. The decision, *Loving v. Virginia*, led to the erosion of the legal force of the antimiscegenation laws in the remaining states.

Impact and Applications

In spite of antimiscegenation laws and societal and cultural taboos, interracial matings have been a frequent occurrence. Many countries around the world, including the United States, are now racially heterogeneous societies. Genetic studies indicate that perhaps 20 to 30 percent of the genes in most African Americans are a result of admixture of white genes from mixed matings since the introduction of slavery to the Americas more than three hundred years ago. Miscegenation has been widespread throughout the world, and there may not even be such a thing as a "pure" race. No adverse biological effects can be attributed to miscegenation.

—Donald J. Nash

See also: Biological Determinism; Eugenics; Eugenics: Nazi Germany; Evolutionary Biology; Genetic Engineering: Social and Ethical Issues; Heredity and Environment; Intelligence; Race; Sociobiology; Sterilization Laws.

Further Reading

Alonso, Karen. *Loving v. Virginia: Interracial Marriage.* Berkeley Heights, N.J.: Enslow, 2000. Covers laws against interracial marriage, the road to the Supreme Court, a look at race-related laws, the Supreme Court's decision, and the impact of the Loving decision. Illustrations, bibliography, index.

Brah, Avtar, and Annie E. Coombes, eds. *Hybridity and Its Discontents: Politics, Science, Culture.* New York: Routledge, 2000. Covers ideas on miscegenation and racial purity, engineering the future, cultural translation, and reconfiguring concepts of nation, community, and belonging. Illustrations, bibliography, index.

Moran, Rachel F. *Interracial Intimacy: The Regulation of Race and Romance.* Chicago: University of Chicago Press, 2001. Discusses antimiscegenation laws and the legal maintenance of racial boundaries; breaking through racial boundaries; judicial review; race and identity; children, custody, and adoption; the new multiracialism; and more.

Sollors, Werner, ed. *Interracialism: Black-White Intermarriage in American History, Literature, and Law.* New York: Oxford University Press, 2000. Collection of foundational writings on interracial marriage and its effects on racial identity and racial relations. Bibliography, index.

Yancey, George. "An Analysis of Resistance to Racial Exogamy." *Journal of Black Studies* 31 (May, 2001). A look at opposition to interracial marriage and at South Carolina's attempt in 1998 to legalize interracial marriage through state referendum.

Web Site of Interest

Cold Spring Harbor Laboratory, Image Archive on the American Eugenics Movement. http://www.eugenicsarchive.org/eugenics. Comprehensive and extensively illustrated site that covers the eugenics movement in the United States, including miscegenation and antimiscegenation laws.

Mitochondrial Diseases

Field of study: Diseases and syndromes
Significance: *Mitochondrial genes are few in number but are necessary for animal cells to grow and survive. Mutations in these genes can result in age-related degenerative disorders and serious diseases of muscles and the central nervous system for which there is no generally effective treatment. Mitochondrial diseases are transmitted maternally and are usually associated with heteroplasmy, a state in which more than one type of gene arrangement, or genotype, occurs in the same individual.*

Key terms

HETEROPLASMY: a mutation in which more than one set of gene products encoded by mitochondrial DNA (mtDNA) can be present in an individual organ or tissue type, a single cell, or a single mitochondrion

MATERNAL INHERITANCE: the transmission pattern characteristically shown by mitochondrial diseases and mutations in mtDNA, where changes that occur in the mother's genetic material are inherited directly by children of both sexes without masking or interference by the mtDNA of the father

MITOCHONDRIA: small structures, or organelles, enclosed by double membranes found outside the nucleus, in the cytoplasm of all higher cells, that produce chemical power for the cells and harbor their own genetic material

MITOCHONDRIAL DNA (mtDNA): genetic material found uniquely in mitochondria, located outside the nucleus and therefore separate from the nuclear DNA

REPLICATIVE SEGREGATION: a mechanism by which individual mtDNAs carrying different mutations can come to predominate in any one mitochondrion

Mitochondrial Genetics and Disease

The unique arrangement of subunits making up individual genes is highly mutable, and thousands of different arrangements, or genotypes, are cataloged in humans. A tiny number of genes in animal cells are strictly inherited from the maternal parent and are found in the mitochondria, located in the cell's cytoplasm, outside the nucleus, where most genetic information resides in nuclear DNA. Some variants in mitochondrial DNA (mtDNA) sequences can cause severe defects in sight, hearing, skeletal muscles, and the central nervous system. Symptoms of these diseases often include great fatigue. The diseases themselves are difficult to diagnose accurately, and they are currently impossible to treat effectively. New genetic screening methods based on polymerase chain reaction (PCR) technologies using muscle biopsies are essential for correct identification of these diseases.

A person normally inherits a single mtDNA type, but families are occasionally found in which multiple mtDNA sequences are present. This condition, called heteroplasmy, is often associated with mitochondrial disease. Heteroplasmy occurs in the major noncoding region of mtDNA without much impact, but if it exists in the genes that control the production of cellular energy, severe consequences result. Weak muscles and multiple organs are involved in most mitochondrial diseases, and there can be variable expression of a particular syndrome within the same family that may either increase or decrease with age. It is easiest to understand this problem by remembering that each cell contains a population of mitochondria, so there is the possibility that some mtDNAs will carry a particular mutation while others do not. Organs also require different amounts of adenosine triphosphate (ATP), the cell's energy source produced in mitochondria. If the population of mutated mitochondria grows to outnumber the unmutated forms, most cells in a particular organ may appear diseased. This process has been called replicative segregation, and a mitochondrial disease is the result. Loss of mtDNA also occurs with increasing age, especially in the brain and heart.

Particular Mitochondrial Diseases

Mitochondrial diseases show a simple pattern of maternal inheritance. The first mitochondrial disease identified was Leber's hereditary optic neuropathy (LHON), a condition associated with the sudden loss of vision when the optic nerve is damaged, usually occurring in a person's early twenties. The damage is not reversible. Biologists now know that LHON is caused by at least four specific mutations that alter the mitochondrial proteins ND1, ND4, and CytB. A second mitochondrial syndrome is myoclonic epilepsy with ragged-red fiber disease (MERRF), which affects the brain and muscles throughout the body. This disease, along with another syndrome called mitochondrial encephalomyopathy, lactic acidosis, and stroke-like episodes (MELAS), is associated with particular mutations in mitochondrial transfer RNA (tRNA) genes that help produce proteins coded for by mtDNA. Finally, deletions and duplications of mtDNA are associated with Kearn-Sayre disease (affecting the heart, other muscles, and the cerebellum), chronic progressive external ophthalmoplegia (CPEO; paralysis of the eye muscles), rare cases of diabetes, heart deficiencies, and certain types of deafness. Some of these conditions have been given specific names, but others have not.

Muscles are often affected by mitochondrial diseases because muscle cells are rich in mitochondria. New treatments for these diseases are based on stimulating undamaged mtDNA in certain muscle precursor cells, called satellite cells, to fuse to damaged muscle cells and regenerate the muscle fibers. Others try to prevent damaged mtDNA genomes from replicating biochemically in order to increase the number of good mtDNAs in any one cell. This last set of experiments has worked on cells in tissue culture but has not been used on humans. These approaches aim to alter the competitive ability of undamaged genes to exist in a cellular environment that normally favors damaged genes. Further advances in treatment will also require better understanding of the natural ability of mtDNA to undergo genetic recombination and DNA repair.

—*Rebecca Cann*

See also: Aging; Extrachromosomal Inheritance; Hereditary Diseases; Human Genetics; Mitochondrial Genes.

Further Reading

Jorde, Lynn, et al. *Medical Genetics.* Rev. 2d ed. St. Louis, Mo.: Mosby, 2000. Presents a simple discussion of these diseases in the context of other genetic syndromes that are sex-linked or sex-limited in their inheritance patterns. Illustrations, bibliography, index.

Lestienne, Patrick, ed. *Mitochondrial Diseases: Models and Methods.* New York: Springer, 1999. Focuses on mitochondrial tRNA structure and its mutations. Illustrated (some color).

Raven, Peter H., and George B. Johnson. *Biology.* 6th ed. New York: W. H. Freeman/Worth, 1999. Helps clarify mitochondria and how they interact with a cell's nucleus. Illustrations, maps, index.

Web Site of Interest

United Mitochondrial Diseases Foundation. http://www.umdf.org. A support organization that promotes research offers support to affected individuals and families; the site explains the genetics of mitochondrial disorders and offers interactive medical advice.

Mitochondrial Genes

Field of study: Molecular genetics

Significance: *Mutations in mitochondrial genes have been shown to cause several human genetic diseases associated with a gradual loss of tissue function. Understanding the functions of mitochondrial genes and their nuclear counterparts may lead to the development of treatments for these debilitating diseases. Analysis of the mitochondrial DNA sequence of different human populations has also provided information relevant to the understanding of human evolution.*

Key terms

ADENOSINE TRIPHOSPHATE (ATP): the molecule that serves as the major source of energy for the cell

ATP SYNTHASE: the enzyme that synthesizes ATP

CYTOCHROMES: proteins found in the electron transport chain

ELECTRON TRANSPORT CHAIN: a series of protein complexes that pump H$^+$ ions out of the mitochondria as a way of storing energy that is then used by ATP synthase to make ATP

MITOCHONDRIAL DNA (mtDNA): genetic material found uniquely in mitochondria, located outside the nucleus and therefore separate from the nuclear DNA

RIBOSOMES: organelles that function in protein synthesis and are made up of a large and a small subunit composed of proteins and ribosomal RNA (rRNA) molecules

SPACERS: long segments of DNA rich in adenine-thymine (A-T) base pairs that separate exons and introns, although most of the spacer DNA is transcribed but is not translated messenger RNA (mRNA)

Mitochondrial Structure and Function

Mitochondria are membrane-bound organelles that exist in the cytoplasm of eukaryotic cells. Structurally, they consist of an outer membrane and a highly folded inner membrane that separate the mitochondria into several compartments. Between the two membranes is the intermembrane space; the innermost compartment bounded by the inner membrane is referred to as the "matrix." In addition to enzymes involved in glucose metabolism, the matrix contains several copies of the mitochondrial chromosome as well as ribosomes, transfer RNA (tRNA), and all the other necessary factors required for protein synthesis. Mitochondrial ribosomes are structurally different from the ribosomes located in the cytoplasm of the eukaryotic cell and, in fact, more closely resemble ribosomes from bacterial cells. This

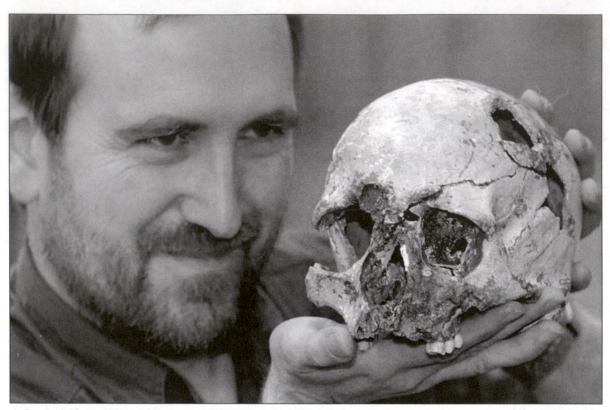

At London's Natural History Museum in 1997, anthropologist Chris Stringer displays the nine-thousand-year-old skull of Cheddar Man (named for the southwestern English town), to whom he traced a modern relative by comparing DNA samples from the skull with samples from a living, forty-two-year-old schoolteacher. This is possible because mitochondrial DNA is passed unchanged from generation to generation down the maternal line. (AP/Wide World Photos)

similarity led to the endosymbiont hypothesis developed by Lynn Margulis, which proposes that mitochondria arose from bacteria that took up residence in the cytoplasm of the ancestor to eukaryotes.

Embedded in the inner mitochondrial membrane is a series of protein complexes that are known collectively as the "electron transport chain." These proteins participate in a defined series of reactions that begin when energy is released from the breakdown of glucose and end when oxygen combines with $2H^+$ ions to produce water. The net result of these reactions is the movement of H^+ ions (also called protons) from the matrix into the intermembrane space. This establishes a proton gradient in which the intermembrane space has a more positive charge and is more acidic than the matrix. Thus mitochondria act as tiny batteries that separate positive and negative charges in order to store energy. Another protein that is embedded in the inner mitochondrial membrane is an enzyme called adenosine triphosphate (ATP) synthase. This enzyme allows the H^+ ions to travel back into the matrix. When this happens, energy is released that is then used by the synthase enzyme to make ATP. Cells use ATP to provide energy for all of the biological work they perform, including movement and synthesis of other molecules. The concept of linking the production of a proton gradient to ATP synthesis was developed by Peter Mitchell in 1976 and is referred to as the chemiosmotic hypothesis.

Mitochondrial Genes

The mitochondrial chromosome is a circular DNA molecule that varies in size from about 16,000 base pairs (bp) in humans to more than 100,000 base pairs in certain species of plants.

Despite these size differences, mitochondrial DNA (mtDNA) contains only a few genes that tend to be similar over a wide range of organisms. This discussion will focus on genes located on the human mitochondrial chromosome that has been completely sequenced. These genes fall into two broad categories: those that play a role in mitochondrial protein synthesis and those involved in electron transport and ATP synthesis.

Mitochondria have their own set of ribosomes that consist of a large and a small subunit. Each ribosomal subunit is a complex of ribosomal RNA (rRNA) and proteins. Genes that play a role in mitochondrial protein synthesis include two rRNA genes designated 16S rRNA and 12S rRNA, indicating the RNA for the large and small subunits respectively. Also in this first category are genes for mitochondrial transfer RNA. Transfer RNA (tRNA) is an *L*-shaped molecule that contains the RNA anticodon at one end and an amino acid attached to the other end. The tRNA anticodon pairs with the codon of the messenger RNA (mRNA) and brings the correct amino acid into position to be added to the growing protein chain. Thus

The Diversity of mtDNA

The mitochondria of plants, animals, and fungi include their own DNA genomes, mitochondrial DNA (mtDNA). The mtDNA genome typically consists of a bacteria-like circular loop of DNA located in highly condensed structures called nucleoids within the mitochondrial matrix. However, the mtDNA of the yeast *Hansenula*, the protozoans *Tetrahymena* and *Paramecium*, and the alga *Chlamydomonas* are chainlike or linear rather than circular, while that of protozoan parasites such as *Trypanosoma*, *Leishmania*, and *Crithidia* is organized into a network of several hundred maxicircles about 21-31 kilobase pairs (kb) long, interlocked with several thousand minicircles, each about 0.5-2.5 kb.

The size of each mtDNA varies greatly among organisms. Most animals have small mtDNA genomes ranging from about 6 to 20 kb, such as the 6-kb mtDNA genome of the protozoan parasite *Plasmodium falciparium*, which causes malaria, and the 14.3-kb mtDNA of free-living *Ascaris* roundworms. The mtDNA genome of humans is about 16.5 kb and comprises about 0.3 percent of the total genome. The mtDNA genomes of most plants and fungi are larger: The mtDNA of the yeast *Saccharomyces cerevisiae* is 86 kb, that of the common pea *Sativa* is 110 kb, that of the liverwort *Marchantia* is 186 kb, and that of the muskmelon *Cucumis melo* is a gigantic 2,400 kb. Much of the size variation is due to the presence of long segments of noncoding sequences embedded within the genome, which seem to be especially abundant in plants and fungi but not in animal mtDNA. More than half of the mtDNA of yeasts, for example, is formed by long segments of spacers, while another quarter consists of introns, intervening sequences between segments consisting of functioning genes.

Despite the size differences, plant and animal mtDNA usually carry the same thirty-seven coding genes: twenty-two genes coding for transfer RNA molecules, two ribosomal RNA genes, and thirteen genes coding for proteins involved in mitochondrial respiration. Again, certain organisms differ. *Marchantia* mtDNA, for example, includes an additional sixteen genes that code for ribosomal proteins and twenty-nine genes that code for proteins of unknown function.

Translation of mtDNA is consistent with the universal genetic code, with notable departures. For example, both AGA and AGG specify the amino acid arginine in the universal genetic code but are stop codes in animal mtDNA. In ciliated protozoans the mtDNA code for glutamine is UAA and UAG, which specifies stop in the universal genetic code. In yeast the mtDNA codes CUU, CUA, CUC, and CUG specify the amino acid threonine instead of leucine, as specified by the universal genetic code. Presumably, all of these mtDNA coding departures from the universal genetic code result from mutations that occurred subsequent to the endosymbiotic incorporation of the original mitochondria into early eukaryotic cells.

Inheritance patterns of mtDNA differ for some plants and animals as well. In animals the mtDNA genome is transmitted primarily through the female egg to the offspring, but in *Chlamydomonas* algae and yeasts male and female gametes are nearly equal in size and contribute mtDNA genome to the offspring.

—*Dwight G. Smith*

the tRNA molecule serves as a bridge between the information in the mRNA molecule and the sequence of amino acids in the protein. Mitochondrial tRNAs are different from those involved in protein synthesis in the cytoplasm. In fact, cytoplasmic tRNAs would not be able to function on mitochondrial ribosomes, nor could mitochondrial tRNAs work with cytoplasmic ribosomes. Thus, mtDNA contains a complete set of twenty-two tRNA genes.

Genes involved in electron transport fall into the second category of mitochondrial genes. The electron transport chain is divided into a series of protein complexes, each of which consists of a number of different proteins, a few of which are encoded by mtDNA. The NADH dehydrogenase complex (called complex I) contains about twenty-two different proteins. In humans, only six of these proteins are encoded by genes located on the mitochondrial chromosome. Cytochrome *c* reductase (complex III) contains about nine proteins, including cytochrome *b*, which is the only one whose gene is located on mtDNA. Cytochrome oxidase (complex IV) contains seven proteins, three of which are encoded by mitochondrial genes. About sixteen different proteins combine to make up the mitochondrial ATP synthase, and only two of these are encoded by mtDNA.

All of the proteins not encoded by mitochondrial genes are encoded by genes located on nuclear chromosomes. In fact, more than 90 percent of the proteins found in the mitochondria are encoded by nuclear genes. These genes must be transcribed into mRNA in the nucleus, then the mRNA must be translated into protein on cytoplasmic ribosomes. Finally, the proteins are transported into the mitochondria where they function. By contrast, genes located on mtDNA are transcribed in the mitochondria and translated on mitochondrial ribosomes.

Impact and Applications

Any mutation occurring in a mitochondrial gene has the potential to reduce or prevent mitochondrial ATP synthesis. Because human cells are dependent upon mitochondria for their energy supply, the effects of these muta-

tions can be wide-ranging and debilitating, if not fatal. If the mutation occurs in a gene that plays a role in mitochondrial protein synthesis, the ability of the mitochondria to perform protein synthesis is affected. Consequently, proteins that are translated on mitochondrial ribosomes such as cytochrome *b* or the NADH dehydrogenase subunits cannot be made, leading to defects in electron transport and ATP synthesis. Mutations in mitochondrial tRNA genes, for example, have been shown to be the cause of several degenerative neuromuscular disorders. Genes involved in electron transport and ATP synthesis have a more directly negative effect when mutated. Douglas C. Wallace and coworkers identified a mutation within the NADH dehydrogenase subunit 4 gene, for example, that was the cause of a maternally inherited form of blindness and was one of the first mitochondrial diseases to be identified.

Of further interest is the study of nuclear genes that contribute to mitochondrial function. Included in this list of nuclear genes are those encoding proteins involved in mtDNA replication, repair, and recombination; enzymes involved in RNA transcription and processing; and ribosomal proteins and the accessory factors required for translation. It is presumed that a mutation in any of these genes could have negative effects upon the ability of the mitochondria to function. Understanding how nuclear genes contribute to mitochondrial activity is an essential part of the search for effective treatments for mitochondrial diseases.

Human evolutionary studies have also been affected by the understanding of mitochondrial genes and their inheritance. Researchers Allan C. Wilson and Rebecca Cann, knowing that mitochondria are inherited exclusively through the female parent, hypothesized that a comparison of mitochondrial DNA sequences in several human populations would enable them to trace the origins of the ancestral human population. These studies led to the conclusion that a female living in Africa about 200,000 years ago was the common ancestor for all modern humans; she is referred to as "mitochondrial Eve."

—Bonnie L. Seidel-Rogol

See also: Aging; Ancient DNA; Extrachromosomal Inheritance; Hereditary Diseases; Human Genetics; Mitochondrial Diseases; RNA World.

Further Reading

Hartwell, L. H., L. Hood, M. L. Goldberg, A. E. Richards, L. M. Silver, and R. C. Veres. *Genetics: From Genes to Genomes.* Boston: McGraw-Hill, 2003. Chapter 15 provides an excellent summary of mitochondrial DNA.

Pon, Liza, and Eric A. Schon, eds. *Mitochondria.* San Diego: Academic Press, 2001. Discusses the effects of impaired mitochondrial function.

Scheffler, Immo E. *Mitochondria.* New York: John Wiley & Sons, 1999. Comprehensive, concise discussion of mitochondria biochemistry, genetics, and pathology.

Wallace, Douglas C. "Mitochondrial DNA in Aging and Disease." *Scientific American,* August, 1997. Gives a detailed explanation of human mitochondrial diseases, aimed at nonspecialists.

Wilson, Allan C., and Rebecca L. Cann. "The Recent African Genesis of Humans." *Scientific American,* April, 1992. Describes how studies of mitochondrial genes have led to information about human origins.

Mitosis and Meiosis

Field of study: Cellular biology

Significance: *Mitosis is the process of cell division in multicellular eukaryotic organisms. Meiosis is the process of cell division that produces haploid gametes in sexually reproducing eukaryotic organisms.*

Key terms

BINARY FISSION: reproduction of a cell by division into two parts

CENTROMERE: a region on the chromosome where chromatids join

CHROMATID: one-half of a replicated chromosome

CYTOKINESIS: division of the cytoplasm to form new cells

DAUGHTER CELLS: cells resulting from the division of a parent cell

DIPLOID CELLS: cells containing two sets of homologous chromosomes

HAPLOID CELLS: cells containing one set of chromosomes; eggs and sperm are haploid cells

Cellular Reproduction

Organisms must be able to grow and reproduce. Prokaryotes, such as bacteria, duplicate DNA and divide by splitting in two, a process called binary fission. Cells of eukaryotes, including those of animals, plants, fungi, and protists, divide by one of two methods: mitosis or meiosis. Mitosis produces two cells, called daughter cells, with the same number of chromosomes as the parent cell, and is used to produce new somatic (body) cells in multicellular eukaryotes or new individuals in single-celled eukaryotes. In sexually reproducing organisms, cells that produce gametes (eggs or sperm) divide by meiosis, producing four cells, each with half the number of chromosomes possessed by the parent cell.

Chromosome Replication

All eukaryotic organisms are composed of cells containing chromosomes in the nucleus. Chromosomes are made of DNA and proteins. Most cells have two complete sets of chromosomes, which occur in pairs. The two chromosomes that make up a pair are homologous, and contain all the same loci (genes controlling the production of a specific type of product). These chromosome pairs are usually referred to as homologous pairs. An individual chromosome from a homologous pair is sometimes called a homolog. For example, typical lily cells contain twelve pairs of homologous chromosomes, for a total of twenty-four chromosomes. Cells that have two homologous chromosomes of each type are called diploid. Some cells, such as eggs and sperm, contain half the normal number of chromosomes (only one of each homolog) and are called haploid. Lily egg and sperm cells each contain twelve chromosomes.

DNA must replicate before mitosis or meiosis can occur. If daughter cells are to receive a

Mitosis

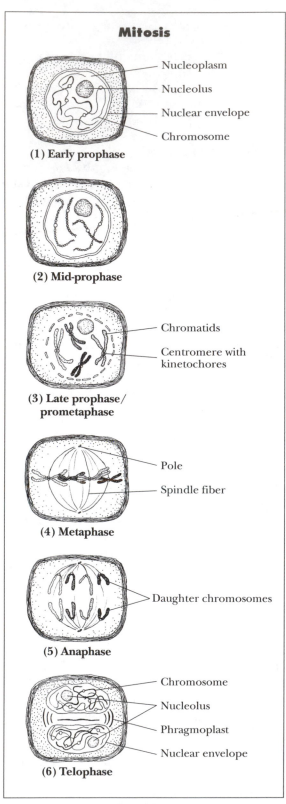

(1) **Early prophase**
- Nucleoplasm
- Nucleolus
- Nuclear envelope
- Chromosome

(2) **Mid-prophase**

(3) **Late prophase/ prometaphase**
- Chromatids
- Centromere with kinetochores

(4) **Metaphase**
- Pole
- Spindle fiber

(5) **Anaphase**
- Daughter chromosomes

(6) **Telophase**
- Chromosome
- Nucleolus
- Phragmoplast
- Nuclear envelope

(Kimberly L. Dawson Kurnizki)

full set of genetic information, a duplicate copy of DNA must be available. Before DNA replication occurs, each chromosome consists of a single long strand of DNA called a chromatid. After DNA replication, each chromosome consists of two chromatids, called sister chromatids. The original chromatid acts as a template for making the second chromatid; the two are therefore identical. Sister chromatids are attached at a special region of the chromosome called the centromere. When mitosis or meiosis starts, each chromosome in the cell consists of two sister chromatids.

Mitosis and meiosis produce daughter cells with different characteristics. When a diploid cell undergoes mitosis, two identical diploid daughter cells are produced. When a diploid cell undergoes meiosis, four unique haploid daughter cells are produced. It is important for gametes to be haploid so that when an egg and sperm fuse, the diploid condition of the mature organism is restored.

Cellular Life Cycles

Mitosis and meiosis occur in the nuclear region of the cell, where all the cell's chromosomes are found. Nuclear control mechanisms begin cell division at the appropriate time. Some cells in an adult organism rarely divide by mitosis in adult organisms, while other cells divide constantly, replacing old cells with new. Meiosis occurs in the nuclei of cells that produce gametes. These specialized cells occur in reproductive organs, such as flower parts in higher plants.

Cells, like organisms, are governed by life cycles. The life cycle of a cell is called the cell cycle. Cells spend most of their time in interphase. Interphase is divided into three stages: first gap (G_1), synthesis (S), and second gap (G_2). During G_1, the cell performs its normal functions and often grows in size. During the S stage, DNA replicates in preparation for cell division. During the G_2 stage, the cell makes materials needed to produce the mitotic apparatus and for division of the cytoplasmic components of the cell. At the end of interphase, the cell is ready to divide. Although each chromosome now consists of two sister chromatids, this is not apparent when viewed through a microscope; all the chromosomes are in a highly relaxed state and simply appear as a diffuse material called chromatin.

Mitosis

Mitosis consists of five stages: prophase, prometaphase, metaphase, anaphase, and telophase. Although certain events identify each stage, mitosis is a continuous process, and each stage gradually passes into the next. Identification of the precise state is therefore difficult at times.

During prophase, the chromatin becomes more tightly coiled and condenses into chromosomes that are clearly visible under a microscope, the nucleolus disappears, and the spindle apparatus begins to form in the cytoplasm. In prometaphase the nuclear envelope breaks down, and the spindle apparatus is now able to invade the nuclear region. Some of the spindle fibers attach themselves to a region near the centromere of each chromosome called the kinetochore. The spindle apparatus is the most obvious structure of the mitotic apparatus. The nuclear region of the cell has opposite poles, like the North and South Poles of the earth. Spindle fibers reach from pole to pole, penetrating the entire nuclear region.

During metaphase, the cell's chromosomes align in a region called the metaphase plate, with the sister chromatids oriented toward opposite poles. The metaphase plate traverses the cell, much like the equator passes through the center of the earth. Sister chromatids separate during anaphase. The sister chromatids of each chromosome split

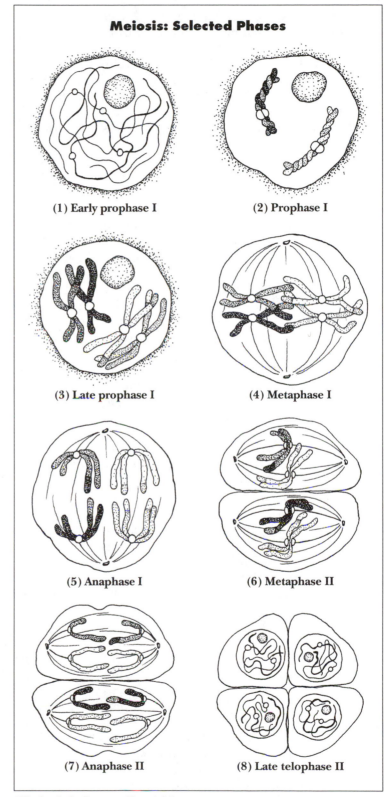

Meiosis: Selected Phases

(1) **Early prophase I**

(2) **Prophase I**

(3) **Late prophase I**

(4) **Metaphase I**

(5) **Anaphase I**

(6) **Metaphase II**

(7) **Anaphase II**

(8) **Late telophase II**

(Kimberly L. Dawson Kurnizki)

apart, and the spindle fibers pull each sister chromatid (now a separate chromosome) from each pair toward opposite poles, much as a rope-tow pulls a skier up a mountain. Telophase begins as sister chromatids reach opposite poles. Once the chromatids have reached opposite poles, the spindle apparatus falls apart, and the nuclear membrane re-forms. Mitosis is complete.

Meiosis

Meiosis is a more complex process than mitosis and is divided into two major stages: meiosis I and meiosis II. As in mitosis, interphase precedes meiosis. Meiosis I consists of prophase I, metaphase I, anaphase I, and telophase I. Meiosis II consists of prophase II, metaphase II, anaphase II, and telophase II. In some cells, an interphase II occurs between meiosis I and meiosis II, but no DNA replication occurs.

During prophase I, the chromosomes condense, the nuclear envelope falls apart, and the spindle apparatus begins to form. Homologous chromosomes come together to form tetrads (a tetrad consists of four chromatids, two sister chromatids for each chromosome). The arms of the sister chromatids of one homolog touch the arms of sister chromatids of the other homolog, the contact points being called chiasmata. Each chiasma represents a place where the arms have the same loci, so-called homologous regions. During this intimate contact, the chromosomes undergo crossover, in which the chromosomes break at the chiasmata and swap homologous pieces. This process results in recombination (the shuffling of linked alleles, the different forms of genes, into new combinations), which results in increased variability in the offspring and the appearance of character combinations not present in either parent.

Tetrads align on the metaphase plate during metaphase I, and one spindle fiber attaches to the kinetochore of each chromosome. In anaphase I, instead of the sister chromatids separating, they remain attached at their centromeres, and the homologous chromosomes separate, each homolog from a tetrad moving toward opposite poles. Telophase I begins as the homologs reach opposite poles, and similar to telophase of mitosis, the spindle apparatus

falls apart, and a nuclear envelope re-forms around each of the two haploid nuclei. Because the number of chromosomes in each of the telophase I nucleus is half the number in the parent nucleus, meiosis I is sometimes called the reductional division.

Meiosis II is essentially the same as mitosis, dividing the two haploid nuclei formed in meiosis I. Prophase II, metaphase II, anaphase II, and telophase II are essentially identical to the stages of mitosis. Meiosis II begins with two haploid cells and ends with four haploid daughter cells.

Nuclear Division and Cytokinesis

Mitosis and meiosis result in the division of the nucleus. Nuclear division is nearly always coordinated with division of the cytoplasm. Cleaving of the cytoplasm to form new cells is called cytokinesis. Cytokinesis begins toward the middle or end of nuclear division and involves not just the division of the cytoplasm but also the organelles. In plants, after nuclear division ends, a new cell wall must be formed between the daughter nuclei. The new cell wall begins when vesicles filled with cell wall material congregate where the metaphase plate was located, producing a structure called the cell plate. When the cell plate is fully formed, cytokinesis is complete. Following cytokinesis, the cell returns to interphase. Mitotic daughter cells enlarge, reproduce organelles, and resume regular activities. Following meiosis, gametes may be modified or transported in the reproductive system.

Alternation of Generations

Meiotic daughter cells continue development only if they fuse during fertilization. Mitosis and meiosis alternate during the life cycles of sexually reproducing organisms. The life-cycle stage following mitosis is diploid, and the stage following meiosis is haploid. This process is called alternation of generations. In plants, the diploid state is referred to as the sporophyte generation, and the haploid stage as the gametophyte generation. In nonvascular plants, the gametophyte generation dominates the life cycle. In other words, the plants normally seen on the forest floor are made of haploid cells.

The sporophytes, which have diploid cells, are small and attached to the body of the gametophyte. In vascular plants, sporophytes are the large, multicellular individuals (such as trees and ferns) whereas gametophytes are very small and either are embedded in the sporophyte or are free-living, as are ferns. The genetic variation introduced by sexual reproduction has a significant impact on the ability of species to survive and adapt to the environment. Alternation of generations allows sexual reproduction to occur without changing the chromosome number characterizing the species.

—*Joyce A. Corban and Randy Moore*

See also: Cell Culture: Animal Cells; Cell Culture: Plant Cells; Cell Cycle, The; Cell Division; Cytokinesis; Gene Regulation: Eukaryotes; Polyploidy; Totipotency.

Further Reading

Alberts, Bruce, Dennis Bray, Julian Lewis, Martin Raff, Keith Roberts, and James D. Watson. *Molecular Biology of the Cell*. 4th ed. New York: Garland, 2002. The chapter "How Cells Are Studied" gives extensive information regarding study methods in cell biology. Light and electron microscopy are discussed as well as staining techniques and tissue culture.

Audesirk, Teresa, Gerald Audesirk, and Bruce E. Myers. *Biology: Life on Earth*. 6th ed. Upper Saddle River, N.J.: Prentice Hall, 2001. The chapter "Cellular Reproduction and the Life Cycles of Organisms" is a brief overview of mitosis, meiosis, and the cell cycle. Includes excellent discussion of alternation of generations.

Campbell, Neil A., and Jane B. Reece. *Biology*. 6th ed. San Francisco: Benjamin Cummings, 2002. The chapter "Reproduction of Cells" provides extensive information regarding mitosis and the cell cycle. The phases of mitosis, the mitotic spindle, cytokinesis, control mechanisms, and abnormal cell division are discussed in detail. The chapter "Meiosis and Sexual Life Cycles" addresses the stages of meiosis, sexual life cycles, and a comparison of mitosis and meiosis. This text is intended for use in introductory biology and is very readable and informative.

John, Bernard. *Meiosis*. New York: Cambridge University Press, 1990. Review and discussion of meiosis, the antithesis of fertilization. Discusses the scheduling, mechanisms, biochemistry, and the genetic control of the events in meiosis.

Keeton, William T., and James L. Gould. *Biological Science*. 5th ed. New York: W. W. Norton, 1993. The chapter "Cellular Reproduction" discusses in detail the stages of mitosis and meiosis. Excellent diagrams allow visualization of cell division.

Model Organism: *Arabidopsis thaliana*

Field of study: Techniques and methodologies

Significance: Arabidopsis thaliana, *also known as mouse-ear cress, can grow from seed to maturity and back to thousands of seeds again in about six weeks. Its short reproduction cycle and simple, low-cost cultivation allow genetic experiments with tens of thousands of plants and make it a popular and convenient organism to use as a model organism.*

Key terms

BRASSICACEAE: the mustard family, a large, cosmopolitan family of plants with many wild species, some of them common weeds, including widely cultivated edible plants like cabbage, cauliflower, radish, rutabaga, turnip, and mustard

GENETIC MAP: a "map" showing distances between genes in terms of recombination frequency

Natural History

Although common as an introduction into America and Australia, *Arabidopsis thaliana* (often referred to simply by its genus name, *Arabidopsis*) is found in the wild throughout Europe, the Mediterranean, the East African highlands, and Eastern and Central Asia (which is probably where it originated). Since *Arabidopsis* is a low winter annual (standing about 1.5 decimeters), it flowers in disturbed habitats from

Two specimens of mouse-ear cress, Arabidopsis thaliana. (AP/Wide World Photos)

March through May. *Arabidopsis* was first described by Johannes Thal (hence the *thaliana* as the specific epithet) in the sixteenth century in Germany's Harz Mountains, but he named it *Pilosella siliquosa* at the time. Undergoing systematic revisions and several name changes, the little plant was finally called *Arabidopsis thaliana* in 1842.

Several characteristics of *Arabidopsis* make it useful as a model organism: First, it has a short life cycle; it goes from seed to seed in only about three months, and each individual plant is prolific, yielding thousands of seeds. Second, the plants are small, easy to grow and to manipulate, so that many genetic screens can be done on petri dishes with a thousand seedlings examined inside just one dish. Also, the genome of *Arabidopsis* is relatively small, with 120 million base pairs (Mbp), 25,000 genes, and five chromosomes containing all the requisite information to encode an entire plant (similar to the functional complexity of the fruit fly *Drosophila melanogaster,* long a favorite model organism among geneticists). Yet in comparison to the genome of corn (*Zea mays*), for example, *Arabidopsis* has a genome that is ten times smaller. Furthermore, *Arabidopsis* is easily transformed using the standard vector *Agrobacterium tumefaciens.* There are also a large number of mutant lines and genomic resources available for *Arabidopsis* at stock centers, and a cooperative multinational research community of academic, government, and industry laboratories exists, all working with *Arabidopsis.*

History of Experimental Work with *Arabidopsis*

The earliest report of a mutant probably was made in 1873 by A. Braun, and Freidrich Laibach first compiled the unique characteristics of *Arabidopsis thaliana* as a model organism for genetics in 1943 (publishing the correct chromosome number of five much earlier, in 1907, later confirmed by other investigators).

Erna Reinholz (a student of Laibach) submitted her thesis in 1945, published in 1947, on the first collection of X-ray-induced mutants. Peter Langridge established the usefulness of *Arabidopsis* in the laboratory in the 1950's, as did George Redei and other researchers, including J. H. van der Veen in the Netherlands, J. Veleminsky in Czechoslovakia, and G. Robbelen in Germany in the 1960's.

Maarten Koorneef and his coworkers published the first detailed genetic map for *Arabidopsis* in 1983. A genetic map allows researchers to observe approximate positions of heritable factors (genes and regulatory elements) on chromosomes. The 1980's saw the first steps in analysis of the genome of *Arabidopsis*. Tagged mutant collections were developed. Physical maps, with distances between genes in terms of DNA length, based on restriction fragment length polymorphisms (RFLPs), were also made in this time period. The physical maps allow genes to be located and characterized, even if their identities remained unknown.

The 1990's saw scientists outlining long-range plans for *Arabidopsis* through the Multinational Coordinated *Arabidopsis* Genome Research Project, which called for genetic and physiological experimentation necessary to identify, isolate, sequence, and understand *Arabidopsis* genes. In the United States, the National Science Foundation (NSF), U.S. Department of Energy (DOE), and Agricultural Research Service (ARS) funded work done at Albany directed by Athanasios Theologis. NSF and DOE funds went also to Stanford, Philadelphia, and four other U.S. laboratories. Worldwide communication among laboratories and the creation of shared databases (particularly in the United States, Europe, and Japan) were established. Transformation methods have become much more efficient, and a large number of *Arabidopsis* mutant lines, gene libraries, and genomic resources have been made and are now available to the scientific community through public stock centers. The expression of multiple genes has been followed, too. Teresa Mozo provided the first comprehensive physical map of the *Arabidopsis* genome, published in 1999; she used overlapping fragments of cloned DNA. These fundamental data provide an important resource for map-based gene cloning and genome analysis. The *Arabidopsis* Genome Initiative, an international effort to sequence the complete *Arabidopsis* genome, was created in the mid-1990's, and the results of this massive undertaking were published on December 14, 2000, in *Nature*.

Comparative Genomics

With full sequencing of the genome of *Arabidopsis* completed, the first catalog of genes involved in the life cycle of a typical plant is now available, and the investigational emphasis has shifted to functional and comparative genomics. Scientists began looking at when and where specific genes are expressed in order to learn more about how plants grow and develop in general, how they survive in the changing environment, and how the gene networks are controlled or regulated. Potentially this research and work can lead to improved crop plants that are more nutritious, more resistant to pests and disease, less vulnerable to crop failure, and capable of producing higher yields with less damage to the natural environment. Since many more people die from malnutrition in the world than from diseases, the *Arabidopsis* genome takes on a much more important consideration than one might think. Of course, plants are fundamental to all ecosystems, and their energy input into those systems is essential and critical.

Already the genetic research on *Arabidopsis* has boosted production of staple crops such as wheat, tomatoes, and rice. The genetic basis for every economically important trait in plants— whether pest resistance, vegetable oil production, or even wood quality in paper products— is under intense scrutiny in *Arabidopsis*.

Although *Arabidopsis* is considered a weed throughout its ecological range on the planet, it is closely related to a number of vegetables, including broccoli, cabbage, brussels sprouts, and cauliflower, which are very important to humans nutritionally and economically. For example, a mutation observed in *Arabidopsis* has resulted in its floral structures assuming the basic shape of a head of cauliflower. This mutation in *Arabidopsis*, not surprisingly, is referred to simply as "cauliflower" and was isolated by

Martin Yanofsky's laboratory. The analogous gene from the cauliflower plant was examined, and it was discovered the cauliflower plant already had a mutation in this gene. From the study of *Arabidopsis*, therefore, researchers have uncovered why a head of cauliflower looks the way it does.

In plants there is an ethylene-signaling pathway (ethylene is a plant hormone) that regulates fruit ripening, plant senescence, and leaf abscission. The genes necessary for the ethylene-signaling pathway have been identified in *Arabidopsis*, including genes coding for the ethylene receptors. As expected, a mutation in these ethylene receptors would also cause the *Arabidopsis* plant to be unable to sense ethylene. Ethylene receptors have now been uncovered from other plant species from the knowledge gained from *Arabidopsis*. Harry Klee's laboratory, for example, has found a tomato mutation in the ethylene receptor, which prevents ripening. When the mutant *Arabidopsis* receptor is expressed in other plants, moreover, the transformed plants also exhibit this insensitivity to ethylene and the lack of ensuing processes associated with it. Therefore, the mechanism of ethylene perception would seem to be conserved in plants, and modifying ethylene receptors can induce change in a plant.

Advances in evolutionary biology and medicine are expected from *Arabidopsis* research, too. Robert Martienssen of Cold Spring Harbor Laboratory has referred to the completion of the *Arabidopsis* genome sequence as having major impact on human health as well as plant biology and agriculture. Surprisingly, some of the newly identified *Arabidopsis* genes are extremely similar or even identical to human genes linked to certain illnesses. No doubt there are many more mysteries to unravel with the proteome analysis of *Arabidopsis* (analysis of how proteins function in the plant), and the biological role of all the twenty-five thousand genes will keep scientists busy for some time to come. For example, this relatively "simple" little plant has surprised workers with its amazing genetic duplication where more than 70 percent of its DNA is copied at least once somewhere else on its genome.

—*F. Christopher Sowers*

See also: Cell Culture: Plant Cells; Extrachromosomal Inheritance; Model Organisms.

Further Reading
Bowman, John L. *Arabidopsis: An Atlas of Morphology and Development.* New York: Springer-Verlag, 1993. Contains images and descriptions of normal and mutant *Arabidopsis* plants.

Russell, Peter J. *Genetics.* San Fransisco, Calif.: Benjamin Cummings, 2002. Good genetic textbook with specific references to genetic duplications, genome sequences, homeotic genes, model organism considerations, and the regulation of development in *Arabidopsis*.

Wilson, Zoe A. *Arabidopsis: A Practical Approach.* New York: Oxford University Press, 2000. Provides an introduction to techniques required for the use of *Arabidopsis* as an experimental system. Provides strategies for the identification, mapping, and characterization of mutants by microscopy, molecular cytogenetics, and gene expression analysis.

Web Site of Interest
The *Arabidopsis* Information Resource (TAIR). http://www.arabidopsis.org. The gateway to the *Arabidopsis* Genome Initiative (AGI), designed for the scientific community, consists of a searchable relational database with many different data types that can be viewed, analyzed, and downloaded. Also has pages for news, lab protocols, and links.

Model Organism: *Caenorhabditis elegans*

Field of study: Techniques and methodologies

Significance: *The roundworm* Caenorhabditis elegans *has helped scientists understand development of multicellular organisms. For their work using* C. elegans *to identify apoptosis, or programmed cell death, three scientists received the Nobel Prize. The* C. elegans *genome project has enabled scientists to develop much of the technology that was used to sequence the human genome. Re-*

search with this organism has also contributed to understanding genetics of the nervous system, aging, and even learning.

Key terms

CELL DIFFERENTIATION: a process during which a cell specifically expresses certain genes, ultimately adopting its final cell fate to become a specific type of cell, such as a neuron, or undergoing programmed cell death (apoptosis)

MODEL ORGANISM: an organism well suited for genetic research because it has a well-known genetic history, a short life cycle, and genetic variation between individuals in the population

The Organism

The nematode *Caenorhabditis elegans* (*C. elegans*) has been the subject of intense analysis by biologists around the world. Nematodes, or roundworms, are simple metazoan animals that have cells specialized to form tissues and organs such as nerve tissue and digestive tissue. Analysis of genetic control of the events that lead to the formation of the tissues in *C. elegans* has revealed biological mechanisms that also control the differentiation of tissues and organs in more complex organisms such as humans.

Caenorhabditis elegans is a microscopic, 1-millimeter-long roundworm that lives in soils and eats bacteria from decaying materials. It belongs to the phylum *Nematoda* (the roundworms), which includes many significant plant and animal parasites. *Caenorhabditis elegans*, however, is free-living (nonparasitic) and does not cause any human diseases. It exists as two sexes, males (containing a single X chromosome) and hermaphrodites (containing two X chromosomes). Both male and hermaphrodite worms have five pairs of autosomal (non-sex) chromosomes. The hermaphrodites are self-fertile. They produce sperm first, which they store, and later "switch" gonads to begin producing eggs. These eggs may be fertilized by the hermaphrodite's own sperm, or if the hermaphrodite mates with a male, sperm from the male will fertilize the eggs. A hermaphrodite that is not mated will lay approximately three hundred fertilized eggs in the first four days of adulthood; hermaphrodites that mate with males will continue to lay eggs as long as sperm are present.

Caenorhabditis elegans eggs begin development within the uterus. They hatch as small L1 larvae and molt four times as they proceed through the easily recognizable larval stages of L2, L3, L4, and adult. The adult hermaphrodite is a little larger than the adult male and can be distinguished by the presence of fertilized eggs lined up in the uterus. The smaller males have specialized tails that contain structures for mating called copulatory spicules.

A Model Organism

Because of its small size and simple diet (bacteria), *C. elegans* is easily adapted to laboratory culture conditions. The worms are grown on small agar-filled petri plates that are seeded with *E. coli*. The worms live comfortably at room temperature, but elevating or lowering the temperature can speed up or slow down development, and changes in temperatures can even reveal conditional phenotypes of some genetic mutations.

One unmated hermaphrodite will produce three hundred progeny over the first four days of adulthood. Additionally, *C. elegans* has a short generation time of approximately three weeks. Obtaining large numbers of progeny allows thorough statistical analysis of the way a mutation is segregated within a population. Because researchers can screen large numbers of worms in a short period of time, extremely rare mutations are likely to be revealed. Genetically "pure" strains are also quickly produced.

Hermaphrodite genetics also provides advantages. Because hermaphrodites are self-fertile, getting homozygous mutations is not difficult. A hermaphrodite that is heterozygous for a given mutation (has one wild-type copy of a gene and one mutated copy of a gene) will produce progeny, one-fourth containing two mutated copies of the gene (homozygotes). Additionally, for researchers studying mutations that affect reproduction or mating behavior, having self-fertile hermaphrodites allows them to maintain mutations that affect processes such as sperm production. A hermaphrodite that cannot make its own sperm can be mated to a wild-type male, and the mutation causing the defect can be maintained. This is not possible in organisms that are strictly male/female or that are strictly hermaphroditic.

Another strength of *C. elegans* is that the genetic strains can be frozen in liquid nitrogen and maintained indefinitely. Even fruit flies have to be constantly mated or "passaged" to maintain the genetic stocks for a laboratory. *Caenorhabditis elegans* strains are maintained in a central location, giving all scientists access to the same well-characterized genetic stocks.

Caenorhabditis elegans is a transparent worm, ideally suited for microscopic analysis. The origin and ultimate fate of every cell in the worm (the cell lineage) has been mapped and traced microscopically. Adult hermaphrodites have 959 somatic (non-sex) cell nuclei, and males

have 1,021. Because the entire cell lineage for the worm is known and the worm is transparent, researchers can use a laser to destroy a single, specific cell and observe how loss of one cell affects development of the worm. These kinds of studies have contributed to the understanding of how neurons find target cells and how one cell can direct the fate of another.

Embryonic Development: Asymmetric Divisions

Research on *C. elegans* has revealed how programmed genetic factors (autonomous development) and cell-cell interactions guide development of an organism from egg to adult. The very first division of the fertilized egg (zygote) in *C. elegans* is asymmetric (uneven) and creates the first difference in the cells of the organism that is reflected in the adult. This division pro-

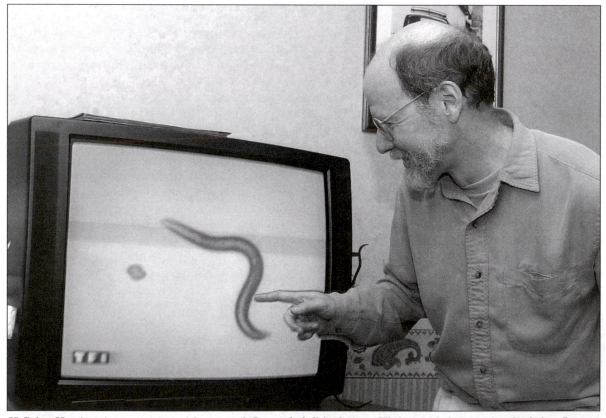

H. Robert Horvitz points to an image of the nematode Caenorhabditis elegans. *Working with this organism, he, Sydney Brenner, and John E. Sulston won the 2002 Nobel Prize in Physiology or Medicine for discovering genes regulating organ development and leading to apoptosis (programmed cell death)—discoveries with significant implications for cancer therapies.* (AP/Wide World Photos)

duces two daughter cells called P and AB. AB is a large cell that gives rise to tissues such as muscle and digestive tract. P is a much smaller cell that ultimately produces the cells that become the gonads (sex cell-producing tissues). The difference in P and AB is determined by the segregation of small P granules in the cell. The location of these granules and the asymmetry of this initial division are determined by the point of entry of the sperm. Until the eight-cell stage, there is no genetic activity by the embryo; the first few divisions are directed by the maternal gene products. This is one example of how maternal gene products can influence the early development of an embryo.

Neural Development

One of the areas of later development that is particularly well understood in *C. elegans* is the development of the nervous system. The nervous system has been completely reconstructed with serial electron micrographs that reveal precisely how one neuron connects to another. Some neurons migrate to assume their final cell fate and function. These migrations are easily studied in the worm because of its transparency, and a single neuron can be visualized by marking it with green fluorescent protein. Many genes and their encoded proteins that have been identified as important for directing the growth, connectivity, and migration of *C. elegans* neurons are highly conserved in evolution and control axon guidance in the vertebrate spinal cord.

Apoptosis: Programmed Cell Death

The 2002 Nobel Prize in Physiology or Medicine was awarded to Sydney Brenner, H. Robert Horvitz, and John E. Sulston for identifying genetically controlled cell death in worms. Cell death is an important part of development in plants and animals. For instance, human embryos have webbing between fingers and toes. This webbing is composed of cells that die in the course of normal development before a human baby is born. The death of these cells occurs because of a genetic program in the cells, apoptosis. The genes that control apoptosis are highly conserved throughout evolution. Apoptosis also plays a role in cancer. Often cancer is

thought of as resulting from uncontrolled proliferation of cells, but it can also result when cells that should die during development fail to die. Scientists are looking at ways to specifically activate apoptosis in tumor cells in order to kill tumors. The clues for what genes to target for such treatments come from studies of the apoptosis pathway in organisms such as *C. elegans*.

A Molecular Tool

The first metazoan genome that was sequenced was *C. elegans*. Many of the technologies (automated machines, chemistries for isolating and preparing DNA) that were developed in the course of the *C. elegans* genome-sequencing project were directly applied to the human genome sequencing project, and many of the scientists involved in sequencing the *C. elegans* genome contributed expertise to the Human Genome Project as well.

The green fluorescent protein, which is a protein that was first used to trace neurons in *C. elegans*, is now used in experiments with a wide variety of organisms or cell cultures to follow specific cells or specific proteins during development. RNA interference, a technique that uses RNA specifically to knock out gene expression of a target gene, was first described in worms. This technique enables scientists to knock out gene expression at the RNA level rather than requiring laborious genetic engineering of DNA. This technique promises to be particularly useful for researchers working with human or other mammal cell culture systems.

Caenorhabditis elegans research identified the first presenilin, a class of proteins later implicated in Alzheimer's disease. Research on the worm has led to a greater understanding of certain proteins that are involved in cellular aging. Studies in *C. elegans* are even contributing to a better understanding of learning and behavior. Most *C. elegans* scientists are studying the worm because it provides a tool for answering many of the hows and whys of biology that cannot be answered easily in more complex systems. The answers to seemingly esoteric questions, such as how *C. elegans* sperm move, will shed light on fundamental biological processes shared by all organisms.

—*Michele Arduengo*

See also: Aging; Antisense RNA; Complementation Testing; Human Genome Project; Model Organism: *Chlamydomonas reinhardtii;* Model Organisms; Noncoding RNA Molecules.

Further Reading

Lewin, Benjamin. *Genes VII.* New York: Oxford University Press, 2001. Contains articles about many of the processes researched in the worm, including apoptosis.

Wood, W. B., et al. *The Nematode Caenorhabditis elegans.* Cold Spring Harbor, N.Y.: Cold Spring Harbor Laboratory Press, 1988. The first "worm book" contains an excellent overview of worm development and an introductory letter from Nobel laureate Sydney Brenner.

Web Site of Interest

Caenorhabditis elegans Web server. http://elegans.swmed.edu. Contains links to major worm labs around the world and to introductory information about the worm. Includes access to WormBase, a "repository of mapping, sequencing and phenotypic information."

Model Organism: *Chlamydomonas reinhardtii*

Field of study: Techniques and methodologies

Significance: Chlamydomonas reinhardtii *is a unicellular green alga that has been extremely useful as a genetics model organism. It has a simple life cycle, is easily mutable, and is accessible for molecular genetic studies.*

Key terms

BACTERIAL ARTIFICIAL CHROMOSOME (BAC): a vector used to clone large fragments of DNA (up to 500 kb) that can be readily inserted in a bacterium, such as *Escherichia coli*

COMPLEMENTARY DNA (cDNA): a DNA molecule that is synthesized using messenger RNA (mRNA) as a template and the enzyme reverse transcriptase; these molecules correspond to genes but lack introns that are present in the actual genome

COSMID: a cloning vector, a hybrid of bacterial plasmid and bacteriophage vectors, that relies on bacteriophage capsules to infect bacteria; these are constructed with selectable markers from plasmids and two regions of lambda phage DNA known as cos (for cohesive end) sites

INSERTIONAL MUTAGENESIS: the generation of a mutant by inserting several nucleotides into a genome

MICROARRAY: a flat surface on which 10,000 to 100,000 tiny spots of DNA molecules fixed on glass or another solid surface are used for hybridization with a probe of fluoresent DNA or RNA

MODEL ORGANISM: an organism well suited for genetic research because it has a well-known genetic history, a short life cycle, and genetic variation between individuals in the population

TRANSFORMATION: a change in both genotype and phenotype resulting from the uptake of exogenous DNA

The Organism

Chlamydomonas reinhardtii is the best-researched member of the green algal genus *Chlamydomonas* (Greek *chlamys,* a cloak, plus *monas,* solitary). *Chlamydomonas reinhardtii* is unicellular with a definite cell wall that consists of glycoproteins rich in the amino acid hydroxyproline. A large, solitary chloroplast folded into a cup shape dominates most of the cytoplasm. The presence of this chloroplast allows autotrophic growth, although *C. reinhardtii* is capable of using acetate as an external carbon source. A circular body that is prominent within the chloroplast is referred to as the pyrenoid. It is the site of carbohydrate synthesis during the light-independent reactions of photosynthesis. The chloroplast also contains a red eyespot with a rhodopsin-like pigmented photoreceptor, called the stigma, that permits phototaxis. *Chlamydomonas reinhardtii* cells display positive phototaxis (that is, swimming toward light) in moderate light and negative phototaxis in intense light.

The cell nucleus is visible with light microscopy and predominates cross-sectional images in electron microscopy, along with the nucleolus. Electron microscopy also indicates sixteen or more chromosomes, which is consistent with the seventeen linkage groups defined by cytogenetic analysis. The cell's anterior end consists of two contractile vacuoles, and mitochondria are dispersed throughout the cytosol. Two long, whiplike flagella extend from basal bodies, which are also located at the anterior end of the cell. *Chlamydomonas reinhardtii* swims using a breaststroke motion. Internally the flagella consist of a central pair of microtubules surrounded by nine doublets. Each doublet consists of arms made of the protein dynein. The dynein interacts with adjacent doublets by pressing and sliding against the neighboring microtubule when adenosin triphosphate (ATP) is hydrolyzed. This brings about the flagellar beat and allows the organism to swim.

Chlamydomonas reinhardtii reproduces asexually by mitotic divisions. Parental cells can produce as many as sixteen progeny cells by successive divisions within the cell wall. Each progeny cell secretes a cell wall and generates flagella. The new cells escape by secreting autolytic enzymes that digest the parental cell wall.

Mating and Laboratory Analysis

The vegetative form of *C. reinhardtii* is haploid and exists as one of two genetically distinct mating types (mt$^+$) and (mt$^-$). When deprived of nitrogen, cells of each mating type differentiate into gametes. Gametes of opposite mating types come into contact with each other by way of their flagella. The gametes fuse, thereby forming a zygote. The zygote secretes a heavy wall and becomes a zygospore. Zygospores can remain dormant and viable in soils for several years. Light and nitrogen can bring about zygospore germination. Four biflagellated cells, known as zoospores, are released. In some strains, meiosis occurs prior to the release of zoospores, followed by a mitotic division. The result is the release of eight zoospores rather than four.

Cells of *C. reinhardtii* are easy to culture. They grow copiously in defined culture media under varying environmental conditions. Mating can be induced when cells of opposite mating types are placed in a nitrogen-free medium. The zygote formed from such a mating can produce four unordered tetrads on appropriate media. Sometimes an additional mitotic event generates eight haploid products that are easy to recover. These features have made *C. reinhardtii* extremely useful as an experimental organism.

Mutagenesis and Transmission Genetics

Research in the 1950's led to the isolation of mutants displaying defects in the ability to photosynthesize. Since then mutants have been developed that affect every structure, function, and behavior of *C. reinhardtii*. Ultraviolet or chemical methods can be used to induce mutants. One of the first mutants isolated was resistant to the antibiotic streptomycin (designated sr). These cells are able to grow on media supplemented with streptomycin as well as media free of streptomycin. Wild-type cells (designated ss) are unable to grow on media containing the antibiotic. Reciprocal crosses with cells of these distinct phenotypes resulted in segregation patterns that departed significantly from Mendelian expectations. The sr phenotype was clearly transmitted only through mt$^+$ cells. Further study has shown that resistance passed through the mt$^+$ chloroplast. The chloroplast contains more than fifty copies of a circular, double-stranded DNA molecule. Uniparental inheritance has been demonstrated for the mitochondrial genome, too. This genome contains fewer genes than the chloroplast, but antibiotic resistant mutations have been generated, along with other types. It is interesting to note that mitochondrial inheritance of antibiotic resistance appears to be transmitted by way of mt$^-$ cells.

Mutational analysis has elucidated aspects of nuclear inheritance, also. The mating type phenotype segregates in a 1:1 ratio in accordance with Mendelian principles. With the advent of molecular techniques, insertional mutagenesis has resulted in a wide array of mutants, including nonphotosynthetic, nonmotile, antibiotic resistant, herbicide resistant, and many more. This type of analysis has resulted in mapping nearly two hundred nuclear loci.

Molecular Analysis

Transformation of *C. reinhardtii* is relatively easy and can be carried out by mixing with DNA-coated glass beads or electroporation, that is, using a current to introduce the DNA into a cell. The frequency of transformation success is highest in wall-less mutants or cells whose walls have been removed prior to transformation. Both nuclear, mitochondrial, and chloroplast transformation studies have been performed, leading to the development of several molecular constructs that have been used to study gene expression. Cosmids and BAC libraries have been created for several markers in order to make the current molecular map of about 240 markers, each having an average spacing of 400 to 500 kb. These markers have been placed on the seventeen linkage groups mentioned previously.

Thus far, the greatest impact these molecular markers are having is in the study of photosynthesis. A chloroplast gene known as *Stt7* has been characterized using these methods. *Stt7* is required for activation of the major light-harvesting protein and interactions between photosystem I and photosystem II when light conditions change. Chloroplast and nuclear transformations have been used in conjunction with developmental mutants to study chloroplast biogenesis. This has increased researchers' understanding of the expression and regulation of many chloroplast genes. A cDNA library composed of many unique chloroplast genes is being constructed and their coding regions sequenced. These cDNAs are called expressed sequence tags (ESTs) and have proven extremely useful for identifying protein-coding genes in other organisms. Thousands of these cDNAs could be placed on pieces of glass the size of a microscope slide using microarray technology to monitor changes in gene expression of virtually the entire genome at the same time. Interactions between the nuclear genome and the chloroplast genome can be assessed in this manner as well.

—*Stephen S. Daggett*

See also: cDNA Libraries; Extrachromosomal Inheritance; Model Organism: *Arabidopsis thaliana*; Model Organism: *Caenorhabditis elegans*; Model Organism: *Drosophila melanogaster*;

Model Organism: *Escherichia coli*; Model Organism: *Mus musculus*; Model Organism: *Neurospora crassa*; Model Organism: *Saccharomyces cerevisiae*; Model Organism: *Xenopus laevis*; Model Organisms.

Further Reading

Graham, Linda E., and Lee W. Wilcox. *Algae.* Upper Saddle River, N.J.: Prentice-Hall, 2000. A textbook for students of introductory phycology that includes a number of chapters dealing with green algae, including members of the genus *Chlamydomonas*.

Harris, Elizabeth H. "*Chlamydomonas* as a Model Organism." *Annual Review of Plant Physiology* 52 (2001): 363-406. A detailed update of what has been learned since the publication of Harris's book in 1989.

_____. *The Chlamydomonas Sourcebook: A Comprehensive Guide to Biology and Laboratory Use.* San Diego, Calif.: Academic Press, 1989. The ultimate guide to working with *Chlamydomonas* species, including a detailed look at the organism, a thorough literature review, and several protocols, some for teaching purposes.

Web Site of Interest

Chlamydomonas Genetics Center. http://www.biology.duke.edu/chlamy. Sponsored by the National Science Foundation, the clearinghouse for data on the genetics of this model organism, including the genome project and the nuclear, chloroplast, and mitochondrial genomes.

Model Organism: *Drosophila melanogaster*

Field of study: Techniques and methodologies

Significance: Drosophila melanogaster *is the scientific name for a species of fruit fly whose study led scientists to discover many of the fundamental principles of the inheritance of traits. The first genetic map that assigned genes to specific chromosomes was developed for* Drosophila. *With advances in molecular technology, continued study* of Drosophila *has led to a greater understanding of genetic control in early embryonic development.*

Key terms

LINKED GENES: genes, and traits they specify, that are situated on the same chromosome and tend to be inherited together

MODEL ORGANISM: an organism well suited for genetic research because it has a well-known genetic history, a short life cycle, and genetic variation between individuals in the population

SEX CHROMOSOMES: The X and Y chromosomes, which determine sex in many organisms; in *Drosophila*, a female carries two X chromosomes and a male carries one X and one Y chromosome

Early Studies of *Drosophila*

By the early 1900's, scientists had discovered chromosomes inside of cells and knew that they occurred in pairs, that one partner of each pair was provided by each parent during reproduction, and that fertilization restored the paired condition. This behavior of chromosomes paralleled the observations of Austrian botanist Gregor Mendel, first published in 1866, which showed that traits in pea plants segregated and were assorted independently during reproduction. This led geneticists Walter Sutton, Theodor Boveri, and their colleagues to propose, in 1902, the "chromosome theory of inheritance," which postulated that Mendel's traits, or "genes," existed on the chromosomes. However, this theory was not accepted by all scientists of the time.

Thomas Hunt Morgan was an embryologist at Columbia University in New York City, and he chose to study the chromosome theory and inheritance in the common fruit fly, *Drosophila melanogaster.* This organism was an ideal one for genetic studies because a single mating could produce hundreds of offspring, it developed from egg to adult in only ten days, it was inexpensively and easily kept in the laboratory, and it had only four pairs of chromosomes that were easily distinguished with a simple microscope. Morgan was the first scientist to keep large numbers of fly "stocks" (organisms with

Thomas Hunt Morgan, one of the most important biologists in classical transmission genetics, established the "Fly Room" at Columbia University in 1910, where for the next quarter century he and his students studied the genetics of the fruit fly. (© The Nobel Foundation)

particular characteristics), and his laboratory became known as the "fly room."

After one year of breeding flies and looking for inherited variations of traits, Morgan found a single male fly with white eyes instead of the usual red, the normal or wild-type color. When he bred this white-eyed male with a red-eyed female, his results were consistent with that expected for a recessive trait, and all the offspring had wild-type eyes. When he mated some of these offspring, he was startled to discover a different inheritance pattern than he expected from Mendel's experiments. In the case of this mating, half of the males and no females had white eyes; Morgan had expected half of all of the males and females to be white-eyed. After many more generations of breeding, Morgan was able to deduce that eye color in a fly was related to its sex, and he located the eye-color gene to the X chromosome of the fruit fly. The X chromosome is one of the sex chromosomes. Because a female fly has two X chromosomes and a male has one X and one Y chromosome, and because the Y chromosome does not carry genes corresponding to those on the X chromosome, any gene on the male's X chromosome is expressed as a trait, even if it is normally recessive. This interesting and unusual example of the first mutant gene in flies was called a "sex-linked" trait because the trait was located on the X chromosome.

This important discovery attracted many students to Morgan's laboratory, and before long they found many other unusual inherited traits in flies and determined their inheritance patterns. One of the next major discoveries by members of the "fly lab" was that of genes existing on the same chromosome, information that was used to map the genes to individual chromosomes.

Linked Genes and Chromosome Maps

Many genes are located on each chromosome. Genes, and the traits they specify, that are situated on the same chromosome tend to be inherited together. Such genes are referred to as "linked" genes. Morgan performed a variety of genetic crosses with linked genes and developed detailed maps of the positions of the genes on the chromosomes based on his results. Morgan did his first experiments with linked genes in *Drosophila* that specified body color and wing type. In fruit flies, a brown body is the wild type and a black body is a mutant type. In wild flies wings are very long, while one mutant variant has short, crinkled wings referred to as "vestigial" wings. When Morgan mated wild-type females with black-bodied, vestigial-winged males, the next generation consisted of all wild-type flies. When he then mated females from this new generation with black-bodied, vestigial-winged males, most of the progeny were either brown and normal winged

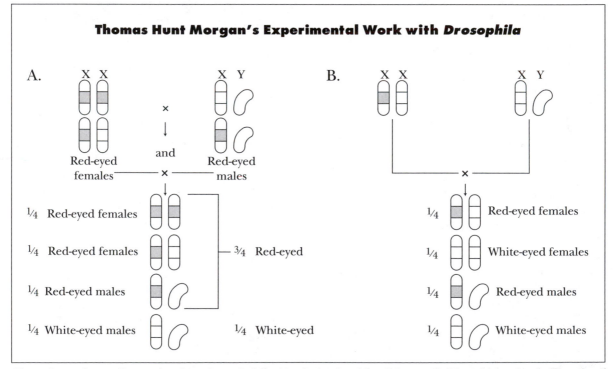

Thomas Hunt Morgan's Experimental Work with *Drosophila*

Morgan's experiments discovered such results as the following: A. A red-eyed female is crossed with a white-eyed male. The red-eyed progeny interbreed to produce offspring in a ¾ red to ¼ white ratio. All the white-eyed flies are male. B. A white-eyed male is crossed with its red-eyed daughter, giving red-eyed and white-eyed males and females in equal proportions. (Electronic Illustrators Group)

or wild-type black and vestigial winged, in about equal proportions. A few of the offspring were either just black bodied (with wild-type wings) or vestigial winged (with wild-type body color), trait combinations found in neither parent. Because of the equal distribution of these mutant traits between males and females, Morgan knew the genes were not sex linked. Because the traits for body color and wing length generally seemed to be inherited together, he deduced that they existed on the same chromosome.

As Morgan and his students and colleagues continued their experiments on the inheritance of body color and wing length, they observed a small but consistent percentage of offspring with trait combinations not observed in either parent, referred to as nonparentals. After repeating these experiments with many different linked genes, Morgan discovered that chromosomes exchange pieces during egg and sperm formation. This exchange of chromosome pieces occurs during a process called meiosis, which occurs in sexually reproducing organisms and results in the production of gametes, generally eggs and sperm. During meiosis, the homologous chromosomes pair tightly and may exchange pieces; since the homologous chromosomes contain genes for the same trait along their length, this exchange does not present any genetic problems. The eggs or sperm produced through meiosis contain one of each pair of chromosomes.

In some of Morgan's genetic crosses, flies carried one chromosome with alleles (alternate forms of a gene at a specific locus) for black bodies and vestigial wings. The homologous chromosome carried wild alleles for both traits. During meiosis, portions of the homologous chromosomes exchanged pieces, resulting in some flies receiving chromosomes carrying genes for black bodies and normal wings or brown bodies and vestigial wings. The exchange of chromosome pieces resulting in new combinations of traits in progeny is referred to as "recombination." Morgan's students and

Alfred H. Sturtevant (California Institute of Technology)

colleagues pursued many different traits that showed genetic recombination. In 1917, one of Morgan's students, Alfred Sturtevant, reasoned that the further apart two genes were on a chromosome, the more likely they were to recombine and the more progeny with new combinations of traits would be observed. Over many years of work, Sturtevant and his colleagues were able to collect recombination data and cluster all the then-known mutant genes into four groupings that corresponded to the four chromosomes of *Drosophila*. They generated the first linkage maps that located genes to chromosomes based on their recombination frequencies.

The chromosomes in the salivary glands of the larval stage of the fruit fly are particularly large. Scientists were able to isolate these chromosomes, stain them with dyes, and observe them under microscopes. Each chromosome had an identifying size and shape and highly detailed banding patterns. X rays and chemicals were used to generate new mutations for study in *Drosophila*, and researchers realized that in many cases they could correlate a particular gene with a physical band along a chromosome. Also noted were chromosome abnormalities, including deletions of pieces, inversions of chromosome sections, and the translocation of a portion of one chromosome onto another chromosome. The pioneering techniques of linkage mapping through recombination of traits and physical mapping of genes to chromosome sections provided detailed genetic maps of *Drosophila*. Similar techniques have been used to construct gene maps of other organisms, including humans.

Control of Genes at the Molecular Level

This seminal genetic work on *Drosophila* was unparalleled in providing insights into the mechanisms of inheritance. Most of the inheritance patterns discovered in the fruit flies were found to be applicable to nearly all organisms. However, the usefulness of *Drosophila* as a research organism did not end with classical transmission genetics; it was found to provide equally valuable insight into the mechanisms of development at the level of DNA.

Drosophila were discovered to be ideal organisms to use in the study of early development. During its development in the egg, the *Drosophila* embryo orchestrates a cascade of events that results in the embryo having a polarity (a head and a tail), with segments between each end defined to become a particular body part in the adult. For example, the second segment of the thorax will support one pair of wings and one of the three pairs of legs. By studying many types of mutants that showed bizarre appearances as adults (for example, two sets of wings or legs replacing the normal antennae on the head), scientists were able to elucidate some of the mechanisms that control development in nearly all animals.

Developmental instructions from the mother fruit fly are sequestered in the egg. When the egg is fertilized, these instructions begin to "turn on" genes within the fertilized eggs that begin to establish the directionality and segment identity within the embryo. So many genes are involved in this process that a defect in a single one will truncate the rest of development, resulting in a severely mutated fly. It was found that conserved regions of DNA outside of the developmental genes received the signals to "turn on." Such sequences were found to be present in all animals studied. These control regions were termed "homeoboxes" after the homeotic genes that control the overall body plan of an organism in early development.

Many other aspects of *Drosophila* were found to be useful in understanding the structure and function of the DNA of all organisms. It was found that in *Drosophila*, large pieces of DNA will, under certain circumstances, pop out of the chromosome and reinsert themselves at another site. One such element, called a P element, was used by scientists to introduce nonfly DNA into the fruit fly embryo, thus providing information on how DNA is expressed in animals. This work also provided early clues into the successful creation of transgenic animals commonly used in research to study cancer and other diseases.

Impact and Applications

Genetic studies of *Drosophila melanogaster* have provided the world with a fundamental understanding of the mechanisms of inheritance. In addition to the inheritance modes shown by Mendel's studies of pea plants, fruit fly genetics revealed that some genes are sex linked in sexually reproducing animals. The research led to the understanding that while many genes are linked to a single chromosome, the linkage is not necessarily static, and that chromosomes can exchange pieces during recombination. The ease with which mutant fruit flies could be generated led to the development of detailed linkage maps for all the chromosomes and ultimately to the localization of genes to specific regions of chromosomes. With the advent of molecular techniques, it was discovered that *Drosophila* again provided a wealth of information in terms of mobile genetic elements and developmental studies. Although all of these breakthroughs were scientifically interesting in terms of the flies themselves, many of the breakthroughs helped identify fundamental principles consistent among all animals. Most of what is known about human genetics and genetic diseases has come from these pioneering studies with *Drosophila*.

Because of the sheer numbers of offspring from any mating of flies, their very short life cycle, and large numbers of traits that are easily observable, fruit flies have become an ideal system to screen for potential chemical carcinogens (cancer-causing agents) or mutagens (agents that cause mutations in DNA) in humans. Flies are exposed to the chemical in question and mated; then their offspring are analyzed for any abnormal appearances or behaviors, or for low numbers of offspring. Should a test substance cause any variation in the expected outcome of a cross, it is then subjected to more rigorous research in other organisms.

The versatile, easy-to-care-for, inexpensive fruit fly is often a fixture in classrooms around the world. Indeed, many geneticists have traced their passion to their first classroom encounters with fruit flies and the excitement of discovering the inheritance patterns for themselves. Flies are also routinely used in the study of neural pathways, learning patterns, behavior, and population genetics. Because of the ease of study and the volumes of information that have been compiled about its genetics, development, and behavior, *Drosophila* will continue to be an important model organism for biological study. The completion of the complete genome sequence of *Drosophila* should greatly increase the usefulness of this model organism, allowing an even more detailed understanding of its genetics.

—*Karen E. Kalumuck*

See also: Aging; Bioinformatics; Biological Clocks; Chemical Mutagens; Chromosome Mutation; Chromosome Theory of Heredity; Developmental Genetics; Genetics, Historical Development of; Homeotic Genes; Human Genome Project; Inbreeding and Assortative Mating; Incomplete Dominance; Lateral Gene Transfer; Linkage Maps; Metafemales; Model Organism: *Arabidopsis thaliana*; Model Organism: *Caenorhabditis elegans*; Model Organism: *Chlamydomonas reinhardtii*; Model Organism: *Escherichia coli*; Model Organism: *Mus musculus*; Model Organism: *Neurospora crassa*; Model Organism: *Saccharomyces cerevisiae*; Model Organism: *Xenopus laevis*; Model Organisms; Mutation and Muta-

genesis; Natural Selection; Noncoding RNA Molecules; Population Genetics.

Further Reading

Abstracts of Papers Presented at the 2001 Meeting on Neurobiology of Drosophila. Arranged by Hugo Bellen and Barbara Taylor. Cold Spring Harbor, N.Y.: Cold Spring Harbor Laboratory Press, 2001. Papers focus on *Drosophila* as a model organism in neurobiology.

Brookes, Martin. *Fly: The Unsung Hero of Twentieth-Century Science*. San Francisco: HarperCollins, 2001. A whimsical history of the fruit fly, *Drosophila melanogaster*, as the star of genetic research, from Thomas Hunt Morgan to DNA sequencing.

Web Sites of Interest

Drosophila Virtual Library. http://www.ceolas.org/VL/fly. Links to databases, labs, and other Web resources of interest to researchers.

FlyBase. http://flybase.bio.indiana.edu:82. A joint venture of the Berkeley and European Drosophila Genome Projects. Includes data from the *Drosophila* genome projects and a vast amount of other information—bibliographies, directories, descriptions of chromosomal aberrations, lists of *Drosophila* stocks, genome project data, images, and more.

Model Organism: *Escherichia coli*

Fields of study: Bacterial genetics; Techniques and methodologies

Significance: *Through the study of the genetics of* Escherichia coli *biologists have come to understand the molecular-level regulation of gene expression and how genes direct routine activities of living cells. This understanding of the genetics of this bacterium has led to the extensive use of this organism in biotechnology. Such technology permits the introduction of foreign genes into the organism's cells, which may result in new bacterial strains capable of solving problems as diverse as environmental pollution, food and energy shortages, and the spread of diseases.*

Key terms

MODEL ORGANISM: an organism well suited for genetic research because it has a well-known genetic history, a short life cycle, and genetic variation between individuals in the population

OPERON: a genetic unit consisting of structural genes coding for amino acid chains; an operator gene controlling the transcriptional (message encoding) activity of the structural genes

REPLICATION: the process of DNA duplication

A Suitable Experimental Organism

Discovered in 1885 by Theodor Escherich, *Escherichia coli* is the most intensely studied bacterium in genetics. In fact, of the earth's living organisms, this bacterium is one of the better understood, and its use as a favorite experimental organism dates to the mid-twentieth century. Even before gaining a rich genetic history, the bacterium was selected for genetic research for several reasons: its ease of handling in experiments, its twenty-minute generation time, its single copy of each gene, and its meager genetic material. Results derived from genetic experiments with *E. coli* have significantly influenced the thinking of biologists, and the genetics of *E. coli* has provided evidence that explains mechanisms underlying important processes: *E. coli* chromosome organization; regulation of gene expression; DNA replication, transcription, and translation; mutation and DNA repair; biotechnology; and evolution.

The *E. coli* cell usually contains a single chromosome, although the cell's actual number of chromosomes depends on the bacterium's growth rate. Fast-growing *E. coli* have two to four chromosome copies per cell, while the slow-growing counterparts have one to two copies per cell. These multiple copies, however, are genetically identical, permitting *E. coli* to behave as haploids (cells containing a single chromosome). This chromosome, a dense cellular structure carrying hereditary information from generation to generation, consists of a single molecule of double-stranded DNA. The DNA, in a closed-circle form, is located in the nucleoid, a central region of the *E. coli* cell. The DNA of *Escherichia coli* is probably made up of four million base pairs and carries 2,800 genes. These genes constitute 75 percent of the DNA molecule; the remaining DNA consists of regions between genes, such as the stretch of DNA acting as the unique origin of replication for the *E. coli* DNA molecule.

Packaging this DNA into the nucleoid is an important concept in *E. coli* genetics because the length of the bacterial chromosome containing the DNA is twelve hundred times that of the *E. coli* cell. The chromosome, therefore, is packaged in a highly compact form. This compact DNA consists of one hundred independent genetic segments, each having forty thousand base pairs (bp) of DNA containing extra twists, with the ends of each genetic segment held, presumably, by proteins. The extra-twisted DNA in one genetic segment is unaffected by events influencing extra twisting of DNA in other genetic segments. Such a structure forms because the DNA, a negatively charged molecule, associates with positively charged structural proteins.

About one-third of the 2,800 genes in *E. coli* have been located on the bacterium's chromo-

A single E. coli *cell.* (AP/Wide World Photos)

some using gene mapping (which determines the locations of genes along the chromosome) and recombinant DNA techniques such as DNA sequencing (which determines the order of the nucleotides in DNA). Of the located genes, 260 of them are organized into seventy-five operons, with the remaining 740 genes scattered, perhaps randomly, around the rest of the DNA molecule.

Regulation of Gene Expression

The genetics of *E. coli* reveals that 26 percent of its mapped genes are organized in transcriptional units (DNA segments containing message-encoding start and stop signals) called operons; these work to regulate gene expression. Operons, coordinately regulated units, often contain genes with related functions. Each regulated unit has a set of adjoining structural genes, a promoter for enzyme binding, and an operator for regulatory protein binding. If the genes encode enzymes involved in an anabolic pathway (in which chemical reactions form larger molecules from smaller ones), they are usually turned off in the presence of the pathway's end product. Alternatively, if the genes encode enzymes involved in a catabolic pathway (in which chemical reactions break down large molecules into smaller ones), they are often expressed in the presence of the enzymes' substrates (molecules whose actions are increased).

The genetics of the lactose (*lac*) and the tryptophan (*trp*) operons were unraveled using *E. coli*. This earned the bacterium a place in history for helping to explain the regulation of gene expression. The *lac* operon consists of three structural genes—*Z, Y,* and *A*—that encode beta-galactosidase, beta-galactoside permease, and beta-galactoside transacetylase, respectively. Other operon components are the promoter and the operator adjoining the *Z* gene. The regulator gene has its own promoter and adjoins the operon. The *lac* operon is an example of an inducible system because the operon's three structural genes are transcribed (put into message code) only in the presence of lactose. In the absence of lactose, the *lac* repressor (a protein product of the regulator gene) binds to the operator and prevents RNA polymerase from initiating operon transcription.

In contrast to the *lac* operon, the *trp* operon is a repressible system, in which the production of an enzyme stops with the addition of the end product of the enzyme reaction. Transcription of the operon's five structural genes, which encode enzymes involved in tryptophan production, is repressed in the presence of tryptophan. A second regulatory mechanism, called attenuation, also controls the system.

Based on the genetics of *E. coli*, biologists know that operon function may change if fused to a new operator. French molecular biologist François Jacob's research team showed this for the structural genes of the *lac* operon. The team used DNA fragments carrying parts of the *lac-pur* region, but with an added deletion that eliminated the *lac* operator and part of the *Z* gene. These modified DNA fragments were inserted into *E. coli* that were unable to produce the enzymes permease and acetylase. The functional lac enzymes produced by the modified particles were no longer activated by lactose. Such enzymes were instead under control of the deactivated purine operator. As a result, excessive purine caused repression of galactoside permease and acetylase. In *E. coli*, gene expression can be regulated at different levels, but transcriptional regulation is the most common.

DNA Replication, Transcription, and Translation

Early in the study of *E. coli*, Matthew Meselson and Franklin Stahl determined how DNA duplicates itself in the bacterium. They grew the organism, across several generations, in culture media containing nucleotides enriched with nitrogen 15 (a heavier isotope of nitrogen), which would be incorporated into all newly synthesized strands of DNA. Then some of these cells with nitrogen 15 enriched DNA were transferred to media containing nucleotides containing normal nitrogen 14, so that all newly synthesized strands of DNA would then contain nitrogen 14 rather than nitrogen 15. After allowing enough time for the cells to divide once, they isolated their DNA and then used cesium chloride density-gradient centrifugation to characterize their results. Their

Sequencing the *E. coli* Genome

As part of the Human Genome Project (begun in 1990), several model organisms were selected for sequencing. Such direct DNA sequence information could be correlated with the extensive data available from classical and molecular genetics. Not only would it provide a means for identifying similar genes in the human genome; it would also provide a means for comparative genomics, that is, to identify similar genes among both model organisms and sequence data from related organisms. The latter is useful to explore the evolution of specific genes and evolutionary relatedness of organisms. Consequently, the sequencing of the *Escherichia coli*, the prokaryotic organism most studied genetically, biochemically, and physiologically, was of high priority. Due to efforts led by Frederick Blattner at the University of Wisconsin, along with colleagues at four other institutions, the six-year project resulted in the complete genomic sequence of *E. coli* K12 (strain MG1655), published on September 5, 1997, in the journal *Science*; the final corrected sequence was updated in October, 2001.

Although there are many different strains of *E. coli*, strain MG1655 was chosen because it is a well-established, stable laboratory strain. The sequencing of a second laboratory strain, W3110, was completed by a consortium of Japanese researchers. The *E. coli* MG1655 genome consists of 4,639,221 base pairs, a number slightly higher than estimated from earlier studies. Of these, 87.8 percent are found in protein-coding genes, 0.8 percent in stable RNA sequences, 0.7 percent in noncoding repeats, and approximately 11 percent in regulatory and other sequences. One difference between eukaryote and prokaryote genomes is the large amount of noncoding sequences in the former and the relative lack of such sequences in the latter. This was borne out by the *E. coli* sequence: The genome analysis indicates that there are 4,405 genes, including 4,286 protein-coding sequences, about 50 percent more than originally predicted. Only about one-third of these represent well-characterized proteins. There are also 7 ribosomal RNA (rRNA) operons and 86 transfer RNA (tRNA) genes.

While *E. coli* is a normal inhabitant of the human gut, the average person associates the name *E. coli* with strain O157:H7, a human pathogen causing intestinal hemorrhaging and resulting in about five hundred deaths per year in the United States. Strain O157:H7 has acquired two toxin genes from a related bacterium, *Shigella dysenteriae*, often found in cattle. The complete sequence of O157:H7 was completed in January, 2001, and provides interesting comparisons. Its genome is 5,528,455 base pairs, with 5,416 genes of which 1,387 are not found in *E. coli* MG1655. These new genes include those for virulence factors, alternative metabolic capacities, and new prophages. Moreover, O157:H7 lacks 528 genes found in *E. coli* MG1655. These marked differences lead some to believe that O157:H7 is actually a different species, having evolutionarily diverged from standard *E. coli* about 4.5 million years ago. This example of comparative genomics illustrates its potential as a powerful tool for medical and other applications.

Sequencing of other strains of *E. coli*, particularly pathogenic strains, is ongoing under the aegis of the *E. coli* Genome Project, based at the University of Wisconsin.

—*Ralph R. Meyer*

findings, verified through autoradiography several years later by John Cairns, showed that in *E. coli*, DNA duplicates itself semiconservatively. This means that in *E. coli*, the strands of the DNA double helix separate and form a Y-shaped replication fork where DNA duplication begins. Proteins stabilize the unwound helix and assist in relaxing the coiling tension created ahead of the duplication activity. A new, complementary strand of DNA, duplicated in *E. coli* at the rate of thirty thousand nucleotides per minute, is produced on each of the two parental template (guide) strands. The DNA duplication process results in two double-stranded DNA molecules, each having one strand from the parent molecule and one newly produced strand. This semiconservative mechanism ensures the faithful copying of the genetic information at each *E. coli* cell division.

During the message-encoding process (transcription), the genic message (RNA transcript) is created step by step, using the DNA template. The template is read in one direction, while RNA is produced in the opposite direction.

The process includes initiation, elongation, and termination phases. The transcription initiation site is signaled by the promoter (a short nucleotide sequence recognized by an RNA polymerase). During elongation, RNA polymerase migrates along the DNA molecule, melting and unwinding the double helix as it moves and sequentially attaching ribonucleotides to one end of the growing RNA molecule. Base pairing to the template strand of the gene determines the identity of the ribonucleotide added to each position. By a complex signal, transcription is terminated shortly after the ends of genes. As a result of the process, a gene-complementary, single-stranded RNA molecule (messenger RNA, or mRNA) is created.

Like the message-encoding process, the message-decoding process (translation) consists of initiation, elongation, and termination. In *E. coli*, the small subunit of a ribosome (the cell's interior structure for protein production) attaches to the ribosome-binding site of an mRNA, resulting in an initiation complex. In elongation, the large subunit of the ribosome attaches to the initiation complex, creating two different binding sites for transfer RNA (tRNA), the amino acid transporter. Ribosomes use mRNA-coded information to take amino acids brought by tRNA and assemble them, on ribosomes, into protein.

Mutation and DNA Repair

In the genetics of *E. coli*, phenotypes resulting from changes in the DNA can occur because of either mutation (a change in the nucleotide sequence of a gene) or recombination (a process leading to new combinations of genes on a chromosome). These new combinations can occur following transfer of chromosomal genes from one bacterial cell to another by transformation (in which a recipient cell acquires genes from free DNA in the medium), transduction (in which a virus carries DNA from donor to recipient cell), or conjugation (in which two bacterial cells make contact and exchange DNA). Transposon and insertion elements, both found in *E. coli*, may also change phenotypes. A transposon is a mobile DNA segment containing genes for inserting DNA into the chromosome and for moving the element to other chromosome locations. Insertion elements, the simplest transposable elements in *E. coli*, contain only genes for mobilizing the elements and inserting them into chromosomes at new locations.

In *E. coli*, as in all other organisms, many chemical and physical agents cause structural changes in DNA. Consequently, mechanisms are needed for repairing such damaged DNA. Such repair mechanisms exist in *E. coli*, although they are complicated and require many different proteins. The three main types of repair mechanisms in *E. coli* are direct repair (the reversal of a structural change), excision repair (in which appropriate enzymes recognize and "label" a damaged nucleotide, excise it, fill in the gap, and seal the strand), and mismatch repair (in which enzymes recognize the mismatch nucleotide and either label it or repair it directly). The parental strand is distinguished from the newly created daughter strand by tagging the parental strand with methyl groups attached to adenines occurring within specific sequences. Such modified adenines act as labels for the parent strand, enabling the repair enzymes to recognize which strand should be repaired at a mismatch position.

Biotechnology

To test a genetic hypothesis, the genetic history of the organism involved must be well known so that the genetic background of the parents used in the experimental crosses is known. The genetics of *E. coli* provides geneticists with such an experimental organism. As a result, *E. coli* is used extensively in biotechnology. In this industry, a foreign gene inserted into the bacterium may be replicated and sometimes translated in the same manner as the native bacterial DNA, producing a foreign gene product. *Escherichia coli* can accept foreign DNA derived from any organism because the genetic code is nearly universal. As an example, genetic mapping of a free-living, nitrogen-fixing bacterium showed that seventeen genes involved in nitrogen fixation are clustered on one portion of the chromosome. Biologists transferred this gene cluster to a plasmid—a circular, independently replicating DNA molecule—and introduced the plasmid into *E. coli* cells, which then produced the enzyme nitrogenase and fixed nitrogen.

A significant breakthrough occurred when a synthetic gene coding for somatostatin, an antigrowth hormone important in the treatment of different human growth disorders, was fused with the start of the *lacZ'* gene contained within a cloning vector (a self-duplicating DNA molecule containing inserted, foreign DNA). Cells of *E. coli* transformed with this recombinant plasmid were able to transcribe the fused gene, recognizing the *lac* promoter as its binding site. The mRNA was then translated by ribosomes that recognized the *lac* ribosome binding sequence. The resulting fused protein was cleaved with cyanogen bromide, which cuts amino acid chains specifically at methionines, resulting in pure-form somatostatin.

Implications for Evolution

The genetics of *E. coli* provides evidence for punctuated equilibrium caused by the appearance of rare, beneficial mutations. This evidence involved studies that measured changes in cell size over three thousand generations of bacteria in a constant environment. During the studies, periods of stagnancy were interrupted by periods of rapid change. The changes in cell size may be the result of direct selection for a rare, beneficial mutation that caused increased cell size. This mutation swept through the population, producing a change in cell size in one hundred generations or less.

Impact and Applications

Geneticists use model organisms for their research. Their favorite organisms, such as *E. coli*, have qualities that make them well suited for genetic experimentation—a rich genetic history, a short life cycle, production of large progeny from a mating, ease in handling, and genetic variation among the individuals in the population. Added to the much that was already known about *E. coli* was the completion of the complete sequence of the chromosome of *E. coli* in 1995. The quantity of genetics involving *E. coli* is a testament to the bacterium's suitability as an experimental organism for testing genetic hypotheses. The hypotheses tested using this experimental organism have contrib-

uted in a revolutionary way to the understanding of significant scientific concepts and to the understanding of the genetics of organisms more complex than bacteria, such as humans.

In addition, recombinant DNA technology (techniques for constructing, studying, and using DNA created in a test tube), which uses *E. coli* extensively, is used in all areas of basic genetics research to investigate genetic circumstances. Many biotechnology companies owe their existence to recombinant DNA technology—and to *E. coli*—as they seek to clone and manipulate genes for the production of commercial products, the improvement of plant and animal agriculture, the development of diagnostic tools for genetic diseases, and the development of new or more effective pharmaceuticals.

—*Robert Haynes*

See also: Antibodies; Archaea; Bacterial Genetics and Cell Structure; Bacterial Resistance and Super Bacteria; Blotting: Southern, Northern, and Western; Chromosome Theory of Heredity; Cloning Vectors; DNA Isolation; DNA Repair; Emerging Diseases; Gene Families; Gene Regulation: *Lac* Operon; Gene Regulation: Viruses; Genetic Code; Genetic Engineering; Genetic Engineering: Historical Development; Genetic Engineering: Industrial Applications; Genetic Engineering: Medical Applications; Genetics, Historical Development of; Human Genome Project; Human Growth Hormone; Model Organisms; Noncoding RNA Molecules; Plasmids; Proteomics; Restriction Enzymes; Shotgun Cloning; Synthetic Genes; Transposable Elements.

Further Reading

Birge, Edward R. *Bacterial and Bacteriophage Genetics.* 4th ed. New York: Springer, 2000. A comprehensive yet concise introductory look at bacterial genetics.

Blattner, Frederick R., et al. "The Complete Genome Sequence of *Escherichia coli* K12." *Science* 277 (1997): 1453-1452. The paper announcing the completion of the *E. coli* genome sequence.

Brown, Terence A. *Genetics: A Molecular Approach.* 3d ed. New York: Chapman & Hall, 1998. A general treatment of the genetics of bacteria, including *E. coli.* Bibliography, index.

Miller, Jeffrey H. *A Short Course in Bacterial Genetics: A Laboratory Manual and Handbook for Escherichia Coli and Related Bacteria.* Cold Spring Harbor, N.Y.: Cold Spring Harbor Laboratory Press, 1999. Summarizes the genes and proteins of *E. coli.*

Parker, James N., and Philip M. Parker, eds. *The Official Patient's Sourcebook on E. Coli.* San Diego: ICON Health, 2002. Discusses topics including the essentials, seeking guidance, the treatment process, and learning more about *E. coli* using the Internet. Includes appendices, glossaries, and an index.

Perna, Nicole T., et al. "Genome Sequence of Enterohaemorrhagic *Escherichia coli* O157: H7." *Nature* 409 (2001): 529-533. Announced the genome sequence for the virulent O157: H7 strain of *E. coli.*

Riley, Monica, and Margrethe Hauge Serres. "Interim Report on Genomics of *Escherichia coli.*" *Annual Review of Microbiology* 54 (2000): 341-411. Updates the genome sequence.

Web Sites of Interest

E. coli Genome Project, University of Wisconsin. http://www.genome.wisc.edu. The genome research center that sequenced the organism's complete K-12 genome now maintains and updates that sequence as well as those of other strains and other pathogenic *Enterobacteriaceae.*

National Institutes of Health, Center for Biotechnology Information. http://www.ncbi .nih.nlm.gov/genbank/genbanksearch .html. For information on *E.coli*, see accession number U00096.

Model Organism: *Mus musculus*

Field of study: Techniques and methodologies

Significance: *Model organisms allow geneticists to investigate how genes affect organismal and cellular function. The mouse is an ideal organism for genetic research because of its size, life span, and litter size. It shares many similarities with humans and is useful for modeling complex phenomena such as cancer and development.*

Key terms

EMBRYONIC STEM CELLS: cultured cells derived from an early embryo

GENOMICS: the study of the entire DNA content of an organism, called its genome

INBREEDING: the process of mating brothers and sisters to create genetically identical offspring

MODEL ORGANISM: an organism well suited for genetic research because it has a well-known genetic history, a short life cycle, and genetic variation between individuals in the population

PHENOTYPE: an observable trait

TRANSGENICS: the technique of modifying an organism by introducing new DNA into its chromosomes

History of Mice in Genetic Research

The use of mice in genetic research had its origin in the efforts of mouse fanciers, who raised mice as pets and developed numerous strains with distinct coat colors. Researchers in the late 1800's who were trying to determine the validity of Gregor Mendel's laws of heredity in animals found the existence of domesticated mice with distinct coat colors to be an ideal choice for their experiments. Through the work of early mouse geneticists such as Lucien Cuénot and others, Mendel's ideas were validated and expanded.

Development of Inbred Strains

As genetic work on mice continued into the 1900's, a number of mouse facilities were created, including the Bussey Institute at Harvard University. One member of the institute, Clarence Little, carried out a set of experiments that would help establish the utility of mice in scientific research. Little mated a pair of mice and then mated the offspring with each other. He continued this process for many generations. After a number of generations of inbreeding, Little's mice lost all genetic variation and became genetically identical. These mice, named DBA mice, became the first strain of inbred mice and marked an important contribution to mouse research. In an experiment using inbred DBA mice, any difference displayed by two mice could not be due to genetic variation and had to be from the result of the experiment. Through inbreeding, genetic variation was removed as a variable. Also, through careful crossing and selection of different inbred strains, populations of mice that differed by only a few genes could be created. Geneticists could then examine the effects of these genes knowing that all other genes were the same. The creation of inbred mice allowed geneticists to study genes in a carefully controlled way.

The first use of inbred mice was in the study of cancer. As inbred strains of mice were created, it was noticed that certain strains had a tendency to develop cancer at a very high frequency. Some of these strains developed tumors that were very similar to those found in human cancers. These mice became some of the first mouse models used to study a human disease.

Unique Aspects of the Mouse Model

The ability of mice to acquire cancer illustrates why the mouse is a unique and valuable tool for research. Although mice are not as easy to maintain as other model organisms, they are vertebrates and thus share a number of physiological and developmental similarities with humans. They can be used to model processes, such as those involved in cancer and skeletal development, that do not exist in simpler organisms. In this capacity, mice represent a balance between the need for an animal with developmental complexity and the need for an animal with a quick generation time that is easily bred and raised. Other organisms, such as chimpanzees, may more closely resemble hu-

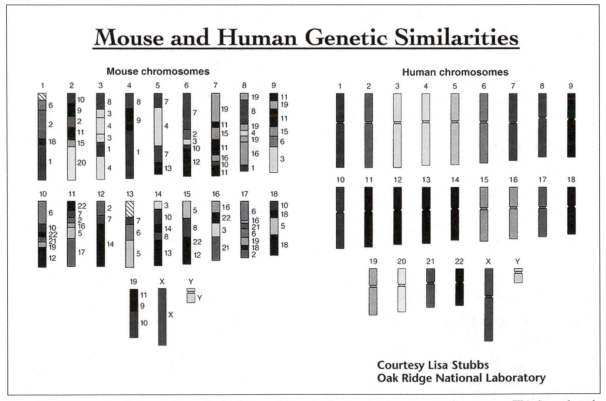

Mouse and Human Genetic Similarities

Courtesy Lisa Stubbs
Oak Ridge National Laboratory

One of the most amazing discoveries in genetics is that very different organisms can have very similar genomes. This figure from the Human Genome Program, for example, shows the similarities between the genes of mice and those of human beings. Approximately 80 to 90 percent of the genes in humans have a counterpart in the mouse. (U.S. Department of Energy Human Genome Program, http://www.ornl.gov/hgmis)

mans, but their lengthy generation time and small litter size make them difficult to use for the many and repeated experiments needed for genetic research.

The use of the mouse model has advanced considerably since the early 1900's. Initially, geneticists relied on the random occurrence of natural mutations to generate mice with traits that mirrored aspects of human biology and disease. Careful crossbreeding and the use of inbred strains allowed the trait to be isolated and maintained. Although this process was slow and tedious, a large number of inbred strains were identified. Later, it was discovered that X rays and other chemicals could increase the rate of mutation, leading to an increase in the rate at which mice with interesting traits could be found. However, the discovery of a mouse strain that modeled a particular human disease was still a matter of chance.

It was the advent of molecular biology that removed this element of chance and brought the mouse to its full prominence as a model organism. Molecular biology provided a mechanistic understanding of gene function and offered tools that allowed for the direct manipulation of genes.

Transgenic Mice

The technique of transgenics allows geneticists to create mice that carry specific mutations in specific genes. Using recombinant DNA technology, a geneticist can construct a piece of DNA containing a mutant form of a chosen gene, then use the mutated gene to modify the existing DNA of mouse embryonic stem cells. These modified embryonic stem cells can be combined with a normal mouse embryo to form a transgenic embryo that can be implanted into the uterus of a female mouse. The transgenic mouse that is born from this process carries in every tissue a mixture of normal cells and cells with the specific DNA alteration introduced by the researcher. Careful crossing of the transgenic mouse with mice of the same inbred strain can then be done to create a new line of mice that carry the DNA alteration in all cells. These mice will then express a phenotype that results directly from the modified gene. Transgenics has allowed geneticists

to custom design mice to display the genetic defects they desire.

In the era of genomics, transgenic mice have become a powerful tool in the effort to understand the function of human genes. Since the complete sequences of the mouse and human genomes are known, it is possible to compare the genes of mice and humans directly. Approximately 80 to 90 percent of the genes in humans have a counterpart in the mouse. Using transgenics to target genes in the mouse that are similar to humans can help geneticists understand their functions. However, care must be used in drawing comparisons. There are a number of examples of mouse genes that carry out functions different from their human counterparts. Despite this concern, comparison of mouse and human genes has provided tremendous insight into the function of the human genome.

Economic and Ethical Considerations

The demand for mice in research has resulted in a $100 million industry devoted to the maintenance and development of mouse models. Companies specializing in mice have developed thousands of inbred strains for use in research. The economic impact of mice has led to patents on transgenic mice and has caused controversy over who has the right to own a particular mouse strain. Also, the extensive use of mice in research (25 million mice in the year 2000) has raised concerns by some for the welfare of mice and questions about the ethics of using them in research.

Research Using the Mouse Model

The study of cancer was the first area of research to benefit from the use of mice. Early mouse geneticists were able to learn about the genetic and environmental factors that influenced the development of cancer. Today's cancer research relies heavily on the mouse model as a way of determining how genes affect the interaction between cancer and the body. Understanding the function of tumor-suppressor genes such as *p53* has come in part from the use of transgenic mice. Mice have also been important in investigating the role of the immune system and angiogenesis in tumor progression.

Mouse work in cancer also made contributions to immunology, which relies heavily on the mouse as a model of an intact immune system. Inbred strains of mice with defective immune systems have been developed to help geneticists understand the role of the immune system in disease progression and transplant rejection. Mice have also been instrumental in studying how genes in pathogenic microorganisms allow the microbes to cause disease. The mouse model has been used to understand how diseases like cholera and anthrax are able to infect and cause damage.

The study of many genetic diseases, such as sickle-cell disease and phenylketonuria (PKU), has benefited from the existence of mouse models that mimic the disease. The genetic components of such complex phenomena as heart disease and obesity are also being elucidated using the mouse model.

Developmental biology has relied heavily on the mouse to determine how gene expression leads to the formation of multicellular organisms. Work that has shown the role homeogenes play in determining mammalian body structure and how genes affect development of organs has been done in mouse models.

The mouse has also proven to be a valuable model in investigating the effects of various genes on brain development and function. Mouse models have provided insights into the way the brain develops and functions, as well as genetic contributions to complex behaviors. Genes have been identified that play roles in complex behaviors such as raising young and predisposition toward addiction.

—*Douglas H. Brown*

See also: Altruism; Chromosome Theory of Heredity; Model Organism: *Arabidopsis thaliana*; Model Organism: *Caenorhabditis elegans*; Model Organism: *Chlamydomonas reinhardtii*; Model Organism: *Drosophila melanogaster*; Model Organism: *Escherichia coli*; Model Organism: *Neurospora crassa*; Model Organism: *Saccharomyces cerevisiae*; Model Organism: *Xenopus laevis*; Model Organisms.

Further Reading

Hartwell, Leland, et al. *Genetics: From Genes to Genomes.* Boston: McGraw-Hill, 2003. An outstanding text that covers the fields of genetics and molecular biology. The reference section in the back contains detailed portraits of model organisms, including the mouse.

Silver, Lee. *Mouse Genetics: Concepts and Applications.* New York: Oxford University Press, 1995. A comprehensive reference providing a thorough explanation of the history and rationale for the use of mice in genetic research. Designed for readers who are new to the field of mouse genetics as well as those with experience.

Web Sites of Interest

Mouse Atlas and Gene Expression Database. http://www3.oup.co.uk/nar/database/summary/20. The Medical Research Council and the University of Edinburgh sponsor the site free on the Web; the data are also available for a fee on CD-ROM. This ongoing project is intended to evolve into the premier source for three-dimensional images on morphology, gene expression, and mutant phenotypes in mouse development. The initial digital embryo images are mounted, accessible through a controlled vocabulary linked to the images. Developmental geneticists will be able to synthesize information from many sources.

Mouse Genome Informatics, Jackson Laboratory, Bar Harbor, Maine. http://www.informatics.jax.org. A center for mutant mouse research, providing access to genetic maps, phenotypes, gene expression data, and sequence information. Includes the Mouse Genome Database, the Gene Expression Database, and the Mouse Genome Sequence Project.

Model Organism: *Neurospora crassa*

Field of study: Techniques and methodologies

Significance: Neurospora crassa *is a bread mold with a relatively small genome, allowing this organism to be studied by causing mutations in its genes and observing the effects of these mutations.*

Such studies are important to the understanding of genetics and genetically related disease, particularly because N. crassa *is eukaryotic and more similar to human DNA than it is to bacteria and viruses.*

Key terms

ASCOMYCETES: organisms of the phylum *Ascomycota*, a group of fungi known as the sac fungi, which are characterized by a saclike structure, the ascus

AUXOTROPHIC STRAIN: a mutant strain of an organism that cannot synthesize a substance required for growth and therefore must have the substance supplied in the growth medium

CYTOGENETICS: the study of normal and mutated chromosomes and their behavior

DIPLOID CELL: a cell that contains two copies of each chromosome

HAPLOID CELL: a cell that contains one copy of each chromosome

MINIMAL MEDIUM: an environment that contains the simplest set of ingredients that the microorganism can use to produce all the substances required for reproduction and growth

MODEL ORGANISM: an organism well suited for genetic research because it has a well-known genetic history, a short life cycle (allowing the production of several generations in a short space of time), and genetic variation between individuals in the population

The Beginning of Biochemical Genetics

Neurospora crassa was first used in genetic experiments by Carl Lindegren in the 1930's. He was able to isolate several morphological mutant strains and create the first "linkage maps" showing where genes are located on chromosomes. This research determined some of the basic principles of "crossing over" during meiosis. Crossing over is the exchange of genes between homologous chromosome pairs by the breaking and reunion of the chromosome. Lindegren was able to show that crossing over occurs before the separation of the homologous pair, between the second and fourth chromatids. *Neurospora crassa* was used as a model organism in the investigation of crossing-over

mechanisms because the four products of meiosis (later duplicated by mitosis to produce eight spores) are arranged in the organism's saclike ascus in a way that exactly reflects the orientation of the four chromatids of each tetrad at the metaphase plate in the first meiotic division. The products of meiosis line up in order and therefore are more easily studied in this organism.

One Gene, One Enzyme

In 1941 George Beadle and Edward Tatum published a paper establishing biochemical genetics as an experimental science. They introduced a procedure for isolating an important class of lethal mutations in an organism, namely, those for blocking the synthesis of essential biological substances. These were expressed in the organism as new nutritional requirements.

By supplying a variety of compounds in the nutrient medium and seeing which allowed various mutant strains to grow and which did not, Beadle and Tatum saw that they could deduce the sequence of biochemical reactions in cells that make necessary compounds, such as amino acids. They concluded that the function of a gene is to direct the formation of a particular enzyme which regulates a chemical event. A mutation can alter a gene so that it no longer produces the normal enzyme, resulting in a physical symptom, such as the need for nutritional supplements. Beadle and Tatum proposed that, in general, each gene directs the formation of one enzyme.

These mutation studies promoted understanding of the biochemistry of gene expression and promoted the use of fungi in genetic experiments. In 1958, Beadle and Tatum were awarded the Nobel Prize in Physiology or Medicine for their discovery that the characteristic function of the gene was to control the synthesis of a particular enzyme.

The Organism

The orange bread mold *Neurospora crassa*, a multicellular lower eukaryote, is the best characterized of the filamentous fungi. Filamentous fungi are a group of fungi with a microscopic, stalklike structure called the mycelium.

They grow on substances of plant or animal origins and reproduce via spores. This group of organisms has importance in agriculture, medicine, and the environment because they are so abundant and are able to proliferate very quickly. It is therefore easy and cheap to reproduce them rapidly. Moreover, the widespread availability of *Neurospora crassa* in nature makes genetic population studies more feasible. Because it can be grown in large quantity, experiments are easier to conduct and their results are more easily analyzed.

Neurospora crassa is a filamentous ascomycete that has asci; an ascus is a saclike structure inside of which four or eight ascospores develop during reproduction. In the *N. crassa* asci, one round of mitosis usually follows meiosis and leaves eight nuclei (new daughter cells). These nuclei eventually become eight ascospores (sexual spores produced by ascomycetes). After the ascospores are formed within the ascus, they are released and germinate to form a new haploid mycelium.

A Model Organism

Geneticists use a variety of organisms in their research. Because it is haploid (containing half the chromosomal material of the parent cell), genotypic changes in *N. crassa* (mutations in genes) are directly observed through the changes in the phenotype (physical characteristics), because only one gene determines physical characteristics. The small size of the genome is a result of a unique feature of *N. crassa:* It has very little repeated DNA. The lack of repetitive DNA is also valuable to researchers when parts of the genome are amplified or sequenced.

Neurospora crassa has been extensively used for genetic research, resulting in hundreds of published articles. They include research on gene expression and effects of external factors, metabolic studies, and genomal mapping experiments. A large number of mutants have been characterized, providing the foundation for many genetic experiments.

Repeat-Induced Point (RIP) Mutations

By using recombinant DNA methods, researchers can study *N. crassa* using a technique known as repeat-induced point (RIP) mutations, the creation of point mutations of a single base pair in specific genes. RIP detects duplications of gene-sized segments and creates repeated point mutations. RIP specifically changes a GC (guanine-cytosine) pair to an AT (adenine-thymine) pair. Repeated sequences are heavily mutated by RIP in the period between fertilization (the time when the sperm comes into contact with the egg) and karyogamy (fusion of the haploid cells to form diploid cells). After the mutation, the altered sequence is methylated (a CH_3, or methyl, group is attached). The methyl group serves as a tag so the mutations can be easily identified. RIP mutations usually indicate a crossing over during meiosis. RIP mutations cause inactivations of duplicate genes, whose functions are then more easily detected.

Sequencing and Linkage

Large-scale sequencing of the *N. crassa* genome has been initiated for several linkage groups (genes that are located on the same chromosomes). Early in the sequencing of the *N. crassa* genome, it became apparent that its genome contains many unique genes. These genes and others have been sorted into linkage groups. There are many maps available for *N. crassa*. The largest group is that at the Whitehead Institute Center for Genome Research under the Fungal Genome Initiative. Restriction fragment length polymorphism (RFLP) maps show the restriction site for a particular restriction endonuclease. Linkage maps show the distribution and linkage of genes throughout the *N. crassa* genome. These maps are particularly important when a researcher is interested in recombinant DNA research.

—*Leah C. Nesbitt, James N. Robinson, and Massimo D. Bezoari*

See also: Chromosome Theory of Heredity; Complementation Testing; Extrachromosomal Inheritance; Genetics, Historical Development of; Model Organism: *Arabidopsis thaliana*; Model Organism: *Caenorhabditis elegans*; Model Organism: *Chlamydomonas reinhardtii*; Model Organism: *Drosophila melanogaster*; Model Organism: *Escherichia coli*; Model Organism: *Mus musculus*; Model Organism: *Saccharomyces cerevisiae*; Model

Organism: *Xenopus laevis*; Model Organisms; One Gene-One Enzyme Hypothesis.

Further Reading

Beadle, G. W., and E. L. Tatum. "Genetic Control of Biochemical Reactions in *Neurospora*." *Proceedings of the National Academy of Sciences* 27 (1941): 499-506. This is the article that made *Neurospora* famous. It lays down the foundations of the one gene, one enzyme hypothesis.

Davis, Rowland H. *Neurospora: Contributions of a Model Organism*. New York: Oxford University Press, 2000. A full account of the organism's history, biology, genome, mitosis, meiosis, metabolism, mutations, and more.

Horowitz, N. H. "Fifty Years Ago: The *Neurospora* Revolution." *Genetics* 127 (1991): 631-636. This article is a brief history of *Neurospora* and its contributions to genetics and biochemistry. The article outlines Beadle's discovery of *Neurospora* as a model organism.

Kinsey, J. A., P. W. Garrett-Engele, E. B. Cambareri, and E. U. Selker. "The *Neurospora* Transposon Tad Is Sensitive to Repeat-Induced Point Mutation (RIP)." *Genetics* 138 (1994): 657-664. This paper describes the RIP mechanisms and the direct effects of RIP on the transposon Tad.

Thancker, Paul D. "Understanding Fungi Through Their Genomes." *Bioscience* 53, no. 1 (January, 2003): 10-15. Useful for students and researchers.

Web Sites of Interest

Neugenesis. http://www.neugenesis.com. Site of a company that produces commercial quantities of monoclonal antibodies (MAbs), generates and screens for new gene sequences specifying commercially valuable products, assembles combinatorial cellular arrays for screening of multicomponent gene and protein variants, and produces cell libraries expressing a wide range of recombinant protein products. Includes a discussion of the repeat-induced point mutation mechanism.

Whitehead Institute for Biomedical Research. http://www-genome.wi.mit.edu. One of the major gateways to genomics research, software, and sequencing databases; provides access to one of the largest collections of linkage maps for *Neurospora* under the Fungal Genome Initiative.

Model Organism: *Saccharomyces cerevisiae*

Field of study: Techniques and methodologies

Significance: *Saccharomyces cerevisiae is a highly tractable yeast organism that was the first eukaryote to have its DNA completely sequenced. Yeast genetic research has been at the forefront of scientists' efforts to identify the genes and processes required for cell growth and division and is now an important tool for nonyeast research to identify proteins that physically interact with one another in the cell.*

Key terms

ASCUS: the cellular structure that results from meiosis in yeast, containing four recombinant spores that are fully capable of growing into haploid yeast cells

BUDDING: the asexual method of duplication used by yeast to create a clone of the original cell

DIPLOID CELL: a cell that contains two copies of each chromosome

HAPLOID CELL: a cell that contains one copy of each chromosome

MATING TYPE: one of two types of yeast cell, depending on a soluble factor that each cell secretes

MODEL ORGANISM: an organism well suited for genetic research because it has a well-known genetic history, a short life cycle, and genetic variation between individuals in the population

The Organism

Saccharomyces cerevisiae (*S. cerevisiae*, or baker's yeast) has been used for millennia to provide leavening to bread products. Yeast is a simple, one-celled eukaryote with six thousand genes on sixteen chromosomes. It was the first eukaryote to have its entire DNA sequenced.

Yeast produce offspring using two different methods, a sexual life cycle and an asexual life cycle. In the asexual life cycle, the yeast cell produces the next generation by a process called budding. All genetic components of the mother cell are duplicated and a small "bud" begins to grow from the mother cell. The bud continues to grow until it is nearly the size of the mother cell. The DNA and other duplicated cellular components are then partitioned into the new bud. The cells undergo cytokinesis and are now separate entities able to grow and continue reproducing independently of one another.

To produce offspring that are not clones of the mother cell, yeast use a sexual life cycle. A yeast cell exists stably as either a diploid or a haploid organism, but only the haploid organism is able to mate and exchange genetic information. Haploid yeast contain either the *MATa* or *MATalpha* gene. These genes produce soluble factors that distinguish them as one of two mating types. An "a" cell (*MATa*) and an "alpha" cell (*MATalpha*) mate by sequentially fusing their cell walls, their cytoplasms, and finally their nuclei. This diploid cell now contains two copies of each chromosome that can undergo recombination during meiosis. When all environmental signals are ideal, the diploid yeast will undergo meiosis, allowing exchange and recombination of genetic information brought to the diploid by both haploid cells. The result of meiosis is an ascus that contains four recombinant spores that will grow into haploid yeast cells when environmental conditions are ideal.

A Model Organism

Researchers choose yeast as a model organism to study specific areas of interest for many different reasons. *Saccharomyces cerevisiae* is nonpathogenic to humans, allowing manipulation in a laboratory with little or no containment required. At a temperature of 30 degrees Celsius (86 degrees Fahrenheit), the yeast population can double in ninety minutes, allowing many experiments to be completed in one day. Among the primary reasons for selection of yeast as a model system is that they offer the possibility of studying the genes and proteins that are required for basic growth functions

and cellular division. Yeast use many of the same genes and proteins to govern the same processes that animal and plant cells use for growth and division. Each single cell has to take in nutrients, grow, and pass along information to its progeny. In many ways, yeast can be considered a simplified version of a plant or animal cell, in that it lacks all the genes that provide the determinants that are expressed as differences between plants and animals. Another important reason for using yeast is that yeast is amenable to investigation using both genetic and biochemical approaches. This allows for correlation of findings from both approaches and a better understanding of a specific process or activity.

Yeast is also ideal for use as a model system due to at least four well-established techniques and procedures. First, genetics in yeast takes advantage of well-established auxotrophic markers. These markers are usually mutations in biosynthetic pathways that are used to synthesize required cellular components such as amino acids and nucleotides. By using these marker genes, researchers can follow genes and their associated chromosomes from one generation to the next.

Second, yeast is readily transformed by plasmids that function as artificial chromosomes. All that is needed is an auxotrophic marker to follow the plasmid through succeeding generations, a yeast origin of replication to allow replication of the plasmid DNA, and a region into which the gene of interest can be inserted in the plasmid DNA. This allows the researcher to move genes easily from yeast strain to yeast strain and quickly examine the effect of the gene in combination with many other genes.

Third, yeast is easily mutated by chemicals and can be grown in a small space, which allows the researcher quickly to identify mutations in genes that result in a specific phenotype. For example, to define all the genes in the adenine biosynthetic pathway, a researcher would mutate a yeast strain with one of many available mutagenic chemicals, resulting in changes within the DNA. The mutated yeast strains would then be checked to see if the strain was able to grow on media lacking adenine. All of the strains mutant for growth on adenine would be collected and could identify a number of genes involved in the adenine biosynthetic pathway. Further research could establish whether each of these mutations in the yeast identified one gene or many genes.

Fourth, yeast is the model system of choice when examining and identifying proteins that interact with one another in the cell. This technique is called the two-hybrid system.

Two-Hybrid System

The two-hybrid system takes advantage of scientists' understanding of transcription at the *GAL1* gene in yeast. The promoter region of *GAL1* contains a binding site for the Gal4p transcription factor. When the cell is grown on the sugar galactose, Gal4p binds to the promoter of *GAL1* and activates transcription of the *GAL1* gene. Gal4p can be essentially divided into two functional regions: one region that binds to DNA and another region that activates transcription.

The two-hybrid system uses the *GAL1*-Gal4p transcription system to identify previously unknown proteins that interact with a protein of interest. The system consists of a reporter gene under the control of the *GAL1* promoter and two plasmids that produce fusions with the Gal4p transcription factor. The first plasmid contains a gene of interest fused to a DNA-binding domain. This plasmid expresses a protein that is able to bind to the DNA-binding site in the *GAL1* promoter of the reporter gene. This plasmid is unable to activate transcription of the reporter gene, since the Gal4p fragment does not contain the information to activate transcription. The second plasmid is provided from a collection of plasmids that consist of unknown or random genes fused to the transcription activation domain of Gal4p. This plasmid by itself is unable to bind to the DNA-binding site in the *GAL1* promoter and thus is unable to activate transcription of the reporter gene. If both plasmids contain genes whose protein products physically interact in the cell, the complex is able to bind to the DNA-binding region of the *GAL1* promoter, and since the activation domain of Gal4p is also present in this complex, activation of the reporter gene will occur. The production of the reporter gene serves as a signal that both of the gene products interact in the cell. The yeast strain containing the active reporter gene is then selected and further examined to determine the unknown DNA that resides on the second plasmid by sequence analysis.

Research and Implications

The years of work on yeast as a model system have provided many insights into how genes and their protein products interact to coordinate the many cellular mechanisms that take place in all cells from simple yeast to complicated humans. It is impossible to exhaustively list the different areas of research currently being examined or completely list the new understandings that have come to light through the use of the *S. cerevisiae* model system. Every major area of cellular research has at one time or another used yeast to ask some of the more difficult questions that could not be asked in other systems. Work in yeast has aided identification of genes and elucidated the mechanism of many different areas of research, including cell cycle regulation, mechanisms of signal transduction, the process of secretion, replication of DNA, transcription of DNA, translation of messenger RNA into proteins, biosynthetic pathways of amino acids and other basic building blocks of cells, and regulation and progression of cells through mitosis and meiosis. Despite all these advances, there is still much to learn from yeast and it will continue to provide information for years to come.

—*John R. Geiser*

See also: Cloning Vectors; Extrachromosomal Inheritance; Linkage Maps; Model Organisms; Noncoding RNA Molecules; Plasmids.

Further Reading

Broach, J., J. Pringle, and E. Jones, eds. *The Molecular and Cellular Biology of the Yeast Saccharomyces.* 3 vols. Cold Spring Harbor, N.Y.: Cold Spring Harbor Laboratory Press, 1991-1997. This comprehensive series is dedicated to reviewing the current understanding in many areas of yeast research. Volume 1 covers genome dynamics, protein synthesis, and energetics; volume 2, gene expression; and volume 3, the cell cycle and cell biology. The individual reviews contain many references to primary literature.

Fields, S., and O. Song. "A Novel Genetic System to Detect Protein-Protein Interactions." *Nature* 340 (1989): 245-246. A seminal article that describes the first use of the two-hybrid system. Contains illustrations and description of how the two-hybrid system functions.

Web Site of Interest

Saccharomyces Genome Database. http://www.yeastgenome.org. The central site for the sequencing projects, with links to data, tables, and much more.

Model Organism: *Xenopus laevis*

Field of study: Techniques and methodologies

Significance: Xenopus laevis, *the African clawed frog, has been used widely in the field of developmental biology. By following the development of this unique organism, scientists have identified and now understand the role of many genes in frog development, providing insight into vertebrate development.*

Key terms

EMBRYOLOGY: the study of developing embryos

FATE MAP: a map created by following the adult fate of embryonic cells

MODEL ORGANISM: an organism well suited for genetic research because it has a well-known genetic history, a short life cycle, and genetic variation between individuals in the population

TRANSGENIC ANIMAL: an animal that contains a gene not normally expressed in its genome

The Organism

The African clawed frog, *Xenopus laevis*, is in the class *Amphibia*, order *Anura*, suborder *Opisthocoela*, family *Pipidae*, and genus *Xenopus*. This genus includes five other species that inhabit silt-filled ponds throughout much of southern Africa. Members of this species share a distinctive habitat and morphology. The organism's name alone provides insight into its structure and habitats: The root *xeno* stems from Greek for "strange," while *pus* is from the Greek for "foot" and *laevis* is Latin for "slippery." *Xenopus laevis* is entirely aquatic, a feature that makes it unique among the other members of the genus, feeding and breeding under water. It is believed that they evolved from terrestrial anurans, organisms that are aquatic as tadpoles but are terrestrial as adults. Migration across land from pond to pond has been observed but is limited by distance and time of year (occurring during the rainy season) because out of water the frogs will dry out and die within a day. In instances of extreme drought, adult frogs will bury themselves in the mud and wait until the next rainfall.

Xenopus laevis is mottled greenish-brown on its dorsal surface and yellowish-white on its ventral surface. In appearance, these frogs are flattened dorsoventrally with dorsally oriented eyes as adults. The members of the genus are collectively known as platannas from the word "plathander," meaning flat-handed. Three toes of the hind limbs are clawed, and a line of specialized sensory organs (the lateral line organs) is found on both the dorsal and ventral surfaces and encircles the eyes. The breeding season for *X. laevis* depends on temperature and rainfall. The tadpoles are herbivorous, feeding on algae, whereas the adults are carnivorous, feeding on worms, crustaceans, and other creatures living in the mud.

A Model Organism

A model organism is defined as one that breeds quickly, is easily managed in the laboratory, and has large numbers of offspring or broods. *Xenopus laevis* meets these requirements nicely. An interesting feature of this organism is its responsiveness to human chorionic gonadotropin, a hormone secreted by the placenta and present in the urine of pregnant women. When exposed to the hormone, female frogs will spawn (lay eggs). As a result of this phenomenon, *X. laevis* was once used as an indicator in human pregnancy tests, whereby the female frogs were injected with human female urine. At present, researchers take advantage of this phenomenon to produce large numbers of offspring by injecting frogs with the hormone. Another characteristic that makes *X. laevis* a good model organism is that it is hardy and can survive in captivity for long periods of time with relatively low mortality rates.

A final requirement for an animal model to be useful is that research on the animal should add to the understanding of biological principles in other organisms. *Xenopus laevis* is widely used in the field of developmental biology. For many decades, amphibian embryologists used salamander embryos, such as *Triturus*, and embryos of the frog *Rana* species. As mentioned above, amphibian embryos have several advantages over other organisms: Amphibian embryos are large, can be obtained in large numbers, and can be maintained easily and inexpensively in the laboratory. However, one disadvantage of traditional amphibian species is that they are seasonal breeders. As a result, investigators cannot conduct experiments throughout the year on most amphibians. *Xenopus laevis* is a notable exception, because it can be induced to breed year-round.

As the fertilized *X. laevis* zygote develops, the yolk-laden cytoplasm, known as the vegetal pole, is oriented downward by gravity. The rest of the cytoplasm, termed the animal pole, orients itself upward. The animal pole is the main portion of the cell, giving rise to the embryo proper. Cell division, or cleavage of cells, in the animal pole increases the number of cells greatly. Movement and migration of these cells, under the influences of interactions with neighboring cells, give rise to a multilaminar embryo that includes the ectoderm (which gives rise to skin and nervous system), the mesoderm (which gives rise to muscle), and the endoderm (which gives rise to many of the "tubes" of the organism, such as the intestines and the respiratory tract).

By following embryos from the very earliest stages, researchers have been able to create "fate maps" of fertilized eggs, which can be used to predict adult derivatives of specific regions in a developing embryo. Early researchers introduced many different techniques to create these kinds of maps. One technique involves destroying single cells during early development and following the development of the embryo to see what tissue is altered. Other methods include transplantation of individual cells or small groups of cells into a host organism and following the fate of the transplanted tissue.

Genetic Manipulation in *Xenopus*

Much of what is known today about the interactions between cells in developing vertebrate embryos has come from *X. laevis*. The early work of embryologists Hans Spemann and Pieter Nieuwkoop has been supported with molecular techniques, and many genes have been identified that control nearly every aspect of *Xenopus* development. A few examples include the *Xenopus Brachury* gene (*Xbra*), which is involved in the establishment of the dorsal-ventral axis; *Xenopus ventral* (*vent1*), which aids in the differentiation of ventral mesoderm and epidermal structures; and *Xenopus nodal-related 1* (*Xnr1*), a gene that is responsible for the specification of the left-right axis.

Xenopus embryos possess a number of advantages that have allowed investigators to study many aspects of developmental biology. One of the struggles that early researchers faced was the lack of dependable techniques for creating transgenic embryos to study the functions and role of individual genes. One can isolate and clone the genes of *Xenopus* and inject RNA into zygotes. RNA, however, is an unstable molecule and relatively short-lived. Therefore, the study of molecular events in the embryo after the period when the embryonic genes are turned on remained problematic. Attempts to inject cloned DNA to be expressed in the embryo were complicated by the fact that it does not integrate into the frog genomic chromosomes during cleavage. Exogenous DNA is then unequally distributed in embryonic cells and, therefore, is always expressed in random patterns. In 1996, Kristen L. Kroll and Enrique Amaya developed a technique to make stable transgenic *Xenopus* embryos. This technique has the potential to boost the utility of *Xenopus* tremendously. One significant advantage of using transgenic frogs over transgenic mice is that one can produce first-generation transgenics, making it unnecessary to wait until the second generation to examine the effects of the exogenous gene on development.

The transgenic technique has several steps, and each step is full of problems. Because exogenous DNA is not incorporated into the zygotic genome, Kroll and Amaya decided to attempt to introduce them into sperm nuclei. Sperm nuclei are treated with the enzyme lysolecithin to remove the plasma membrane prior to incubation with the linearized DNA plasmid containing the exogenous gene. The sperm nuclei are then incubated with restriction enzyme to introduce nicks in the nuclear DNA. The nicks facilitate incorporation of the plasmid DNA. The nuclei are then placed in an interphase egg extract, which causes the nuclei to swell as if they were male pronuclei. This technique has been used in many laboratories to introduce into the frog genes that are not normally expressed, allowing the researcher to study the function of these genes.

The National Institutes of Health is supporting the Trans-NIH *Xenopus* Initiative, specifically developed to support research in the areas of genomics and genetics in *Xenopus* research. While there is still much to be learned from this unique organism, it is clear that the advantages of this animal model far outweigh the disadvantages. With continued work in laboratories around the world, scientists may soon fully understand the genetics involved in vertebrate development. *Xenopus laevis* is ideally suited to provide critical breakthroughs in embryonic body patterning and cell fate determination, later development and the formation of organs, and cell biological and biochemical processes.

—*Steven D. Wilt*

See also: Model Organism: *Arabidopsis thaliana*; Model Organism: *Caenorhabditis elegans*; Model Organism: *Chlamydomonas reinhardtii*; Model Organism: *Drosophila melanogaster*; Model Organism: *Escherichia coli*; Model Organism: *Mus musculus*; Model Organism: *Neurospora crassa*; Model Organism: *Saccharomyces cerevisiae*; Model Organisms; Noncoding RNA Molecules; Totipotency.

Further Reading

Brown, A. L. *The African Clawed Toad Xenopus laevis: A Guide for Laboratory Practical Work.* London: Butterworths, 1970. A useful, introductory-level text describing the anatomy, behavior, and maintenance of *X. laevis*. Illustrations.

Gurdon, J. B., C. D. Lane, H. R. Woodland, and G. Marbaix. "Use of Frog Eggs and Oocytes for the Study of Messenger RNA and Its Translation in Living Cells." *Nature* 233 (1971): 177-182. Describes early work in the field of developmental biology and the functions of messenger RNA in protein translation.

Kroll, K. L., and E. Amaya. "Transgenic *Xenopus* Embryos from Sperm Nuclear Transplantations Reveal FGF Signaling Requirements During Gastrulation." *Development* 12 (1996): 3173-3183. A seminal research article that describes the methods of creating transgenic *X. laevis* embryos.

Nieuwkoop, P. D., J. Faber, and M. W. Kirschner. *Normal Table of Xenopus laevis (Daudin): A Systematical and Chronological Survey of the Development from the Fertilized Egg Till the End of Metamorphosis.* New York: Garland, 1994. Excellent reference on the stages of embryological development in *X. laevis*. Illustrations.

Seidman, S., and H. Soreq. *Transgenic Xenopus: Microinjection Methods and Developmental Neurobiology.* Totowa, N.J.: Humana Press, 1997. Explains basic background and protocols for transgenic frog research. Illustrations.

Web Sites of Interest

National Institutes of Health. Trans-NIH *Xenopus* Initiative. http://www.nih.gov/science/models/xenopus. This site keeps researchers aware of NIH's plans regarding support of the genomic and genetic needs for *Xenopus* research.

Xenbase: A *Xenopus* Web Resource. http://www.xenbase.org. A "database of information pertaining to the cell and developmental biology of the frog, *Xenopus*" with genomic information, directories, methods, links to databases and electronic journals, conference announcements.

Model Organisms

Field of study: Techniques and methodologies

Significance: *Due to evolutionary relationships between organisms, different organisms share similar, evolutionarily conserved genes and mechanisms of inheritance. This similarity between different species allows researchers to use model organisms to examine general genetic principles that are applicable to a wide variety of living organisms, including human beings. Findings from studies on model organisms not only reveal information about the influence of genetics on basic biology but also provide important insights into the role of genetics in human health and disease.*

Key terms

HOMOLOGY: similarity resulting from descent from a common evolutionary ancestor

MODEL ORGANISM: a species used for genetic analysis because of characteristics that make it desirable as a research organism and because of similarity to other organisms

Why Models?

Genetics research seeks to understand how genetic information is transmitted from one generation to the next and how this information influences the structure, function, development, and behavior of cells and organisms. However, the sheer number of different species and even greater diversity of cell types make the examination of every organism or type of cell impossible. Instead, researchers choose to investigate how genes influence function in a relatively small number of species. They then apply what they learn from these species to other organisms. Those species that are most commonly studied are called model organisms because they serve as models for researchers' understanding of gene function in other organisms.

Basic activities required for cells to survive are retained in virtually all organisms. Genes that have a common evolutionary origin and thus carry out a similar function are said to have homology. For example, many of the same genes used to repair damaged DNA molecules in the bacterial cell *Escherichia coli* are retained in multicellular, eukaryotic organisms. Thus, much of what is known about genetic control of DNA repair in human cells has been learned by studying homologous genes in the relatively simple *E. coli*. Model organisms provide practical systems in which to ask important genetic questions.

Selection of Model Organisms

Scientific researchers choose which model organisms to study based on the presence of characteristics that make an organism useful for investigating a particular question. Because of the extensive number of questions being asked in biological research, a tremendous number of species are used as model organisms. However, virtually all model organisms fulfill three basic criteria:

(1) they are relatively easy to grow and maintain
(2) they reproduce rapidly
(3) they are of reasonably small size

Geneticists add other criteria to their selection of model organisms, including the use of species for which many mutant forms have been isolated, into which mutations can be easily introduced, and for which techniques have been developed that allow for DNA introduction, isolation, and manipulation. Increasingly, model organisms are those whose genomes have been or will be completely sequenced, allowing for easier isolation and characterization of selected genes and subsequent analysis of gene function. Finally, the model organism must have enough similarity to other organisms that it can be used to ask interesting questions. Many model organisms are used to address questions that help scientists to better understand human cellular and genetic activities. Other model organisms are selected because they provide important information about pathogenic organisms, such as bacteria or viruses, or about economically significant organisms, such as agriculturally important species.

Some Commonly Used Model Organisms

Arguably the first model organism utilized by a geneticist was the garden pea, used by

Gregor Mendel to elucidate how particular traits are transmitted from generation to generation. The patterns of inheritance described by Mendel for the garden pea are applicable to all diploid, sexually reproducing organisms, making the pea a model organism for studying gene transmission. Many other organisms have subsequently been exploited to investigate all aspects of genetic influence on cell function. Prokaryotic cells, particularly the intestinal bacterium *Escherichia coli*, have provided important insights into basic cellular activities, ranging from DNA synthesis to protein translation to secretion of extracellular material. As unicellular eukaryotic cells, the brewer's yeast *Saccharomyces cerevisiae* and fission yeast *Schizosaccharomyces pombe* have provided models for eukaryotic cell function, including how genes regulate cell division, how proteins are targeted to particular locations in cells, and how specific genes are turned on and off under specific conditions.

Multicellular model species are used to reveal how genes influence the interactions between cells, as well as the organization and function of the whole organism. The fruit fly *Drosophila melanogaster* has been used since the early twentieth century to investigate the association of particular traits with specific chromosomes and was the first organism in which sex-linked inheritance was described. *Drosophila* has also been used to study developmental and behavioral genetics, providing important insights into the role genes play in determining the organizational pattern of developing embryos and in influencing how organisms behave.

More recently, genetic examination of the roundworm *Caenorhabditis elegans* has provided further insights into the role of genes in generating developmental patterns. Some of these insights resulted in the awarding of the 2002 Nobel Prize in Physiology or Medicine to Sydney Brenner, H. Robert Horvitz, and John E. Sulston for their work on apoptosis, or "programmed cell death," in *C. elegans* and its applicability to investigations of apoptosis in other organisms, including humans.

Genetic analysis of plants is also performed using model organisms, the most important of which is the mustard plant *Arabidopsis thaliana*, whose small genome, rapid generation time, and prolific seed production make it useful for studying plant inheritance patterns, flower generation, genetic responses to stress and pathogen attack, and developmental patterning, among other important plant activities.

Model organisms are also critical for enhancing our understanding of vertebrate genetics. The African clawed frog *Xenopus laevis* and zebrafish *Danio rerio* are used to study basic vertebrate developmental patterns and the organization of specific cell types into tissues and organs. The primary model organism for analysis of mammalian gene function is the house mouse, *Mus musculus*. The generation of thousands of mouse mutants, the ability to perform targeted knockouts of specific mouse genes, and the completion of DNA sequencing of the mouse genome have made the mouse a useful model for examining the role of genes in virtually all aspects of mammalian biology. In addition, the regions of DNA encoding genes in mice and humans are approximately 85 percent identical, making the mouse important not only for studying basic human biology but also as a model for understanding genetic influences on human health and disease.

—*Kenneth D. Belanger*

See also: Model Organism: *Arabidopsis thaliana*; Model Organism: *Caenorhabditis elegans*; Model Organism: *Chlamydomonas reinhardtii*; Model Organism: *Drosophila melanogaster*; Model Organism: *Escherichia coli*; Model Organism: *Mus musculus*; Model Organism: *Neurospora crassa*; Model Organism: *Saccharomyces cerevisiae*; Model Organism: *Xenopus laevis*.

Further Reading

Brookes, M. *Fly: The Unsung Hero of Twentieth Century Science.* New York: Ecco Press, 2001. A descriptive history and analysis of the use of *Drosophila melanogaster* to study biological principles, from inheritance and development to aging and alcohol tolerance.

Malakoff, D. "The Rise of the Mouse: Biomedicine's Model Mammal." *Science* 288 (2000): 248-253. Describes the use of the mouse in enhancing scientists' understanding of human biology, including the role of genes in disease and the development of new biomedical treatments.

Moore, J. A. *Science as a Way of Knowing: The Foundations of Modern Biology*. Cambridge, Mass.: Harvard University Press, 1993. A biologist describes the history of biological research from Aristotle to modern molecular analysis. Contains several outstanding chapters on the use of model organisms to understand fundamental genetic concepts.

Pennisi, E. "*Arabidopsis* Comes of Age." *Science* 290 (2000): 32-35. Insightfully reviews the role of *Arabidopsis thaliana* in elucidating plant biology.

Web Site of Interest

Genetics Society of America. http://www.genetics-gsa.org. Click on "Model Organisms" for links to Web pages on more than two dozen model organisms.

Molecular Clock Hypothesis

Fields of study: Evolutionary biology; Molecular genetics

Significance: *The molecular clock hypothesis (MCH) predicts that amino acid changes in proteins and nucleotide changes in DNA are approximately constant over time. When first proposed, it was immediately embraced by many evolutionists as a way to determine the absolute age of evolutionary lineages. After more protein sequences were analyzed, however, many examples were inconsistent with the MCH. The theory has generated a great deal of controversy among evolutionists, and although it is now generally accepted that many genes do not change at constant rates, methods are still being developed to determine the ages of lineages based on amino acid and nucleotide substitutions.*

Key terms

CODON: a three-letter nucleotide sequence in RNA or DNA that codes for a specific amino acid; a gene is composed of a long string of codons

INTRON: an intervening sequence in a eukaryotic gene (generally there are several to many

per gene) that must be removed when it is transcribed into messenger RNA (mRNA); introns are assumed to have no function and therefore mutations in them are often considered neutral

NEUTRAL MUTATION: a mutation in a gene which is considered to have no effect on the fitness of the organism

PHYLOGENY: often called an evolutionary tree, the branching patterns that show evolutionary relationships, with the taxa on the ends of the branches

TAXON (*pl.* TAXA): a general term used by evolutionists to refer to a type of organism at any taxonomic rank in a classification of organisms

History

In 1962 Émile Zuckerkandl and Linus Pauling published evidence that the rate of amino acid substitution in proteins is constant over time. In 1965, after several protein sequences (cytochrome c, hemoglobin, and fibrinopeptides) seemed to show this pattern, they proposed the molecular clock hypothesis (MCH). According to their hypothesis, mutations leading to changes in the amino acid sequence of a protein should occur at a constant rate over time, rather than per generation, as previously assumed. In other words, if the sequence of cytochrome c were determined 1,000,000 years ago, 500,000 years ago, and in the present, the rate of amino acid substitution would be the same between the first two samples as it would be between the second and third. To state this more accurately, they considered the rate approximately constant, which means that one protein may display some variation, but if the average rates of change for several were considered as a group, they would be constant.

Importance of the Molecular Clock Hypothesis

The evolutionary importance of the MCH was almost immediately apparent. Paleontologists had long determined the ages of fossils using radioactive dating techniques, but determining the date of a fossil was not the same as determining how long ago flowering plants diverged (evolved from) the other vascular plants, for example. Using the MCH, researchers could compare the amino acid sequences of a protein in a flowering plant and another vascular plant, and if the substitution rate (that is, substitutions per unit of time) was known, they could determine how long ago these two plants diverged. The MCH held great promise for solving many of the questions about when various groups of organisms diverged from their common ancestors. To "calibrate" the clock—that is, to determine the rate of amino acid substitutions—all that was needed were the sequences of some taxa and a reliable age for fossils considered to represent the common ancestor to the taxa. Once this clock had been calibrated, other taxa that might not be as well represented in the fossil record could be studied, and their time of divergence could be determined as well.

As more data accumulated through the next twenty years, it was discovered that amino acid substitutions in many proteins were not as clocklike as hoped. Rates over time seemed to slow down and speed up, and there was no predictable pattern to the changes. In fact, the same proteins in different evolutionary lineages often "ticked" at a different rate.

The Neutral Theory

During the time that more and more proteins were being sequenced, DNA sequencing gradually began to dominate. One of the theories about why the MCH did not seem to be working was that protein sequences were constrained by natural selection. The intensity of natural selection has always been assumed to vary over time, and if this is true, then amino acid substitution rates should also increase and decrease as some kind of function of the pressure exerted by natural selection. DNA sequences were quickly hailed as the solution to this problem. In 1968, Motoo Kimura proposed the neutral theory, in which he proposed that any nucleotide substitution in DNA that occurred in a noncoding region, or that did not change the amino acid sequence in the gene's product, would be unaffected by natural selection. He suggested that because of this, neutral mutations (nucleotide substitutions) would be free to take place without being weeded out by selection.

The strength of the neutral theory was that, unlike mutations that affect the amino acid sequence, neutral mutations should occur at a constant rate over time. Therefore, Kimura predicted that the MCH would be valid for neutral mutations. Most eukaryotic genomes are riddled with sequences, like introns or highly repetitive DNA, that have no apparent function and can therefore be assumed to be prone to neutral mutations. Even within the coding regions (exons) of expressed genes, the third position of many codons can be changed without affecting the amino acid for which it codes. A number of evolutionists expressed skepticism concerning the neutral theory, arguing that there is probably no truly neutral mutation.

As DNA sequences poured in, much the same story emerged as for protein sequences. Whether or not neutral mutations exist, nucleotide substitutions that were assumed to be neutral turned out to tick no better. In the 1980's the controversy over the MCH reached its height, and most evolutionists were forced to conclude that very few genes, or neutral sequences, behaved like a clock. Even those that did behave like clocks did not tick at the same rate in all lineages, and even worse, some genes ticked more or less steadily in some lineages and very erratically in others. Comparisons among the many amino acid and nucleotide sequences revealed another surprise: Amino acid sequences tended, on average, to be more reliable than nucleotide sequences.

Beyond the Molecular Clock

Since the 1980's, the MCH has fallen into disfavor among most evolutionists, but attempts to use amino acid and nucleotide sequences to estimate evolutionary ages are still being made. In a few cases, often in closely related taxa, the MCH works, but other approaches are used more often. Many of these approaches attempt to take into account the highly variable substitution rates among different lineages and over time. Rather than using a single protein or DNA sequence, as was attempted when the MCH was first developed, they use several in the same analysis. Data analysis relies on complex, and sometimes esoteric, statistical algorithms that often require considerable computational power.

In some ways, the research community is in disarray when it comes to post-MCH methods. There are several alternative approaches, and some that represent blended approaches, and agreement is far from being achieved. It is hoped that as more data are collected and analyzed, a coherent approach will be developed.

See also: Ancient DNA; DNA Sequencing Technology; Evolutionary Biology; Natural Selection; Punctuated Equilibrium; Repetitive DNA.

—*Bryan Ness*

Further Reading

Ayala, Francisco J. "Vagaries of the Molecular Clock." *Proceedings of the National Academy of Science USA* 94 (1997): 7776-7783. A somewhat technical overview of the molecular clock hypothesis in relation to two specific genes in fruit flies.

Benton, Michael J., and Francisco J. Ayala. "Dating the Tree of Life." *Science* 300 (2003): 1698-1700. An overview of the current debate on the use of molecular dating techniques.

Nei, Masatoshi, and Sudhir Kumar. *Molecular Evolution and Phylogenetics.* New York: Oxford University Press, 2000. Textbook-type coverage of a variety of topics, with one complete chapter on the molecular clock hypothesis.

Pagel, Mark. "Inferring the Historical Patterns of Biological Evolution." *Nature* 401 (1999): 877-884. An overview of phylogenies and how they are constructed, including a discussion of the molecular clock hypothesis.

Molecular Genetics

Field of study: Molecular genetics

Significance: *Molecular genetics is the branch of genetics concerned with the central role that molecules, particularly the nucleic acids DNA and RNA, play in heredity. The understanding of molecular genetics is at the heart of biotechnology, which has had a tremendous impact on medicine, agriculture, forensics, and many other fields.*

Key terms

DNA: dexoyribonucleic acid, a long-chain macromolecule, made of units called nucleotides and structured as a double helix joined by weak hydrogen bonds, which forms genetic material for most organisms

GENOME: the assemblage of the genetic information of an organism or of one of its organelles

REPLICATION: the process by which one DNA molecule is converted to two DNA molecules identical to the first

RNA: ribonucleic acid, the macromolecule in the cell that acts as an intermediary between the genetic information stored as DNA and the manifestation of that genetic information as proteins

TRANSCRIPTION: the process of forming an RNA molecule according to instructions contained in DNA

TRANSLATION: the process of forming proteins according to instructions contained in an RNA molecule

Identity and Structure of Genetic Material

Molecular genetics is the branch of genetics that deals with the identity of the molecules of heredity, their structure and organization, how these molecules are copied and transmitted, how the information encrypted in them is decoded, and how the information can change from generation to generation. In the late 1940's and early 1950's, scientists realized that the materials of heredity were nucleic acids. DNA was implicated as the substance extracted from a deadly strain of pneumococcal bacteria that could transform a mild strain into a lethal one and as the substance injected into bacteria by viruses as they start an infection. RNA was shown to be the component of a virus that determined what kind of symptoms of infection appeared on tobacco leaves.

The nucleic acids are made up of nucleotides linked end to end to produce very long molecules. Each nucleotide has sugar and phosphate parts and a nitrogen-rich part called a base. Four bases are commonly found in each DNA and RNA. Three, adenine (A), guanine (G), and cytosine (C), are found in both DNA and RNA, while thymine (T) is normally found only in DNA and uracil (U) only in RNA. In the double-helical DNA molecule, two strands are helically intertwined in opposite directions. The nucleotide strands are held together in part by interactions specific to the bases, which "pair" perpendicular to the sugar-phosphate strands. The structure can be envisioned as a ladder. The A and T bases pair with each other, and G and C bases pair with each other, forming "rungs"; the sugar-phosphates, joined end to end, form the "sides" of the ladder. The entire molecule twists and bends in on itself to form a compact whole. An RNA molecule is essentially "half" of this ladder, split down the middle. RNA molecules generally adopt less regular structures but may also require pairing between bases.

DNA and RNA, in various forms, serve as the molecules of heredity. RNA is the genetic material that some viruses package in viral particles. One or several molecules of RNA may make up the viral information. The genetic material of most bacteria is a single circle of double-helical DNA, the circle consisting of from slightly more than 500,000 to about 5 million nucleotide pairs. In eukaryotes such as humans, the DNA genetic material is organized into multiple linear DNA molecules, each one the essence of a morphologically recognizable and genetically identifiable structure called a chromosome.

In each organism, the DNA is closely associated with proteins. Proteins are made of one or more polypeptides. Polypeptides are linear polymers, like nucleic acids, but the units linked end to end are amino acids rather than nucleotides. More than twenty kinds of amino acids make up polypeptides. Proteins are generally smaller than DNA molecules and assume a variety of shapes. Proteins contribute to the biological characteristics of an organism in many ways: They are major components of structures both inside (membranes and fibers) and outside (hair and nails) the cell; as enzymes, they initiate the thousands of chemical reactions that cells use to get energy and build new cells; and they regulate the activities of cells. Histone proteins pack eukaryotic nuclear DNA into tight bundles called nucleosomes. Further coiling and looping of nucleosomes results in the compact structure of chromosomes. These can be seen with help of a microscope. The complex of DNA and protein is called chromatin.

The term "genome" denotes the roster of genes and other DNA of an organism. Most eukaryotes have more than one genome. The principal genome is the genome of the nucleus that controls most of the activities of cells. Two organelles, the mitochondria (which produce energy by oxidizing chemicals) and the plastids (such as chloroplasts, which convert light to chemical energy in photosynthesis) have their own genomes. The organelle genomes have only some of the genes needed for their functioning. The others are present in the nuclear genome. Nuclear genomes have many copies of some genes. Some repeated sequences are organized tandemly, one after the other, while others are interspersed with unique sequences. Some repeated sequences are genes present in many copies, while others are DNAs of unknown function.

Copying and Transmission of Genetic Nucleic Acids

James Watson and Francis Crick's double-helical structure for DNA suggested to them how a faithful copy of a DNA could be made. The strands would pull apart. One by one, the new nucleotide units would then arrange themselves by pairing with the correct base on the exposed strands. When zipped together, the new units make a new strand of DNA. The process, called DNA replication, makes two double-helical DNAs from one original one. Each daughter double-helical DNA has one old and one new strand. This kind of replication, called semiconservative replication, was confirmed by an experiment by Matthew Meselson and Franklin Stahl.

Enzymes cannot copy DNA of eukaryotic chromosomes completely to each end of the DNA strands. This is not a problem for bacteria, whose circular genomes do not have ends. To keep the ends from getting shorter with each cycle of replication, eukaryotic chromosomes have special structures called telomeres at their ends that are targets of a special DNA synthesis enzyme.

When a cell divides, each daughter cell must get one and only one complete copy of the mother cell's DNA. In most bacterial chromosomes, this DNA synthesis starts at only one place, and that starting point is controlled so that the number of starts equals the number of cell fissions. In eukaryotes, DNA synthesis begins at multiple sites, and each site, once it has begun synthesis, does not begin another round until after cell division. When DNA has been completely copied, the chromosomes line up for distribution to the daughter cells. Protein complexes called kinetochores bind to a special region of each chromosome's DNA called the "centromere." Kinetochores attach to microtubules, fibers that provide the tracks along which the chromosomes move during their segregation into daughter cells.

Gene Expression, Transcription, and Translation

DNA is often dubbed the blueprint of life. It is more accurate to describe DNA as the computer tape of life's instructions because the DNA information is a linear, one-dimensional series of units rather than a two-dimensional diagram. In the flow of information from the DNA tape to what is recognized as life, two steps require the decoding of nucleotide sequence information. The first step, the copying of the DNA information into RNA, is called transcription, an analogy to medieval monks sitting in their cells copying, letter by letter, old Latin manuscripts. The letters and words in the new version are the same as in the old but are written with a different hand and thus have a slightly different appearance. The second step, in which amino acids are polymerized in response to the RNA information, is called translation. Here, the monks take the Latin words and find English, German, or French equivalents. The product is not in the nucleotide language but in the language of polypeptide sequences. The RNAs that direct the order of amino acids are called messenger RNAs (mRNAs) because they bring instructions from the DNA to the ribosome, the site of translation.

Multicellular organisms consist of a variety of cells, each with a particular function. Cells also respond to changes in their environment. The differences among cell types and among cells in different environmental conditions are caused by the synthesis of different proteins. For the most part, regulation of which proteins are synthesized and which are not occurs by controlling the synthesis of the mRNAs for these proteins. Genes can have their transcription switched on or switched off by the binding of protein factors to a segment of the gene that determines whether transcription will start or not. An important part of this gene segment is the promoter. It tells the transcription apparatus to start RNA synthesis only at a particular point in the gene.

Not all RNAs are ready to function the moment their synthesis is over. Many RNA transcripts have alternating exon and intron segments. The intron segments are taken out with splicing of the end of one exon to the beginning of the next. Other transcripts are cut at several specific places so that several functional RNAs arise from one transcript. Eukaryotic mRNAs get poly-A tails (about two hundred nucleotide units in which every base is an A) added after transcription. A few RNAs are edited after transcription, some extensively by adding or removing U nucleotides in the middle of the RNA, others by changing specific bases.

Translation occurs on particles called ribosomes and converts the sequence of nucleotide residues in mRNA into the sequence of amino acid residues in a polypeptide. Since protein is created as a consequence of translation, the process is also called protein synthesis. The mRNA carries the code for the order of insertion of amino acids in three nucleotide units called codons. Failure of the ribosome to read nucleotides three at a time leads to shifts in the frame of reading the mRNA message. The frame of reading mRNA is set by starting translation only at a special codon.

Transfer RNA (tRNA) molecules actually do the translating. There is at least one tRNA for each of the twenty common amino acids. Anticodon regions of the tRNAs each specifically pair with only a specific subset of mRNA codons. For each amino acid there is at least one enzyme that attaches the amino acid to the correct tRNA. These enzymes are thus at the center of translation, recognizing both amino acid and nucleotide residues.

The ribosomes have sites for binding of mRNA, tRNA, and a variety of protein factors. Ribosomes also catalyze the joining of amino acids to the growing polypeptide chain. The protein factors, usually loosely bound to ribosomes, assist in the proper initiation of polypeptide chains, in the binding of amino acid-bearing tRNA to the ribosome, and in moving the ribosome relative to the mRNA after each additional step. Three steps in translation use biochemical energy: attaching the amino acid to the tRNA, binding the amino acyl tRNA to the ribosome-mRNA complex, and moving the ribosome relative to the mRNA.

Protein Processing and DNA Mutation

The completed polypeptide chain is processed in one or more ways before it assumes its role as a mature protein. The linear string of amino acid units folds into a complex, three-dimensional structure, sometimes with the help of other proteins. Signals in some proteins' amino acid sequences direct them to their proper destinations after they leave the ribosomes. Some signals are removable, while others remain part of the protein. Some newly synthesized proteins are called polyproteins because they are snipped at specific sites, giving several proteins from one translation product. Finally, individual amino acid units may get other groups attached to them or be modified in other ways.

The DNA information can be corrupted by reaction with certain chemicals, some of which are naturally occurring while others are present in the environment. Ultraviolet and ionizing radiation can also damage DNA. In addition, the apparatus that replicates DNA will make a mistake at low frequency and insert the wrong nucleotide.

Collectively, these changes in DNA are called DNA damage. When DNA damage goes unrepaired before the next round of copying of the DNA, mutations (inherited changes in nucleotide sequence) result. Mutations may be substitutions, in which one base replaces another. They may also be insertions or deletions of one or more nucleotides. Mutations may be beneficial, neutral, or harmful. They are the targets of the natural selection that drives evolution. Since some mutations are harmful, survival of the species requires that they be kept to a low level.

Systems that repair DNA are thus very important for the accurate transmission of the DNA information tape. Several kinds of systems have evolved to repair damaged DNA before it can be copied. In one, enzymes directly reverse the damage to DNA. In a second, the damaged base is removed, and the nucleotide chain is split to allow its repair by a limited resynthesis. In a third, a protein complex recognizes the DNA damage, which results in incisions in the DNA backbone on both sides of the damage. The segment containing the damage is removed, and the gap is filled by a limited resynthesis. In still another, mismatched base pairs, such as those that result from errors in replication, are recognized, and an incision is made some distance away from the mismatch. The entire stretch from the incision point to past the mismatch is then resynthesized. Finally, the molecular machinery that exchanges DNA segments, the recombination machinery, may be mobilized to repair damage that cannot be handled by the other systems.

Invasion and Amplification of Genes

Mutation is only one way that genomes change from generation to generation. Another way is via the invasion of an organism's genome by other genomes or genome segments. Bacteria have evolved restriction modification systems to protect themselves from such invasions. The gene for restriction encodes an enzyme that cleaves DNA whenever a particular short sequence of nucleotides is present. It does not recognize that sequence when it has been modified with a methyl group on one of its bases. The gene for modification encodes the enzyme that adds the methyl group. Thus the bacterium's own DNA is protected. However, DNA that enters the cell from outside, such as by phage infection or by direct DNA uptake, is not so protected and will be targeted for degradation by the restriction enzyme. Despite restriction, transfer of genes from one species to another (horizontal, or lateral, gene transfer) has occurred.

As far as is known, restriction modification systems are unique to bacteria. Gene transfer from bacteria to plants occurs naturally in diseases caused by bacteria of the *Agrobacterium* genus. As part of the infection process, these bacteria transfer a part of their DNA containing genes, only active in plants, into the plant genome. Studies with fungi and higher plants suggest that eukaryotes cope with gene invasion by inactivating the genes (gene silencing) or their transcripts (cosuppression).

Another way that genomes change is by duplications of gene-sized DNA segments. When the environment is such that the extra copy is advantageous, the cell with the duplication survives better than one without the duplication. Thus genes can be amplified under selective pressure. In some tissues, such as salivary glands of dipteran insects and parts of higher plant embryos, there is replication of large segments of chromosomes without cell division. Monster chromosomes result.

Genomes also change because of movable genetic elements. Inversions of genome segments occur in bacteria and eukaryotes. Other segments can move from one location in the genome to another. Some of these movements appear to be rare, random events. Others serve particular functions and are programmed to occur under certain conditions. One kind of mobile element, the retrotransposon, moves into new locations via an RNA intermediate. The element encodes an enzyme that makes a DNA copy of the element's RNA transcript. That copy inserts itself into other genome locations. The process is similar to that used by retroviruses to establish infection in cells. Other mobile elements, called transposons or transposable elements, encode a transposase enzyme that inserts the element sequence, or a copy of it, into a new location. When that new location is in or near a gene, normal functioning of that gene is disturbed.

The production of genes for antibodies (an important part of a human's immune defense system) is a biological function that requires gene rearrangements. Antibody molecules consist of two polypeptides called light and heavy chains. In most cells in the body, the genes for light chains are in two separated segments, and those for heavy chains are in three. During the maturation of cells that make antibodies, the genes are rearranged, bringing these segments together. The joining of segments is not precise. The imprecision contributes to the diversity of possible antibody molecules.

Cells of baker's or brewer's yeast (*Saccharomyces cerevisiae*) have genes specifying their sex, or mating type, in three locations. The information at one location, the expression locus, is the one that determines the mating type of the cell. A copy of this information is in one of the other two sites, while the third has the information specifying the opposite mating type. Yeast cells switch mating types by replacing the information at the expression locus with information from a storage locus. Mating-type switching and antibody gene maturation are only two examples of programmed gene rearrangements known to occur in a variety of organisms.

Genetic Recombination

Recombination occurs when DNA information from one chromosome becomes attached to the DNA of another. When participating chromosomes are equivalent, the recombination is called homologous. Homologous recombination in bacteria mainly serves a repair function for extreme DNA damage. In many eukaryotes, recombination is essential for the segregation of chromosomes into gamete cells during meiosis. Nevertheless, aspects of the process are common between bacteria and eukaryotes. Starting recombination requires a break in at least one strand of the double-helical DNA. In the well-studied yeast cells, a double-strand break is required. Free DNA ends generated by breaks invade the double-helical DNA of the homologous chromosome. Further invasion and DNA synthesis result in a structure in which the chromosomes are linked to one another. This structure, called a half-chiasma, is recognized and resolved by an enzyme system. Resolution can result in exchange so that one end of one chromosome is linked to the other end of the other chromosome and vice versa. Resolution can also result in restoration of the original linkage. In the latter case, the DNA around the exchange point may be that of the other DNA. This is known as gene conversion.

Impact and Applications

Molecular genetics is at the heart of biotechnology, or genetic engineering. Its fundamental investigation of biological processes has provided tools for biotechnologists. Molecular cloning and gene manipulation in the test tube rely heavily on restriction enzymes, other nucleic-acid-modifying enzymes, and extrachromosomal DNA, all discovered during molecular genetic investigation. The development of nucleic acid hybridization, which allows the identification of specific molecular clones in a pool of others, required an understanding of DNA structure and dynamics. The widely used polymerase chain reaction (PCR), which can amplify minute quantities of DNA, would not have been possible without discoveries in DNA replication. Genetic mapping, a prelude to the isolation of many genes, was sped along by molecular markers detectable with restriction enzymes or the PCR. Transposable elements and the transferred DNA of *Agrobacterium*, because they often inactivate genes when they insert in them, were used to isolate the genes they inactivate. The inserted elements served as tags or handles by which the modified genes were pulled out of a collection of genes.

The knowledge of the molecular workings of genes gained by curious scientists has allowed other scientists to intervene in many disease situations, provide effective therapies, and improve biological production. Late twentieth century scientists rapidly developed an understanding of the infection process of the acquired immunodeficiency syndrome (AIDS) virus. The understanding, built on the skeleton of existing knowledge, has helped combat this debilitating disease. Molecular genetics has also led to the safe and less expensive production of proteins of industrial, agricultural, and pharmacological importance. The transfer of DNA from *Agrobacterium* to plants has been exploited in the creation of transgenic plants. These plants offer a new form of pest protection that provides an alternative to objectionable pesticidal sprays and protects against pathogens for which no other protection is available. Recombinant insulin and recombinant growth hormone are routinely given to those whose conditions demand them. Through molecular genetics, doctors have diagnostic kits that can, with greater rapidity, greater specificity, and lower cost, determine whether a pathogen is present. Finally, molecular genetics has been used to identify genes responsible for many inherited diseases of humankind. Someday medicine may correct some of these diseases by providing a good copy of the gene, a strategy called gene therapy.

—*Ulrich Melcher*

See also: Ancient DNA; Antisense RNA; Biochemical Mutations; Central Dogma of Molecular Biology; Chemical Mutagens; Chloroplast Genes; Chromatin Packaging; DNA Isolation; DNA Repair; DNA Structure and Function; Gene Families; Genetic Code; Genetic Code, Cracking of; Genome Size; Genomics; Molecular Clock Hypothesis; Mutation and Mutagenesis; Noncoding RNA Molecules; Oncogenes; One Gene-One Enzyme Hypothesis; Protein Structure; Protein Synthesis; Proteomics; Pseudogenes; Repetitive DNA; Restriction Enzymes; Reverse Transcriptase; RNA Isolation; RNA Structure and Function; RNA Transcription and mRNA Processing; RNA World; Signal Transduction; Steroid Hormones; Telomeres; Transposable Elements; Tumor-Suppressor Genes.

Further Reading

Brown, Terence A. *Genetics: A Molecular Approach.* 3d ed. New York: Chapman & Hall, 1998. Solid text with bibliography, index.

Carroll, Sean B., Jennifer K. Grenier, and Scott D. Weatherbee. *From DNA to Diversity: Molecular Genetics and the Evolution of Animal Design.* Malden, Mass.: Blackwell, 2001. Discusses morphology and its genetic basis, and evolutionary biology's synthesis with genetics and embryology. Illustrations (some color), figures, tables, glossary, bibliography.

Clark, David P., and Lonnie D. Russell. *Molecular Biology Made Simple and Fun.* 2d ed. Vienna, Ill.: Cache River Press, 2000. A detailed and entertaining account of molecular genetics. Bibliography, index.

Hancock, John T. *Molecular Genetics.* Boston: Butterworth-Heinemann, 1999. Covers the basics of molecular genetics, especially for advanced high school and beginning-level college students. Illustrations, bibliography, summaries of key chapter concepts.

Hartl, D. L. *Genetics: Analysis of Genes and Genomes.* 5th ed. Boston: Jones and Bartlett, 2001. An excellent introductory genetics textbook.

Hartwell, L. H., L. Hood, M. L. Goldberg, A. E. Reynolds, L. M. Silber, and R. C. Veres. *Genetics: From Genes to Genomes.* 2d ed. New York: McGraw-Hill, 2003. A comprehensive textbook on genetics, by the 2001 Nobel laureate in physiology or medicine. Available as an e-book.

Lewin, Benjamin. *Genes VII.* New York: Oxford University Press, 2001. Covers structure, function, and molecular processes of genes.

Miesfeld, Roger L. *Applied Molecular Genetics.* New York: John Wiley, 1999. Presents an overview of the practical implications of molecular genetics in modern biotechnology. Illustrations (mostly color), appendices, bibliography, Web resources.

Russell, Peter J. *Genetics.* San Fransisco, Calif.: Benjamin Cummings, 2002. Good genetic textbook with basic coverage of molecular genetics.

Strachan, T., and Andrew P. Read. *Human Molecular Genetics 2.* 2d ed. New York: Wiley-Liss, 1999. Introductory discussion of DNA, chromosomes, and the Human Genome Project. Illustrated.

Watson, James, et al. *Molecular Biology of the Gene.* 5th ed. 2 vols. Menlo Park, Calif.: Benjamin Cummings, 2003. A widely used textbook by the co-discoverer of DNA's helical structure. Bibliography, index.

Web Sites of Interest

Human Molecular Genetics. http://hmg.oup journals.org. The Web site for the online journal, with abstracts of articles available online and full text available for a fee.

Max Planck Institute for Molecular Genetics. http://www.molgen.mpg.de. Research institute focuses on molecular mechanisms of DNA replication, recombination, protein synthesis, and ribosome structure, and offers educational information and history.

Monohybrid Inheritance

Field of study: Classical transmission genetics

Significance: *Humans and other organisms show a number of different patterns in the inheritance and expression of traits. For many inherited characteristics, the pattern of transmission is monohybrid inheritance, in which a trait is determined by one pair of alleles at a single locus. An understanding of monohybrid inheritance is critical for understanding the genetics of many medically significant traits in humans and economically significant traits in domestic plants and animals.*

Key terms

ALLELE: one of the pair of possible alternative forms of a gene that occurs at a given site or locus on a chromosome

DOMINANT GENE: the controlling member of a pair of alleles that is expressed to the exclusion of the expression of the recessive member

RECESSIVE GENE: an allele that can only be expressed when the controlling or dominant allele is not present

Mendel and Monohybrid Inheritance

The basic genetic principles first worked out and described by Gregor Mendel in his classic experiments on the common garden pea have been found to apply to many inherited traits in all sexually reproducing organisms, including humans. Until the work of Mendel, plant and animal breeders tried to formulate laws of inheritance based upon the principle that characteristics of parents would be blended in their offspring. Mendel's success came about because he studied the inheritance of contrasting or alternative forms of one phenotypic trait at a time. The phenotype of any organism includes not only all of its external characteristics but also all of its internal structures, extending even into all of its chemical and metabolic functions. Human phenotypes would include characteristics such as eye color, hair color, skin color, hearing and visual abnormalities, blood disorders, susceptibility to various diseases, and muscular and skeletal disorders.

Mendel experimented with seven contrasting traits in peas: stem height (tall vs. dwarf), seed form (smooth vs. wrinkled), seed color (yellow vs. green), pod form (inflated vs. constricted), pod color (green vs. yellow), flower color (red vs. white), and flower position (axial vs. terminal). Within each of the seven sets, there was no overlap between the traits and thus no problem in classifying a plant as one or the other. For example, although there was some variation in height among the tall plants and some variation among the dwarf plants, there was no overlap between the tall and dwarf plants.

Mendel's first experiments crossed parents that differed in only one trait. Matings of this type are known as monohybrid crosses, and the rules of inheritance derived from such matings yield examples of monohybrid inheritance. These first experiments provided the evidence for the principle of segregation and the principle of dominance. The principle of segregation refers to the separation of members of a gene pair from each other during the formation of gametes (the reproductive cells: sperm in males and eggs in females). It was Mendel who first used the terms "dominant" and "recessive." It is of interest to examine his words and to realize how appropriate his definitions are today: "Those characters which are transmitted entire, or almost unchanged by hybridization, and therefore in themselves constitute the characters of the hybrid, are termed the dominant and those which become latent in the process recessive." The terms dominant and recessive are used to describe the characteristics of a phenotype, and they may depend on the level at which a phenotype is described. A gene that acts as a recessive for a particular external trait may turn out not to be so when its effect is measured at the biochemical or molecular level.

An Example of Monohybrid Inheritance

The best way of describing monohybrid inheritance is by working through an example. Although any two people obviously differ in many genetic characteristics, it is possible, as Mendel did with his pea plants, to follow one trait governed by a single gene pair that is separate and independent of all other traits. In effect, by doing this, the investigator is working with the equivalent of a monohybrid cross. In selecting an example, it is best to choose a trait that does not produce a major health or clinical effect; otherwise, the clear-cut segregation ratios expected under monohybrid inheritance might not be seen in the matings.

Consider the trait of albinism, a phenotype caused by a recessive gene. Albinism is the absence of pigment in the hair, skin, and eyes. Similar albino genes have been found in many animals, including mice, buffalo, bats, frogs, and rattlesnakes. Since the albino gene is recessive, the gene may be designated with the symbol c and the gene for normal pigmentation as C. Thus a mating between a homozygous normal person (CC) and a homozygous albino person (cc) would be expected to produce children who are heterozygous (Cc) but phenotypically normal, since the normal gene is dominant to the albino gene. Only normal genes, C, would be passed on by the normally pigmented parent, and only albino genes, c, would be passed on by the albino parent. If there was a mating between two heterozygous people (Cc and Cc), the law of segregation would predict that each parent would produce two kinds of gametes: C and c. The resulting progeny would be expected to appear at a ratio of $1\,CC\!:\!2\,Cc\!:\!1\,cc$. Since C is dominant to c, $\frac{3}{4}$ of the progeny would be expected to have normal pigment, and $\frac{1}{4}$ would be expected to be albino. There are three genotypes (CC, Cc, and cc) and two phenotypes (normal pigmentation and albino). By following the law of segregation and taking account of the dominant gene, it is possible to determine the types of matings that might occur and to predict the types of children that would be expected (see the table "Phenotype Predictions: Albino Children").

Because of dominance, it is not always possible to tell what type of mating has occurred. For example, in matings 1, 2, and 4 in the table, the parents are both normal in each case. Yet in mating 4, $\frac{1}{4}$ of the offspring are expected to be albino. A complication arises when it is realized that in mating 4 the couple might not produce any offspring that are cc; in that case, all offspring would be normal. Often, because of the small number of offspring in humans and

Phenotype Predictions: Albino Children		
Parents	**Phenotypes**	**Offspring Expected**
1. AA × AA	Normal × Normal	All AA (Normal)
2. AA × Aa	Normal × Normal	½ AA, ½ Aa (All Normal)
3. AA × aa	Normal × Albino	All Aa (Normal)
4. Aa × Aa	Normal × Normal	¼ AA, ½ Aa, ¼ aa (¾ Normal, ¼ Albino)
5. Aa × aa	Normal × Albino	½ Aa, ½ aa (½ Normal, ½ Albino)
6. aa × aa	Albino × Albino	All aa (Albino)

other animals, the ratios of offspring expected under monohybrid inheritance might not be realized. Looking at the different matings and the progeny that are expected, it is easy to see how genetics can help to explain not only why children resemble their parents but also why children do not resemble their parents.

Modification of Basic Mendelian Inheritance

After Mendel's work was rediscovered early in the twentieth century, it soon became apparent that there were variations in monohybrid inheritance that apparently were not known to Mendel. Mendel studied seven pairs of contrasting traits, and in each case, one gene was dominant and one gene was recessive. For each trait, there were only two variants of the gene. It is now known that other possibilities exist. For example, other types of monohybrid inheritance include codominance (in which both genes are expressed in the heterozygote) and sex linkage (an association of a trait with a gene on the X chromosome). Nevertheless, the law of segregation operates in these cases as well, making it possible to understand inheritance of the traits.

Within a cell, genes are found on chromosomes in the nucleus. Humans have forty-six chromosomes. Each person receives half of the chromosomes from each parent, and it is convenient to think of the chromosomes in pairs. Examination of the chromosomes in males and females reveals an interesting difference. Both sexes have twenty-two pairs of what are termed "autosomes" or "body chromosomes." The difference in chromosomes between the two sexes occurs in the remaining two chromosomes. The two chromosomes are known as the sex chromosomes. Males have an unlike pair of sex chromosomes, one designated the X chromosome and the other, smaller one designated the Y chromosome. Females, on the other hand, have a pair of like sex chromosomes, and these are similar to the X chromosome of the male. Although the Y chromosome does not contain many genes, it is responsible for male development. A person without a Y chromosome would undergo female development. Since genes are located on chromosomes, the pattern of transmission of the genes demonstrates some striking differences from that of genes located on any of the autosomes. For practical purposes, "sex linked" usually refers to genes found on the X chromosome since the Y chromosome contains few genes. Although X-linked traits do not follow the simple pattern of transmission of simple monohybrid inheritance as first described by Mendel, they still conform to his law of segregation. Examination of a specific example is useful to understand the principle.

The red-green color-blind gene is X-linked and recessive, since females must have the gene on both X chromosomes in order to exhibit the trait. For males, the terms "recessive" and "dominant" really do not apply since the male has only one X chromosome (the Y chromosome does not contain any corresponding genes) and will express the trait whether the gene is recessive or dominant. An important implication of this is that X-linked traits appear more often in males than in females. In general, the more severe the X-linked recessive trait is from a health point of view, the greater the proportion of affected males to affected females.

If the color-blind gene is designated *cb* and the normal gene *Cb*, the types of mating and offspring expected may be set up as they were for the autosomal recessive albino gene. In the present situation, the X and Y chromosomes will also be included, remembering that the *Cb* and *cb* genes will be found only on the X chromosome and that any genotype with a Y chromosome will result in a male. (See the table "Phenotype Predictions: Color Blindness.")

"Carrier" females are heterozygous females who have normal vision but are expected to pass the gene to half their sons, who would be color blind. Presumably, the carrier female would have inherited the gene from her father, who would have been color blind. Thus, in some families the trait has a peculiar pattern of transmission in which the trait appears in a woman's father, but not her, and then may appear again in her sons.

Impact and Applications

The number of single genes known in humans has grown dramatically since Victor McKusick published the first *Mendelian Inheritance in Man* catalog in 1966. In the first catalog, there were 1,487 entries representing loci identified by Mendelizing phenotypes or by cellular and molecular genetic methods. In the 1994 catalog, the number of entries had grown to 6,459. Scarcely a day goes by without a news re-

port or story in the media involving an example of monohybrid inheritance. Furthermore, genetic conditions or disorders regularly appear as the theme of a movie or play. An understanding of the principles of genetics and monohybrid inheritance provides a greater appreciation of what is taking place in the world, whether it is in the application of DNA fingerprinting in the courtroom, the introduction of disease-resistant genes in plants and animals, the use of genetics in paternity cases, or the description of new inherited diseases.

Perhaps it is in the area of genetic diseases that knowledge of monohybrid inheritance offers the most significant personal applications. Single-gene disorders usually fall into one of the four common modes of inheritance: autosomal dominant, autosomal recessive, sex-linked dominant, and sex-linked recessive. Examination of individual phenotypes and family histories allows geneticists to determine which mode of inheritance is likely to be present for a specific disorder. Once the mode of inheritance has been identified, it becomes possible to determine the likelihood or the risk of occurrence of the disorder in the children. Since the laws governing the transmission of Mendelian traits are so well known, it is possible to pre-

Phenotype Predictions: Color Blindness

Parents	Phenotypes	Offspring Expected
1. $X^{Cb}X^{Cb} \times X^{Cb}Y$	Normal × Normal	$X^{Cb}X^{Cb}$ normal female $X^{Cb}Y$ normal male
2. $X^{Cb}X^{Cb} \times X^{cb}Y$	Normal × Color blind	$X^{Cb}X^{cb}$ normal female $X^{Cb}Y$ normal male
3. $X^{Cb}X^{cb} \times X^{Cb}Y$	Normal × Normal	$X^{Cb}X^{Cb}$ $X^{Cb}X^{cb}$ ½ normal females, ½ carrier females $X^{Cb}Y$ $X^{cb}Y$ ½ normal males, ½ color-blind males
4. $X^{Cb}X^{cb} \times X^{cb}Y$	Normal × Color blind	$X^{Cb}X^{cb}$ $X^{cb}X^{cb}$ ½ carrier females, ½ color-blind females $X^{Cb}Y$ $X^{cb}Y$ ½ normal males, ½ color-blind males
5. $X^{cb}X^{cb} \times X^{Cb}Y$	Color blind × Normal	$X^{Cb}X^{cb}$ carrier females $X^{cb}Y$ color-blind males
6. $X^{cb}X^{cb} \times X^{cb}Y$	Color blind × Color blind	$X^{cb}X^{cb}$ color-blind females $X^{cb}Y$ color-blind males

dict with great accuracy when a genetic condition will affect a specific family member. In many cases, testing may be done prenatally or in individuals before symptoms appear. As knowledge of the human genetic makeup increases, it will become even more essential for people to have a basic knowledge of how Mendelian traits are inherited.

—*Donald J. Nash*

See also: Albinism; Classical Transmission Genetics; Complete Dominance; Dihybrid Inheritance; Epistasis; Hereditary Diseases; Incomplete Dominance; Mendelian Genetics; Multiple Alleles.

Further Reading

Cooke, K. J. "Twisting the Ladder of Science: Pure and Practical Goals in Twentieth-Century Studies of Inheritance." *Endeavour* 22, no. 1 (1998). Author argues that genetics is powerfully entwined with, and thus affected by, social, individual, and commercial factors.

Derr, Mark. "The Making of a Marathon Mutt." *Natural History* (March, 1966). The principles of Mendelian inheritance are applied to the world of champion sled dogs.

McKusick, Victor A., comp. *Mendelian Inheritance in Man: A Catalog of Human Genes and Genetic Disorders.* 12th ed. Baltimore: Johns Hopkins University Press, 1998. A comprehensive catalog of Mendelian traits in humans. Although it is filled with medical terminology and clinical descriptions, there are interesting family histories and fascinating accounts of many of the traits. Bibliography, index.

Pierce, Benjamin A. *The Family Genetic Sourcebook.* New York: John Wiley & Sons, 1990. An introduction to the principles of heredity and a catalog of more than one hundred human traits. Topics include heredity, inheritance patterns, chromosomes and chromosomal abnormalities, genetic risks, and family history. Suggested readings, appendices, glossary, and index.

Wexler, Alice. *Mapping Fate: A Memoir of Family, Risk, and Genetic Research.* Berkeley: University of California Press, 1996. Intimate story of one family's struggles with the inheritability of Huntington's disease.

Multiple Alleles

Field of study: Classical transmission genetics

Significance: *Alleles are alternate forms of genes at the same locus. When three or more variations of a gene exist in a population, they are referred to as multiple alleles. The human ABO blood groups provide an example of multiple alleles.*

Key terms

BLOOD TYPE: one of the several groups into which blood can be classified based on the presence or absence of certain molecules called antigens on the red blood cells

CODOMINANT ALLELES: two contrasting alleles that are both fully functional and fully expressed when present in an individual

DOMINANT ALLELE: an allele that masks the expression of another allele that is considered recessive to it

RECESSIVE ALLELE: an allele that will be exhibited only if two copies of if are present

The Discovery of Alleles and Multiple Alleles

Although Gregor Mendel, considered to be the father of genetics, did not discover multiple alleles, an understanding of his work is necessary to understand their role in genetics. In the 1860's, Mendel formulated the earliest concepts of how traits or characteristics are passed from parents to their offspring. His work on pea plants led him to propose that there are two factors, since renamed "genes," that cause each trait that an individual possesses. A particular form of the gene, called the "dominant" form, will enable the characteristic to occur whether the offspring inherits one or two copies of that allele. The alternate form of the gene, or allele, will be exhibited only if two copies of this allele, called the "recessive" form, are present. For example, pea seeds will be yellow if two copies of the dominant, yellow-causing gene are present and will be green if two copies of the recessive gene are present. However, since yellow is dominant to green, an individual plant with one copy of each allele will be as yellow as a plant possessing two yellow genes. Mendel discovered only two alternate appearances, called phenotypes, for each trait he studied. He found that violet is the allele dominant to white in causing flower color, while tall is the allele dominant to short in creating stem length.

Early in the twentieth century, examples of traits with more than an either/or phenotype caused by only two possible alleles were found in a variety of organisms. Coat color in rabbits is a well-documented example of multiple alleles. Not two but four alternative forms of the gene for coat color exist in rabbit populations, with different letters used to designate those colors. The gene producing color is labeled c; thus, c^+ produces full, dark color, c^{ch} produces mixed colored and white hairs, c^h produces white on the body but black on the paws, and c creates a pure white rabbit. It is important to note that although three or more alternative forms can exist in a population, each individual organism can only possess two, acquiring only one from each of its parents. What, then, of Mendel's principle of one allele being dominant to the other? In the rabbit color trait, c^+ is dominant to the c^{ch}, which is dominant to c^h, with c, the gene for pure white, recessive to the other three.

If mutation can create four possible color alleles, is it not also possible that successive mutations might cause a much larger number of multiple alleles? Numerous examples exist of genes with many alleles. For example, sickle-cell disease, and related diseases called thalassemias, are all caused by mutations in one of the two genes that code for the two protein subunits of hemoglobin, the protein that carries oxygen in the blood. Dozens of different types of thalassemia exist, all caused by mutations in the same gene.

Blood Types

One of the earliest examples of multiple alleles discovered in humans concerns the ABO blood type system. In 1900, the existence of four blood types (A, B, AB, and O) was discovered. The study of pedigrees (the family histo-

The Relationship Between Genotype and Blood Type

Genotype	Blood Type	Comments
AA	A	These two genotypes produce
AO	A	identical blood types.
BB	B	These two genotypes produce
BO	B	identical blood types.
AB	AB	Both dominant alleles are expressed.
OO	O	With no dominant alleles, the recessive allele is expressed.

ries of many individuals) revealed by 1925 that these four blood types were caused by multiple alleles. The alleles are named I^A, I^B, and I^O, or simply A, B, and O. Both A and B are dominant to O. However, A and B are codominant to each other. Thus, if both are present, both are equally seen in the individual. A person with two A alleles or an A and an O has type A blood. Someone with two B alleles or a B and an O has type B. Two O alleles result in type O blood. Because A and B are codominant, the individual with one of each allele is said to have type AB blood.

To say people are "type A" means that they have an antigen (a glycoprotein or protein-sugar molecule) of a particular type embedded in the membrane of all red blood cells. The presence of an A allele causes the production of an enzyme that transfers the sugar galactosamine to the glycoprotein. The B allele produces an enzyme that attaches a different sugar, called galactose, and the O allele produces a defective enzyme that cannot add any sugar. Because of codominance, people with type AB blood have both antigens on their red blood cells.

Transfusion with blood from a donor with a different blood type than the recipient can cause death, due to the potential presence of A or B antibodies in the recipient's blood. Antibodies are chemical molecules in the plasma (the liquid portion of the blood). If, by error, type A blood is given to a person with type B blood, the recipient will produce antibodies against the type A red blood cells, which will attach to them, causing them to agglutinate, or form clumps. By this principle, a person with type O blood can donate it to people with any blood type, because their blood cells have neither an A nor a B antigen. Thus, people with type O blood are often referred to as universal donors because no antibodies will be formed against type O blood red blood cells. Likewise, people with type AB blood are often referred to as universal recipients because they have both types of antigens and therefore will not produce antibodies against any of the blood types. Medical personnel must carefully check the blood type of both the recipient and the donated blood to avoid agglutination and subsequent death.

Blood types have been used to establish paternity because a child's blood type can be used to determine what the parents' blood types could and could not be. Since a child receives one allele from each parent, certain men can be eliminated as a child's potential father if the alleles they possess could not produce the combination found in the child. However, this proves only that a particular person could be the father, as could millions of others who possess that blood type; it does not prove that a particular man is the father. Modern methods of analyzing the DNA in many of the individual's genes now make the establishment of paternity a more exact science.

Impact and Applications

The topic of multiple alleles has implications for many human disease conditions. One of these is cystic fibrosis (CF), the most common deadly inherited disease afflicting Caucasians. Characterized by a thick mucus buildup in lungs, pancreas, and intestines, it frequently brings about death by age twenty. Soon after the gene that causes CF was found in 1989, geneticists realized there may be as many as one hundred multiple alleles for this gene. The extent of the mutation in these alternate genes apparently causes the great variation in the severity of symptoms from one patient to another.

The successful transplantation of organs is also closely linked to the existence of multiple alleles. A transplanted organ has antigens on its cells that will be recognized as foreign and destroyed by the recipient's antibodies. The genes that build these cell-surface antigens, called human leukocyte antigen (HLA), occur in two main forms. HLA-A has nearly twenty different alleles, and HLA-B has more than thirty. Since any individual can only have two of each type, there are an enormous number of possible combinations in the population. Finding donors and recipients with the same or a very close combination of HLA alleles is a very difficult task for those arranging successful organ transplantation.

Geneticists are coming to suspect that multiple alleles, once thought to be the exception to the rule, may exist for the majority of human genes. If this is so, the study of multiple alleles for many disease-producing genes should shed more light on why the severity of so many genetic diseases varies so widely from person to person.

—Grace D. Matzen, updated by Bryan Ness

See also: Complementation Testing; Cystic Fibrosis; Organ Transplants and HLA Genes; Population Genetics.

Further Reading

Klug, William S. *Essentials of Genetics.* 3d ed. Upper Saddle River, N.J.: Prentice Hall, 1999. Supplies a solid explanation of multiple alleles. Bibliography, index.

Mutation and Mutagenesis

Field of study: Molecular genetics

Significance: *A mutation is a heritable change in the structure or composition of DNA. Depending on the function of the altered DNA segment, the effect of a mutation can range from undetectable to causing major deformities and even death. Mutation is a natural process by which new genetic diversity is produced. However, chemical pollutants and radiation can increase mutation rates and have a serious effect on health.*

Key terms

GENE POOL: all of the genes carried by all members of a population of organisms; the genetic diversity in the gene pool provides the variation that allows adaptation to new conditions

GERMINAL MUTATION: a mutation in gamete-forming (germinal) tissue, which can be passed from a parent to its offspring

MUTAGEN: a chemical or physical agent that causes an increased rate of mutation

MUTAGENESIS: the process of a heritable change occurring in a gene, either spontaneously or in response to a mutagen

MUTATION RATE: the probability of a heritable change occurring in the genetic material over a given time period, such as a cell division cycle or a generation

PHENOTYPE: the observable effects of a gene; phenotypes include physical appearance, biochemical activity, cell function, or any other measurable factor

SOMATIC MUTATION: a mutation that occurs in a body cell and produces a group of mutant cells but is not transmitted to the next generation

WILD TYPE: the normal genetic makeup of an organism, as it occurs in nature (the wild); a mutation alters the phenotype of a wild-type trait to produce a mutant phenotype

Definitions

A mutation is any change in the genetic material that can be inherited by the next generation of cells or progeny. A mutation can occur at any time in the life of any cell in the body. If a mutation occurs in the reproductive tissue, the change can be passed to an offspring in the egg or sperm. That new mutation may then affect the development of the offspring and be passed on to later generations. However, if the mutation occurs in cells of the skin, muscle, blood, or other body (somatic) tissue, the new mutation will only be passed on to other body cells when that cell divides. This can produce a mosaic patch of cells carrying the new genetic change. Most of these are undetectable and have no effect on the carrier. An important exception is a somatic mutation that causes the affected cell to lose control of the cell cycle and divide uncontrollably, resulting in cancer. Many environmental chemicals and agents that cause mutations (such as X rays and ultraviolet

radiation) therefore also cause cancer.

Mutation also has an important, beneficial role in natural populations of all organisms. The ability of a species to adapt to changes in its environment, combat new diseases, or respond to new competitors is dependent on genetic diversity in the population's gene pool. Without sufficient resources of variability, a species faced by a serious new stress can become extinct. The reduced population sizes in rare and endangered species will result in reduced genetic diversity and a loss of the capacity to respond to selection pressures. Zoo breeding programs often take data on genetic diversity into account when planning the captive breeding of endangered species. The creation of new agricultural crops or of animal breeds with economically desirable traits also depends on mutations that alter development in a useful way. Therefore, mutation can have both damaging and beneficial effects.

The Role of Mutations in Cell Activity and Development

The genetic information in a cell is encoded in the sequence of subunits, the nucleotides, that make up the DNA molecule. A mutation is a change in the cell's genetic makeup, and it can range from changing just a single nucleotide in the DNA molecule to altering long pieces of DNA. To appreciate how such changes can affect an organism, it is important to understand how information is encoded in DNA and how it is translated to produce a specific protein. There are four different nucleotides in the DNA molecule: adenine (A), guanine (G), thymine (T), and cytosine (C). The DNA molecule is composed of two strands linked together by a sequence of base pairs (bp). An adenine on one strand pairs with a thymine on the other (A-T), and a guanine on one strand pairs with a cytosine on the other (G-C). When a gene is activated, one of the two strands is used as a model, or template, for the synthesis of a single-stranded molecule called messenger RNA (mRNA). The completed mRNA molecule is then transported out of the nucleus, and it binds with ribosomes (small structures in the cytoplasm of the cell), where a protein is made using the mRNA's nucleotide sequence as its coded message. The nucleotides are read on the ribosome in triplets, with three adjacent nucleotides (called a codon) corresponding to one of the twenty amino acids found in protein.

Thus the sequence of nucleotides eventually determines the order of amino acids that are linked together to form a specific protein. The amino acid sequence in turn determines how the protein will work, either as a structural part of a cell or as an enzyme that will catalyze a specific biochemical reaction. A gene is often 1,000 bp or longer, so there are many points at which a genetic change can occur. If a mutation occurs in an important part of the gene, even the change of a single amino acid can cause a major change in protein function. Sickle-cell disease is a good example of this. In sickle-cell disease a base-pair substitution in the DNA causes the sixth codon in the mRNA to change from GAG to GUG. When this modified mRNA is used to create a protein, the amino acid valine is substituted for the normal glutamic acid in the sixth position in a string of 146 amino acids. This small change causes the protein to form crystals and thus deform cells when the amount of available oxygen is low. Since this protein is one of the parts of the oxygen-carrying hemoglobin molecule in red blood cells, this single DNA nucleotide change has potentially severe consequences for an affected individual.

Types of Mutation

Because they can be so diverse, one way to organize mutations is to describe the kind of molecular or structural change that has occurred. There are three broad classes of mutation. "Genomic mutations" are changes in the number of chromosomes in a cell. Inheriting an extra chromosome, as in Down syndrome, is an example of a genomic mutation. "Chromosome mutations" are changes in the structure of a chromosome and can include the loss, gain, or altered order of a series of genes. "Gene mutations" are genetic changes limited to an individual gene or the adjacent regions that control its activity during development. Thus, the amount of genetic information affected by a mutation can vary from a single gene to hundreds of them. Genes also vary in the severity of their effects. Some are undetectable in the carrier, some cause small defects or even beneficial changes in the function of a protein, while others can produce major changes in several different developmental processes at the same time.

Gene mutations are sometimes called point mutations because their genetic effects are limited to a single point, or gene, on a chromosome that can carry up to several thousand different genes. The simplest kind of point mutation is a base substitution, in which one base pair is replaced by another (for example, the replacement of an A-T base pair at one point in the DNA molecule by a C-G base pair). This can change a codon triplet so that a different amino acid is placed in the protein at that point. This often changes the function of the protein, at least in minor ways. However, some base substitutions are silent. Since several different triplets can code for the same amino acid, not all base changes will result in an amino acid substitution.

Another common kind of gene mutation called a "frame shift" can have a much larger effect on protein structure. A frame-shift mutation occurs when a nucleotide is added to, or lost from, the DNA strand when it is duplicated during cell division. Since translation of the mRNA is done by the ribosomes adding one amino acid to the growing protein for every three adjacent nucleotides, adding or deleting one nucleotide will effectively shift that reading frame so that all following triplets are different. By analogy, one can consider the following sentence of three-letter words: THE BIG DOG CAN RUN FAR. If a base (for example, a letter *X*, in this analogy) is added at the end of the second triplet, the "sentence" will still read three letters at a time during translation and the meaning will be completely altered. THE BIX GDO GCA NRU NFA R. In a cell, a nonfunctional protein is produced unless the frame shift is near the terminal end of the gene.

Environmental agents such as ultraviolet (UV) radiation can affect DNA and base pairing. Certain UV wavelengths, for example, cause some DNA nucleotides to pair abnormally. Gene mutations have also been traced to the movement of transposable DNA elements.

Transposable elements were first discovered by Barbara McClintock while studying chromosome breakage and kernel traits in maize. Now they are known from many organisms, including humans. Transposable elements are small DNA segments that can become inserted into a chromosome and later excised and change their position. If one becomes inserted in the middle of a gene, it effectively separates the gene into two widely spaced fragments. In the fruit fly (*Drosophila melanogaster*), in which spontaneous mutations have been studied in detail at the DNA level, as many as half of the spontaneous mutations in certain genes have been traced to transposable elements.

There are four major kinds of chromosome mutations. A chromosome deletion or deficiency is produced when two breaks occur in the chromosome but are repaired by leaving out the middle section. For example, if the sections of a chromosome are labeled with the letters ABCDEFGHIJKLMN and chromosome breaks occur at F-G and at K-L, the broken chromosome can be erroneously repaired by enzymes that link the ABCDEF fragment to the LMN fragment. The genes in the unattached middle segment, GHIJK, will be lost from the chromosome. Losing these gene copies can affect many different developmental processes and even cause the death of the organism. Chromosome breaks and other processes can also cause some genes to be duplicated in the chromosome (for example, ABCDEFGHDEFG HIJKLMN). A third kind of chromosome mutation, an inversion, changes the order of the genes when the segment between two chromosomal breaks is reattached backward (for example, ABCDJIHGFEKLMN). Finally, chromosome segments can be moved from one kind of chromosome to another in a structural change called a translocation. Some examples of heritable Down syndrome are caused by this type of chromosome mutation.

Genomic mutations are a large factor in the genetic damage that occurs in humans. Whole chromosomes can be lost or gained by errors during cell division. In animals, almost all examples of chromosome loss are so developmentally severe that the individual cannot survive to birth. On the other hand, since extra chromosomes provide an extra copy of each of their genes, the amount of each protein they code for is unusually high, and this, too, can create biochemical abnormalities for the organism. In humans, an interesting exception is changes in chromosome number that involve the sex-determining chromosomes, especially the X chromosome (the Y is relatively silent in development). Since normal males have one X and females have two, the cells in females inactivate one of the X chromosomes to balance gene dosage. This dosage compensation mechanism can, therefore, also come into operation when one of the X chromosomes is lost or an extra one is inherited because of an error in cell division. The resulting conditions, such as Turner syndrome and Klinefelter syndrome, are much less severe than the developmental problems associated with other changes in chromosome number.

Mutation Rate

A mutation is any heritable change in the genetic material, but there are several different ways one can look at genetic change. For example, errors can occur when the DNA molecule is being duplicated during cell division. In simple organisms such as bacteria, about one thousand nucleotides are added to the duplicating DNA molecule each second. The speed is not as great in plants and animals, but errors still occur when mispairing between A and T or between C and G nucleotides occurs. DNA breaks are also common. These kinds of genetic change can be classified as "genetic damage." Some mutations are spontaneous, caused by changes that occur in the process of normal cell biochemistry. Other damage that can be traced to environmental factors changes bases, causes mispairing, or breaks DNA strands. Fortunately, almost all of this initial genetic damage is repaired by enzymes that recognize and correct errors in nucleotide pairing or DNA strand breaks. It is the unrepaired genetic damage that appears as new mutations. One of the first geneticists to design experiments to measure mutation rate was Hermann Müller, who received the Nobel Prize for his work on mutagenesis, including the discovery that X rays cause mutations.

Induction of Mutations by X Rays

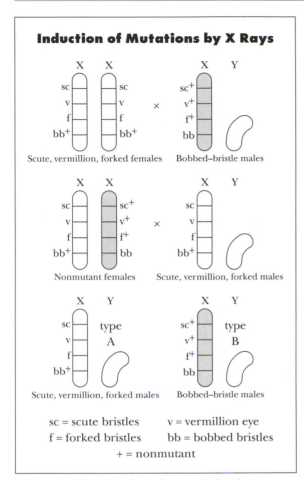

sc = scute bristles   v = vermillion eye

f = forked bristles   bb = bobbed bristles

+ = nonmutant

If, for a given fly and its descendants, an induced or spontaneous lethal mutation occurs in the paternal X chromosome (shaded), no third-generation males of type B will result. If a spontaneous lethal mutation occurs in an original maternal X chromosome, then no third-generation males of type A will result.

The experiments by Müller provide a useful example of the kind of experimental design that can be used to measure mutation rates. Müller focused on new mutations (lethals) on the X chromosome of *Drosophila* that could cause the carrier to die. Since a male has only one X chromosome, a lethal mutation on that chromosome causes death. A living male *Drosophila* must, therefore, have no lethal mutations on his X chromosome. If a male *Drosophila* is treated with an agent such as X irradiation or certain chemicals, new lethal mutations can be detected when he is mated with special genetic strains of females. His X chromosomes are

eventually passed on to male descendants. If a new lethal mutation exists on a specific X chromosome, all males that inherit that chromosome copy will die during development. Spontaneous mutation rates measured by this technique average about 1×10^{-5} for each gene. In other words, there is a probability of about 1 in 100,000 that a mutation will occur in a particular gene each generation. This is a very low probability for a specific gene, but when it is multiplied for all of the genes in an animal or plant, it is likely that a new mutation has occurred somewhere on the chromosomes of an organism each generation.

Spontaneous mutation rates vary to some extent from one gene to another and from one organism to another, but one major source of variation in mutation rate comes from external agents that act on the DNA to increase damage or inhibit repair. One of the most widely used techniques for measuring the mutagenic activity of a chemical was developed in the 1970's by Bruce Ames. The Ames test uses bacteria that have a mutation that makes them unable to produce the amino acid histidine. These bacteria cannot survive in culture unless they are given histidine in the medium. To test whether a chemical increases the mutation rate, it is mixed with a sample of these bacteria, and they are placed on a medium without histidine. Any colonies that survive represent bacteria in which a new mutation has occurred to reverse the original defect (a back-mutation). Since many chemicals that cause mutations also cause cancer, this quick and inexpensive test is now used worldwide to screen potential carcinogenic, or cancer-causing, agents.

Mutation rates in mice are measured by use of the specific-locus test. In this test, wild-type male mice are mated with females that are homozygous for up to seven visible, recessive mutations that cause changes in coat color, eye color, and shape of the ear. If no mutations occur in any of the seven genes in the germ cells of the male, the male offspring will all be wild type in appearance. However, a new mutation in any of the seven genes will yield a progeny with a mutant phenotype (for example, a new coat color). The same cross can also be used to identify new mutations in females. Since mice are mammals, they are a close model system to humans. Thus, results from mutation studies in mice have helped identify agents that are likely to be mutagenic in humans.

The Use of Mutations to Study Development

Mutations offer geneticists a powerful tool to analyze development. By understanding the way development is changed by a mutation, one can determine the role the normal gene plays. Although most people tend to think of mutations as causing some easily visible change in the appearance of a plant or animal (such as wrinkled pea seeds or white mouse fur), most mutations are actually lethal when present in two copies (homozygous). These lethal mutations affect some critical aspect of cell structure or other fundamental aspect of development or function. Genes turn on and off at specific times during development, and by studying the abnormalities that begin to show when a lethal

mutation carrier dies, a geneticist can piece together a picture of the timing and role of important gene functions.

Another useful insight comes from mutations with effects that vary. For example, many mutations have phenotypic effects that depend on the conditions, such as temperature, in which the individual develops. An interesting example of such temperature sensitivity is the fur color of Siamese cats. The biochemical pathway for pigmentation is active in cool temperatures but is inactivated at warmer body temperature. For this reason, a Siamese cat will only be pigmented in the cooler parts such as the tips of the ears and tail. Gene interactions like this allow geneticists to study the conditions under which the protein coded by a mutant gene works.

It would be a mistake, however, to think that all mutations have large phenotypic effects. Many complex traits are produced by many genes working together and are affected by

The coats of Siamese cats are darker at their extremities as the result of a mutation that is affected by body temperature. (AP/Wide World Photos)

environmental variables such as temperature. These are called quantitative traits because they are measured on some kind of scale, such as size, number, or intensity. The mutations that affect quantitative traits are not different, except perhaps in the magnitude of their individual effects, from other kinds of gene mutation. Mutations in quantitative traits are a major source of heritable variation on which natural and artificial selection can act to change a phenotype.

Impact and Applications

It will probably never be possible to eliminate all mutation events because many mutations are caused by small errors in normal DNA duplication when cells divide. Learning how mutations affect cell division and cell function can help one to understand processes such as cancer and birth defects that can often be traced to genetic change. Some explanations of processes such as aging have focused on mutation in somatic cells. Mutation is also the source of genetic variation in natural populations, and the long-term survival of a species depends on its ability to draw on this variation to adapt to new environmental conditions.

Two aspects of mutagenesis will continue to grow in importance. First, environmental and human-made mutagens will continue to be a source of concern as technological advances occur. Many scientists are working to monitor and correct potential mutagenic hazards. Second, geneticists are beginning to use molecular tools, such as transposable elements and the techniques of genetic engineering, to produce preplanned genetic changes. Directed mutagenesis of DNA offers a way to correct preexisting genetic defects or alter phenotypes in planned ways. Mutation is, therefore, both a source of problems and a source of promise.

—*James N. Thompson, Jr.*
—*R. C. Woodruff*

See also: Biochemical Mutations; Cancer; Cell Cycle, The; Central Dogma of Molecular Biology; Chemical Mutagens; Chromosome Mutation; Classical Transmission Genetics; Complementation Testing; Congenital Disorders; Consanguinity and Genetic Disease; Cystic Fibrosis; Extrachromosomal Inheritance; Genetic Load; Hereditary Diseases; Huntington's Disease; Inborn Errors of Metabolism; Mitochondrial Genes; Molecular Genetics; Oncogenes; Phenylketonuria (PKU); Transposable Elements; Tumor-Suppressor Genes.

Further Reading

Braman, Jeff, ed. *In Vitro Mutagenesis Protocols.* 2d ed. Totowa, N.J.: Humana Press, 2002. Presents advanced mutagenesis techniques. Illustrated.

Friedberg, Errol C., et al., eds. *DNA Repair and Mutagenesis.* Washington, D.C.: ASM Press, 1995. An accessible, comprehensive look at how living cells respond to genomic injury and alterations, covering mutagenesis and other forms of DNA damage tolerance. Includes illustrations and more than four thousand references.

Radman, Miroslav, and Robert Wagner. "The High Fidelity of DNA Duplication." *Scientific American* 259 (August, 1988). Discusses the high degree of accuracy in the process of DNA duplication.

Smith, Paul J., and Christopher J. Jones, eds. *DNA Recombination and Repair.* New York: Oxford University Press, 2000. Addresses the integrity of genomes for good health and how DNA repair and recombination relates to illness, especially cancer. Illustrated.

Sobti, R. C., G. Obe, and P. Quillardet, eds. *Trends in Environmental Mutagenesis.* New Delhi: Tausco, 1999. Discusses genetic toxicology, environmental mutagenic microbes, asbestos genotoxicity, and more. Illustrations, bibliography.

Natural Selection

Fields of study: Evolutionary biology; Population genetics

Significance: *Natural selection is the mechanism proposed by Charles Darwin to account for biological evolutionary change. Using examples of artificial selection as analogies, he suggested that any heritable traits that allow an advantage in survival or reproduction to an individual organism would be "naturally selected" and increase in frequency until the entire population had the trait. Selection, along with other evolutionary forces, influences the changes in genetic and morphological variation that characterize biological evolution.*

Key terms

ADAPTATION: the evolution of a trait by natural selection, or a trait that has evolved as a result of natural selection

ARTIFICIAL SELECTION: selective breeding of desirable traits, typically in domesticated organisms

FITNESS: an individual's potential for natural selection as measured by the number of offspring of that individual relative to those of others

GROUP SELECTION: selection in which characteristics of a group not attributable to the individuals making up the group are favored

Natural Selection and Evolution

In 1859, English naturalist Charles Darwin published *On the Origin of Species by Means of Natural Selection*, in which he made two significant contributions to the field of biology: First, he proposed that biological evolution can occur by "descent with modification," with a succession of minor inherited changes in a lineage leading to significant change over many generations; and second, he proposed natural selection as the primary mechanism for such change. (This was also proposed independently by Alfred R. Wallace and was presented with Darwin in the form of a joint research paper some years earlier.) Darwin reasoned that if an individual organism carried traits that allowed it to have some advantage in survival or reproduction, then those traits would be carried by its offspring, which would be better represented in future generations. In other words, the individuals carrying those traits would be "naturally selected" because of the advantages of the traits. For example, if a small mammal happened to have a color pattern that made it more difficult for predators to see, it would have a better chance of surviving and reproducing. The mammal's offspring would share the color pattern and the advantage over differently patterned members of the same species. Over many generations, the proportion of individuals with the selected pattern would increase until it was present in every member of the species, and the species would be said to have evolved the color pattern trait.

Natural selection is commonly defined as "survival of the fittest," although this is often misinterpreted to mean that individuals who are somehow better than others will survive while the others will not. As long as the traits convey some advantage in reproduction so that the individual's offspring are better represented in the next generation, then natural selection is occurring. The advantage may be a better ability to survive, or it may be something else, such as the ability to produce more offspring.

For natural selection to lead to evolutionary change, the traits under selection must be heritable, and there must be some forms of the traits that have advantages over other forms (variation). If the trait is not inherited by offspring, it cannot persist and become more common in later generations. Darwin recognized this, even though in his time the mechanisms of heredity and the sources of new genetic variation were not understood. After the rediscovery of Gregor Mendel's principles of genetics in the early years of the twentieth century, there was not an immediate integration of genetics into evolutionary biology. In fact, it was suggested that genetic mutation might be the major mechanism of evolution. This belief, known as Mendelism, was at odds with Darwinism, in which natural selection was the primary force of evolution. However, with the "modern synthesis" of genetics and evolutionary theory in the 1940's and 1950's, Mendelian genetics was shown to be entirely compatible with Darwinian evolution. With this recognition, the role of mutation in evolution was relegated to the source of variation in traits upon which natural selection can act.

The potential for natural selection of an organism is measured by its "fitness." In practice, the fitness of an individual is some measure of the representation of its own offspring in the next generation, often relative to other individuals. If a trait has evolved as a result of natural selection, it is said to be an "adaptation." The term "adaptation" can also refer to the process of natural selection driving the evolution of such a trait. There are several evolutionary forces in addition to selection (for example, genetic drift, migration, and mutation) that can influence the evolution of a trait, though the process is called adaptation as long as selection is involved.

Population Genetics and Natural Selection

Population geneticists explore the actual and theoretical changes in the genetic composition of natural or hypothetical populations. Not surprisingly, a large part of the theoretical and empirical work in the field has concentrated on the action of natural selection on genetic variation in a population. Ronald A. Fisher and J. B. S. Haldane were the primary architects of selection theory beginning in the 1930's, and Theodosius Dobzhansky was a pioneer in the detection of natural selection acting on genetic variants in populations of *Drosophila melanogaster* (fruit flies).

The most basic mathematical model of genes in a population led to the Hardy-Weinberg law, which predicts that there would be no change in the genetic composition of a population in the absence of any evolutionary forces such as natural selection. However, models that include selection show that it can have specific influences on a population's genetic variation. In such models, the fitness of an organism's genotype is represented by a fitness coefficient (or the related selection coefficient), in which the genotype with the highest fitness is assigned a value of 1, and the remaining genotypes are assigned values relative to the highest fitness. A fitness coefficient of 0 represents a lethal genotype (or, equivalently, one that is incapable of reproduction).

The simplest models of selection include the assumption that a genotype's fitness does not change with time or context and demonstrate three basic types of selection, defined by how selection acts on a distribution of varying forms of a trait (where extreme forms are rare and average forms are common). These three types are directional selection (in which one extreme is favored), disruptive selection (in which both extremes are favored), and stabilizing selection (in which average forms are favored). The first two types (with the first probably being the most common) can lead to substantial genetic change and thus evolution, though in the process genetic variation is depleted. The third type maintains variation but does not result in much genetic change. These results create a problem: Natural populations generally have substantial genetic variation, but most selection is expected to deplete it. The problem has led population geneticists to explore the role of other forces working in place of, or in conjunction with, natural selection and to study more complex models of selection. Examples include models that allow a genotype to be more or less fit if it is more common (frequency-dependent selection) or that allow many genes to interact in determining a genotype's fitness (multilocus selection). Despite the role of other forces, selection is considered an important and perhaps complex mechanism of genetic change.

Detecting and Measuring Fitness

Although a great amount of theoretical work on the effects of selection has been done, it is also important to relate theoretical results to actual populations. Accordingly, there has been a substantial amount of research on natural and laboratory populations to measure the presence and strength of natural selection. In practice, selection must be fairly strong for it to be distinguished from the small random effects that are inherent in natural processes.

Ideally, a researcher would measure the total selection on organisms over their entire life cycles, but in some cases this may be too difficult or time-consuming. Also, a researcher may be interested in discovering what specific parts of the life cycle selection influences. For these reasons, many workers choose to measure components of fitness by breaking down the life cycle into phases and looking for fitness differences among individuals at some or all of them. These components can differ with different species but often include fertility selection (differences in the number of gametes produced), fecundity selection (differences in the number of offspring produced), viability selection (differences in the ability to survive to reproductive age), and mating success (differences in the ability to successfully reproduce). It is often found in such studies that total lifetime fitness is caused primarily by fitness in one of these components, but not all. In fact, it may be that genotypes can have a disadvantage in one component but still be selected with a higher overall fitness because of greater advantages in other components.

There are several empirical methods for detection and measurement of fitness. One relatively simple way is to observe changes in gene or genotype frequencies in a population and fit the data on the rate of change to a model of gene-frequency change under selection to yield an estimate of the fitness of the gene or genotype. The estimate is more accurate if the rate of mutation of the genes in question is taken into account. In the famous example of "industrial melanism," it was observed that melanic (dark-colored) individuals of the peppered moth *Biston betularia* became more common in Great Britain in the late nineteenth century, corresponding to the increase in pollution that came with the Industrial Revolution. It was suggested that the melanic moths were favored over the lighter moths because they were camouflaged on tree trunks where soot had killed the lichen and were therefore less conspicuous to bird predators. Although it is now known that the genetics of melanism are more complex, early experiments suggested that there was a single locus with a dominant melanic allele and a recessive light allele; the data from one hundred years of moth samples were used to infer that light moths have two-thirds the survival ability of melanic moths. Later studies also showed that peppered moths do not rest on tree trunks, calling into question

the role of bird predation in the selection process. Nevertheless, selection of some sort is still considered the best explanation for the changes observed in peppered moth populations, even though the selective factor responsible is not known.

Later, a second method of fitness measurement was applied to the peppered moth using a mark-recapture experiment. In such an experiment, known quantities of marked genotypes are released into nature and collected again some time later. The change in the proportion of genotypes in the recaptured sample provides a way to estimate their relative fitnesses. In practice, this method has a number of difficulties associated with making accurate and complete collections of organisms in nature, but the fitness measure of melanic moths by this method was in general agreement with that of the first method. A third method of measuring fitness is to measure deviations from the genotype proportions expected if a population is in Hardy-Weinberg equilibrium. This method can be very unreliable if deviations are the result of something other than selection.

Units of Selection

Darwin envisioned evolution by selection on individual organisms, but he also considered the possibility that there could be forms of selection that would not favor the survival of the individual. He noted that in many sexual species, one sex often has traits that are seemingly disadvantageous but may provide some advantage in attracting or competing for mates. For instance, peacocks have a large, elaborately decorated tail that is energetically costly to grow and maintain and might be a burden when fleeing from predators. However, it seems to be necessary to attract and secure a mate. Darwin, and later Fisher, described how such a trait could evolve by sexual selection if the female evolves a preference for it, even if natural selection would tend to eliminate it.

Other researchers have suggested that in some cases selection may act on biological units other than the individual. Richard Dawkins's *The Selfish Gene* (1976) popularized the idea that selection may be acting directly on genes and only indirectly on the organisms that carry them. This distinction is perhaps only a philosophical one, but there are specific cases in which genes are favored over the organism, such as the "segregation distorter" allele in *Drosophila* that is overrepresented in offspring of heterozygotes but lethal in homozygous conditions.

The theory of kin selection was developed to explain the evolution of altruistic behavior such as self-sacrifice. In some bird species, for example, an individual will issue a warning call against predators and subsequently be targeted by the predator. Such behavior, while bad for the individual, can be favored if those benefiting from it are close relatives. While the individual may perish, relatives that carry the genes for the behavior survive and altruism can evolve. Kin selection is a specific type of group selection in which selection favors attributes of a group rather than an individual. It is not clear whether group selection is common in evolution or limited to altruistic behavior.

Impact and Applications

The development of theories of selection and the experimental investigation of selection have always been intertwined with the field of evolutionary biology and have led to a better understanding of the history of biological change in nature. More recently, there have been medical applications of this knowledge, particularly in epidemiology. The specific mode of action of a disease organism or other parasite is shaped by the selection pressures of the host it infects. Selection theory can aid in the understanding of cycles of diseases and the response of parasite populations to antibiotic or vaccination programs used to combat them.

Although the idea of natural selection as a mechanism of biological change was suggested in the nineteenth century, artificial selection in the form of domestication of plants and animals has been practiced by humans for many thousands of years. Early plant and animal breeders recognized that there was variation in many traits, with some variations being more desirable than others. Without a formal understanding of genetics, they found that by choosing and breeding individuals with the desired traits, they could gradually improve the lineage. Darwin used numerous examples of artificial selection to illustrate biological change and argued that natural selection, while not necessarily as strong or directed, would influence change in much the same way. It is important to make a clear distinction between the two processes: Breeders have clear, long-term goals in mind in their breeding programs, but there are no such goals in nature. There is only the immediate advantage of the trait to the continuation of the lineage. The application of selection theory to more recent breeding programs has benefited human populations in the form of new and better food supplies.

—*Stephen T. Kilpatrick*

See also: Altruism; Ancient DNA; Artificial Selection; Classical Transmission Genetics; Evolutionary Biology; Genetic Code; Genetic Code, Cracking of; Genetics, Historical Development of; Hardy-Weinberg Law; Human Genetics; Lamarckianism; Mendelian Genetics; Molecular Clock Hypothesis; Mutation and Mutagenesis; Population Genetics; Punctuated Equilibrium; Repetitive DNA; RNA World; Sociobiology; Speciation; Transposable Elements.

Further Reading

Dawkins, Richard. *Extended Phenotype: The Long Reach of the Gene.* Rev. 2d ed. Afterword by Daniel Dennett. New York: Oxford University Press, 1999. Argues that the selfish (individual) gene extends to making artifacts, such as birds' nests, and to manipulative, persuasive behavior for survival. Bibliography, index.

_____. *The Selfish Gene.* New York: Oxford University Press, 1989. Argues that the world of the selfish gene revolves around competition and exploitation and yet acts of apparent altruism do exist in nature. A popular account of sociobiological theories that revitalized Darwin's natural selection theory.

Dover, Gabriel A. *Dear Mr. Darwin: Letters on the Evolution of Life and Human Nature.* Berkeley: University of California Press, 2000. A fictional tale of correspondence with Charles Darwin. Illustrated.

Fisher, Ronald Aylmer. *The Genetical Theory of*

Natural Selection: A Complete Variorum Edition. Edited with a foreword and notes by J. H. Bennett. New York: Oxford University Press, 1999. Facsimile of the 1930 edition. Illustrated.

Gould, Stephen Jay. *The Structure of Evolutionary Theory.* Cambridge, Mass.: Harvard University Press, 2002. Gould considers this book on natural selection his major work, a collection of twenty-five years of study exploring the history and future of evolutionary theory. Includes a chapter on punctuated equilibrium. Illustrations, bibliography, and index.

Keller, Laurent, ed. *Levels of Selection in Evolution.* Princeton, N.J.: Princeton University Press, 1999. Addresses the question of what keeps competition between various levels of natural selection from destroying the common interests to be gained from cooperation between members of a species. Illustrated.

Levy, Charles K. *Evolutionary Wars: A Three-Billion-Year Arms Race: The Battle of Species on Land, at Sea, and in the Air.* Illustrations by Trudy Nicholson. New York: W. H. Freeman, 1999. Discusses the often violent nature of natural selection and adaptation, including the survival skills and mechanisms of dragonflies, frogs, viruses, poison-filled jellyfish, and beetles, and the tongues of woodpeckers and anteaters. Ninety-four illustrations, index.

Lynch, John M., ed. *Darwin's Theory of Natural Selection: British Responses, 1859-1871.* 4 vols. Bristol, England: Thoemmes Press, 2002. A collection of rare, primary sources by scientists, theologians, and others on Darwin's theory, including the 1867 critical review by Fleeming Jenkin that Darwin thought best summarized his work on natural selection. Bibliography, index.

Magurran, Anne E., and Robert M. May, eds. *Evolution of Biological Diversity: From Population Differentiation to Speciation.* New York: Oxford University Press, 1999. Discusses species variation as theorized by proponents of natural selection, ecological, and behavioral models. Looks at fossil records for empirical data. Illustrations, bibliography, index.

Michod, Richard E. *Darwinian Dynamics: Evolutionary Transitions in Fitness and Individuality.* Princeton, N.J.: Princeton University Press, 1999. Argues that cooperation instead of competition and violence accounts for species survival and fitness, and that evolution occurs through genetic change instead of the more common theory of endurance. Illustrations, bibliography, index.

Ryan, Frank. *Darwin's Blind Spot: Evolution Beyond Natural Selection.* Boston: Houghton Mifflin, 2002. Argues for a symbiotic instead of the most widely accepted competitive and survival-based theory of evolution. Bibliography, index.

Shermer, Michael. *In Darwin's Shadow: The Life and Science of Alfred Russel Wallace: A Biographical Study on the Psychology of History.* New York: Oxford University Press, 2002. The unsung contemporary of Darwin deserves equal credit for developing early evolutionary theory. Illustrated.

Williams, George C. *Adaptation and Natural Selection: A Critique of Some Current Evolutionary Thought.* 1966. Reprint. Princeton, N.J.: Princeton University Press, 1996. A good introduction to adaptation and units of selection. New preface. Bibliography, index.

Web Site of Interest

Writings of Charles Darwin. http://pages.britishlibrary.net/charles.darwin. Resource for Darwin's writings, a bibliography, and biographical material. Includes online version of Darwin's *On the Origin of Species by Means of Natural Selection* (1859).

Neural Tube Defects

Field of study: Diseases and syndromes
Significance: *Neural tube defects are a category of birth defects that usually result from the failure of the neural tube to close properly during gestational development. Many neural tube defects can be prevented through folic acid supplementation and avoidance of other risk factors. However, because the neural tube closes during the first gestational*

month, preventive measures must be instituted prior to pregnancy. Therefore, prevention of neural tube defects depends on the planning or expectation of pregnancies while initiating positive lifestyle changes.

Key terms

ANENCEPHALUS: a neural tube defect characterized by the failure of the cerebral hemispheres of the brain and the cranium to develop normally

ETIOLOGY: the cause or causes of a disease or disorder

MULTIFACTORIAL: characterized by a complex interaction of genetic and environmental factors

NEURAL TUBE: the embryonic precursor to the spinal cord and brain that normally closes at small openings, or neuropores, by the twenty-eighth day of gestation

SPINA BIFIDA: a neural tube defect that usually results from the failure of the posterior neuropore to close properly during gestation

Formation of the Neural Tube

Neural tube defects represent congenital defects that have long been prevalent in human populations. Documented cases of anencephalus and spina bifida have been found among the skeletal remains of the ancient Egyptians and prehistoric Native Americans. In the contemporary world, spina bifida remains one of the most common birth defects. Yet despite their prevalence and antiquity, some questions concerning the causes of neural tube defects remain unanswered.

Neural tube defects result from a disruption in the formation or closure of the neural tube, which, during embryonic development, differentiates into the brain and spinal cord. The neural tube develops first out of the neural plate. The borders of the neural plate are folded, forming the neural groove. The neural groove becomes progressively deeper, placing the two folds in opposition. Final development of the neural tube occurs as the dorsal folds fuse along the midline. Closure of this structure begins around the third gestational week,

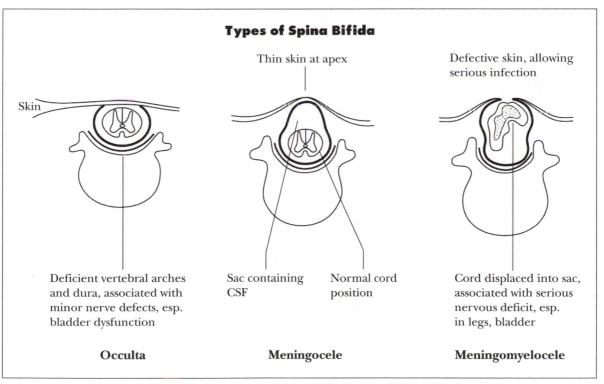

Types of Spina Bifida

Thin skin at apex

Defective skin, allowing serious infection

Skin

Deficient vertebral arches and dura, associated with minor nerve defects, esp. bladder dysfunction

Sac containing CSF

Normal cord position

Cord displaced into sac, associated with serious nervous deficit, esp. in legs, bladder

Occulta

Meningocele

Meningomyelocele

Spina bifida is among the most common neural tube disorders. (Hans & Cassidy, Inc.)

beginning at its midportion and ending at the anterior and posterior neuropores around the twenty-fifth and twenty-seventh gestational days, respectively.

Classification of Neural Tube Defects

Disruption in the formation and closure of the posterior neuropores is associated with spina bifida. Spina bifida occulta is generally unaccompanied by protrusion of the spinal cord or its coverings through the open, unfused arches of the vertebrae. On the other hand, much more severe conditions, categorized as spina bifida cystica, result in the herniation of neural tissues and the formation of cystic swelling. One form of spina bifida cystica, meningomyelocele, is marked by the protrusion of both the meninges and the spinal cord through the unfused vertebral arch. In the most severe cases, portions of the spinal cord and nerve roots are encased in the walls of the sac, damaging and hindering normal neurological functioning and development. In such instances, the severity of the neurological dysfunction depends on the location of the lesion along the vertebral column, as nerves below the defect are adversely affected. Meningocele, a more moderate manifestation of spina bifida cystica, is encountered four to five times less frequently than meningomyelocele. Unlike the latter condition, the cystic sacs of meningoceles are made up solely of meninges and spinal fluid. This factor, coupled with the lack of involvement of the spinal cord, generally affords a more favorable prognosis, although some sensory and motor deficits may persist after surgery.

Anencephalus, which results from the disruption of the anterior neuropore, is the most devastating and severe of all the neural tube defects. Infants born with this birth defect are lacking significant areas of their brain and skull. The region normally occupied by the cerebral hemispheres consists of a formless mass of highly vascular connective tissue, while most of the bones of the skull are simply absent. Many anencephalic infants are stillborn; most die soon after birth.

Encephalocele, like anencephalus, is believed to result from defective closure of the anterior neuropore. In these conditions, a saclike protrusion of neural tissue occurs through an opening along the midline of the skull. The prognosis and outcome of infants born with encephalocele depends upon the size of the lesion and the extent to which neural tissues are involved.

Prevalence-at-Birth Rates and Causes

Prevalence-at-birth rates of neural tube defects show substantial geographic and temporal variation. Historically, some of the highest prevalence-at-birth rates have been documented in the British Isles and range from as high as 4.5 in 1,000 births in Belfast, Ireland, to as low as 1.5 in 1,000 births in London, England. In the United States, the highest rates of neural tube defects have historically occurred in northeastern states. Rates of neural tube defects are declining in most areas of the world, although regional outbreaks, marked by higher birth prevalence rates, have been reported and are generally unexplained. Typically, rates in the contemporary United States average around 1 to 2 in 1,000 births, and the risk of having an infant with a neural tube defect increases by about 2 percent if a couple has previously had a child with such a defect.

Among the most important risk factors are those relating to the diet and health status of prospective mothers. Also, there are indications that excessive elevation of a woman's body temperature during early pregnancy, through hot baths or recreational hot tubs, may increase her chances of having an infant with a neural tube defect. A number of studies have suggested that women who give birth to infants with neural tube defects have lower health status and poorer diets than other women. Inadequate levels of folate appear to place women at greater risk of having an infant with a neural tube defect. Doctors now recommend that women planning a pregnancy supplement their diets with folate, although any woman planning to become pregnant should first consult her doctor before taking any supplement. Tests for alpha-fetoprotein in the mother's blood during the prenatal period can help detect the presence of a neural tube defect in the developing fetus.

—*Mary K. Sandford*

See also: Amniocentesis and Chorionic Villus Sampling; Congenital Defects; Developmental Genetics; Prenatal Diagnosis.

Further Reading

Bock, Gregory, and Joan Marsh, eds. *Neural Tube Defects.* New York: Wiley, 1994. Discusses prenatal screening, before- and after-birth treatment, and the genetic and environmental causes of congenital malformations. Illustrations, bibliography, index.

Evans, Mark I., ed. *Metabolic and Genetic Screening.* Philadelphia: W. B. Saunders, 2001. Covers principles of screening, neural tube defects, prenatal genetic screening in the Ashkenazi Jewish population, cystic fibrosis, and identifying and managing hereditary risk of breast and ovarian cancer.

Massaro, Edward J., and John M. Rogers, eds. *Folate and Human Development.* Totowa, N.J.: Humana Press, 2002. Focuses on how folate could help prevent human developmental disorders, including neural tube defects. Illustrations, bibliography, index.

Web Sites of Interest

Medline Plus. http://www.nlm.nih.gov/medlineplus/neuraltubedefects.html. Medline, sponsored by the National Institutes of Health, is one of the first stops for any medical question; this page provides descriptions and links to clinical trials and other resources for neural tube defects.

Spina Bifida Association of America. http://www.sbaa.org. Offers information, fact sheets, testimonials, a clinic directory, copious links to other resources, and more.

Noncoding RNA Molecules

Field of study: Molecular genetics
Significance: *Although less familiar than mRNA, tRNA, and rRNA, noncoding RNA molecules (ncRNAs) play many roles—including some that have not yet been elucidated—in normal cellular functions. The existence of ncRNAs has been known since the 1960's, but it was not until the last decade of the twentieth century that their sig-nificance and functions began to be understood. The functions of the many ncRNAs so far discovered include roles in DNA replication, post-transcriptional control of gene expression, processing of other RNAs, and mRNA stability.*

Key terms

cDNA LIBRARY: a collection of clones produced from all the RNA molecules in the cells of a particular organism, often from a single tissue

CLONE: a culture of bacteria, usually *Escherichia coli*, whose cells contain a recombinant plasmid

CODON: a three-letter nucleotide sequence in RNA or DNA that codes for a specific amino acid; a gene is composed of a long string of codons

INTRON: an intervening sequence in a eukaryotic gene (generally there are several to many per gene) which must be removed when it is transcribed into messenger RNA (mRNA); introns are assumed to have no function and therefore mutations in them are often considered neutral

SPLICEOSOME: a complex assemblage of proteins and RNA in the nucleus of cells that cuts out introns and splices the exons of a maturing mRNA

Definition

Noncoding RNAs (ncRNAs) include any RNA that is not messenger RNA (mRNA), ribosomal RNA (rRNA), or transfer RNA (tRNA). The discovery of the first ncRNAs in the 1960's occurred because they were expressed in such high numbers. At the time, RNA was considered to function only as a means to express a gene, all three of the main types of RNA being intimately involved in this process. Many of the ncRNAs discovered over the next twenty years were also discovered fortuitously, before any speculation about their possible functions was even considered. Once transcription and processing of mRNAs was elucidated, many of the ncRNAs were considered leftover fragments representing the introns that had been cut out of pre-mRNAs. At the same time it was discovered that some of the ncRNAs were involved in the process of intron removal and exon splicing. Systematic searches for ncRNAs did not begin until the later 1990's and, once undertaken, revealed a veritable universe of ncRNAs, ranging from very short sequences of less than 100 nucleotides to some around 100,000 nucleotides, and possibly more. For a system considered so well understood, the entry of so many new players has added a whole new layer of complexity to the study of genetics.

Types of ncRNA and Their Occurrence

Researchers have now identified ncRNAs in essentially all organisms, from bacteria to humans. In bacteria they tend to be smaller and in most cases are called small RNA (sRNA). Although the most common name for noncoding RNAs in other organisms is ncRNA, they also have gone by the name "small non-messenger RNA" (snmRNA). After these general names, there is a collection of names for ncRNAs that have particular characteristics or functions, and the list of names will probably grow as new ncRNAs are discovered. Some newly discovered ncRNAs cannot be assigned a function.

ncRNAs Involved in RNA Processing

Because almost all eukaryotic genes contain intervening sequences, called introns, that in-

Types of ncRNA

Type of ncRNA	Abbreviation
guide RNA	gRN
heterogeneous nuclear RNA	hnRNA
micro-RNA	miRNA
small cytoplasmic RNA	scRN
small interfering RNA	siRNA
small non-messenger RNA	snmRNA
small nuclear RNA	snRNA
small nucleolar RNA	snoRNA
small temporal RNA	stRNA
transfer messenger RNA	tmRNA

—Bryan Ness

terrupt the coding sequence of the gene, when an RNA is first transcribed it cannot be translated without being processed. Processing involves removal of the introns and the splicing of the remaining fragments, the exons, which contain the coding sequence of the gene. The cellular "machine" that does this job is the spliceosome. It is a complex assemblage of proteins and small RNAs. The proteins and RNAs are grouped together into several particles called small nuclear ribonucleoproteins (snRNPs, pronounced "snurps" by geneticists). The RNA component of snRNPs are small nuclear RNAs (snRNAs), the best known being U1 snRNA. Several different snRNAs are now known, and they are components of the several snRNPs that come together to make a functional spliceosome.

Subsequent discoveries revealed that snRNPs, and thus snRNAs, were involved in other types of RNA processing. Some are involved in polyadenylation, the addition of adenine nucleotides to the 3′ end of mRNAs to make what is called a poly-A tail. Histone protein mRNAs are known to lack poly-A tails, but in *Xenopus* (the African clawed frog), snRNPs are still involved in properly finishing the 3′ end. A final role for some is maturation of rRNA transcripts, whose spacer RNA sequences must be removed. All of these functions verge on being enzyme-like.

A complex related to snRNPs was first found in bacteria and has now been found in all groups of organisms. It contains proteins and RNA and is called ribonuclease P (RNase P); it is involved in the processing of tRNA and some rRNAs. Experiments have shown that the RNA component can catalyze the required reactions, even without the protein component, making it the first clear-cut "ribozyme," an RNA with catalytic properties. Several types of ncRNA are now known to act as ribozymes, and this ability prompted the evolutionary community to propose that early "life" was RNA-based rather than protein and DNA-based.

ncRNAs Involved in RNA Modification

RNA modification by small nucleolar RNAs (snoRNAs) has been best studied in *Saccharomyces cerevisiae* (yeast). Mature rRNAs must have some of their ribose sugars methylated, and although their exact role in the process has not yet been completely defined, snoRNAs are involved. They bind to rRNAs in small regions where they have complementary base sequences and somehow direct methylation. Other snoRNAs are involved in pseudouridylation (that is, conversion of some of the uracil nucleotides in rRNA to pseudouracil, a modified nucleotide) of rRNA. The enzyme that actually performs the pseudouridylation is not known. Many eukaryotes have snoRNAs, and recently snoRNA homologs (a homolog is a molecule that is similar to another) have been found in *Archaea*, but not yet in *Bacteria*.

Not as well known are guide RNAs (gRNAs), discovered in some protists. They also modify rRNA, by guiding the insertion or deletion of uracil nucleotides. The details of the process are not well understood, but the mechanism involves complementary base pairing between the rRNA and a gRNA, much like that seen with snoRNAs. It is possible that, as more studies are undertaken, gRNAs will be found in other types of organisms.

ncRNAs Affecting mRNA Stability and Translation

Another type of ncRNA that has been known for some time is small interfering RNA (siRNA). A type of antisense RNA, siRNAs have base sequences complementary to the coding, or "sense," region of an mRNA. By binding to an mRNA, a siRNA is able to block translation, and there is evidence that it also tags the mRNA for degradation. This type of genetic control is often called post-transcriptional gene silencing, or RNAi (RNA interference). Another ncRNA, micro-RNA (miRNA), also seems to target specific mRNAs for degradation. Both of these types represent very small RNA molecules of generally fewer than thirty nucleotides.

Apparently targeting specific mRNAs for degradation in bacteria are sRNAs, the mechanism being somewhat uncertain. Another function of sRNAs in bacteria is activation of certain mRNAs by preventing formation of an inhibitory structure in the mRNA. Another ncRNA, simply called OxyS RNA, represses translation by interfering with ribosome binding.

A final ncRNA, originally believed to be found only in the nematode *Caenorhabditis elegans*, is small temporal RNA (stRNA). The stRNAs block translation of specific mRNAs after translation has begun, but apparently they do not tag the mRNA for degradation. They are now believed to represent a subset of miRNA with slightly different properties. Their size is typically between twenty-one and twenty-five nucleotides, and screening of a variety of other organisms suggests that they may be more widespread than first assumed.

Other Specialized ncRNAs

A variety of other ncRNAs carry out more specialized functions, some just beginning to be understood. Gene silencing is a very important component of normal development. As cells become differentiated and specialized, they must express certain genes, and the remaining genes must be silenced. One form of silencing is called imprinting, whereby certain alleles from an allele pair are silenced, often those received from only one sex. A large ncRNA (a little longer than 100,000 nucleotides) called *Air* is responsible for silencing the paternal alleles in a small autosomal gene cluster. How it does this is still being studied.

In human females, one of the X chromosomes (females have two) must be inactivated so the genes on it will not be expressed. This inactivation, called Lyonization after the discoverer of the phenomenon, Mary Lyon, occurs during development on a random basis in each cell, so that the X chromosome subjected to deactivation is randomly determined. An ncRNA called *Xist* plays a central part in this process. It is a large RNA of 16,500 nucleotides and is transcribed from genes on both X chromosomes. It is inherently unstable but somehow becomes stable and binds all over one of the two X chromosomes. The X chromosome that gets coated with *Xist* is then inactivated, and the only gene it transcribes thereafter is the *Xist* gene. Transcription of *Xist* ceases on the active X chromosome.

A type of ncRNA called transfer messenger RNA (tmRNA) is involved in resuming translation at ribosomes that have stalled. When a stalled ribosome is encountered, a tmRNA first acts as a tRNA charged with the amino acid alanine. The stalled polypeptide is transferred to the alanine on the tmRNA. Then translation continues, but now the tmRNA acts as the mRNA, instead of the mRNA the ribosome was initially translating. A termination codon is soon reached and the amino acids that were added based on the tmRNA code act as a tag for enzymes in the cytoplasm to break it down. This allows those ribosomes that would normally remain tied up with an mRNA they cannot complete translating to be recycled for translating another mRNA.

The Future of ncRNA Research

Most of the ncRNAs described above were unknown until the 1980's, and some of them were only discovered in the 1990's. What appeared to be a relatively simple picture of genetic control in cells has now gained many, previously hidden, layers involving all manner of RNAs, ranging from a mere 20 nucleotides to 100,000 nucleotides or so in length. Some are suggesting that this glimpse is just the tip of the iceberg and that continued research will reap increasingly complex interactions among RNAs and between RNAs and proteins. Genomics, the study of the DNA sequence of genomes, has been a hot field for some time, but now it looks as if "RNomics" is beginning to steal the show.

Some strides have already been made in RNomics with surveys of cDNA libraries for ncRNA sequences, especially some of the smaller ones that were long thought merely to be leftover scraps from other processes. For example, one study in 2001, which included a survey of a mouse-brain cDNA library, revealed 201 potential novel, small ncRNAs. In a 2003 survey of a cDNA library from *Drosophila melanogaster* (fruit fly), sixty-six potential novel ncRNAs were discovered. Judging by the large numbers of candidate ncRNAs showing up in what are essentially first-time surveys, many more may remain to be found. There could potentially be thousands of ncRNA genes. What is surprising is that many of these ncRNA genes are being found in spacer regions and introns, places that were once considered useless junk. With so much now being found in these regions, many geneticists have become ever more cautious in calling any DNA sequence junk DNA.

Because the field of RNomics is in its infancy and the functions of many of the ncRNAs are just barely understood, it may be premature to predict specific medical applications, but certainly the potential is there. The population of ncRNAs in a cell, in some sense, resembles a complex set of switches that turn genes on and off—before they are transcribed, while they are being transcribed, or even once translation has begun. Once these switches are better understood, researchers may be able to exploit the system with artificially produced RNAs. Geneticists will probably also discover that a number of diseases that appeared to have unexplained genetic behavior will find the solutions in ncRNA.

—Bryan Ness

See also: cDNA Libraries; Central Dogma of Molecular Biology; DNA Structure and Function; RNA Structure and Function; RNA Transcription and mRNA Processing.

Further Reading

Bass, Brenda L. "The Short Answer." *Nature* 411 (2001): 428-429. A look at RNA interference (RNAi) and the role of ncRNAs.

Gottesman, Susan. "Stealth Regulation: Biological Circuits with Small RNA Switches." *Genes and Development* 16 (2002): 2829-2842. A relatively comprehensive overview of what is known about ncRNA. Contains some helpful figures showing some of the mechanisms of action.

Grosshans, Helge, and Frank J. Slack. "MicroRNAs: Small Is Plentiful." *The Journal of Cell Biology* 156, no. 1 (2002): 17-21. Overviews stRNA and miRNA and their functions.

Hentze, Matthias W., Elisa Izaurralde, and Bertrand Séraphin. "A New Era for the RNA World." *EMBO Reports* 1, no. 5 (2000): 394-398. A report on the RNA 2000 Conference, hosted by the RNA Society. Focuses on certain ncRNAs, such as those in spliceosomes and in the brain.

Lewin, Benjamin. *Genes VII.* New York: Oxford University Press, 2001. An upper-division college textbook that is better than many other textbooks. Various chapters include discussion of ncRNAs.

Storz, Gisela. "An Expanding Universe of Noncoding RNAs." *Science* 296 (2002): 1260-1263. A fairly complete overview of the various kinds of ncRNA, along with as much as is known about many of them.

Nondisjunction and Aneuploidy

Fields of study: Cellular biology; Diseases and syndromes

Significance: *Nondisjunction is the faulty disjoining of replicated chromosomes during mitosis or meiosis, which causes an alteration in the normal number of chromosomes (aneuploidy). Nondisjunction is a major cause of Down syndrome and various sex chromosome anomalies. Understanding the mechanisms associated with cell division may provide new insight into the occurrence of these aneuploid conditions.*

Key terms

MEIOSIS: a series of two nuclear divisions that occur in gamete formation in sexually reproducing organisms

MITOSIS: nuclear division of chromosomes, usually accompanied by cytoplasmic division; two daughter cells are formed with identical genetic material

Background

Each cell in multicellular organisms contains all the hereditary information for that individual, in the form of DNA. In eukaryotes, DNA is packaged in rodlike structures called chromosomes, and any given species has a characteristic chromosome number. There are typically two of each kind of chromosome, which is referred to as being diploid. In humans (*Homo sapiens*), there are forty-six chromosomes; in corn (*Zea mays*), there are twenty chromosomes. A haploid cell has half the number of chromosomes as a diploid cell of the same species, which constitutes one of each kind of chromosome. One set of chromosomes is contributed to a new individual by each parent in sexual reproduction through the egg and sperm, which are both haploid. Thus, a fertilized egg will contain two sets of chromosomes and will be diploid.

A karyotype is a drawing or picture that displays the number and physical appearance of the chromosomes from a single cell. A normal human karyotype contains twenty-two pairs of autosomes (chromosomes that are not sex chromosomes) and one pair of sex chromosomes. Females normally possess two X chromosomes in their cells, one inherited from each parent. Males have a single X chromosome, inherited from the mother, and a Y chromosome, inherited from the father.

The many cells of a multicellular organism are created as the fertilized egg undergoes a series of cell divisions. In each cell division cycle, the chromosomes are replicated, and, subsequently, one copy of each chromosome is distributed to two daughter cells through a process called mitosis. When gametes (eggs or sperm) are produced in a mature organism, a different type of nuclear division occurs called meiosis. Gametes contain one set of chromosomes instead of two. When two gametes join (when a sperm cell fertilizes an egg cell), the diploid chromosome number for the species is restored, and, potentially, a new individual will form with repeated cell divisions.

When replicated chromosomes are distributed to daughter cells during mitosis or meiosis, each pair of chromosomes is said to disjoin from one another (disjunction). Occasionally, this process fails. When faulty disjoining (nondisjunction) of replicated chromosomes occurs, a daughter cell may result with one or more chromosomes than normal or one or more fewer than normal. This alteration in the normal number of chromosomes is called aneuploidy. One chromosome more than normal is referred to as a "trisomy." For example, Down syndrome is caused by trisomy 21 in humans. One chromosome fewer than normal is called monosomy. Turner syndrome in humans is an example of monosomy. Turner's individuals are women who have only one X chromosome in their cells, whereas human females normally have two X chromosomes. When

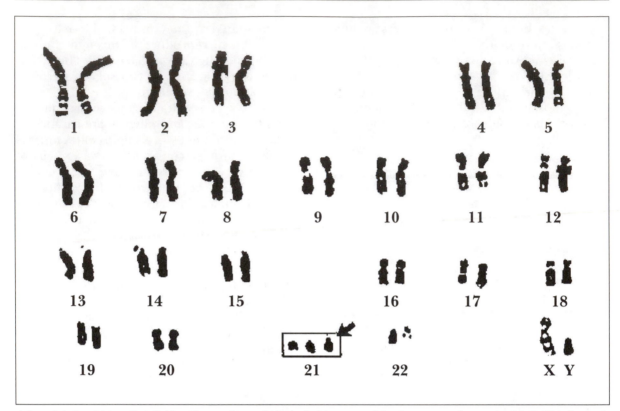

A karyotype is a picture that displays the number and physical appearance of the chromosomes from a single cell. This karyotype shows the trisomy at chomosome 21 that results in Down syndrome. (U.S. Department of Energy Human Genome Program, http://www.ornl.gov/hgmis)

nondisjunction occurs in the dividing cells of a mature organism or a developing organism, a portion of the cells of the organism may be aneuploid. If nondisjunction occurs in meiosis during gamete formation, then a gamete will not have the correct haploid chromosome number. If that gamete joins with another, the resulting embryo will be aneuploid. Examples of human aneuploid conditions occurring in live births include Down syndrome (trisomy 21), Edwards' syndrome (trisomy 18), Patau syndrome (trisomy 13), metafemale (more than two X chromosomes), Klinefelter syndrome (XXY), and Turner syndrome (XO). Most aneuploid embryos do not survive to birth.

Causes of Nondisjunction

There are both environmental and genetic factors associated with nondisjunction in plants and animals. Environmental factors that may induce nondisjunction include physical factors such as heat, cold, maternal age, and ionizing radiation, in addition to a wide variety of chemical agents.

In humans, it is well established that increased maternal age is a cause of nondisjunction associated with the occurrence of Down syndrome. For mothers who are twenty years of age, the incidence of newborns with Down syndrome is 0.4 in 1,000 newborns. For mothers over forty-five years of age, the incidence of newborns with Down syndrome is 17 in 1,000 newborns. While it is clear that increased maternal age is linked to nondisjunction, it is not known what specific physiological, cellular, or molecular mechanisms or processes are associated with this increased nondisjunction. While nondisjunction in maternal meiosis may be the major source of trisomy 21 in humans, paternal nondisjunction in sperm formation does occur and may result in aneuploidy.

In a study conducted by Karl Sperling and colleagues published in the *British Medical Journal* (July 16, 1994), low-dose radiation in the form of radioactive fallout from the Chernobyl nuclear accident (April, 1986) was linked to a significant increase in trisomy 21 in West Berlin in January, 1987: twelve births of trisomy 21 compared to the expected two or three births. This study suggests that, at least under certain circumstances, ionizing radiation may affect the occurrence of nondisjunction. Researchers have shown that ethanol (the alcohol in alcoholic beverages) causes nondisjunction in mouse-egg formation, suggesting a similar possibility in humans. Other researchers have found that human cells in tissue culture (cells growing on nutrient media) had an increased occurrence of nondisjunction if the media was deficient in folic acid. This implies that folic acid may be necessary for normal chromosome segregation or distribution during cell division.

Scientists know from genetics research that mutations (changes in specific genes) in the fruit fly result in the occurrence of nondisjunction. This genetic component of nondisjunction is further supported by the observation that an occasional family gives birth to more than one child with an aneuploid condition. In these instances, it is likely that genetic factors are contributing to repeated nondisjunction.

Impact and Applications

There are several reasons scientists are devoting research efforts to understanding the consequences of nondisjunction and aneuploidy. First, at least 15 to 20 percent of all recognized human pregnancies end in spontaneous abortions. Of these aborted fetuses, between 50 and 60 percent are aneuploid. Second, of live births, 1 in 700 is an individual with Down syndrome. Mental retardation is a major symptom in individuals with Down syndrome. Thus, nondisjunction is one cause of mental retardation. Finally, aneuploidy is common in cancerous cells. Scientists do not know whether nondisjunction is part of the multistep process of tumor formation or whether aneuploidy is a consequence of tumor growth. Continued research into the mechanics of cell division and the various factors that influence that process will increase the understanding of the consequences of nondisjunction and possibly provide the means to prevent its occurrence.

—*Jennifer Spies Davis*

See also: Chromosome Theory of Heredity; Down Syndrome; Hereditary Diseases; Klinefelter Syndrome; Metafemales; Polyploidy; Turner Syndrome; XYY Syndrome.

Further Reading

Bender, Bruce G., and Robert J. Harmon. "Psychosocial Adaptation of Thirty Nine Students with Sex Chromosome Abnormalities." *Pediatrics* 96 (August, 1995). Evaluates the risks for problems with cognitive skills, learning abilities, and psychosocial adaptation in adolescents.

Berch, Daniel B., and Bruce G. Bender, eds. *Sex Chromosome Abnormalities and Human Behavior.* Boulder, Colo.: Westview Press, 1990. Explores the cognitive, emotional, and psychosocial skills of those with sex chromosome abnormalities.

Cunningham, Cliff. *Understanding Down Syndrome: An Introduction for Parents.* 1988. Reprint. Cambridge, Mass.: Brookline Books, 1999. Expanded and updated to reflect changes in the education and care of children and adults with Down syndrome. Also provides in-depth advice and information for parents, including issues of professional guidance, treatment, and prenatal testing.

Massimini, Kathy, ed. *Genetic Disorders Sourcebook: Basic Consumer Information About Hereditary Diseases and Disorders.* 2d ed. Detroit, Mich.: Omnigraphics, 2001. Discusses the ethics of gene testing, the causes of and treatments for genetic disorders, and includes a section on chromosomal disorders. Glossary, resource directory, charts, tables, and index.

Orr-Weaver, Terry L., and Robert A. Weinberg. "A Checkpoint on the Road to Cancer." *Nature* (March 19, 1998). Examines the possible role of aneuploidy in tumor progression.

Pai, G. Shashidhar, Raymond C. Lewandowski, Digamber S. Borgaonkar. *Handbook of Chromosomal Syndromes.* New York: John Wiley &

Sons, 2002. Covers two hundred chromosomal aneuploidy syndromes, including information on diagnosis, behavior, and life expectancy. Illustrated.

Patterson, D. "The Causes of Down Syndrome." *Scientific American* 257 (August, 1987). Discusses nondisjunction, the ongoing research into what genes occur on chromosome 21 and how they contribute to Down syndrome, the history of the syndrome, and associated disorders.

Vig, Baldev K., ed. *Chromosome Segregation and Aneuploidy.* New York: Springer-Verlag, 1993. A comprehensive collection of research into the beginning stages of aneuploidy, the malsegregation of chromosomes, and environmental mutagenesis. Illustrations, bibliography, index.

Oncogenes

Fields of study: Molecular genetics; Viral genetics

Significance: *Oncogenes are a group of genes origi-nally identified in RNA tumor viruses and later identified in many types of human tumors. The discovery of oncogenes has revolutionized the un-derstanding of cancer genetics and contributed to the development of a model of cancer as a multi-stage genetic disorder. The identification of these abnormally functioning genes in many types of human cancer has also provided new molecular targets for therapeutic intervention.*

Key terms

PROTO-ONCOGENES: cellular genes that carry out specific steps in the process of cellular proliferation; as a consequence of mutation or deregulation, these genes may be con-verted into cancer-causing genes

RETROVIRUS: a virus that converts its RNA ge-nome into a DNA copy that integrates into the host chromosome

The Discovery of Oncogenes

The discovery of oncogenes has been closely linked to the study of the role of a group of RNA tumor viruses, retroviruses (*Retroviridae*), in the etiology of many animal cancers. In the early part of the twentieth century, Peyton Rous identified a virus (called Rous sarcoma vi-rus after its discoverer) capable of inducing tu-mors called sarcomas in chickens. Many other RNA tumor viruses capable of causing tumor formation in animals or experimental systems were later discovered, which led to a search for specific viral genes responsible for the cancer-causing properties of these viruses.

The identification of these cancer-causing genes (oncogenes) awaited developments in the area of recombinant DNA technology and molecular genetics, which ultimately facili-tated the molecular analysis of this group of genes. These analyses revealed that viral onco-genes were actually cellular genes that were in-corporated into the genetic material of the RNA tumor virus during the process of infec-tion. The acquisition of these host-cell genes was responsible for the cancer-causing proper-ties of these viruses. The first oncogene discov-ered was the *src* gene of the Rous sarcoma virus. Subsequently, at least thirty different oncogenes were discovered in avian and mammalian RNA tumor viruses. Each of these oncogenes has a cellular counterpart that is the presumed ori-gin of the viral gene; with the exception of the Rous sarcoma virus, the incorporation of the host-cell gene into the virus, involving a process called transduction, results in the loss of viral genes, generating a defective virus.

In addition to the oncogenes originally iden-tified in viruses, more than fifty oncogenes have been identified in malignant tumors as part of chromosomal rearrangements or the amplifica-tion or mutations of specific genes. The first ge-netic rearrangement linked to a specific type of human malignancy involved the "Philadelphia" chromosome in patients with chronic myelog-enous leukemia (CML). This chromosome rep-resents a shortened version of chromosome 22, which results from an exchange of genetic mate-rial between chromosomes 9 and 22 (called a re-ciprocal translocation). Subsequent molecular analyses showed that the oncogene *abl*, origi-nally identified in a mammalian RNA tumor virus, was translocated to chromosome 22 in CML patients. Additional human malignancies involving translocated oncogenes previously identified in RNA tumor viruses have been iden-tified, notably the oncogene *myc* in patients with Burkitt's lymphoma, a disease primarily found in parts of Africa.

Additional genetic rearrangements repre-sent amplification of existing oncogenes (seg-ments of genetic material duplicated many times in genetically unstable tumor cells). These gene amplifications may be associated with the presence of multiple copies of genetic segments along a chromosome, designated as homoge-neously staining regions (HSRs), or may appear in the form of minichromosomes containing the amplified genes, termed double-minutes (DMs). For example, late-stage neuroblastomas often contain numerous double-minute chro-mosomes containing amplified copies of the *N-myc* gene.

The Properties of Oncogenes

The first dramatic evidence linking onco-genes with cancer was provided by studies of the *sis* oncogene of simian sarcoma virus, which proved to be an altered form of the mammalian platelet derived growth factor (PDGF). Growth factors are proteins that bind to receptors on target cells to initiate an intracellular signaling cascade, which results in cellular proliferation. This seminal discovery led to the development of the proto-oncogene model. This model states that oncogenes are derived from normal host genes called proto-oncogenes, which encode gene products involved in controlling cell division. If proto-oncogene expression is altered by mutation or deregulation, they may disrupt the normal control of cell division, resulting in unregulated cellular proliferation, a hallmark of malignancy.

Subsequent analyses of oncogene activities and the structure and function of the cellular proto-oncogenes from which they are derived have provided strong evidence for this model. Viral and cellular oncogenes with functions affecting every step in the control of the cell have been identified. In addition to altered growth factors such as *sis*, researchers have also identified altered growth factor receptors such as the epidermal growth factor receptor (*erb-b*), elements of the intracellular signal cascade (*src* and *ras*), nuclear transcriptional activators (*myc*), cell-cycle regulators called cyclin-dependent kinases (cdk's), and cell death inhibitors (*bc12*) in human tumors of diverse tissue origin. Each of these oncogenic gene products represents an altered form of normal cellular genes that participate in cell-division pathways. Numerous mutations in proto-oncogenes have

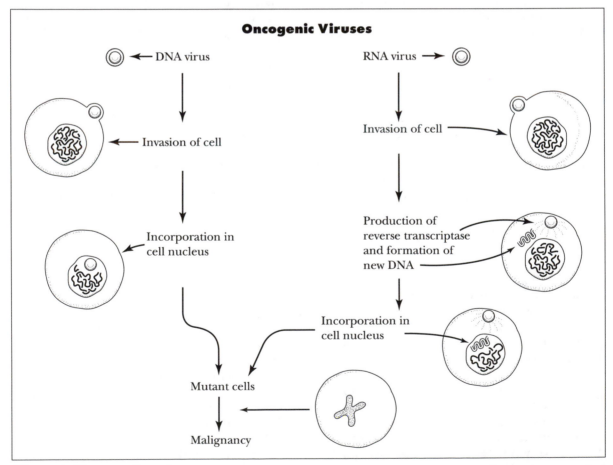

Oncogenic Viruses

DNA virus

RNA virus

Invasion of cell

Invasion of cell

Incorporation in cell nucleus

Production of reverse transcriptase and formation of new DNA

Incorporation in cell nucleus

Mutant cells

Malignancy

(Electronic Illustrators Group)

been identified, including single base changes (point mutations), gene truncations, gene amplifications, and gene rearrangements resulting from exchanges between different chromosomes called translocations.

One of the most dramatic discoveries involved a comparative analysis of the structures of the normal and oncogene forms of the *ras* proto-oncogene isolated from human bladder carcinomas. Surprisingly, a single-base change was sufficient to convert a normal cellular gene to a cancer-causing gene. The observed mutations were localized to regulatory regions of the *ras* gene product, resulting in its permanent activation. Molecular analyses of many other oncogenes have shown that the observed mutations fall into several categories: nucleotide base changes that result in gene products whose functions are not subject to normal inhibitory processes, overproduction of gene products caused by gene amplification or translocation, and loss of regulatory components caused by gene translocation or truncation. The generalized consequence of these mutations is to convert normal cellular gene products important in cell division to dominant, unregulated gene products that cause the inappropriate stimulation of cell division.

Interestingly, most tumors analyzed show the involvement of multiple oncogenes and tumor-suppressor genes (another class of otherwise normal genes that are modified in some manner). Studies of tumor development in human colorectal carcinomas in which it is possible to identify discrete stages of tumor development have indicated a progressive increase in the number and types of cellular oncogenes at successive stages of tumor development. From these studies, a model of oncogenesis has emerged in the form of a multistage disorder characterized by the successive accumulation of mutations in specific cellular oncogenes and tumor-suppressor genes, which results in the inability to regulate cellular proliferation.

Impact and Applications

The identification of oncogenes has provided enormous amounts of information on the cellular mechanisms responsible for the loss of growth control in cancer cells. In addition, these dysfunctional gene products represent potential targets for therapeutic applications. Research studies have been directed toward the design of inhibitors of specific oncogenes such as *ras* and *erb-b* in order to block the effects of oncogenes in malignant cells. Additional molecular targets include overexpressed oncogenes that stimulate cellular proliferation or blood vessel formation (angiogenesis), processes critical to tumor establishment. The advantages of these approaches include better targeting of cancer cells, as well as a potential decrease in side effects as compared to conventional chemotherapy. Structural abnormalities in oncogene products may be used in the development of monoclonal antibodies directed against these dysfunctional proteins. Toxins may also be linked to the antibodies to generate immunotoxins whose cell-killing activities directly target malignant cells. Malignant melanoma (skin cancer) has been the focus of many of these targeted approaches directed against specific abnormal gene products. Successful clinical applications will most likely combine approaches involving cytotoxic drugs and inhibitors targeting multiple sites of oncogene dysfunction in the cancer cell.

—*Sarah Crawford Martinelli*

See also: Aging; Blotting: Southern, Northern, and Western; Breast Cancer; Burkitt's Lymphoma; Cancer; Cell Culture: Animal Cells; Cell Cycle, The; Gene Therapy; Genetics, Historical Development of; Human Genetics; Hybridomas and Monoclonal Antibodies; Repetitive DNA; RNA Transcription and mRNA Processing; Tumor-Suppressor Genes.

Further Reading

Angier, Natalie. *Natural Obsessions: Striving to Unlock the Deepest Secrets of the Cancer Cell.* Boston: Mariner Books/Houghton Mifflin, 1999. Explores mutant-gene research and laboratory work to find the essence of the human cancer cell.

Cooper, Geoffrey M. *Oncogenes.* 2d ed. Boston: Jones and Bartlett, 1995. Provides a framework for studying oncogenes and tumor-suppressor genes and discusses advances in the field, including knowledge of signal transduction pathways, which lead to cell proliferation.

Ehrlich, Melanie, ed. *DNA Alterations in Cancer: Genetic and Epigenetic Changes*. Natick, Mass.: Eaton, 2000. Provides an introduction to cancer genes, tumor-suppressor genes, inherited mutations, and more. Illustrations, bibliography, index.

Hartwell, Leland, et al. "Cell Cycle Control and Cancer." *Science* 266 (1994). Provides a clear description of the role of oncogenes in cell-cycle dysregulation.

La Thangue, Nicholas B., and Lasantha R. Bandara, eds. *Targets for Cancer Chemotherapy: Transcription Factors and Other Nuclear Proteins*. Totowa, N.J.: Humana Press, 2002. Discusses research on protein targets for cancer drugs. Illustrations, bibliography, index.

Mulvihill, John J. *Catalog of Human Cancer Genes: McKusick's Mendelian Inheritance in Man for Clinical and Research Oncologists*. Foreword by Victor A. McKusick. Baltimore: Johns Hopkins University Press, 1999. Discusses the hereditary traits and genes that lead to susceptibility or resistance to cancer. Includes seven hundred entries grouped according to body organ.

Varmus, Harold. "The Molecular Genetics of Cellular Oncogenes." *Annual Review of Genetics* 18 (1994). Nobel laureate Varmus details the structure and function of oncogenes.

Web Sites of Interest

American Cancer Society. http://www.cancer.org. Site has searchable information on oncogenes and tumor suppressor genes.

American Society of Clinical Oncology. http://www.asco.org. Searchable site on oncogenes and molecular oncology.

One Gene-One Enzyme Hypothesis

Field of study: History of Genetics; Molecular genetics

Significance: *The formulation of the one gene-one enzyme hypothesis in 1941, which simply states that each gene gives rise to one enzyme, was foundational to understanding the molecular basis of gene action. Today, with a more detailed understanding of how genes work, geneticists consider the original hypothesis an oversimplification and have reformulated it as the "one gene-one polypeptide" hypothesis. Even in its new form, however, there are exceptions.*

Key terms

MESSENGER RNA (mRNA) PROCESSING: chemical modifications that alter messenger RNAs, often resulting in more than one gene product formed from the same gene

METABOLIC PATHWAY: a series of enzyme-catalyzed reactions leading to the complete breakdown or synthesis of a particular biological molecule

POLYPEPTIDE: a complex molecule encoded by the genetic code and composed of amino acids; one or more of which compose a protein

POST-TRANSLATIONAL MODIFICATION: chemical alterations to proteins that alter their properties as enzymes

Genetics Meets Biochemistry

In the early part of the twentieth century, genetics was becoming an established discipline, but the relationship between genes and how they are expressed as phenotypes was not yet understood. Biochemistry was also in its infancy, particularly the study of the enzyme-catalyzed chemical reactions of metabolic pathways. In 1902, a British medical doctor named Archibald Garrod brought genetics and biochemistry together in the discovery that a human disease called alkaptonuria, which causes individuals with the disease to accumulate a black pigment in their urine—was inherited as a recessive trait. Equally important, however, was Garrod's observation that alkaptonurics were unable to metabolize alkapton, the molecule responsible for the black pigmentation, an intermediate in the degradation of amino acids. Garrod's conclusion was that people with alkaptonuria lack the enzyme that normally degrades alkapton. Because it thus appeared that a defective gene led to an enzyme deficiency, Garrod predicted that genes form enzymes. This statement was the precursor of what came to be known as the one gene-one enzyme hypothesis.

George Wells Beadle. (© The Nobel Foundation)

Formation of the Hypothesis

Garrod's work went largely ignored until 1941, when George Beadle and Edward Tatum, geneticists at Stanford University, used bread mold (*Neurospora crassa*) to test and refine Garrod's theory. Wild-type *Neurospora* grows well on minimal media containing only sugar, ammonia, salts, and biotin, because it can biosynthesize all other necessary biochemicals. Beadle and Tatum generated mutants that did not grow on minimal media but instead grew only when some other factor, such as an amino acid, was included. They surmised that the mutant molds lacked specific enzymes involved in biosynthesis. With several such mutants, Beadle and Tatum demonstrated that mutations in single genes often corresponded to disruptions of single enzymatic steps in biosynthetic metabolic pathways. They concluded that each enzyme is controlled by one gene, a relationship they called the "one gene-one enzyme hypothesis." This time, the scientific community

took notice, awarding a Nobel Prize in Physiology or Medicine to Beadle and Tatum in 1958, and the hypothesis served as the basis for biochemical genetics for the next several years.

Modifications to the Hypothesis

The one gene-one enzyme hypothesis was accurate in predicting many of the findings in biochemical genetics after 1941. It is now known that DNA genes are often transcribed into messenger RNAs (mRNAs), which in turn are translated into polypeptides, many of which form enzymes. Thus, the basic premise that genes encode enzymes still holds. On the other hand, Beadle and Tatum had several of the details wrong, and today the hypothesis should be restated as follows: Most genes encode information for making one polypeptide.

There are at least three reasons that the original one gene-one enzyme hypothesis does not accurately explain biologists' current understanding of gene expression. First of all, enzymes are often formed from more than one

Edward Lawrie Tatum. (© The Nobel Foundation)

polypeptide, each of which is the product of a different gene. For example, the enzyme ATP synthase is composed of at least seven different polypeptides, all encoded by separate genes. Thus, the one-to-one ratio of genes to enzymes implied by the hypothesis is clearly incorrect. This fact was recognized early and led to the theory's reformulation as the "one gene-one polypeptide" hypothesis. However, even this newer version of the hypothesis has since been shown to be inaccurate.

Second, several important genes do not encode enzymes. For example, some genes encode transfer RNAs (tRNAs), which are required for translating mRNAs. Thus, clearly even the one gene-one polypeptide hypothesis is insufficient, since tRNAs are not polypeptides.

Finally, further deviation from the original one gene-one enzyme hypothesis is required when one considers that several modifications to RNAs and polypeptides occur after gene transcription, and can do so in more than one way. Thus, a single gene can give rise to more than one mRNA, and potentially to numerous different polypeptides with varying properties. Post-transcriptional variation in gene expression occurs first during RNA processing, when the polypeptide-encoding regions of mRNA are spliced together. It is important to note that the exact splicing pattern can vary depending on the exact needs of the cell. One example of a gene that undergoes differential mRNA processing leading to two dramatically different phenotypes is the fruit fly gene sex-lethal (*sxl*). A long version of *sxl* mRNA is generated in developing male flies and a shorter one in female flies. Because the *sxl* protein regulates sexual development, mutant female flies that mistakenly splice *sxl* mRNA display male sexual characteristics.

Like differential mRNA processing, post-translational protein modification varies by cellular context, allowing a single gene to generate more than one kind of enzyme. However, unlike mRNA processing, protein modification is often reversible. For example, liver cells responding to insulin will chemically modify some of their enzymes by way of a process called signal transduction, thereby changing their en-zymatic properties, often essentially making them into different enzymes. Once insulin is no longer present, the cell can undo the modifications, returning the enzymes back to their original forms.

—Stephen Cessna

See also: Complementation Testing; Genetics, Historical Development of; Model Organism: *Neurospora crassa*; Signal Transduction.

Further Reading

Beadle, G. W., and E. L. Tatum. "Genetic Control of Biochemical Reactions in *Neurospora.*" *Proceedings of the National Academy of Sciences* 27 (1941): 499-506. The original research article that postulated the one gene-one enzyme hypothesis.

Davis, Rowland H. *Neurospora: Contributions of a Model Organism.* New York: Oxford University Press, 2000. A full account of the organism's history, biology, genome, mitosis, meiosis, metabolism, mutations, and more.

Science 291 (February, 2001). A special issue on the human genome. Articles estimate the number of genes in the human genome and guess at the corresponding number of active gene products.

Weaver, Robert F. *Molecular Biology.* 2d ed. New York: McGraw-Hill, 2002. Gives a modern overview of gene expression, including differential mRNA processing, and explains the original work of Beadle and Tatum in detail.

Organ Transplants and HLA Genes

Field of study: Immunogenetics

Significance: *Organ transplantation has saved the lives of countless people. Although the success rate for organ transplantation continues to improve, many barriers remain, including an inadequate supply of donor organs and the phenomenon of transplant rejection. Transplant rejection is caused by an immune response by the recipient to molecules on the transplanted organs that are coded for by the human leukocyte antigen (HLA) gene complex.*

Key terms

ALLELES: the two alternate forms of a gene at the same locus on a pair of homologous chromosomes

ANTIGENS: molecules recognized as foreign to the body by the immune system, including molecules associated with disease-causing organisms (pathogens)

HISTOCOMPATIBILITY ANTIGENS: molecules expressed on transplanted tissues that are recognized as foreign by the immune system, causing rejection of the transplant; the most important histocompatibility antigens in vertebrates are coded for by a cluster of genes called the major histocompatibility complex (MHC)

LOCUS (*pl.* LOCI): the location of a gene on a chromosome

POLYMORPHISM: the presence of many different alleles for a particular locus in individuals of the same species

Transplantation

The replacement of damaged organs by transplantation was one of the great success stories of modern medicine in the latter decades of the twentieth century. During the 1980's, the success rates for heart and kidney transplants showed marked improvement and, most notably, the one-year survival for pancreas and liver transplants rose from 20 percent and 30 percent to 70 percent and 75 percent, respectively. These increases in organ survival were largely attributable to improvements in two aspects of the transplantation protocol that directly reduced tissue rejection: the development of more accurate methods of tissue typing that allowed better tissue matching of donor and recipient, and the discovery of more effective and less toxic antirejection drugs. In fact, these changes helped make transplantation procedures so common by the 1990's that the low number of donor organs became a major limiting factor in the number of lives saved by this procedure.

Rejection and the Immune Response

The rejection of transplanted tissues is associated with genetic differences between the donor and recipient. Transplants of tissue within the same individual, called autografts, are never rejected. Thus the grafting of blood vessels transplanted from the leg to an individual's heart during bypass operations are never in danger of being rejected. On the other hand, organs transplanted between genetically distinct humans tend to undergo clinical rejection within a few days to a few weeks after the procedure. During the rejection process, the transplanted tissue is gradually destroyed and loses its function. When examined under the microscope, tissue undergoing rejection is observed to be infiltrated with a variety of cells, causing its destruction. These infiltrating cells are part of the recipient's immune system, which recognizes molecules on the transplant as foreign to the body and responds to them as they would to a disease-causing, pathogenic organism.

The human immune response is a complex system of cells and secreted proteins that has evolved to protect the body from invasion by pathogens. Immune mechanisms are directed against molecules or parts of molecules called antigens. The ultimate function of the immune response is to recognize pathogen-associated antigens as foreign to the body and to eliminate and destroy the organism, thus resolving the disease. On the other hand, the immune response is prevented, under most circumstances, from attacking the antigens expressed on the tissues of the body in which they originate. The ability to distinguish between self and foreign antigens is critical to protecting the body from pathogens and to the maintenance of good health.

A negative consequence of the ability of the immune system to discriminate between self and foreign antigens is the recognition and destruction of transplants. The antigens associated with transplants are recognized as foreign in the same fashion as pathogen-associated antigens, and many of the same immune mechanisms used to kill pathogens are responsible for the destruction of the transplant. The molecules on the transplanted tissues recognized by the immune system are called histocompatibility antigens. The term "histocompatibility" refers to the fact that transplanted organs are often not compatible with the body of a genetically distinct recipient. All vertebrate animals have a cluster of genes that code for the most

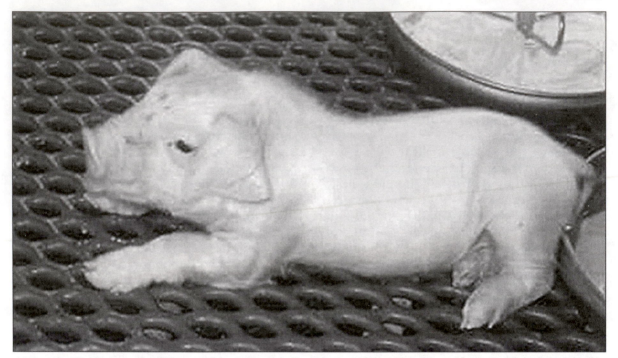

A two-week-old piglet in April, 2002, one of three that were the first to be cloned from both human and pig cells. Normal pigs have been sources of human "replacement parts" (such as heart valves) from some time. The hope is that organs from pigs with human genes will be more easily accepted by the human body after transplantion. (AP/Wide World Photos)

important histocompatibility antigens, called the major histocompatibility complex (MHC).

MHC Polymorphism, HLA Genes, and Tissue Typing

Each MHC locus is highly polymorphic, meaning that many different alleles exist within a population (members of a species sharing a habitat). The explanation for the polymorphism of histocompatibility antigens is related to the actual function of these molecules within the body. Clearly, histocompatibility molecules did not evolve to induce the rejection of transplants, despite the fact that this characteristic led to their discovery and name.

Histocompatibility molecules function by regulating immunity against foreign antigens. Each allele codes for a protein that allows the immune response to recognize a different set of antigens. Many pathogens, including the viruses associated with influenza and acquired immunodeficiency syndrome (AIDS), undergo genetic mutations that lead to changes in their antigens, making it more difficult for the body

to make an immune response to the virus. The existence of multiple MHC alleles in a population, therefore, ensures that some individuals will have MHC alleles allowing them to mount an immune response against a particular pathogen. If an entire population lacked these alleles, their inability to respond to certain pathogens could threaten the very existence of the species. The disadvantage of MHC polymorphism, however, is the immune response to the donor's histocompatibility antigens that causes organ rejection.

The human leukocyte antigen (HLA) gene complex is located on chromosome 6 in humans. Six important histocompatibility antigens are coded for by the HLA complex: the *A, B, C, DR, DP,* and *DQ* alleles. Differences in HLA antigens between the donor and recipient are determined by tissue typing. For many years, tissue typing was performed using antibodies specific to different HLA alleles. Antibodies are proteins secreted by the cells of the immune system that are used in the laboratory to identify specific antigens. As scientists began to clone the genes for the most common HLA alleles in the 1980's and 1990's, however, it appeared that direct genetic analysis would eventually replace or at least supplement these procedures.

Fewer differences in these antigens between donor organ and recipient mean a better prognosis for transplant survival. Therefore, closely related individuals who share many of their histocompatibility alleles are usually preferred as donors. When a family member is not available, the process of finding a donor is problematic. Worldwide computer databases are used to match potential donors with recipients, who are placed on a waiting list based on the severity of their disease.

Immunosuppressive Antirejection Drugs

Perhaps the most important medical breakthrough responsible for the increased success of organ transplantation occurred in the last two decades of the twentieth century. This breakthrough involved the discovery and successful use of antirejection drugs, most of which act by suppressing the immune response to the transplanted tissue. Immunosuppressive drugs are usually given in high doses for the first few weeks after transplantation or during a rejection crisis, but the dosage of these drugs is usually reduced thereafter to avoid their toxic effects.

Cyclosporine is by far the most effective of these drugs and has largely been responsible for the increased efficacy of liver, pancreas, lung, and heart transplantation procedures. In spite of its successes, cyclosporine has limitations in that it can cause kidney damage when given in high doses. Azathioprine, associated with bone marrow toxicity, was largely supplanted by the introduction of the less toxic cyclosporine. However, azathioprine has been used as part of a combined cyclosporine-azathioprine regimen. This practice allows the reduction of both the cyclosporine and azathioprine dosages, reducing the toxicity of both drugs. The search for more effective and less toxic antirejection drugs continues. Individuals receiving immunosuppressive therapy have other concerns in addition to the toxicity of the drugs themselves. As these individuals will have an impaired ability to mount an immune response to pathogens, their susceptibility to a variety of diseases will be increased. Thus transplant recipients must take special precautions to avoid exposure to potential pathogens, especially when receiving high doses of the drugs.

—James A. Wise

See also: Animal Cloning; Bacterial Genetics and Cell Structure; Bioethics; Biological Weapons; Cancer; Cloning; Cloning: Ethical Issues; Diabetes; Gene Therapy: Ethical and Economic Issues; Genetic Engineering: Historical Development; Genetics, Historical Development of; Heart Disease; Huntington's Disease; Hybridomas and Monoclonal Antibodies; Immunogenetics; In Vitro Fertilization and Embryo Transfer; Model Organism: *Mus musculus*; Model Organism: *Xenopus laevis*; Multiple Alleles; Paternity Tests; Polymerase Chain Reaction; Prion Diseases: Kuru and Creutzfeldt-Jakob Syndrome; Race; Sickle-Cell Disease; Stem Cells; Synthetic Antibodies; Totipotency; Transgenic Organisms; Xenotransplants.

Further Reading

Browning, Michael, and Andrew McMichael, eds. *HLA and MHC: Genes, Molecules, and Function.* New York: Academic Press, 1999. A review of molecular genetics of MHC, the structure and function of MHC-encoded molecules, and how they factor in health and disease. Illustrations, bibliography, index.

Janeway, Charles A., Paul Travers, et al. *Immunobiology: The Immune System in Health and Disease.* 5th rev. ed. Philadelphia: Taylor & Francis, 2001. Provides an excellent review of the HLA complex.

Lechler, Robert I., et al. "The Molecular Basis of Alloreactivity." *Immunology Today* 11 (March, 1990). Discusses the molecular basis of transplantation rejection.

Sasaki, Mutsuo, et al., eds. *New Directions for Cellular and Organ Transplantation.* New York: Elsevier Science, 2000. A collection of conference papers on organ transplantation and organ donation. Illustrations, bibliography, index.

Scientific American 269 (September, 1993). A special issue devoted to "Life, Death, and the Immune System," providing an excellent overview of the immune system.

Parthenogenesis

Field of study: Genetic engineering and biotechnology

Significance: *Parthenogenesis is the development of unfertilized eggs, which produces individuals that are genetically alike and allows rapid expansion of a population of well-adapted individuals into a rich environment. This clonal reproduction strategy is used by a number of species for rapid reproduction under very favorable conditions, and it appears to offer a selective advantage to individuals living in disturbed habitats.*

Key terms

ADAPTIVE ADVANTAGE: increased fertility in offspring as a result of passing on favorable genetic information

DIPLOID: having two sets of homologous chromosomes

FERTILIZATION: the fusion of two cells (egg and sperm) in sexual reproduction

HAPLOID: having one set of chromosomes

MEIOSIS: nuclear division that reduces the chromosome number from diploid to haploid in the production of the sperm and the egg

ZYGOTE: the product of fertilization in sexually reproducing organisms

The Nature of Parthenogenesis

Parthenogenesis is derived from two Greek words that mean "virgin" (*parthenos*) and "origin" (*genesis*) and describes a form of reproduction in which females lay diploid eggs (containing two sets of chromosomes) that develop into new individuals without fertilization—there is no fusion of a sperm nucleus with the egg nucleus to produce the new diploid individual. This is a form of clonal reproduction because all of the individuals are genetically identical to the mother and to each other. The mechanisms of parthenogenesis do not show any single pattern and have evolved independently in different groups of organisms. In some organisms, such as rotifers and aphids, parthenogenesis alternates with normal sexual reproduction. When there is a rich food source, such as new rose bushes emerging in the early spring, aphids re-

produce by parthenogenesis; late in the summer, however, as the food source is decreasing, sexually reproducing females appear. The same pattern has been observed in rotifers, in which a decrease in the quality of the food supply leads to the appearance of females that produce haploid eggs by normal meiosis that require fertilization for development. The strategy appears to involve the clonal production of large numbers of genetically identical individuals that are well suited to the environment when the conditions are favorable and the production of a variety of different types, by the recombination that occurs during normal meiosis and the mixing of alleles from two individuals in sexual reproduction, when the conditions are less favorable. In social insects, such as bees, wasps, and ants, parthenogenesis is a major factor in sex determination, although it may not be the only factor. In these insects, eggs that develop by parthenogenesis remain haploid and develop into males, while fertilized eggs develop into diploid, sexually reproducing females.

In algae and some forms of plants, parthenogenesis also allows rapid reproduction when conditions are favorable. In citrus, seed development by parthenogenesis maintains the favorable characteristics of each plant. For this reason, most commercial citrus plants are propagated by asexual means, such as grafting. Parthenogenesis has also been induced in organisms that do not show the process in natural populations. In sea urchins, for example, development can be induced by mechanical stimulation of the egg or by changes in the chemistry of the medium. Even some vertebrate eggs have shown signs of early development when artificially stimulated, but haploid vertebrate cells lack all of the information required for normal development, so such "zygotes" cease development very early.

Parthenogenesis in Vertebrates

Parthenogenesis has been observed in vertebrates such as fish, frogs, and lizards. In these parthenogenetic populations, all the individuals are females, so reproduction of the clone is restricted to parthenogenesis. Parthenogenetic fish often occur in populations along with sexually reproducing individuals. The parthenogenetic forms produce diploid eggs that develop without fertilization; in rare cases, however, fertilization of a parthenogenetic egg gives rise to a triploid individual that has three sets of chromosomes rather than the normal two sets (two from the diploid egg and one from the sperm). In some groups, penetration of a sperm is necessary to activate development of the zygote, but the sperm nucleus is not incorporated into the zygote.

Evidence indicates that in each of these vertebrate situations, the parthenogenetic populations have resulted from a hybridization between two different species. The parthenogenetic forms always occur in regions where the two parental species overlap in their distribution, often an area that is not the most favorable habitat for either species. The hybrid origin has been confirmed by the demonstration that the animals have two different forms of an enzyme that have been derived from the two different species in the region. Genetic identity has also been confirmed using skin graft studies. In unrelated organisms, skin grafts are quickly rejected because of genetic incompatibilities; clonal animals, on the other hand, readily accept grafts from related donors. Parthenogenetic fish from the same clone accept grafts that confirm their genetic identity, but rejection of grafts by other parthenogenetic forms from different populations shows that they are different clones and must have a different origin. This makes it possible to better understand the structure of the populations and helps in the study of the origins of parthenogenesis within those populations. Comparisons using nuclear and mitochondrial DNA also allow the determination of species origin and the maternal species of the parthenogenetic form since the mitochondria are almost exclusively transmitted through the vertebrate egg. Within the hybrid, a mechanism has originated that allows the egg to develop without fertilization, although, as already noted, penetration by a sperm may be required to activate development in some of the species.

The advantage of parthenogenesis appears to be the production of individuals that are genetically identical. Since the parthenogenetic form may, at least in vertebrates, be a hybrid, it is heterozygous at most of its genetic loci. This provides greater variation that may provide the animal with a greater range of responses to the environment. Maintaining this heterozygous genotype may give the animals an advantage in environments where the parental species are not able to reproduce successfully and may be a major reason for the persistence of this form of reproduction. Many vertebrate parthenogenetic populations are found in disturbed habitats, so their unique genetic composition may allow for adaptation to these unusual conditions.

Mechanisms of Development

The mechanisms of diploid egg development are as diverse as the organisms in which this form of reproduction is found. In normal meiosis, the like chromosomes of each pair separate at the first division and the copies of each chromosome separate at the second division (producing four haploid cells). During the meiotic process in the egg, three small cells (the polar bodies), each with one set of chromosomes, are produced, and one set of chromosomes remains as the egg nucleus. In parthenogenetic organisms, some modification of this process occurs that results in an egg nucleus with two sets of chromosomes—the diploid state. In some forms, the first meiotic division does not occur, so two chromosome sets remain in the egg following the second division. In other forms, one of the polar bodies fuses back into the cell so that there are two sets of chromosomes in the final egg. In another variation, there is a replication of chromosomes after the first division, but no second division takes place in the egg, so the chromosome number is again diploid. In all of these mechanisms, the genetic content of the egg is derived from the mother's genetic content, and there is no contribution to the genetic content from male material.

The situation may be even more complex, however, because some hybrid individuals may retain the chromosomal identity of one species by a selective loss of the chromosomes of the other species during meiosis. The eggs may carry the chromosomes of one species but the mitochondria of the other species. The haploid eggs must be fertilized, so these individuals are not parthenogenetic, but their presence in the population shows how complex reproductive strategies can be and how important it is to study the entire population in order to understand its dynamics fully: A single population may contain individuals of the two sexual species, true parthenogenetic individuals, and triploid individuals resulting from fertilization of a diploid egg.

—D. B. Benner

See also: Totipotency.

Further Reading

Beatty, Richard Alan. *Parthenogenesis and Polyploidy in Mammalian Development.* Cambridge, England: Cambridge University Press, 1957. An early but still useful study.

Kaufman, Matthew H. *Early Mammalian Development: Parthenogenetic Studies.* New York: Cambridge University Press, 1983. By a well-known expert in mouse studies.

Patents on Life-Forms

Field of study: Bioethics; Human genetics and social issues

Significance: *In 1980, the U.S. Supreme Court upheld the right to patent a live, genetically altered organism. The decision was opposed by many scientists and theologians who believed that such organisms would pose a threat to the future of humanity. Although "legally" settled, the debate has continued, opponents arguing that patenting life-forms and DNA sequences imposes too great a cost and greatly inconveniences genetic research.*

Key term

PATENT: a grant made by the government that gives the creator or inventor the sole right to make, use, or sell that invention for a specific period of time, usually seventeen years in the United States

Patent on Life-Form Upheld

On June 16, 1980, the U.S. Supreme Court voted 5 to 4 that living organisms could be patented under federal law. The case involved Ananda M. Chakrabarty, a scientist who, while working for General Electric in 1972, had created a new form of bacteria, *Pseudomona originosa*, which could break down crude oil, and, therefore, could be used to clean up oil spills. Chakrabarty filed for a patent, but an examiner for the Patent Office rejected the application on the ground that living things are not patentable subject matter under existing patent law. Commissioner of Patents and Trademarks Sidney A. Diamond supported this view. Federal patent law provided that a patent could be issued only to a person who invented or discovered any new and useful "manufacture" or "composition of matter." The U.S. Court of Customs and Patent Appeals reversed that decision in 1979, concluding that the fact that microorganisms are alive has no legal significance. It held that a live, human-made bacterium is a patentable item since the microorganism was manufactured by crossbreeding four existing strains of bacteria and had never existed in nature.

Writing for the majority, Chief Justice Warren Burger upheld the patent appeals court judgment, making a distinction between the new bacterium and "laws of nature, physical phenomena and abstract ideas," which are not patentable. In the Court majority's view, Chakrabarty had invented a form of life that did not exist in the natural world, so it could not be considered part of nature. Instead, it was a product of human "ingenuity and research" that deserved patent protection. Items not patentable include new minerals that are discovered in the earth or a new species of plant found in a distant forest. These things occur naturally and are not created by humans. Burger also stressed that physicist Albert Einstein could not have patented his formula $E = mc^2$, since it is a law of nature, nor could Sir Isaac Newton have received a patent for the law of gravity. Discoveries such as these are part of the natural world and cannot be owned by a single individual.

Chakrabarty, on the other hand, had not found an unknown, natural species, nor had he discovered a law of nature. His new bacterium had a distinctive name and was developed in the laboratory for a specific purpose. None of the characteristics of the new organism could be found in nature. His discovery, Burger re-emphasized, was patentable because he had created it.

Opposition to the Ruling

The Court majority refused to consider arguments made in friend-of-the-court briefs filed by opponents of genetic engineering. The briefs were presented by groups representing scientists, including several Nobel Prize winners, and religious organizations. One brief suggested that genetic research posed a dangerous and serious threat to the future of humanity and should, therefore, be prohibited. Possible dangers included the spread of pollution and disease by newly created bacteria, none of which would have any natural enemies. Other threats involved the possible loss of genetic diversity, if, for instance, only the "best" form of laboratory-created plant seeds were grown. Research into human genetics could lead to newly designed gene material that could be used to build a "master race," thereby devaluing other human lives. Justice Burger concluded, however, that humans could be trusted not to create such horrible things. Quoting William Shakespeare's *Hamlet*, the chief justice asserted that it is sometimes better "to bear those ills we have than to fly to others that we know not of." People can try to guess what genetic manipulation could lead to, but it would also be a good idea to expect good things from science rather than "a gruesome parade of horribles." Besides, he then said, it did not matter whether a patent was granted in this case; in either case, scientific research would continue into the nature of genes.

The People's Business Commission, a non-profit educational foundation, had argued that granting General Electric and Chakrabarty a patent would give corporations the right "to own the processes of life in the centuries to come" through genetic manipulation. Chief Justice Burger wrote that the Court was "without competence to entertain these arguments." They did not have enough information available to determine whether to ignore such fears "as fantasies generated by fear of the unknown"

or accept them. Such a determination was not the responsibility of the Court, however. Questions of the morality of genetic research and manipulation were better left to Congress and the political process. How to proceed in these matters could only be resolved "after the kind of investigation, examination, and study that legislative bodies can provide and courts cannot."

Justice William J. Brennan, Jr., presented a brief dissenting opinion. He noted that Congress had twice, in 1930 and 1970, permitted new types of plants to be patented. However, those laws made no mention of bacteria. Thus, Brennan argued, Congress had indicated that only plants could receive patents and that the legislators had thus clearly indicated that other life-forms were excluded from the patent process. The Court majority rejected this view, arguing that Congress had not specifically excluded other life-forms.

Developments Since 1980

Since the patenting of the petroleum-eating bacteria, a variety of other genetically modified (GM) organisms have been patented, including pest-resistant crop plants and numerous types of "knockout" mice used by many researchers. The controversy around such patents initially calmed, but more recent developments have rekindled the flames. Since the advent of the Human Genome Project, the sequencing of genomes has accelerated exponentially. Because many sequences might contain valuable genes or markers, companies and nonprofit organizations began patenting the sequences.

After much debate, the patenting of DNA sequences has been allowed, and although the guidelines are still being fine-tuned, the general rule is that any "distinctive" DNA sequence can be patented. Those opposed to the patenting of DNA sequences say that it will impede research and even the development of useful medical applications. Ethicists argue that no one should have a right to patent DNA sequences, which represent the very basis of life. Some scientists have pushed for more restrictive rules, such as that a sequence cannot be patented unless there is clear evidence that the sequence codes for a useful product or would likely lead to a specific application.

At present it is too early to predict the final outcome of the push for patenting DNA sequences. There is still debate and there is a large backlog of sequences for which patents are still pending. A survey of medical testing laboratories in 2003 found that a number of labs either no longer used certain tests or did not plan to develop them when licensing fees were required for permission to use a relevant DNA sequence. Overwhelmingly, the labs surveyed saw patenting of DNA sequences as having a negative effect on the development of affordable clinical genetic tests.

—Leslie V. Tischauser, updated by Bryan Ness
See also: Genetic Engineering: Historical Development; Genetic Engineering: Industrial Applications; Genetic Engineering: Social and Ethical Issues; Human Genetics; Human Genome Project; Hybridization and Introgression; Model Organism: *Mus musculus*; Transgenic Organisms.

Further Reading

Chapman, Audrey R., ed. *Perspectives on Genetic Patenting: Religion, Science, and Industry in Dialogue.* Washington, D.C.: American Association for the Advancement of Science, 1999. Discusses questions such as, Should products of nature be patentable? Are genes or gene fragments discoveries or inventions? Should patenting of genes, cell lines, or genetically modified organisms be equated with ownership of them? Is the DNA in genes just a complex molecule or is it sacred? Does patenting human DNA and tissue demean human life and human dignity?

Diamond, Commissioner of Patents and Trademarks v. Chakrabarty (1980), 447 U.S. 303. The official citation of the Supreme Court decision.

Doll, John. "Talking Gene Patents." *Scientific American*, August, 2001. A brief interview with the director of biotechnology for the U.S. Patent and Trademark Office on what makes a gene eligible for a chemical compound patent and the number of genes patented.

Hanson, Mark J. "Religious Voices in Biotechnology: The Case of Gene Patenting." *Hastings Center Law Report*, November/December, 1997. Discusses the religious, legal, moral, and scientific concerns about patenting human genetic material, DNA and patents, and the biotechnology industry.

Vogel, Fredrich, and Reinhard Grunwald, eds. *Patenting of Human Genes and Living Organisms.* New York: Springer, 1994. Provides an overview of patent acquisition and legal concerns. Illustrated.

Paternity Tests

Field of study: Human genetics and social issues

Significance: *Establishing paternity can be important for establishing legal responsibility for child support, health insurance, veterans' and social security benefits, and legal access to medical records. It may also affect a child's future as it relates to inherited diseases.*

Key terms

FORENSIC GENETICS: the use of genetic tests and principles to resolve legal questions

HUMAN LEUKOCYTE ANTIGENS (HLA): antigens produced by a cluster of genes that play a critical role in the outcome of transplants; because they are made up of a large number of genes, they are used in individual identification and the matching of parents and offspring

PATERNITY EXCLUSION: the indication, through genetic testing, that a particular man is not the biological father of a particular child

Genetic Principles of Paternity Testing

The basic genetic principles utilized in paternity testing have remained the same from the first applications of ABO blood groups to applications of DNA fingerprinting. Available tests may positively exclude a man from being a child's biological father. Evidence supporting paternity, however, cannot be considered conclusive. Ultimately, a court must decide whether a man is determined to be the legal father based on all lines of evidence.

The genetic principles can be illustrated with a very simple example that uses ABO blood types. The four blood groups (A, B, AB, and O) are controlled by three pairs of genes. In the example, however, only three of the blood groups will be used to demonstrate the range of matings with the possible children for each of them (see the table headed "Blood Types, Genes, and Possible Offspring").

> Example 1: A man is not excluded.
> Mother: A
> Child: A
> Putative Father: AB

It can be seen that the mothers in matings 1 and 4 satisfy the condition of the mother being A and possibly having a child being A. Mating 4 satisfies the condition of a father being AB, the mother A, and a possible child being A. Results indicate that the putative father could be the father. He is not excluded.

> Example 2: A man is excluded.
> Mother: A
> Child: A
> Putative Father: B

Again, it is seen that the mothers in matings 1, 4, and 7 satisfy the condition of the mother being A and possibly having a child being A.

Mating 7 satisfies the condition of a father being B and the mother A, but mating 7 cannot produce a child being A. The putative father cannot be the father, and he is excluded.

DNA Fingerprinting

After the initial use of ABO blood groups in paternity testing, it became apparent that there were many cases in which the ABO phenotypes did not permit exclusion. Other blood group systems have also been used, including the MN and Rh groups. As more blood groups are utilized, the probability of exclusion (or nonexclusion) increases. Paternity tests have not been restricted to blood groups alone; tissue types and serum enzymes have also been used.

The most powerful tool developed has been DNA testing. DNA fingerprinting was developed in England by Sir Alec Jeffreys. DNA is extracted from white blood cells and broken down into fragments by bacterial enzymes (restriction endonucleases). The fragments are separated by size, and specific fragments are identified. Each individual has a different DNA profile, but the profiles of parents and children have similarities in greater proportion than those between unrelated people. Also, frequencies of different fragments tend to vary among ethnic groups. It is possible not only to exclude

Blood Types, Genes, and Possible Offspring

Mating Number	Genes of Parents		Blood Type of Parents		Possible Children	
	Father	Mother	Father	Mother	Genes	Blood Type
1	AA	AA	A	A	AA	A
2	AA	AB	A	AB	AA or AB	A or AB
3	AA	BB	A	B	AB	AB
4	AB	AA	AB	A	AA or AB	A or AB
5	AB	AB	AB	AB	AA, AB, or BB	A, AB, or B
6	AB	BB	AB	B	AB or BB	AB or B
7	BB	AA	B	A	AB	AB
8	BB	AB	B	AB	AB or BB	AB or B
9	BB	BB	B	B	BB	B

someone who is not the biological father but also to determine actual paternity with a probability approaching 100 percent.

Impact and Applications

The personal, social, and economic implications involved in paternity testing have far-reaching consequences. Blood-group analysis is cheaper but less consistent than DNA testing. Paternity can often be excluded but rarely proven with the same degree of accuracy that DNA testing provides. Human leukocyte antigen (HLA) testing can also be used but suffers from many of the same problems as blood-group analysis. The development of DNA testing after 1984 revolutionized the field of paternity testing. DNA fingerprinting has made decisions on paternity assignments virtually 100 percent accurate. The same technique has also been applied in cases of individual identification, and results have helped to release people who have been falsely imprisoned as well as convict other people with the analysis of trace evidence.

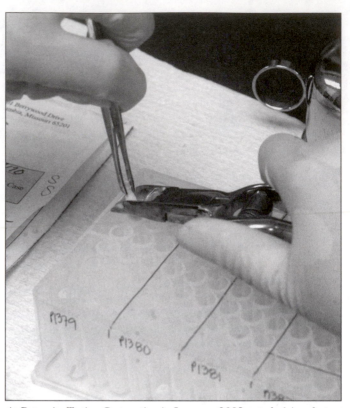

At Paternity Testing Corporation in January, 2003, a technician drops a specimen into a dish with lysis solution to extract DNA for a paternity test. (AP/Wide World Photos)

—*Donald J. Nash*

See also: DNA Fingerprinting; Forensic Genetics; Repetitive DNA.

Further Reading

Anderlik, Mary R., and Mark A. Rothstein. "DNA-Based Identity Testing and the Future of the Family: A Research Agenda." *American Journal of Law and Medicine* 28 (2002). Covers DNA-based identity testing, misattributed paternity, legal issues, and more.

Cohen, Warren. "Kid Looks Like the Mailman? Genetic Labs Boom as the Nation Wonders Who's Daddy." *U.S. News and World Report* 122 (January 27, 1997). Discusses paternity testing at genetic laboratories.

Goodman, Christi. *Paternity, Marriage, and DNA.* Denver, Colo.: National Conference of State Legislatures, 2001. A concise discussion of critical policy issues surrounding paternity, DNA, and marriage.

Office of the Inspector General, Department of Health and Human Services. *Eight Reports on Paternity Establishment.* 1997-2000 (OEI-06-98-0054). Report no. 5 covers genetic testing for paternity at the state level.

Sonenstein, Freya L., Pamela A. Holcomb, and Kristin S. Seefeldt. *Promising Approaches to Improving Paternity Establishment Rates at the Local Level.* Washington, D.C.: Urban Institute, 1993. Reviews paternity establishment procedures.

Weir, Bruce S. *Human Identification: The Use of DNA Markers.* New York: Kluwer Academic, 1995. Discussion includes the debates over using DNA profiles to identify paternity. Bibliography.

Web Sites of Interest

Earl's Forensic Page. http://members.aol.com/EarlNMeyer/DNA.html. Summarizes how DNA fingerprinting works and its use in crime investigations and in determining paternity.

National Newborn Screening and Genetics Resource Center. http://genes-r-us.uthscsa.edu. Site serves as a resource for information on genetic screening, including paternity testing.

Pedigree Analysis

Fields of study: Population genetics; Techniques and methodologies

Significance: *Charts called pedigrees are used to represent the members of a family and to indicate which individuals have particular inherited traits. A pedigree is built of shapes connected by lines. Pedigrees are used by genetic counselors to help families determine the risk of genetic disease and are used by research scientists in determining how traits are inherited.*

Key terms

ALLELES: alternate forms of a gene locus, some of which may cause disease

AUTOSOMAL TRAIT: a trait that typically appears just as frequently in either sex because an autosomal chromosome, rather than a sex chromosome, carries the gene

DOMINANT ALLELE: an allele that is expressed even when only one copy (instead of two) is present

HEMIZYGOUS: the human male is considered to be hemizygous for X-linked traits, because he has only one copy of X-linked genes

HETEROZYGOUS CARRIERS: individuals who have one copy of a particular recessive allele that is expressed only when present in two copies

HOMOZYGOTE: an organism that has identical alleles at the same locus

RECESSIVE ALLELE: an allele that is expressed only when there are two copies present

X-LINKED TRAIT: a trait caused by a gene carried on the X chromosome, which has different patterns of inheritance in females and males because females have two X chromosomes while males have only one

Overview and Definition

Pedigree analysis involves the construction of family trees that can be used to trace inheritance of a trait over several generations. It is a graphical representation of the appearance of a particular trait or disease in related individuals along with the nature of the relationships.

Standardized symbols are used in pedigree charts. Males are designated by squares, females by circles. Symbols for individuals affected by a trait are shaded, while symbols for unaffected individuals are not. Heterozygous carriers are indicated by shading of half of the symbol, while carriers of X-linked recessive traits have a dot in the middle of the symbol. Matings are indicated by horizontal lines linking the mated individuals. The symbols of the individuals who are offspring of the mated individuals are linked to their parents by a vertical line intersecting with the horizontal mating line.

The classic way to determine the mode of inheritance of a trait is to conduct experimental matings of large numbers of individuals. Such experimental matings between humans are not possible, so it is necessary to infer the mode of inheritance of traits in humans through the use of pedigrees. Large families with good historical records are the easiest to analyze. Once a pedigree is established, it can be used to determine the likely mode of inheritance of a particular trait and, if the mode of inheritance can be determined with certainty, to determine the risk of the trait's appearing in offspring.

Typical Pedigrees

There are four common modes of inheritance detected using pedigree analysis: autosomal dominant, autosomal recessive, X-linked dominant, and X-linked recessive. Autosomal traits are governed by genes found on one of the autosomes (chromosomes 1-22), while the genes that cause X-linked traits are found on the X chromosome. Males and females are equally likely to be affected by autosomal traits, whereas X-linked traits are never passed on from father to son and all affected males in a family received the mutant allele from their mothers.

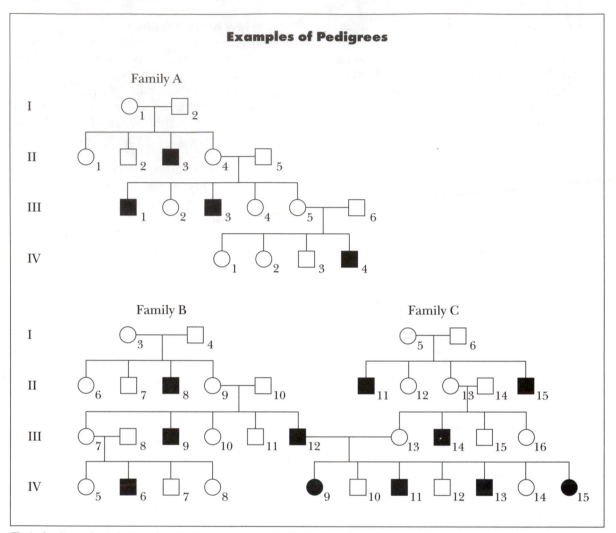

Examples of Pedigrees

Typical pedigree charts for three families: Roman numerals indicate generations. Squares denote male individuals; circles, female individuals; white or blank individuals are "normal" phenotype; black denotes "affected" phenotype. The charts read like a family tree, with "mother" and "father" at the top and vertical lines denoting offspring; individuals connected only by horizontal lines are mates that have entered the genetic line from outside ("in-laws" in the case of humans). Family A provides an example of a sex-linked recessive trait. Families B and C (joined at 12 and 13) give examples of autosomal traits and how they can resemble sex-linked recessive traits sometimes—and hence the reason for using large families when constructing pedigrees. (Bryan Ness)

The pattern of autosomal dominant inheritance is perhaps the easiest type of Mendelian inheritance to recognize in a pedigree. A trait that appears in successive generations, and is found only among offspring where at least one of the parents is affected, is normally due to a dominant allele.

If neither parent has the characteristic phenotype displayed by the child, the trait is recessive. For recessive traits, on average, the recurrence risk to the unborn sibling of an affected individual is one in four. The majority of X-linked traits are recessive. The hallmark of X-linked recessive inheritance is that males are much more likely to be affected than females, because males are hemizygous, that is, they possess only one X chromosome, while females have two X chromosomes. Therefore, a recessive trait on the X chromosome will be expressed in all males who possess that X chromosome, while females with one affected X chromosome will be asymptomatic carriers unless their other X chromosome also carries the recessive trait. The trait or disease is typically passed from an affected grandfather through his carrier daughters to half of his grandsons.

X-linked dominant traits are rare but distinctive. All daughters of an affected male and a normal female are affected, while all sons of an affected male and a normal female are normal. For matings between affected females and normal males, the risk of having an affected child is one in two, regardless of the sex of the child. Males are usually more severely affected than females. The trait may be lethal in males. In the general population, females are more likely to be affected than males, even if the disease is not lethal in males.

Usefulness

Pedigrees are important both for helping families identify the risk of transmitting an inherited disease and as starting points for searching for the genes responsible for inherited diseases. Mendelian ratios do not apply in individual human families because of the small size. Pooling of families is possible; in the United States, the Mormons and the Amish have kept good records that have aided genetic studies.

However, even using large, carefully constructed records, pedigrees can be difficult to construct and interpret for several reasons. Tracing family relationships can be complicated by adoption, children born out of wedlock, blended families, and assisted reproductive technologies that result in children who may not be genetically related to their parents. Additionally, people are sometimes hesitant to supply information because they are embarrassed by genetic conditions that affect behavior or mental stability.

Many traits do not follow clear-cut Mendelian ratios. Extensions and exceptions to Mendel's laws that can confound efforts to develop a useful pedigree are numerous. In diseases with variable expressivity, some of the symptoms of the disease are always expressed but may range from very mild to severe. In autosomal dominant diseases with incomplete penetrance, some individuals who possess the dominant allele may not express the disease phenotype at all. Some traits have a high recurrent mutation rate. An example is achondroplasia (a type of dwarfism), in which 85 percent of cases are due to new mutations, where both parents have a normal phenotype. Traits due to multifactorial inheritance have variable expression as a result of interactions of the genes involved with the environment. Early-acting lethal alleles can lead to embryonic death and a resulting dearth of expected affected individuals. Pleiotropy is the situation in which a single gene controls several functions and therefore has several effects; it can result in different symptoms in different affected individuals. Finally, one trait can have a different basis of inheritance in different families. For example, mutations in any one of more than four hundred different genes can result in hereditary deafness.

Modern Applications

Genetic counseling is one of the key areas in which pedigrees are employed. A genetic counseling session usually begins with the counselor taking a family history and sketching a pedigree with paper and pencil, followed by use of a computer program to create an accurate pedigree. The Human Genome Project has accelerated the number of genetic disorders that can be detected by heterozygote and prenatal screening. A large part of the genetic counselor's job is to determine for whom specific genetic tests are appropriate.

Although genetic tests for many disorders are now available, the genes involved in many other disorders have yet to be identified. Therefore most human gene mapping utilizes molecular DNA markers, which reflect variation at noncoding regions of the DNA near the affected gene, rather than biochemical, morphological, or behavioral traits. A DNA marker is a piece of DNA of known size, representing a specific locus, that comes in identifiable variations. These allelic variations segregate according to Mendel's laws, which means it is possible to follow their transmission as one would any gene's transmission. If a particular allelic variant of the DNA marker is found in individuals with a particular phenotype, the DNA marker can be used to develop a pedigree. The DNA from all available family members is examined and the pedigree is constructed using the presence of the DNA marker rather than phenotypic categories. This method is particularly useful for late-onset diseases such as Huntington's disease, whose victims may not know they carry the deleterious allele until they are in their forties or fifties, well past reproductive years. Although using DNA markers is a powerful method, crossover in the chromosome between the marker and the gene can cause an individual to be normal but still have the marker that suggests presence of the mutant allele. Thus, for all genetic tests there is a small percentage of false positive and false negative results, which must be factored into the advice given during genetics counseling.

—*Lisa M. Sardinia*

See also: Artificial Selection; Classical Transmission Genetics; Complete Dominance; Eugenics; Genetic Counseling; Homosexuality; Incomplete Dominance; Multiple Alleles.

Further Reading

Bennett, Robin L. *The Practical Guide to the Genetic Family History.* New York: Wiley-Liss, 1999. Designed for primary care physicians, this practical book provides the foundation in human genetics necessary to recognize inherited disorders and familial disease susceptibility. Shows how to create a family pedigree.

Bennett, Robin L., et al. "Recommendations for Standardized Human Pedigree Nomenclature." *American Journal of Human Genetics* 56, no. 3 (1995): 745-752. A report from the Pedigree Standardization Task Force that addresses current usage, consistency among symbols, computer compatibility, and the adaptability of symbols to reflect the rapid technical advances in human genetics.

Cummings, Michael R. "Pedigree Analysis in Human Genetics." In *Human Heredity,* edited by Cummings. Pacific Grove, Calif.: Brooks/Cole, 2003. Textbook designed for an introductory human genetics course for non-science majors. This chapter contains many useful diagrams and pictures, ending with several case studies and numerous problems.

Wolff, G., T. F. Wienker, and H. Sander. "On the Genetics of Mandibular Prognathism: Analysis of Large European Noble Families." *Journal of Medical Genetics* 30, no. 2 (1993): 112-116. Good, not overly technical, example of the use of human pedigrees to determine mode of inheritance.

Penetrance

Field of study: Population genetics
Significance: *Penetrance is a measure of how frequently a specific genotype results in the same, predictable phenotype. Such variable expression of the same genotype is the result of different genetic backgrounds and the effects of variations in the environment. Geneticists desire 100 percent penetrance for desirable genes that offer disease resistance but reduced penetrance and low expressivity for others that may contribute to human diseases.*

Key terms

EXPRESSIVITY: the degree to which a phenotype is expressed, or the extent of expression of a phenotype
PHENOTYPE: the physical appearance or biochemical and physiological characteristics of an individual, which is determined by both heredity and environment

Gene Expression and Environment

Gene expression results in a chemical product (protein) with a specific function. The genotype (genetic makeup, or gene) and environmental conditions determine the phenotype of an individual.

Penetrance and Expressivity

Gene expression is dependent upon environmental factors and may be modified, enhanced, silenced, and/or timed by the regulatory mechanisms of the cell in response to internal and external forces. A range of phenotypes can result from a genotype in response to different environments; the phenomenon is called "norms of reaction" or "phenotypic plasticity." Norms of reaction represent the expression of phenotypic variability in individuals of a single genotype.

The question of which is more important in the formation of an organism, nature (genotype) or nurture (environment), has been debated for centuries. The answer is that it depends. The genotype defines phenotypic potential. The environment works on the plasticity of expression to produce different phenotypes from similar genotypes.

Penetrance is the proportion of individuals with a specific genotype who display a defined phenotype. Some individuals may not express a gene if modifiers, epistatic genes, or suppressors are also present in the genome. Penetrance is the likelihood, or probability, that a condition or disease phenotype will, in fact, appear when a given genotype is present. If every person carrying a gene for a dominantly inherited disorder has the mutant phenotype, then the gene is said to have 100 percent penetrance. If

only 30 percent of those carrying the mutant allele exhibit the mutant phenotype, the penetrance is 30 percent. Sometimes an individual with a certain genotype fails to express the expected phenotype, and then the allele is said to be nonpenetrant in the individual. If the phenotype is expressed to any degree, the genotype is penetrant.

Given a particular phenotypic trait and a genotype, penetrance can be expressed as the probability of the phenotype given the genotype. For example, penetrance can be the probability of round seeds, a phenotype, given the genotype G; it can also be the probability of wrinkled seeds, another phenotype, given the genotype G. One could label the specific phenotype of interest as P_i (P_i might refer to either the round or wrinkled seeds) and the specific genotype among many possibilities as G_j. The penetrance would then be the probability of P_i given G_j. These penetrances can all be ex-

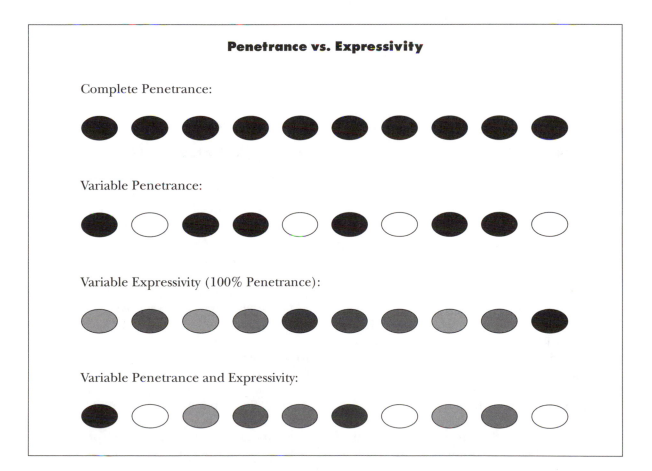

Penetrance vs. Expressivity

Complete Penetrance:

Variable Penetrance:

Variable Expressivity (100% Penetrance):

Variable Penetrance and Expressivity:

pressed using the mathematical notation of conditional probabilities as follows:

Case 1: $\Pr(\text{round}|G)$
Case 2: $\Pr(\text{wrinkled}|G)$
Case 3: $\Pr(P_i|G_j)$

A 100 percent penetrance means that all individuals who possess a particular genotype express the phenotype (common in all homozygous lethal genes). Tay-Sachs disease shows complete, or 100 percent, penetrance, as all homozygotes for this allele develop the disease and die.

An allele, *Fu*, in mice causes fusion in the tail in heterozygotes, *Fufu*, and extremely fused and abnormal tails in the homozygotes, *FuFu*. From testcross matings of *Fufu* × *fufu*, 87 fused-tailed mice and 129 nonfused-tailed mice resulted. Genetic analyses of the 129 nonfused-tailed mice revealed that 22 were genotypically *Fufu*. The number of fused-tailed mice was 87 and the number of mice with the *Fufu* genotype but nonfused tails was 22. The total number of fused-tailed mice expected was (87 + 22) = 109. Therefore, penetrance was calculated at 87/109 = 0.798

Expressivity

Whereas penetrance describes the frequency that a genotype is expressed as a specified phenotype, expressivity describes the range of variation in the phenotype when expression is observed. Expressivity is variation in allelic expression when the allele is penetrant. Not all traits are expressed 100 percent of the time even though the allele is present. Expressivity is the range of variation in a phenotype; it refers to the degree of expression of a given trait or combination of traits that is associated with a gene. Affected individuals may have severe or mild symptoms; they may have symptoms that show up in one organ or combination of organs in one individual but not in the same locations in other individuals.

Phenotype may be altered by heterogeneity of other genes that affect the expression of a particular locus in question, or by environmental influence. Variable expressivity is a common feature of a variety of cancers. The lower the penetrance, the fewer number of individuals will be affected. In humans, the dominant allele *P* produces polydactyly—extra toes and/or fingers. Matings between two normal appearing parents sometimes produce offspring with polydactyly. The parent with the *Pp* genotype exhibits reduced penetrance for the *P* allele.

—Manjit S. Kang

See also: Hereditary Diseases; Pedigree Analysis.

Further Reading

Fairbanks, Daniel J., and W. Ralph Anderson. *Genetics: The Continuity of Life.* New York: Brooks/Cole, 1999. This is one of the rare books that contains a good discussion, in Chapter 13, of the concepts of penetrance and expressivity.

Kang, Manjit S. "Using Genotype-by-Environment Interaction for Crop Cultivar Development." *Advances in Agronomy* 62 (November, 1997): 199-252. This paper discusses environmental influences on heredity.

Kang, Manjit S., and Hugh G. Gauch, Jr. *Genotype-by-Environment Interaction.* Boca Raton, Fla.: CRC Press, 1996. For those interested in in-depth treatment of the interactions between genotypes and environments.

Phenylketonuria (PKU)

Field of study: Diseases and syndromes

Significance: *Phenylketonuria is a relatively common genetic disease affecting about one in every ten thousand newborn babies. If the disease is not detected and treatment is not begun within the first few weeks of life, the child will develop various neurological symptoms including retardation. If the disease is detected shortly after birth and dietary treatment is instituted, symptoms characteristic of the disease usually will not develop.*

Key terms

PHENYLALANINE: an essential amino acid that accumulates in those affected by PKU; phenylalanine and tyrosine, another essential amino acid, can be converted into various compounds such as melanin, epinephrine, norepinephrine, and dopamine

PHENYLALANINE HYDROXYLASE: the enzyme that converts phenylalanine into tyrosine; in those affected by PKU, this enzyme is defective or missing

PHENYLPYRUVIC ACID: a compound derived from phenylalanine that accumulates in those affected by PKU

TYROSINE: an essential amino acid that can be derived from phenylalanine; it can be converted into various compounds such as melanin, epinephrine, norepinephrine, and dopamine

Discovery of PKU

Phenylketonuria, or PKU, was discovered in 1934 by Asbjørn Følling in Norway. Følling discovered that the urine of retarded children turned green when ferric chloride, a chemical used to detect ketones in the urine of diabetics, was added. The urine of diabetics normally turns purple or burgundy with the addition of ferric chloride. Følling conducted further investigations and discovered that the substance responsible for turning urine green upon addition of ferric chloride was phenylpyruvic acid. Følling discovered that the origin of phenylpyruvic acid was the amino acid phenylalanine.

Symptoms and Effects on Metabolism

Common characteristics of untreated patients with PKU are mental retardation, light-colored skin, hyperactivity, schizophrenia, tremors, and eczema.

PKU also has major metabolic effects. In people with normal metabolisms, phenylalanine, an essential amino acid, must be consumed in the diet. Phenylalanine is either incorporated into the body's proteins or converted by the enzyme phenylalanine hydroxylase into tyrosine, another amino acid. Tyrosine is either incorporated into protein or converted into other important biological molecules, such as dopamine, epinephrine, norepinephrine, and melanin. Alternatively, tyrosine can be completely metabolized and eliminated from the body.

People with PKU cannot metabolize phenylalanine into tyrosine at normal rates. Normally, blood phenylalanine concentrations are between 2 and 6 milligrams per deciliter (mg/dl), but in PKU phenylalanine accumulates to 20 mg/dl or more. Since phenylalanine cannot be properly converted into tyrosine, melanin, dopamine, norepinephrine, and epinephrine, there is a deficiency of those important compounds, which probably contributes to the development of symptoms characteristic of the disease. The high levels of phenylalanine may also interfere with the transport of other important amino acids into the brain. Since several amino acids use the same transport system as phenylalanine, phenylalanine is preferentially transported at the expense of the others. This may also contribute to the development of symptoms characteristic of PKU. Additionally, since phenylalanine cannot be metabolized normally, it is metabolized into abnormal compounds such as phenylpyruvic acid, which further contributes to the development of PKU symptoms.

The PKU Gene

The gene responsible for PKU encodes the information for the liver enzyme phenylalanine hydroxylase (PAH), which catalyzes the conversion of phenylalanine to tyrosine. The disease-causing mutant PKU gene is recessive. Thus, in order for a person to have PKU he or she must inherit two copies of the mutant gene. Approximately one in every fifty people in the United States is a heterozygous carrier for the disease. About one in every ten thousand newborn babies has the disease. African Americans have a much lower incidence of PKU than do Caucasian Americans. In certain other populations, such as in Ireland, the incidence of the disease is much higher.

The PKU gene was isolated in 1992, and soon afterward it was discovered that there is no one type of PKU mutation. Instead, the disease can be caused by a variety of defects affecting the PKU gene. Many of these defects are "point" mutations resulting in single base-pair changes in the DNA which lead to amino acid substitutions in the *PAH* gene. Other defects include base-pair changes leading to splicing defects in *PAH* messenger RNA (mRNA), deletions resulting in one or more missing amino acids in PAH, and insertions resulting in mRNA reading frame shifts. More than four hundred mutations have been found in the PKU gene.

The variety of different defects in the PKU gene leads to variability in the activity of PAH and the severity of the disease.

PKU Screening

In 1957 Willard Centerwall introduced ferric chloride as a screening technique by impregnating babies' diapers with ferric chloride. If the babies' urine contained phenylpyruvic acid, the diaper would turn green. Since the test was reliable only after the baby was several weeks old and after brain damage may already have occurred, a new, more reliable and more sensitive test was needed.

Robert Guthrie developed a more sensitive test. In the Guthrie test, bacteria are grown on an agar medium that contains an inhibitor of growth that can be overcome by exogenously added phenylalanine. If a small piece of filter paper containing blood is placed on the agar medium with the bacteria, the phenylalanine in the blood leaches out of the filter paper and stimulates growth of the bacteria. The extent of the growth around the filter paper is directly proportional to the amount of phenylalanine in the blood. Guthrie published his procedure in 1961. In 1963 Massachusetts became the first state to legislate mandatory PKU screening of all newborns. It is now mandatory in all fifty states.

Treatment

The treatment of choice for PKU is dietary or nutritional intervention. PKU babies placed on very low phenylalanine diets show normal cognitive development. The PKU diet eliminates high-protein foods, which are replaced with low-phenylalanine foods and supplemented with a nutritional formula. In 1954, Horst Bickel was the first to treat PKU with diet therapy.

It is recommended that dietary intervention begin as soon as possible after birth and continue for life. It is especially important that pregnant PKU women adhere closely to the diet, or their babies will be mentally retarded. Studies have shown that if children or adults are taken off the diet, some PKU symptoms may develop.

—*Charles L. Vigue*

See also: Biochemical Mutations; Genetic Screening; Genetic Testing; Hereditary Diseases; Inborn Errors of Metabolism; Model Organism: *Mus musculus*.

Further Reading

Koch, Jean Holt. *Robert Guthrie, the PKU Story: A Crusade Against Mental Retardation.* Pasadena, Calif.: Hope, 1997. A longtime friend profiles the scientific work and personal life work of Robert Guthrie.

National PKU News. This newsletter, published in Seattle, Washington, three times per year, provides the latest information about PKU.

Parker, James N. *The Official Parent's Sourcebook on Phenylketonuria.* San Diego, Calif.: ICON Health Publications, 2002. This resource, created for parents with PKU children, tells parents how and where to look for information about PKU.

Web Site of Interest

National Organization for Rare Disorders. http://www.rarediseases.org. Searchable site by type of disorder. Includes background information on PKU, a list of other names for the disease, and a list of related organizations.

Plasmids

Field of study: Molecular genetics

Significance: *Plasmids are DNA molecules that exist separately from the chromosome. Plasmids exist in a commensal relationship with their host and may provide the host with new abilities. They are used in genetic research as vehicles for carrying genes. In the wild, they promote the exchange of genes and contribute to the problem of antibiotic resistance.*

Key terms

COMMENSALISM: a relationship in which two organisms rely on each other for survival

GENE: a region of DNA containing instructions for the manufacture of a protein

TRANSPOSON: a piece of DNA that can copy itself from one location to another

Plasmid Structure

The structure of plasmids is usually circular, although linear forms do exist. Their size ranges from a few thousand base pairs to hundreds of thousands of base pairs. They are found primarily in bacteria but have also been found in fungi, plants, and even humans.

In its commensal relationship with its host, the plasmid can be thought of as a molecular parasite whose primary function is to maintain itself within its host and to spread itself as widely as possible to other hosts. The majority of genes that are present on a plasmid will be dedicated to this function. Researchers have discovered that despite the great diversity of plasmids, most of them have similar genes, dedicated to this function. This relative simplicity of plasmids makes them ideal models of gene function, as well as useful tools for molecular biology. Genes of interest can be placed on a plasmid, which can easily be moved in and out of cells. Using plasmids isolated from the wild, molecular biologists have designed many varieties of artificial plasmids, which have greatly facilitated research in molecular biology.

Plasmid Replication

To survive and propagate, a plasmid must be able to copy itself, or replicate. The genes that direct this process are known as the replication genes. These genes do not carry out all the functions of replication, but instead coopt the host's replication machinery to replicate the plasmid. Replication allows the plasmid to propagate by creating copies of itself that can be passed to each daughter cell when the host divides. In this manner, the plasmid propagates along with the host.

A second function of the replication genes is to control the copy number of the plasmid. The number of copies of a plasmid that exist inside a host can vary considerably. Plasmids can exist at a very low copy number (one or two copies per cell) or at a higher copy number, with dozens of copies per cell. Adjusting the copy number is an important consideration for a plasmid. Plasmid replication is an expensive process that consumes energy and resources of the host cell. A plasmid with a high copy number can place a significant energy drain on its host cell. In environments where the nutrient supply is low, a plasmid-bearing cell may not be able to compete successfully with other, non-plasmid-containing cells. Wild plasmids often exist at a low copy number, or create a high copy number for only a brief period of time.

Plasmid Partitioning

Because the presence of a plasmid is expensive in terms of energy, a cell harboring a plasmid will grow more slowly than a similar cell with no plasmid. This can cause a problem for a plasmid if it fails to partition properly during its host's division. If the plasmid does not partition properly, then one of the host's daughter cells will not contain a plasmid. Since this cell does not have to spend energy replicating a plasmid, it will gain an ability to grow faster, as will all of its offspring. In such a situation, the population of non-plasmid-containing cells could outgrow the population of plasmid-containing cells and use up all the nutrients in the environment. To avoid this problem, plasmids have evolved strategies to prevent improper partitioning. One strategy is for the plasmid to contain partitioning genes. Partitioning genes encode proteins that actively partition plasmids into each daughter cell during the host cell's division. Active partitioning greatly reduces the errors in partitioning that might occur if partitioning were left to chance.

A second strategy that plasmids use to prevent partitioning errors is the plasmid addition system. In this strategy, genes on the plasmid direct the production of both a toxin and an antidote. The antidote protein is very unstable and degrades quickly, but the toxin is quite stable. As long as the plasmid is present, the cytoplasm of the cell will be full of toxin and antidote. Should a daughter cell fail to receive a plasmid during division, the residual antidote and toxin present in the cytoplasm from the mother cell will begin to degrade, since there is no longer a plasmid present to direct the synthesis of either toxin or antidote. Since the antidote is very unstable, it will degrade first, leaving only toxin, which will kill the cell.

Plasmid Transfer Between Cells

Propagation of plasmids can occur through the spread of plasmids from parent cells to their offspring (referred to as vertical transfer), but propagation can also occur between two different cells (referred to as horizontal transfer). Many plasmids are able to transfer themselves from one host to another through the process of conjugation. Conjugal plasmids contain a collection of genes that direct the host cell that contains them to attach to other cells and transfer a copy of the plasmid. In this manner, the plasmid can spread itself to other hosts and is not limited to spreading itself only to the descendants of the original host cell.

One of the first plasmids to be identified was discovered because of its ability to conjugate. This plasmid, known as the F plasmid, or F factor, is a plasmid found in the bacterium *Escherichia coli*. Cells harboring the F plasmid are designated F^+ cells and can transfer their plasmid to other *E. coli* cells that do not contain the F plasmid (called F^- cells).

Conjugal plasmids can be very specific and transfer only between closely related members of the same species (such as the F plasmid), or they can be very promiscuous and allow transfer between unrelated species. An extreme example of cross-species transfer is the Ti plasmid of the bacterial species *Agrobacterium tumefaciens*. The Ti plasmid is capable of transferring part of itself from *A. tumefaciens* into the cells of dicotyledonous plants. Plant cells that receive parts of the Ti plasmid are induced to grow and form a tumorlike structure, called a gall, that provides a hospitable environment for *A. tumefaciens*.

Host Benefits from Plasmids

In most commensal relationships, there is an exchange of benefits between the two partners. The same is true for plasmids and their hosts. In many cases, plasmids provide their host cells with a collection of genes that enhance the ability of the host cell to survive. Enhancements include the ability to metabolize a wider range of materials for food and the ability to survive in hostile environments. One particular hostile environment in which plasmids can provide the ability to survive is the human body. A number of pathogenic microorganisms gain their ability to inhabit the human body, and thus cause disease, from genes contained on plasmids. An example of this is *Bacillus anthracis*, the agent that causes anthrax. Many of the genes that allow this organism to cause disease are contained on one of two plasmids, called pXO1 and pXO2. *Yersinia pestis*, the causative agent of bubonic plague, also gains its disease-causing ability from plasmids.

R Factors

Another example of plasmids conferring on their hosts the ability to survive in a hostile environment is antibiotic resistance. Plasmids known as R factors contain genes that make their bacterial hosts resistant to antibiotics. These R factors are usually conjugal plasmids, so they can move easily from cell to cell. Because the antibiotic resistance genes they carry are usually parts of transposons, they can readily copy themselves from one piece of DNA to another. Two different R factors that happened to be together in one cell could exchange copies of each other's antibiotic resistance genes. A number of R factors exist that contain multiple antibiotic resistance genes. Such plasmids can result in the formation of "multi-drug resistant" (MDR) strains of pathogenic bacteria, which are difficult to treat. There is much evidence to suggest that the widespread use of antibiotics has contributed to the development of MDR pathogens, which are emerging as an important health concern.

Role of Plasmids in Evolution

Through conjugation, plasmids can transfer genetic information from one species of bacterial cell to another. During its stay in a particular host, a plasmid may acquire some of the chromosomal genes of the host, which it then carries to a new host by conjugation. These genes can then be transferred from the plasmid to the chromosome of the new host. If the new host and the old host are different species, this gene transfer can result in the introduction of new genes, and thus new traits, into a cell. Bacteria, being asexual, produce daughter cells that are genetically identical to their parent. The existence of conjugal plasmids, which allow for the transfer of genes between bacterial species, may represent an important mechanism by which bacteria generate diversity and create new species.

—*Douglas H. Brown*

See also: Anthrax; Antisense RNA; Archaea; Bacterial Genetics and Cell Structure; Bacterial Resistance and Super Bacteria; Biopesticides; Biopharmaceuticals; Blotting: Southern, Northern, and Western; Cloning; Cloning Vectors; DNA Sequencing Technology; Emerging Diseases; Extrachromosomal Inheritance; Gene Regulation: Bacteria; Genetic Engineering; Genetic Engineering: Agricultural Applications; Genetic Engineering: Historical Development; Genetic Engineering: Industrial Applications; Genome Size; Genomics; High-Yield Crops; Human Growth Hormone; Immunogenetics; Model Organism: *Chlamydomonas reinhardtii*; Model Organism: *Escherichia coli*; Model Organism: *Saccharomyces cerevisiae*; Model Organism: *Xenopus laevis*; Noncoding RNA Molecules; Polymerase Chain Reaction; Proteomics; Shotgun Cloning; Transgenic Organisms; Transposable Elements.

Further Reading

Levy, Stuart B. "The Challenge of Antibiotic Resistance." *Scientific American* 278 (1998): 46-53. A discussion on the growing problem of antibiotic resistance. Written by one of the experts in the field.

Summers, David K. *The Biology of Plasmids.* Malden, Mass.: Blackwell, 1996. A comprehensive book on plasmid biology written for college undergraduates.

Thomas, Christopher M. "Paradigms of Plasmid Organization." *Molecular Microbiology* 37, no. 3 (2000): 485-491. A review that clearly discusses the evolution and organization of plasmid genes.

Polygenic Inheritance

Fields of study: Classical transmission genetics

Significance: *Polygenically inherited traits—characterized by the amount of some attribute that they possess but not by their presence or absence—are central to plant and animal breeding, medicine, and evolutionary biology. Most of the economically important traits in plants and animals—for example, yield and meat production—are polygenic in nature. Quantitative genetic principles are applied to improve such traits.*

Key terms

HERITABILITY: the proportion of the total observed variation for a trait attributable to heredity or genes

MERISTIC TRAIT: traits that are counted, such as number of trichomes or bristles

QUANTITATIVE TRAIT: a trait, such as human height or weight, that shows continuous variation in a population and can be measured; also called a metric trait

QUANTITATIVE TRAIT LOCI (QTLs): genomic regions that condition a quantitative trait, generally identified via DNA-based markers

THRESHOLD TRAITS: characterized by discrete classes at an outer scale but exhibiting continuous variation at an underlying scale; for example, diabetes, schizophrenia, and cancer

Discovery of Polygenic Inheritance

Soon after the rediscovery of Gregor Mendel's laws of inheritance in 1900, Herman Nilsson-Ehle, a Swedish geneticist, showed in 1909 how multiple genes with small effects could collectively affect a continuously varying character. He crossed dark, red-grained wheat with white-grained wheat and found the progeny with an intermediate shade of red. Upon crossing the progeny among themselves, he obtained grain colors ranging from dark red to white. He could classify the grains into five groups in a symmetric ratio of 1:4:6:4:1, with the extreme phenotypes being one-sixteenth dark red and one-sixteenth white. This suggested two-gene segregation. For a two-gene

($n = 2$) model, the number and frequency of phenotypic classes ($2n + 1 = 5$) can be determined by expanding the binomial $(a + b)4$, where a represents number of favorable alleles and b represents number of nonfavorable alleles.

Subsequently, Nilsson-Ehle crossed a different variety of red-grained wheat with white-grained wheat. He found that one-sixty-fourth of the plants produced dark red kernels and one-sixty-fourth produced white kernels. There were a total of seven phenotypic (color) classes instead of five. The segregation ratio corresponded to three genes: $(a + b)^6 = 1a^6 + 6a^5b^1 + 15a^4b^2 + 20a^3b^3 + 15a^2b^4 + 6a^1b^5 + 1b^6$. Here, a^6 means that one of sixty-four individuals possessed six favorable alleles, $20a^3b^3$ means that twenty of sixty-four individuals had three favorable and three nonfavorable alleles, and b^6 means that one individual had six nonfavorable alleles. An assumption was that each of the alleles had an equal, additive effect. These experiments led to what is known as the multiple-factor hypothesis, or polygenic inheritance (Kenneth Mather coined the terms "polygenes" and "polygenic traits"). Around 1920, Ronald Aylmer Fisher, Sewall Green Wright, and John Burdon Sanderson Haldane developed methods of quantitative analysis of genetic effects.

Polygenic traits are characterized by the amount of some attribute that they possess but not by presence or absence, as is the case with qualitative traits that are controlled by one or two major genes. Environmental factors generally have little or no effect on the expression of a gene or genes controlling a qualitative trait, whereas quantitative traits are highly influenced by the environment and genotype is poorly represented by phenotype. Genes controlling polygenic traits are sometimes called minor genes.

Examples and Characteristics of Polygenic Traits

Quantitative genetics encompasses analyses of traits that exhibit continuous variation caused by polygenes and their interactions among themselves and with environmental factors. Such traits include height, weight, and some genetic defects.

Diabetes and cancer are considered to be threshold traits because all individuals can be classified as affected or unaffected (qualitative). They are also continuous traits because severity varies from nearly undetectable to extremely severe (quantitative). Because it is virtually impossible to determine the exact genotype for such traits, it is difficult to control defects with a polygenic mode of inheritance.

Detection of Genes Controlling Polygenic Traits

The detection of genes controlling polygenic traits is challenging and complex because:

(1) The expression of genes controlling such traits is modified by fluctuations in environmental and/or management factors.
(2) A quantitative trait is usually a composite of many other traits, each influenced by many genes with variable effects.
(3) Effects of allele substitution are small because many genes control the trait.
(4) Expression of an individual gene may be modified by the expression of other genes and environment.

Polygenic traits are best analyzed with statistical methods, the simplest of which are estimation of arithmetic mean, standard error, variance, and standard deviation. Two populations can have the same mean, but their distribution may be different. Thus, one needs information on variances for describing the two populations more fully. From variances, effects of genes can be ascertained in the aggregate rather than as individual genes.

The issues in quantitative genetics are not only how many and which genes control a trait but also how much of what is observed (phenotype) is attributable to genes (heritability) and how much to the environment. The concept of heritability in the broad sense is useful for quantitative traits, but heritability itself does not give any clues to the total number of genes involved. If heritability is close to 1.0, the variance for a trait is attributable entirely to genetics, and when it is close to zero, the population's phenotype is due entirely to the variation in the underlying environment. Environmental effects mask or modify genetic effects.

Distribution or frequency of different classes in segregating populations—for example, F_2—may provide an idea about the number of genes, particularly if the gene number is small (say, three to four). Formulas have been devised to estimate the number of genes conditioning a trait, but these estimates are not highly reliable. Genes controlling quantitative traits can be estimated via use of chromosomal translocations or other cytogenetic procedures. The advent of molecular markers, such as restriction fragment length polymorphisms, has made it easier and more reliable to pinpoint the location of genes on chromosomes of a species of interest. With much work in a well-characterized organism, these polygenes can be mapped to chromosomes as quantitative trait loci.

—*Manjit S. Kang*

See also: Congenital Defects; Genetic Engineering; Hereditary Diseases; Neural Tube Defects; Pedigree Analysis; Plasmids; Quantitative Inheritance.

Further Reading

Kang, Manjit S. *Quantitative Genetics, Genomics, and Plant Breeding.* Wallingford, Oxon, England: CABI, 2002. Provides various methods of studying metric or quantitative traits, especially with DNA-based markers.

Lynch, Michael, and Bruce Walsh. *Genetics and Analysis of Quantitative Traits.* Sunderland, Mass.: Sinauer Associates, 1998. Gives an overview of the history of quantitative genetics and covers evolutionary genetics.

Polymerase Chain Reaction

Fields of study: Genetic engineering and biotechnology; Molecular genetics; Techniques and methodologies

Significance: *Polymerase chain reaction (PCR) is the in vitro (in the test tube) amplification of specific nucleic acid sequences. In a few hours, a single piece of DNA can be copied one billion times. Because this technique is simple, rapid, and very sensitive, it is used in a very wide range of applications, including forensics, disease diagnosis, molecular genetics, and nucleic acid sequencing.*

Key terms

DNA POLYMERASE: an enzyme that copies or replicates DNA; it uses a single-stranded DNA as a template for synthesis of a complementary new strand and requires an RNA primer or a small section of double-stranded DNA to initiate synthesis

MOLECULAR CLONING: the process of splicing a piece of DNA into a plasmid, virus, or phage vector to obtain many identical copies of that DNA

The Development of the Polymerase Chain Reaction

The polymerase chain reaction (PCR) was developed by Kary B. Mullis in the mid-1980's. The technique revolutionized molecular genetics and the study of genes. One of the difficulties in studying genes is that a specific gene can be one of approximately twenty-one thousand genes in a complex genome. To obtain the number of copies of a specific gene needed for accurate analysis required the time-consuming techniques of molecular cloning and detection of specific DNA sequences. The polymerase chain reaction changed the science of molecular genetics by allowing huge numbers of copies of a specific DNA sequence to be produced without the use of molecular cloning. The tremendous significance of this discovery was recognized by the awarding of the 1993 Nobel Prize in Chemistry to Mullis for the invention of the PCR method. (The 1993 prize was also awarded to Michael Smith, for work on oligonucleotide-based, site-directed mutagenesis and its development for protein studies.)

How Polymerase Chain Reaction Works

PCR begins with the creation of a single-stranded DNA template to be copied. This is done by heating double-stranded DNA to temperatures near boiling (about 94 to 99 degrees Celsius, or about 210 degrees Fahrenheit). This is followed by the annealing (binding of a complementary sequence) of pairs of oligonucleotides (short nucleic acid molecules about ten to twenty nucleotides long) called primers. Because DNA polymerase requires a double-stranded region to prime (initiate) DNA synthesis, the starting point for DNA synthesis is specified by the location at which the primer anneals to the template. The primers are chosen to flank the DNA to be amplified. This annealing is done at a lower temperature (about 30-65 degrees Celsius, or about 86-149 degrees Fahrenheit). The final step is the synthesis by DNA polymerase of a new strand of DNA complementary to the template starting from the primers. This step is carried out at temperatures about 65-75 degrees Celsius (149-167 degrees Fahrenheit). These three steps are repeated many times (for many cycles) to amplify the template DNA. The time for each of the three steps is typically one to two minutes. If, in each cycle, one copy is made of each of the strands of the template, the number of DNA molecules produced doubles each cycle. Because of this doubling, more than one million copies of the template DNA are made at the end of twenty cycles.

The PCR reaction is made more efficient by the use of heat-stable DNA polymerases, isolated from bacteria that live at very high temperatures in hot springs or deep-sea vents, and by the use of a programmable water bath (called a thermal cycler) to change the temperatures of samples quickly to each of the temperatures needed in each of the steps of a cycle.

Impact and Applications

PCR is extremely rapid. One billion copies of a specific DNA can be made in a few hours. It is also extremely sensitive. It is possible to copy a single DNA molecule. Great care must be taken to avoid contamination, however, for even trace contaminants can readily be amplified by this method.

PCR is a useful tool for many different applications. It is used in basic research to obtain DNA for sequencing and other analyses. PCR is

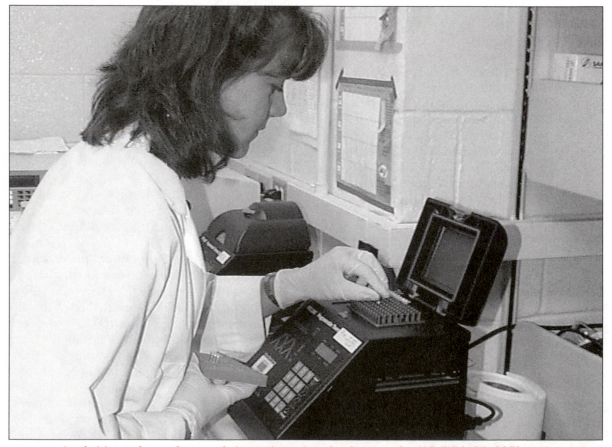

A technician performs polymerase chain reaction testing of anthrax samples. (AP/Wide World Photos)

used in disease diagnosis, in prenatal diagnosis, and to match donor and recipient tissues for organ transplants. Because a specific sequence can be amplified greatly, much less clinical material is needed to make a diagnosis. The assay is also rapid, so results are available sooner. PCR is used to detect pathogens, such as the causative agents for Lyme disease or for acquired immunodeficiency syndrome (AIDS), that are difficult to culture. PCR can even be used to amplify DNA from ancient sources such as mummies, bones, and other museum specimens. PCR is an important tool in forensic investigations. Target DNA from trace amounts of biological material such as semen, blood, and hair roots can be amplified. There are probes for regions of human DNA that show hypervariability in the population and therefore make good markers to identify the source of the DNA. PCR can therefore be used to evaluate evidence at the scene of a crime, help identify missing people, and resolve paternity cases.

—*Susan J. Karcher*

See also: Ancient DNA; Anthrax; Bioinformatics; Blotting: Southern, Northern, and Western; Central Dogma of Molecular Biology; Cloning Vectors; DNA Fingerprinting; DNA Sequencing Technology; Forensic Genetics; Genetic Engineering: Historical Development; Human Genome Project; In Vitro Fertilization and Embryo Transfer; Mitochondrial Diseases; Molecular Genetics; Paternity Tests; Repetitive DNA; RFLP Analysis; RNA Isolation; Swine Flu.

Further Reading

Budowle, Bruce, et al. *DNA Typing Protocols: Molecular Biology and Forensic Analysis.* Natick, Mass.: Eaton, 2000. Discussion includes DNA extraction and PCR-based analyses. Illustrations, bibliography, index.

Chen, Bing-Yuan, and Harry W. Janes, eds. *PCR Cloning Protocols.* Rev. 2d ed. Totowa, N.J.: Humana Press, 2002. Presents helpful introductory chapters with each section and guidelines for PCR cloning. Illustrations, bibliographies, index.

Guyer, Ruth L., and Daniel E. Koshland, Jr. "The Molecule of the Year." *Science* 246 (December 22, 1989): 1543-1546. Reviews the "major scientific development of the year," the polymerase chain reaction, noting that the technique, although introduced earlier, "truly burgeoned" in 1989.

Innis, Michael A., David H. Gelfand, and John J. Sninsky, eds. *PCR Applications: Protocols for Functional Genomics.* San Diego: Academic Press, 1999. Discusses gene discovery, genomics, and DNA array technology. Entries on nomenclature, expression, sequence analysis, structure and function, electrophysiology, parmacology, and information retrieval. Illustrations, bibliography, index.

Kochanowski, Bernd, and Udo Reischl, eds. *Quantitative PCR Protocols.* Methods in Molecular Medicine 26. Totowa, N.J.: Humana Press, 1999. Outlines protocols and includes methodological and process notes. Illustrations, bibliography, index.

Lloyd, Ricardo V., ed. *Morphology Methods: Cell and Molecular Biology Techniques.* Totowa, N.J.: Humana Press, 2001. Includes an overview of PCR. Black-and-white and color illustrations, bibliography, index.

McPherson, M. J., and S. G. Møller. *PCR Basics.* Oxford, England: BIOS Scientific, 2000. Provides introductory information about PCR theory, background, and protocols. Illustrations, bibliography, index.

Mullis, Kary B. "The Unusual Origin of the Polymerase Chain Reaction." *Scientific American* 262 (April, 1990). Nobel laureate Mullis describes the initial development of the technique for the general audience.

Watson, James D., et al. *Recombinant DNA.* New York: Scientific American Books, 1992. Summarizes polymerase chain reaction and its applications. Full-color illustrations, diagrams, bibliography, index.

Polyploidy

Field of study: Population genetics
Significance: *Polyploids have three or more complete sets of chromosomes in their nuclei instead of the two sets found in diploids. Polyploids are especially common in plants, with some examples also existing in animals, and have a prominent role in*

the evolution of species. Some tissues of diploid organisms are polyploid, while the remaining cells in the organism are diploid.

Key terms

ALLOPOLYPLOID: a type of polyploid species that contains genomes from more than one ancestral species

ANEUPLOID: a cell or an organism with one or more missing or extra chromosomes; the opposite is "euploid," a cell with the normal chromosome number

AUTOPOLYPLOID: a type of polyploid species that contains more than two sets of chromosomes from the same species

HOMOLOGOUS CHROMOSOMES: chromosomes that are structurally the same and have the same gene loci, although they may have different alleles (alternative forms of a gene) at many of their shared loci

The Formation of Polyploidy

Most animals are diploid, meaning that they have two homologous sets of chromosomes in their cells; and their gametes (eggs and sperm) are haploid, that is, having one set of chromosomes. Plants, a variety of single-celled eukaryotes, and some insects have individual or parts of an individual's life cycle when they are haploid. In any case, when there are more than two sets of homologous chromosomes, the cell or organism is considered polyploid. A triploid organism has three sets of homologous chromosomes, a tetraploid has four sets, a dodecaploid has twelve sets, and there are organisms known to have many more than a dozen sets of homologous chromosomes.

How polyploids are formed in nature is still debated. Regardless of what theory is accepted, the first step certainly involves a failure during cell division, in either meiosis or mitosis. For example, if cytokinesis (division of the cytoplasm) fails at the conclusion of meiosis II, the daughter cells will be diploid. If, by chance, a diploid sperm fertilizes a diploid egg, the resulting zygote will be tetraploid. Although polyploidy might occur this way, biologists have proposed an alternative model involving a triploid intermediate stage.

The triploid intermediate model has been applied primarily to plants, in which polyploidy is better studied. Hybrids between two species are often sterile, but occasionally a diploid gamete from one of the species joins with a normal haploid gamete from the other species, which produces a triploid hybrid. Triploids are also sterile, for the most part, but do produce a small number of gametes, many of which are diploid. This makes the probability that two diploid gametes will join, to form a tetraploid, much higher. This hypothesis is supported by the discovery of triploid hybrid plants that do produce a small number of viable gametes. This type of polyploid, formed as a result of hybridization between two species, is called an allopolyploid. Allopolyploids are typically fertile and represent a new species.

Polyploidy can also occur within a single species, without hybridization, in which case it is called an autopolyploid. Autopolyploids can form in the same way as allopolyploids, but they can also occur as the result of a failure in cell division in a bud. If a cell in the meristematic region (a rapidly dividing group of cells at the tip of a bud) completes mitosis but not cytokinesis, it will be a tetraploid cell. All daughter cells from this cell will also be tetraploid, so that any flowers borne on this branch will produce diploid gametes. If the plant is self-compatible, it can then produce tetraploid offspring from these flowers. Autopolyploids are often a little larger and more robust than the diploids that produce them, but they are often so similar they cannot be easily distinguished. An autopolyploid, when formed, represents a new species but is not generally recognized as such unless it looks different enough physically from diploids.

The Genetics of Polyploids

A polyploid has more copies of each gene than a diploid. For example, a tetraploid has four alleles at each locus, which means tetraploids can contain much more individual variability than diploids. This has led some evolutionists to suggest that polyploids should have higher fitness than the diploids from which they came. With more variation, the individual would be preadapted to a much wider range of conditions. Because there are so many extra

Wheat is one of many important polyploid crops. (AP/Wide World Photos)

copies of genes, a certain amount of gene silencing (loss of genes through mutation or other processes) occurs, with no apparent detriment to the plant.

The pairing behavior of chromosomes in polyploids is also unique. In a diploid, during meiosis, homologous chromosomes associate in pairs. In an autotetraploid there are four homologous chromosomes of each type which associate together in groups of four. In an allotetraploid, the chromosomes from the two species from which they are derived are commonly not completely homologous and do not associate together. Consequently, the pairs of homologous chromosomes from one parent species associate together in pairs, as do the chromosomes from the other parent species. For this reason, sometimes allopolyploids are referred to as amphidiploids, because their pairing behavior looks the same as it does in a diploid. This is also why an allopolyploid is fertile (because meiosis occurs normally), but a hybrid between two diploids commonly is not, because the chromosomes from the two species are unable to pair properly.

Polyploid Plants and Animals

In the plant kingdom, it is estimated by some that 95 percent of pteridophytes (plants, including ferns, that reproduce by spores) and perhaps as many as 80 percent of angiosperms (flowering plants that form seeds inside an ovary) are polyploid, although there is high variability in its occurrence among families of angiosperms. In contrast, polyploidy is uncommon in gymnosperms (plants that have naked seeds that are not within specialized structures). Extensive polyploidy is observed in chrysanthemums, in which chromosome numbers range from 18 to 198. The basic chromosome number (haploid or gamete number of chromosomes) is 9. Polyploids from triploids (with 27 chromosomes) to 22-ploids (198 chromosomes) are observed. The stonecrop *Sedum suaveolens*, which has the highest chromosome number of any angiosperm, is believed to be about 80-ploid (720 chromosomes). Many important agricultural crops, including wheat, corn, sugarcane, potatoes, coffee, apples, and cotton, are polyploid.

Polyploid animals are less common than polyploid plants but are found among some groups, including crustaceans, earthworms, flatworms, and insects such as weevils, sawflies, and moths. Polyploidy has also been observed in some vertebrates, including tree frogs, lizards, salamanders, and fish. It has been suggested that the genetic redundancy observed in vertebrates may be caused by ancestral polyploidy.

Polyploidy in Tissues

Most plants and animals contain particular tissues that are polyploid or polytene, while the rest of the organism is diploid. Polyploidy is observed in multinucleate cells and in cells that have undergone endomitosis, in which the chromosomes condense but the cell does not undergo nuclear or cellular division. For example, in vertebrates, liver cells are binucleate and therefore tetraploid. In addition, in humans, megakaryocytes can have polyploidy levels of up to sixty-four. A megakaryocyte is a giant bone-marrow cell with a large, irregularly lobed nucleus that is the precursor to blood platelets. A megakaryocyte does not circulate, but forms platelets by budding. A single megakaryocyte can produce three thousand to four thousand platelets. A platelet is an enucleated, disk-shaped cell in the blood that has a role in blood coagulation. In polytene cells, the replicated copies of the chromosomal DNA remain associated to produce giant chromosomes that have a continuously visible banding pattern. The trophoblast cells of the mammalian placenta are polytene.

Importance of Polyploids to Humans

Most human polyploids die as embryos or fetuses. In a few rare cases, a polyploid infant is born that lives for a few days. In fact, polyploidy is not tolerated in most animal systems. Plants, on the other hand, show none of these problems with polyploidy. Some crop plants are much more productive because they are polyploid. For example, wheat (*Triticum aestivum*) is

an allohexaploid and contains chromosome sets that are derived from three different ancient types. Compared to the species from which it evolved, *T. aestivum* is far more productive and produces larger grains of wheat. *Triticum aestivum* was not developed by humans but appears to have arisen by a series of chance events in the past, humans simply recognizing the better qualities of *T. aestivum*. Another fortuitous example involves three species of mustard that have given rise to black mustard, turnips, cabbage, broccoli, and several other related crops, all of which are allotetraploids.

Polyploids may be induced by the use of drugs such as colchicine, which halts cell division. Because of the advantages of the natural polyploids used in agriculture, many geneticists have experimented with artificially producing polyploids to improve crop yields. One prime example of this approach is *Triticale*, which represents an allopolyploid produced by hybridizing wheat and rye. Producing artificial polyploids often produces a new variety that has unexpected negative characteristics, so that only a few such polyploids have been successful. Nevertheless, research on polyploidy continues.

—Susan J. Karcher, updated by Bryan Ness

See also: Cell Division; Cytokinesis; High-Yield Crops; Gene Families; Genome Size; Hereditary Diseases; Nondisjunction and Aneuploidy.

Further Reading

Adams, Keith L., Richard Cronn, Ryan Percifield, and Jonathan F. Wendel. "Genes Duplicated by Polyploidy Show Unequal Contributions to the Transcriptome and Organ-Specific Reciprocal Silencing." *Proceedings of the National Academy of Sciences* 100 (April 15, 2003): 4649-4654. This article shows that with multiple copies of a gene due to polypoidy, some of the copies are silenced.

Hunter, Kimberley L., et al. "Investigating Polyploidy: Using Marigold Stomates and Fingernail Polish." *American Biology Teacher* 64 (May, 2002). A guide to exploring polyploidy through hands-on learning. Experiment supports National Science Education Standards.

Leitch, Illia J., and Michael D. Bennett. "Polyploidy in Angiosperms." *Trends in Plant Science* 2 (December, 1997). Describes the role of polyploidy in the evolution of higher plants.

Lewis, Ricki. *Human Genetics: Concepts and Applications.* 5th ed. Boston: McGraw-Hill, 2003. Gives an overview of polyploidy and aneuploidy in humans. Color ilustrations, maps, and CD-ROM with this edition.

Population Genetics

Field of study: Population genetics

Significance: *Population genetics is the study of how genes behave in populations. It is concerned with both theoretical and experimental investigations of changes in genetic variation caused by various forces; therefore, the field has close ties to evolutionary biology. Population genetics models can be used to explore the evolutionary histories of species, make predictions about future evolution, and predict the behavior of genetic diseases in human populations.*

Key terms

ALLELE: one of the different forms of a particular gene (locus)

FITNESS: a measure of the ability of a genotype or individual to survive and reproduce compared to other genotypes or individuals

GENE POOL: all of the alleles in all the gametes of all the individuals in a population

GENETIC DRIFT: random changes in genetic variation caused by sampling error in small populations

GENOTYPE: the pair of alleles carried by an individual for a specific gene locus

HARDY-WEINBERG LAW: a mathematical model that predicts, under particular conditions, that allele frequencies will remain constant over time, with genotypes in specific predictable proportions

MODERN SYNTHESIS: the merging of the Darwinian mechanisms for evolution with Mendelian genetics to form the modern fields of population genetics and evolutionary biology

NEUTRAL THEORY OF EVOLUTION: Motoo Kimura's theory that nucleotide substitutions in the DNA often have no effect on fitness, and thus changes in allele frequencies in populations are caused primarily by genetic drift

The Hardy-Weinberg Law

The branch of genetics called population genetics is based on the application of nineteenth century Austrian botanist Gregor Mendel's principles of inheritance to genes in a population. (Although, for some species, "population" can be difficult to define, the term generally refers to a geographic group of interbreeding individuals of the same species.) Mendel's principles can be used to predict the expected proportions of offspring in a cross between two individuals of known genotypes, where the genotype describes the genetic content of an individual for one or more genes. An individual carries two copies of all chromosomes (except perhaps for the sex chromosomes, as in human males) and therefore has two copies of each gene. These two copies may be identical or somewhat different. Different forms of the same gene are called alleles. A genotype in which both alleles are the same is called a homozygote, while one in which the two alleles are different is a heterozygote. Although a single individual can carry no more than two alleles for a particular gene, there may be many alleles of a gene present in a population.

It would be essentially impossible to track the inheritance patterns of every single mating pair in a population, in essence tracking all the alleles in the gene pool. However, by making some simplifying assumptions about a population, it is possible to predict what will happen to the gene pool over time. Working independently in 1908, the British mathematician Godfrey Hardy and the German physiologist Wilhelm Weinberg were the first to formulate a simple mathematical model describing the behavior of a gene (locus) with two alleles in a population. In this model, the numbers of each allele and of each genotype are not represented as actual numbers but as proportions (known as allele frequencies and genotype frequencies, respectively) so that the model can be applied to any population regardless of its size. By assuming Mendelian inheritance of alleles, Hardy and Weinberg showed that allele frequencies in a population do not change over time and that genotype frequencies will change to specific proportions, determined by the allele frequencies, within one generation and remain at those proportions in future generations. This result is known as the Hardy-Weinberg law, and the stable genotype proportions predicted by the law are known as Hardy-Weinberg equilibrium. It was shown in subsequent work by others that the Hardy-Weinberg law remains true in more complex models with more than two alleles and more than one locus.

In order for the Hardy-Weinberg law to work, certain assumptions about a population must be true:

(1) the gene pool must be infinite in size;
(2) mating among individuals (or the fusion of gametes) must be completely random;
(3) there must be no new mutations;
(4) there must be no gene flow (that is, no alleles should enter or leave the population; and
(5) there should be no natural selection.

Since real populations cannot meet these conditions, it may seem that the Hardy-Weinberg model is too unrealistic to be useful, but, in fact, it can be useful. First, the conditions of a natural population may be very close to Hardy-Weinberg assumptions, so the Hardy-Weinberg law may be approximately true for at least some populations. Second, if genotypes in a population are not in Hardy-Weinberg equilibrium, it is an indication that one or more of these assumptions is not met. The Hardy-Weinberg law has been broadly expanded, using sophisticated mathematical modeling, and with adequate data can be used to determine why a population's allele and genotype frequencies are out of Hardy-Weinberg equilibrium.

Genetic Variation and Mathematical Modeling

Sampling and genetic analyses of real populations of many different types of organisms reveal that there is usually a substantial amount of genetic variation, meaning that for a fairly large proportion of genes (loci) that are analyzed, there are multiple alleles, and therefore multiple genotypes, within populations. For example, in the common fruit fly *Drosophila melanogaster* (an organism that has been well studied genetically since the very early 1900's), between one-third and two-thirds of the genes that have been examined by protein electrophoresis have been found to be variable. Genetic variation can be measured as allele frequencies (allelic variation) or genotype frequencies (genotypic variation). A major task of population geneticists has been to describe such variation, to try to explain why it exists, and to predict its behavior over time.

The Hardy-Weinberg law predicts that if genetic variation exists in a population, it will remain constant over time, with genotypes in specific proportions. However, the law cannot begin to explain natural variation, since genotypes are not always found in Hardy-Weinberg proportions, and studies that involve sampling populations over time often show that genetic variation can be changing. The historical approach to explaining these observations has been to formulate more complex mathematical models based on the simple Hardy-Weinberg model that violate one or more of the implicit Hardy-Weinberg conditions.

Beginning in the 1920's and 1930's, a group of population geneticists, working independently, began exploring the effects of violating Hardy-Weinberg assumptions on genetic variation in populations. In what has become known as the "modern synthesis," Ronald A. Fisher, J. B. S. Haldane, and Sewall Wright merged Darwin's theory of natural selection with Mendel's theory of genetic inheritance to create a field of population genetics that allows for genetic change. They applied mathematics to the problem of variation in populations and were eventually able to incorporate what happens when each, or combinations, of the Hardy-Weinberg assumptions are violated.

Assortative Mating and Inbreeding

One of the implicit conditions of the Hardy-Weinberg model is that genotypes form mating pairs at random. In most cases mates are not selected based on genotype. Unless the gene in question has some direct effect on mate choice, mating with respect to that gene is random. However, there are conditions in natural populations in which mating is not random. For example, if a gene controls fur color and mates are chosen by appropriate fur color, then the genotype of an individual with respect to that gene will determine mating success. For this gene, then, mating is not random but rather "assortative." Positive assortative mating means that individuals tend to choose mates with genotypes like their own, while negative assortative mating means that individuals tend to choose genotypes different than their own.

Variation in a population for a gene subject to assortative mating is altered from Hardy-Weinberg expectations. Although allele frequencies do not change, genotype frequencies are altered. With positive assortative mating, the result is higher proportions of homozygotes and fewer heterozygotes, while the opposite is true when assortative mating is negative. Sometimes random mating in a population is not possible because of the geographic organization of the population or general mating habits. Truly random mating would mean that any individual can mate with any other, but this is nearly impossible because of gender differences and practical limitations. In natural populations, it is often the case that mates are somewhat related, even closely related, because the population is organized into extended family groups whose members do not (or cannot, as in plants) disperse to mate with members of other groups. Mating between relatives is called inbreeding. Because related individuals tend to have similar genotypes for many genes, the effects of inbreeding are much like those of positive assortative mating for many genes. The proportions of homozygotes for many genes tend to increase. Again, this situation has no effect on allelic variation, only genotypic variation. Clearly, the presence of nonrandom mating patterns cannot by themselves explain the majority of patterns of genetic variation in natural populations but can contribute to the action of other forces, such as natural selection.

Migration and Mutation

In the theoretical Hardy-Weinberg population, there are no sources of new genetic variation. In real populations, alleles may enter or leave the population, a process called migration or "gene flow" (a more accurate term, since migration in this context means not only movement between populations but also successful reproduction to introduce alleles in the new population). Also, new alleles may be introduced by mutation, the change in the DNA sequence of an existing allele to create a new one, as a result of errors during DNA replication or the inexact repair of DNA damage from environmental influences such as radiation or mutagenic chemicals. Both of these processes can change both genotype frequencies and allele frequencies in a population. If the tendency to migrate is associated with particular genotypes, a long period of continued migration tends to push genotype and allele frequencies toward higher proportions of one type (in general, more homozygotes) so that the overall effect is to reduce genetic variation. However, in the short term, migration may enhance genetic variation by allowing new alleles and genotypes to enter. The importance of migration depends on the particular population. Some populations may be relatively isolated from others so that migration is a relatively weak force affecting genetic variation, or there may be frequent migration among geographic populations. There are many factors involved, not the least of which is the ability of members of the particular species to move over some distance.

Mutation, because it introduces new alleles into a population, acts to increase genetic variation. Before the modern synthesis, one school of thought was that mutation might be the driving force of evolution, since genetic change over time coming about from continual introduction of new forms of genes seemed possible. In fact, it is possible to develop simple mathematical models of mutation that show resulting patterns of genetic variation that resemble those found in nature. However, to account for the rates of evolution that are commonly observed, very high rates of mutation are required. In general, mutation tends to be quite rare, making the hypothesis of evolution by mutation alone unsatisfactory.

The action of mutation in conjunction with other forces, such as selection, may account for the low-frequency persistence of clearly harmful alleles in populations. For example, one might expect that alleles that can result in genetic diseases (such as cystic fibrosis) would be quickly eliminated from human populations by natural selection. However, low rates of mutation can continually introduce these alleles into populations. In this "mutation-selection balance," mutation tends to introduce alleles while selection tends to eliminate them, with a net result of continuing low frequencies in the population.

Genetic Drift

Real populations are not, of course, infinite in size, though some are large enough that this Hardy-Weinberg condition is a useful approximation. However, many natural populations are small, and any population with less than about one thousand individuals will vary randomly in the pattern of genetic variation from generation to generation. These random changes in allele and genotype frequencies are called genetic drift. The situation is analogous to coin tossing. With a fair coin, the expectation is that half of the tosses will result in heads and half in tails. On average, this will be true, but in practice a small sample will not show the expectation. For example, if a coin is tossed ten times, it is unlikely that the result will be exactly five heads and five tails. On the other hand, with a thousand tosses, the results will be closer to half and half. This higher deviation from the expected result in small samples is called a sampling error. In a small population, there is an expectation of the pattern of genetic variation based on the Hardy-Weinberg law, but sampling error during the union of sex cells to form offspring genotypes will result in random deviations from that expectation. The effect is that allele frequencies increase or decrease randomly, with corresponding changes in genotype frequencies. The smaller the population, the greater the sam-

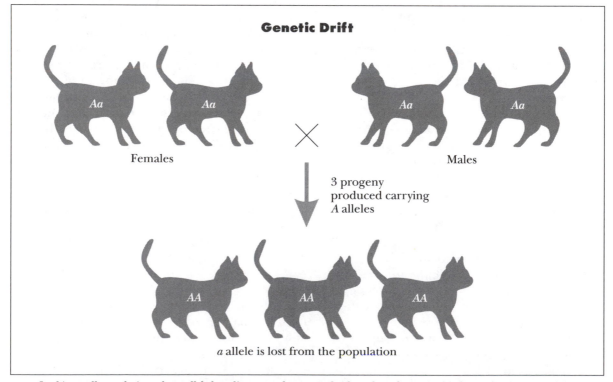

Genetic Drift

Aa   Aa   ×   Aa   Aa

Females   Males

3 progeny
produced carrying
A alleles

AA   AA   AA

a allele is lost from the population

In this small population, the a *allele has disappeared as a result of random chance and is lost to future generations.*

pling error and the more pronounced genetic drift will be.

Genetic drift has an effect on genetic variation that is similar to that of other factors. Over the long term, allele frequencies will drift until all alleles have been eliminated but one, eliminating variation. (For the moment, ignore the action of other forces that increase variation.) Over a period of dozens of generations, however, drift can allow variation to be maintained, especially in larger populations in which drift is minimal.

In the early days of population genetics, the possibility of genetic drift was recognized but often considered to be a minor consideration, with natural selection as a dominant force. Fisher in particular dismissed the importance of genetic drift, engaging over a number of years in a published debate with Wright, who always felt that drift would be important in small populations. Beginning in the 1960's with the acquisition of data on DNA-level population variation, the role of drift in natural populations became more recognized. It appears to be an especially strong force in cases in which a small number of individuals leave the population and migrate to a new area where they establish a new population. Large changes can occur, especially if the number of migrants is only ten or twenty. This type of situation is now referred to as a founder effect.

Natural Selection

Natural selection in a simple model of a gene with two alleles in a population can be easily represented by assuming that genotypes differ in their ability to survive and produce offspring. This ability is called fitness. In applying natural selection to a theoretical population, each genotype is assigned a fitness value between zero and one. Typically, the genotype in a population that is best able to survive and can, on average, produce more offspring than other genotypes is assigned a fitness value of one, and genotypes with lower fitness are assigned fitnesses with fractional values relative to the high-fitness genotype.

The study of this simple model of natural selection has revealed that it can alter genetic variation in different ways, depending on which genotype has the highest fitness. In the simple one-gene, two-allele model, there are three possible genotypes: two homozygotes and one heterozygote. If one homozygote has the highest fitness, it will be favored, and the genetic composition of the population will gradually shift toward more of that genotype (and its corresponding allele). This is called directional selection. If both homozygotes have higher fitness than the heterozygote (disruptive selection), one or the other will be favored, depending on the starting conditions. Both of these situations will decrease genetic variation in the population, because eventually one allele will prevail. Although each of these types of selection (particularly directional) may be found for genes in natural populations, they cannot explain why genetic variation is present, and is perhaps increasing, in nature.

Heterozygote advantage, in which the heterozygote has higher fitness than either homozygote, is the other possible situation in this model. In this case, because the heterozygote carries both alleles, both are expected to be favored together and therefore maintained. This is the only condition in this simple model in which genetic variation may be maintained or increased over time. Although this seems like a plausible explanation for the observed levels of natural variation, studies in which fitness values are measured almost never show heterozygote advantage in genes from natural populations. As a general explanation for the presence of genetic variation, this simple model of selection is unsatisfactory.

Studies of more complex theoretical models of selection (for example, those with many genes and different forms of selection) have revealed conditions that allow patterns of variation very similar to those observed in natural populations, and in some cases it seems clear that natural selection is a major factor determining patterns of genetic change. However, in many cases, selection does not seem to be the most important factor or even a factor at all.

Experimental Population Genetics and the Neutral Theory

Population genetics has always been a field in which the understanding of theory is ahead of empirical observation and experimental testing, but these have not been neglected. Although Fisher, Haldane, and Wright were mainly theorists, there were other architects of the modern synthesis who concentrated on testing theoretical predictions in natural populations. Beginning in the 1940's, for example, Theodosius Dobzhansky showed in natural and experimental populations of *Drosophila* species that frequency changes and geographic patterns of variation in chromosome variants are consistent with the effects of natural selection.

Natural selection was the dominant hypothesis for genetic changes in natural populations for the first several decades of the modern synthesis. In the 1960's, new techniques of molecular biology allowed population geneticists to examine molecular variation, first in proteins and later, with the use of restriction enzymes in the 1970's and DNA sequencing in the 1980's and 1990's, in DNA sequences. These types of studies only confirmed that there is a large amount of genetic variation in natural populations, much more than can be attributed only to natural selection. As a result, Motoo Kimura proposed the "neutral theory of evolution," the idea that most DNA sequence differences do not have fitness differences and that population changes in DNA sequences are governed mainly by genetic drift, with selection playing a minor role. This view, although still debated by some, was mostly accepted by the 1990's, although it was recognized that evolution of proteins and physical traits may be governed by selection to a greater extent.

Impact and Applications

The field of population genetics is a fundamental part of the modern field of evolutionary biology. One possible definition of evolution would be "genetic change in a population over time," and population geneticists try to describe patterns of genetic variation, document changes in variation, determine their theoretical causes, and predict future patterns. These types of research have been valuable in studying the evolutionary histories of organisms for which there are living representatives, including humans.

In addition to the scientific value of understanding evolutionary history better, there are more immediate applications of such work. In conservation biology, data about genetic variation in a population can help to assess its ability to survive in the future. Data on genetic similarities between populations can aid in decisions about whether they can be considered as the same species or are unique enough to merit preservation.

Population genetics has had an influence on medicine, particularly in understanding why "disease genes," while clearly harmful, persist in human populations. The field has also affected the planning of vaccination protocols to maximize their effectiveness against parasites, since a vaccine-resistant strain is a result of a rare allele in the parasite population. In the 1990's it began to be recognized that effective treatments for medical conditions would need to take into account genetic variation in human populations, since different individuals might respond differently to the same treatment.

—Stephen T. Kilpatrick, updated by Bryan Ness

See also: Artificial Selection; Behavior; Consanguinity and Genetic Disease; Emerging Diseases; Evolutionary Biology; Genetic Load; Genetics, Historical Development of; Hardy-Weinberg Law; Heredity and Environment; Hybridization and Introgression; Inbreeding and Assortative Mating; Lateral Gene Transfer; Natural Selection; Polyploidy; Punctuated Equilibrium; Quantitative Inheritance; Sociobiology; Speciation.

Further Reading

Christiansen, Freddy B. *Population Genetics of Multiple Loci.* New York: Wiley, 2000. Reinterprets classical population genetics to include the mixture of genes not only from one generation to the next but also within existing populations. Illustrations, map, bibliography, index.

Dobzhansky, Theodosius. *Genetics and the Origin of Species.* 3d ed. New York: Columbia University Press, 1951. A classic treatment of population genetics and evolution.

Gillespie, John H. *Population Genetics: A Concise Guide.* Baltimore: Johns Hopkins University

Press, 1997. Boils down the basics to less than two hundred pages.

Hartl, Daniel L. *A Primer of Population Genetics.* Rev. 3d ed. Sunderland, Mass.: Sinauer Associates, 2000. Sections cover genetic variation, the causes of evolution, molecular population genetics, and the genetic architecture of complex traits. Illustrations, bibliography, index.

Hedrick, Philip W. *Genetics of Populations.* 2d ed. Boston: Jones and Bartlett, 2000. Quantitative analysis. Illustrations, bibliography.

Landweber, Laura F., and Andrew P. Dobson, eds. *Genetics and the Extinction of Species: DNA and the Conservation of Biodiversity.* Princeton, N.J.: Princeton University Press, 1999. Offers theories on and methods for maintaining biodiversity and for preventing species' extinction. Illustratations, bibliography, index.

Lewontin, Richard C. *The Genetic Basis of Evolutionary Change.* New York: Columbia University Press, 1974. Discusses genetic variation in populations. Bibliography.

Papiha, Surinder S., Ranjan Deka, and Ranajit Chakraborty, eds. *Genomic Diversity: Applications in Human Population Genetics.* New York: Kluwer Academic/Plenum, 1999. Emphasis is on genetic variation and the application of molecular markers. Illustrations (some color), bibliography, index.

Provine, William B. *The Origins of Theoretical Population Genetics.* 2d ed. Chicago: University of Chicago Press, 2001. An account of the early history of the field. Illustrated, bibliography, index.

Slatkin, Montgomery, and Michel Veuille, eds. *Modern Developments in Theoretical Population Genetics: The Legacy of Gustave Malécot.* New York: Oxford University Press, 2002. Discusses the work of the late cofounder of population genetics. Focuses on the theory of coalescents. Illustrations, bibliography, index.

Prader-Willi and Angelman Syndromes

Field of study: Diseases and syndromes

Significance: *Both Prader-Willi and Angelman syndromes are caused by errors at the same site in the long arm of chromosome 15 (15q11-q13), but the clinical outcomes of these errors are markedly different, because the chromosomes containing the errors come from different parents. Thus, these syndromes offer a striking example of the concept known as parental imprinting.*

Key terms

DELETION: loss of a portion of a chromosome, which may be very small or very large

DISOMY: a case in which both copies of a chromosome come from a single parent, rather than (as is usual) one being maternal and one being paternal

FLUORESCENT IN SITU HYBRIDIZATION (FISH): an extremely sensitive assay for determining the presence of deletions on chromosomes, which uses a fluorescence-tagged segment of DNA that binds to the DNA region being studied

IMPRINTING: marking a chromosome so that there are differences between maternal and paternal inheritance

TRANSLOCATION: the removal of a portion of a chromosome which is then attached to the end of another chromosome; may involve loss of control for several genes

Symptoms

Angelman syndrome (AS) was first described in 1965 by Dr. Harry Angelman, who described three children with a stiff, jerky gait, absent speech, excessive laughter, and seizures. Newer reports include severe mental retardation and a characteristic face that is small with a large mouth and prominent chin. These characteristics give rise to the alternate name for the syndrome, that being "happy puppet syndrome." The syndrome is fairly rare, with an incidence estimated to be between one in fifteen thousand to one in thirty thousand. It is usually not recognized at birth or in infancy, since the developmental problems are nonspecific during this period.

Prader-Willi syndrome (PWS), by comparison, is characterized by mental retardation, hypotonia (decreased muscle tone), skin picking, short stature, crytorchidism (small or undescended testes), and hyperphagia (overeating leading to severe obesity). Delayed motor and language development are common, as is intellectual impairment (the average IQ is about 70). The syndrome was first described by Doctors Andrea Prader, Alexis Labhart, and Heinrich Willi in 1956. Like Angelman syndrome, PWS has a fairly low incidence, estimated at one in fifteen thousand. Neither condition is race-specific, and neither is considered to be a familial disease.

The primary cause of both syndromes appears to be a small deletion on the long arm of chromosome 15 (del 15q11-q13). The deleted area is estimated to be about 4 million base pairs (bp), small by molecular standards but large enough to contain several genes. This area of chromosome 15 is known to contain several genes that are activated or inactivated depending on the chromosome's parent of origin (that is, a gene may be turned on in the chromosome inherited from the mother but turned off in the chromosome inherited from the father). This parent-specific activation is referred to as genetic imprinting. It is now known that the deletions causing AS appear in the chromosome inherited from the mother, while those causing PWS occur in the chromosome inherited from the father. Since the genes of only one chromosome are active at a time, any disruption (deletion) in the active chromosome will lead to the effects seen in one of these syndromes.

Genetic Basis of AS

In 1997 a gene within the AS deletion region called *UBE3A* was found to be mutated in approximately 5 percent of AS individuals. These mutations can be as small as a single base pair. This gene codes for a protein/enzyme called a ubiquitin protein ligase, and *UBE3A* is believed to be the causative gene in AS. All mechanisms known to cause AS appear to cause inactivation or absence of this gene. *UBE3A* is an enzymatic component of a complex protein degradation system termed the ubiquitin-proteasome pathway. This pathway is located in the cytoplasm of all cells. The pathway involves a small protein molecule (ubiquitin) that can be attached to proteins, thereby causing them to be degraded. In the normal brain, *UBE3A* inherited from the father is almost completely inactive, so the maternal copy performs most of the ubiquitin-producing function. Inheritance of a *UBE3A* mutation from the mother causes AS; inheritance of the mutation from the father has no apparent effect on the child. In some families, AS caused by a *UBE3A* mutation can occur in more than one family member.

Another cause of AS (3 percent of cases) is paternal uniparental disomy (UPD). In this case a child inherits both copies of chromosome 15 from the father, with no copy inherited from the mother. Even though there is no deletion or mutation, the child is still missing the active *UBE3A* gene because the paternally derived chromosomes only have brain-inactivated *UBE3A* genes.

A fourth class of AS individuals (3-5 percent) have chromosome 15 copies inherited from both parents, but the copy inherited from the mother functions in the same way as a paternally inherited one would. This is referred to as an "imprinting defect." Some individuals may have a very small deletion of a region known as the imprinting center (IC), which regulates the activity of *UBE3A* from a distant location. The mechanism for this is not yet known.

While there are several genetic mechanisms for AS, all of them lead to the typical clinical features found in AS individuals, although minor differences in incidence of features may occur between each group.

Genetic Basis of PWS

The primary genes involved in PWS are *SNRPN*, a gene that encodes the small ribonucleotide polypeptide SmN that is found in the fetal and adult brain, and *ZFN127*, a gene that encodes a zinc-finger protein of unknown function. *SNRPN* is involved in messenger RNA (mRNA) processing, an intermediate step between DNA transcription and protein formation. A mouse model of PWS has been developed with a large deletion that includes the *SNRPN* region and the PWS imprinting center and shows a phenotype similar to that of infants with PWS.

It is probable that the hypothalamic problems (such as overeating) associated with PWS might result from a loss of *SNRPN*. The production of this protein is found mainly in the hypothalamic regions of the brain and in the olfactory cortex. Thus, disruption of hypothalamic functions such as satiety are a likely result of this defect. Prader-Willi syndrome is the most common genetic cause of obesity. In addition to its role in satiety, the hypothalamus regulates growth, sexual development, metabolism, body temperature, pigmentation, and mood—all functions that are affected in those with PWS.

PWS may also be caused by uniparental disomy, as seen in AS. However, in PWS both copies of chromosome 15 are derived from the mother instead of from the father.

As mentioned above, the imprinting center may be involved in at least some cases of both syndromes. This chromosome 15 IC is about 100 kilobase pairs (kb) long and includes exon 1 of the *SNRPN* gene. Mutations in this area appear to prevent the paternal-to-maternal imprinting switch in the AS families and prevents the maternal-to-paternal switch in PWS families. Therefore, it is possible that the IC is needed to regulate alternate RNA splicing in the *SNRPN* gene transcripts.

Genetic Diagnosis

The usual chromosome studies carried out during prenatal diagnosis are interpreted as normal in fetuses with AS and PWS syndromes, since the small abnormalities on chromosome 15 are not detected by this type of study. Likewise, fetal ultrasound offers no help in detecting physical abnormalities related to AS or PW, since the affected fetus is well formed. Amniotic fluid volume and alpha-feto protein levels also appear normal.

Specialized chromosome 15 FISH studies are needed to determine the presence of either syndrome resulting from chromosomal deletions. Testing for parent-specific methylation imprints at the 15q11-q13 locus detects more than 95 percent of cases. For cases caused by uniparental disomy, polymerase chain reaction (PCR) testing can be used.

Relevance to Geneticists

Few examples of known parental imprinting occur in the human, so AS and PWS provide rare opportunities for geneticists and biologists to study this important phenomenon. Examples of nonhuman parental imprinting are well known, but the genetic and biochemical mechanisms have not been established. Detailing the IC for chromosome 15 will be key to understanding how imprinting occurs and how the effects of AS and PWS are manifested.

The suggestion has been made that PWS (and therefore disruption of the IC) may also, at least in some cases, have an environmental trigger. A high association of PWS with fathers employed in hydrocarbon-related occupations (such as factory workers, lumbermen, machinists, chemists, and mechanics) at the time of conception has been reported by one investigative team. This is an area that needs further exploration.

—*Kerry L. Cheesman*

See also: Amniocentesis and Chorionic Villus Sampling; Chromosome Structure; Congenital Defects; Down Syndrome; Fragile X Syndrome; Hereditary Diseases; Human Growth Hormone; Huntington's Disease; Intelligence; Polymerase Chain Reaction; Prenatal Diagnosis.

Further Reading

Cassidy, S. B. "Genetics of Prader-Willi Syndrome." In *Management of Prader-Willi Syndrome*, edited by Louise R. Greenswag and Randell C. Alexander. 2d ed. New York: Springer-Verlag, 1995. An in-depth chapter providing details of PWS genetics; written with the assumption that readers understand the basics of human genetics.

Cassidy, S. B., and S. Schwartz. "Prader-Willi and Angelman Syndromes: Disorders of Genomic Imprinting." *Medicine* (Baltimore) 77 (1998): 140-151. A review of these syndromes written for health care professionals.

Hall, J. G. "Genomic Imprinting: Nature and Clinical Relevance." *Annual Review of Medi-*

cine 48 (1997): 35-44. A well-documented review that includes discussion of both AS and PWS as examples of human genomic imprinting.

Lai, L. W., R. P. Erickson, and S. B. Cassidy. "Clinical Correlates of Chromosome Fifteen Deletions and Maternal Disomy in Prader-Willi Syndrome." *American Journal of Diseases of Children* 147 (1993): 1217-1223. Discusses the signs and symptoms of PWS, including those that are needed to make a clinical diagnosis in children.

Lalalande, M. "Parental Imprinting and Human Disease." *Annual Review of Genetics* 30 (1997): 173-195. A well-written, well-documented review of imprinting and how it relates to both AS and PWS. Good reading for undergraduate students.

Mann, M. R., and M. S. Bartolomei. "Towards a Molecular Understanding of Prader-Willi and Angelman Syndromes." *Human Molecular Genetics* 8, no. 10 (1999): 1867-1873. Details the molecular mechanisms, rather than the clinical correlates, of AS and PWS; fairly technical but readable by students of biology.

Nicholls, R. D. "Genomic Imprinting and Uniparental Disomy in Angelman and Prader-Willi Syndromes: A Review." *American Journal of Medical Genetics* 46 (1993): 16-25. Not as far-reaching as Lalalande, but a well-written review of the genetic aspects of both AS and PWS.

Web Sites of Interest

National Organization for Rare Disorders. http://www.rarediseases.org. Searchable site by type of disorder. Includes background information on Prader-Willi and Angelman syndrome, a list of other names for the disorder, and a list of related organizations.

Prader-Willi Syndrome Association. http://www.pwsausa.org. This site offers background information on the syndrome, a research/medical section, links, and more.

Prenatal Diagnosis

Field of study: Human genetics and social issues

Significance: *Tests ranging from ultrasound and maternal blood tests to testing fetal cells from the amniotic fluid or placenta are performed to detect genetic disorders that the fetus may have. Although tests may show the absence of specific genetic defects, the detection of a genetic defect can produce an ethical dilemma for the parents and their physician.*

Key terms

AMNIOTIC FLUID: the liquid that surrounds the developing fetus

NEURAL TUBE: the embryonic structure that becomes the brain and spinal cord

PLACENTA: an organ composed of both fetal and maternal tissue through which the fetus is nourished

TRISOMY: the presence of three copies (instead of two) of a particular chromosome in a cell

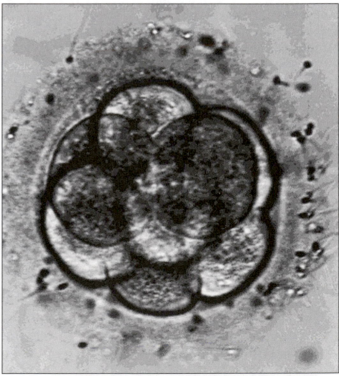

An eight-cell human embryo. (AP/Wide World Photos)

The Eight-Cell Stage

Preimplantation genetic diagnosis (PGD) has been used since 1988 to screen for genetic disorders. The most common type of PGD involves embryo biopsy at the 6-8 cell stage after fertilization has occurred in vitro. This early form of prenatal diagnosis is typically performed on day 3 embryos. One to two blastomeres (cells) are removed from the embryo (either by aspiration or by extrusion) using a fine glass needle. The biopsied embryo is then returned to culture, where the lost cells are replenished. Genetic testing is then carried out on the biopsied cells using either a technique known as fluorescent in situ hybridization (FISH) or a second technique known as fluorescent polymerase chain reaction (PCR). The FISH technique can be used to determine the presence of chromosomes 13, 16, 18, 21, 22, X, and Y. Aneuploidies (abnormal numbers of chromosomes) involving these chromosomes account for the majority of first-trimester miscarriages and for 95 percent of all postnatal chromosomal abnormalities. PCR involves amplification of DNA and allows diagnosis of single-gene diseases.

By enabling very early diagnosis of these abnormalities, PGD allows physicians to determine which embryos are most likely to be chromosomally normal prior to placement in the uterus. This increases the probability of a successful pregnancy and a healthy baby. Genetic testing generally takes only six to eight hours to complete, so that intrauterine transfer of the chromosomally normal embryo can take place within one day. If more normal embryos are obtained than one wishes to implant, the extra embryos may be preserved for future use by cryopreservation. The survival rate of frozen embryos is thought to be about 50 percent.

While this early form of prenatal diagnosis allows the elimination of many embryos carrying major genetic defects prior to implantation, it is still recommended that followup prenatal diagnosis (using chorionic villus sampling or amniocentesis) be done on resulting pregnancies. It must be realized that in order to employ preimplantation diagnostic testing, a couple must undergo in vitro fertilization even if they are fertile. This is an expensive, time-consuming process, and generally results in only a 20 percent pregnancy rate per cycle. For couples with fertility problems, this is an easy path to choose as they try to ensure implantation of normal, healthy embryos leading to healthy babies.

This technique has been used increasingly by couples who have a history of genetic disorders. In the past, couples who had a history of genetic abnormalities could decide (1) not to have children, (2) become pregnant and knowingly risk and accept abnormalities, or (3) become pregnant and rely on chorionic villus sampling or amniocentesis to diagnose genetic problems and terminate (abort) problem pregnancies.

To date, PGD has been used to detect cases of cystic fibrosis, Tay-Sachs disease, beta-thalassemia, Huntington's disease, myotonic dystrophy, X-linked disorders, and aneuploidies such as trisomy 13, 18, or 21, Turner syndrome, or Klinefelter syndrome. The number of detectable genetic defects has greatly increased since 1988. In that time, hundreds of healthy children have been born to parents undergoing preimplantation diagnosis.

—*Robin Kamienny Montvilo*

Prenatal Testing

Prenatal testing is administered to a large number of women, and the tests are becoming more informative. Some of the tests are only mildly invasive to the mother, but others involve obtaining fetal cells. Some are becoming routine for all pregnant women; others are offered only when an expectant mother meets a certain set of criteria. Some physicians will not offer the testing (especially the more invasive procedures) unless the parents have agreed that they will abort the fetus if the testing reveals a major developmental problem, such as Down syndrome or Tay-Sachs disease. Others will order testing without any such guarantees, believing that test results will give the parents time to prepare themselves for a special-needs baby. The test results are also used to determine if additional medical teams should be present at the delivery to deal with a newborn who is not normal and healthy. Most often, prenatal testing is offered if the mother is age thirty-five or older, if a particular disorder is present in relatives on one or both sides of the family, or if the parents have already produced one child with a genetic disorder.

Maternal Blood Tests and Ultrasound

Screening maternal blood for the presence of alpha fetoprotein (AFP) is offered to pregnant women who are about eighteen weeks into a pregnancy. Although AFP is produced by the fetal liver, some will cross the placenta into the mother's blood. Elevated levels of AFP can indicate an open neural tube defect (such as spina bifida), although it can also indicate twins. Unusual AFP findings are usually followed up by ultrasound examination of the fetus.

Other tests of maternal blood measure the amounts of two substances that are produced by the fetal part of the placenta: hCG and UE3. Lower-than-average levels of AFP and UE3, combined with a higher-than-average amount of hCG, increases the risk that the woman is carrying a Down syndrome (trisomy 21) fetus. For example, a nineteen-year-old woman has a baseline risk of conceiving a fetus with Down syndrome of 1 in 1,193. When blood-test results show low AFP and UE3 along with high hCG, the probability of Down syndrome rises to 1 in 145.

During an ultrasound examination, harmless sound waves are bounced off the fetus from an emitter placed on the surface of the mother's abdomen or in her vagina. They are used to make a picture of the fetus on a television monitor. Measurements on the monitor can often be used to determine the overall size, the head size, and the sex of the fetus, and whether all the arms and legs are formed and of the proper length. Successive ultrasound tests will indicate if the fetus is growing normally. Certain ultrasound findings, such as shortened long bones, may indicate an increased probability for a Down syndrome baby. Because Down syndrome is a highly variable condition, normal ultrasound findings do not guarantee that the child will be born without Down syndrome. Only a chromosome analysis can determine this for certain.

Amniocentesis, Karyotyping, and FISH

Amniocentesis is the process of collecting fetal cells from the amniotic fluid. Fetal cells collected by amniocentesis can be grown in culture; then the fluid around the cells is collected and analyzed for enzymes produced by the cells. If an enzyme is missing (as in the case of Tay-Sachs disease), the fetus may be diagnosed with the disorder before it is born. Because disorders such as Tay-Sachs disease are untreatable and fatal, a woman who has had one Tay-Sachs child may not wish to give birth to another. Early diagnosis of a second Tay-Sachs fetus would permit her to have a therapeutic abortion.

Chromosomes in the cells obtained by amniocentesis may be stained to produce a karyotype. In a normal karyotype, the chromosomes will be present in pairs. If the fetus has Down syndrome (trisomy 21), there will be three copies of chromosome 21. Other types of chromosome abnormalities that also appear in karyotypes are changes within a single chromosome. If a chromosome has lost a piece, it is said to contain a deletion. Large deletions will be obvious when a karyotype is analyzed because the chromosome will appear smaller than normal. Sometimes the deletion is so small that it is not visible on a karyotype.

If chromosome analysis is needed early in pregnancy before the volume of amniotic fluid is large enough to permit amniocentesis, the mother and doctor may opt for chorionic villus sampling (CVS). The embryo produces finger-like projections (villi) into the uterine lining. Because these projections are produced by the embryo, their cells will have the same chromosome number as the rest of the embryonic cells. After growing in culture, the cells may be karyotyped in the same way as those obtained by amniocentesis. Both amniocentesis and CVS carry risks of infection and miscarriage. Normally these procedures are not offered unless the risk of having an affected child is found to be greater than the risk of complications from the procedures.

If the doctor is convinced that the fetus has a tiny chromosomal defect that is not visible on a karyotype, it will then be necessary to probe (or "FISH") the fetal chromosomes; the initials "FISH" stand for "fluorescent in situ hybridization." A chromosome probe is a piece of DNA that is complementary to DNA within a gene. Complementary pieces of DNA will stick together (hybridize) when they come in contact. The probe also has an attached molecule that will glow when viewed under fluorescent light. A probe for a particular gene will stick to the part of the chromosome where the gene is located and make a glowing spot. If the gene is not present because it has been lost, no spot will appear. Probes have been developed for many individual genes that cause developmental abnormalities when they are deleted from the chromosomes.

Cells obtained by amniocentesis can be probed in less time than it takes to grow and prepare them for karyotyping. Probes have been developed for the centromeres of the chromosomes that are frequently present in extra copies, such as 13, 18, 21, X, and Y. Y chromosomes that have been probed appear as red spots, X chromosomes as green spots, and number 18 chromosomes as aqua spots. A second set of probes attached to other cells from the same fetus will cause number 13 chromo-

somes to appear as green spots and number 21 chromosomes to appear as red spots. Cells from a girl with trisomy 21 would have two green spots and two aqua spots, but no red spot when the first set of probes is used. Some other cells from the same girl will show two red spots, but three green ones, when the second set of probes is used.

More recently, tests for many more genetic defects using advanced molecular genetics tests have been developed. DNA can be isolated from fetal cells, obtained by one of the methods already described, which is then probed for single gene defects. Hundreds of potential genetic defects can be detected in this way, although only a few such tests are generally available. Another barrier to their use is their high cost. Costs will likely drop in the future as the tests are perfected and are used more widely. These same tests may be performed on the parents to determine whether they are carriers of certain genetic diseases.

Yury Verlinsky (right), known for his cutting-edge work in prenatal testing, and Ridvan Seckin Ozen examine human chromosomes at the Reproductive Genetics Institute in Chicago. (AP/Wide World Photos)

Impact and Applications

Until the development of prenatal techniques, pregnant women had to wait until delivery day to find out the sex of their child and whether or not the baby was normal. Now much more information is available to both the woman and her doctor weeks before the baby is due. Even though tests are not available for all possible birth defects, normal blood tests, karyotypes, or FISH can be very comforting. On the other hand, abnormal test results give the parents definite information about birth defects, as opposed to the possibilities inherent in a statement of risk. The parents must decide whether to continue the pregnancy. If they do, they must then cope with the fact that they are not going to have a normal child. When properly administered, the test results are explained by a genetic counselor who is also equipped to help the parents deal with the strong emotions that bad news can produce. Genetic testing also has far-reaching implications. If insurance companies pay for the prenatal testing, they receive copies of the results. Information about genetic abnormalities could cause the insurance companies to deny claims arising from treatment of the newborn or to deny insurance to the individual later in life.

—*Nancy N. Shontz, updated by Bryan Ness*

See also: Albinism; Amniocentesis and Chorionic Villus Sampling; Burkitt's Lymphoma; Color Blindness; Congenital Defects; Consanguinity and Genetic Disease; Cystic Fibrosis; Down Syndrome; Dwarfism; Fragile X Syndrome; Gender Identity; Genetic Counseling; Genetic Screening; Genetic Testing; Genetic Testing: Ethical and Economic Issues; Heart Disease; Hemophilia; Hereditary Diseases; Hermaphrodites; Human Genetics; Human Genome Project; Huntington's Disease; In Vitro Fertilization and Embryo Transfer; Inborn Errors of Metabolism; Klinefelter Syndrome; Metafemales; Monohybrid Inheritance; Neural Tube Defects; Pedigree Analysis; Phenylketonuria (PKU); Polymerase Chain Reaction; Prader-Willi and Angelman Syndromes; Pseudohermaphrodites; RFLP Analysis; Sickle-Cell Disease; Tay-Sachs Disease; Testicular Feminization Syndrome; Thalidomide and Other Teratogens; Turner Syndrome; XYY Syndrome.

Further Reading

Bianchi, Diana W., Timothy M. Crombleholme, and Mary E. D'Alton. *Fetology: Diagnosis and Management of the Fetal Patient.* New York: McGraw-Hill, 2000. A resource for practitioners and a guide for parents. Illustrations, bibliography, index.

Hadley, Andrew G., and Peter Soothill, eds. *Alloimmune Disorders of Pregnancy: Anaemia, Thrombocytopenia, and Neutropenia in the Fetus and Newborn.* New York: Cambridge University Press, 2002. Discusses an often overlooked subject. Written especially for practitioners and students of pediatrics, obstetrics, fetal-maternal medicine, blood banking, and immunology. Illustrations, bibliography, index.

Heyman, Bob, and Mette Henriksen. *Risk, Age, and Pregnancy: A Case Study of Prenatal Genetic Screening and Testing.* New York: Palgrave, 2001. Provides a detailed case study of a prenatal genetic screening and testing system in a British hospital, giving perspectives of pregnant women, hospital doctors, and midwives, and elucidating the communication between women and the hospital doctors who advise them.

McConkey, Edwin H. *Human Genetics: The Molecular Revolution.* Boston: Jones & Bartlett, 1993. Contains additional information on FISH. Illustrations, bibliography, index.

New, Maria I., ed. *Diagnosis and Treatment of the Unborn Child.* Reddick, Fla.: Idelson-Gnocchi, 1999. Provides an overview of prenatal testing and treatment. Illustrations (some color), bibliography.

Petrikovsky, Boris M., ed. *Fetal Disorders: Diagnosis and Management.* New York: Wiley-Liss, 1999. Discusses testing and treatment protocols, prognoses, counseling, and more. Illustrations (some color), bibliography, index.

Pilu, Gianluigi, and Kypros H. Nicolaides. *Diagnosis of Fetal Abnormalities: The 18-23-Week Scan.* New York: Parthenon Group, 1999. Covers mid-trimester fetal ultrasound. Illustrations, index.

Rodeck, Charles H., and Martin J. Whittle, eds. *Fetal Medicine: Basic Science and Clinical Practice.* New York: Churchill Livingstone, 1999. Good general reference for residents and fellows in neonatal and maternal-fetal medicine. Illustrations (some color), bibliography, index.

Twining, Peter, Josephine M. McHugo, and David W. Pilling, eds. *Textbook of Fetal Abnormalities*. New York: Churchill Livingstone, 2000. Covers the safety of ultrasound; the routine fetal anomaly scan; disorders of amniotic fluid; cranial, spinal, cardiac, pulmonary, abdominal, skeletal, urinary tract, and chromosomal abnormalities; intra-uterine therapy; and counseling. Illustrations (some color), bibliography, index.

Weaver, David D., with the assistance of Ira K. Brandt. *Catalog of Prenatally Diagnosed Conditions*. 3d ed. Baltimore: Johns Hopkins University Press, 1999. Covers about eight hundred conditions and has 1,221 literature references. Bibliography, index.

Web Sites of Interest

Association of Women's Health, Obstetric, and Neonatal Nurses. http://www.awhonn.org. Offers pages for education and practice resources as well as legal policy.

March of Dimes. http://www.marchofdimes.com. This site is searchable by keyword and includes information on the basics of amniocentesis and chorionic villus sampling and articles on how the two procedures relate to genetics.

National Newborn Screening and Genetics Resource Center. http://genes-r-us.uthscsa.edu. Site serves as a resource for information on genetic screening.

Prion Diseases: Kuru and Creutzfeldt-Jakob Syndrome

Field of study: Diseases and syndromes

Significance: *Kuru and Creutzfeldt-Jakob syndrome are rare, fatal diseases of the brain and spinal cord. Nerve cell death is caused by the accumulation of a protein called a "prion" that appears to be a new infectious agent that interferes with gene expression in nerve cells. Understanding these diseases has far-reaching implications for the study of other degenerative mental disorders.*

Key terms

DEMENTIA: mental deterioration ranging from forgetfulness and disorientation to complete unresponsiveness

PRION: short for "proteinaceous infectious particle," an element consisting mainly of protein and generally lacking nucleic acid (DNA and RNA), which is often the causative agent behind various spongiform encephalopathies

Causes, Symptoms, and Treatment

Kuru and Creutzfeldt-Jakob syndrome, degenerative diseases of the human central nervous system, are among a group of diseases that also affect cattle (mad cow disease) and sheep (scrapie). They have been classified in several ways, including "slow-virus" infections (because of the extremely long incubation period between contact and illness) and "spongiform encephalopathies" (because of the large holes seen in the brain after death). However, a virus that may cause such a disease has never been found, and the body does not respond to the disease as an infection. The only clue to the cause is the accumulation of a transmissible, toxic protein known as a prion; therefore, these disorders are now known simply as "prion diseases."

Creutzfeldt-Jakob syndrome is rare: Approximately 250 people die from it yearly in the United States. It usually begins in middle age with symptoms that include rapidly progressing dementia, jerking spastic movements, and visual problems. Within one year after the symptoms begin, the patient is comatose and paralyzed, and powerful seizures affect the entire body. Death occurs shortly thereafter. The initial symptom of the illness (rapid mental deterioration) is similar to other disorders; therefore, diagnosis is difficult. No typical infectious agent (bacteria or viruses) can be found in the blood or in the fluid that surrounds the brain and spinal cord. X rays and other scans are normal. There is no inflammation, fever, or antibody production. Brain wave studies are, however, abnormal, and at autopsy, the brain is found to have large holes and massive protein deposits in it.

Kuru is found among the Fore tribe of Papua New Guinea. Until the early 1960's, more than

The Discovery of Prions

In 1972, Stanley B. Prusiner, then a resident in neurology at the University of California School of Medicine at San Francisco, lost a patient to Creutzfeldt-Jakob disease. He resolved to learn more about the condition. He read that it and related diseases, scrapie and kuru, could be transmitted by injecting extracts from diseased brain into the brains of healthy animals. At the time, the diseases were thought to be caused by a slow-acting virus, but it had not been identified. He was intrigued by a study from the laboratory of Tikvah Alper that suggested that the scrapie agent lacked nucleic acid. When he started his own lab in 1974, Prusiner decided to pursue the nature of the infectious agent.

He and his associates determined to purify the causative agent in scrapie-infected brains and, by 1982, had a highly purified preparation. They subjected it to extensive analysis, and all of their results indicated that it indeed lacked DNA or RNA and that it consisted mainly, if not exclusively, of protein. The infectivity was lost when treated with procedures that denatured protein, but not when treated with those detrimental to nucleic acids. He named the agent a "prion," an abbreviation for "proteinaceous infectious particle." Shortly afterward, he showed that it consisted of a single protein. This was a highly unorthodox discovery because all pathogens studied to date contained nucleic acid. Skeptics were convinced that a very small amount of nucleic acid must be contaminating the prions, although the limits on detection showed that it contained fewer than one hundred nucleotides and would have to be smaller than any known virus.

Prusiner and his collaborators subsequently learned that the gene for the prion protein was found in chromosomes of hamsters, mice, humans, and all other mammals that have been examined. Furthermore, most of the time these animals make the prion protein without getting sick—a startling observation. Prusiner and his team subsequently showed that the prion protein existed in two forms, one harmless and the other leading to disease. The latter proved to be highly resistant to degradation by proteolytic enzymes and accumulated in the brain tissue of affected animals and people. In infectious disease, the harmful form of the prions appears to convert the harmless form to the harmful form, although the mechanism is not understood. In inherited disease, mutations in the prion may cause it to adopt the harmful form spontaneously or after some unknown signal, leading eventually to the disease state. While questions remain, research since the 1980's has established the involvement of prions in various spongiform encephalopathies.

In 1997, Prusiner was awarded the Nobel Prize in Physiology or Medicine for his pioneering discovery of prions and their role in various neurological diseases. The Nobel Committee also noted his perseverance in pursuing an unorthodox hypothesis in the face of major skepticism.

—*James L. Robinson*

one thousand Fore died of Kuru each year. Anthropologists recording their customs described their practice of eating the brains of their dead relatives in order to gain the knowledge they contained. Clearly, some infectious agent was being transmitted during this ritual. Such cannibalism has since stopped, and Kuru has declined markedly. Kuru, like Creutzfeldt-Jakob syndrome, shows the same spongiform changes and protein deposits in the brain after death. Similarly, early symptoms include intellectual deterioration, spastic movements, and visual problems. Within a year, the patient becomes unresponsive and dies.

The outbreak of "mad cow" disease in the mid-1990's in Great Britain led to widespread fear. Thousands of cattle were killed to prevent human consumption of contaminated beef. The cows were infected by supplemental feedings tainted by infected sheep meat. Animal-to-human transmission of these diseases appears to occur, and research has shown that human-to-animal infection is possible as well.

Both Kuru and Creutzfeldt-Jakob syndrome, as well as the animal forms, have no known treatment or cure. Because of the long incubation period, decades may pass before symptoms appear, but once they do, the central nervous system is rapidly destroyed, and death comes quickly. It is likely that many more people die of these disorders than is known because they are so rarely diagnosed.

Properties of Prions

Most of the research on prion diseases has focused on scrapie in sheep. It became clear that the infectious particle had novel properties: It was not a virus as had been suspected, nor did the body react to it as an invader. It was discovered that this transmissible agent was an abnormal version of a common protein, which defied medical understanding. This protein is normally secreted by nerve cells and is found on their outer membranes. Its gene is on chromosome 20 in humans. The transmissible, infectious fragment of the prion somehow disrupts the nerve cell, causing it to produce the abnormal fragment instead of the normal protein. This product accumulates to toxic levels in the tissue and fluid of the brain and spinal cord over many years, finally destroying the central nervous system.

Prion infection appears to occur from exposure to infected tissues or fluids. Transmission has occurred accidentally through nerve tissue transplants and neurosurgical instruments. Prions are not affected by standard sterilization techniques; prevention requires careful handling of infected materials and extended autoclaving of surgical instruments (for at least one hour) or thorough rinsing in chlorine bleach. The agent is not spread by casual contact or air, and isolating the patient is not necessary.

Other human degenerative nervous system diseases whose causes remain unclear also show accumulations of proteins to toxic levels. Alzheimer's disease is the best-studied example, and it is possible that a process similar to that in prion diseases is at work. The discovery of prions has far-reaching implications for genetic and cellular research. Scientists have already learned a startling fact: Substances as inert as proteins and far smaller than viruses can act as agents of infections.

—*Connie Rizzo*

See also: Alzheimer's Disease; Huntington's Disease.

Further Reading

Aguzzi, Adriano. "A Brief History of Prions." *Nature* 389 (October 23, 1997). An overview of the history of research into prions and prion diseases.

Baker, Harry F., ed. *Molecular Pathology of the Prions.* Totowa, N.J.: Humana Press, 2001. Overview of research on prion diseases. Illustrated, bibliography, index.

Ferry, Georgina. "Mad Brains and the Prion Heresy." *New Scientist* 142, no. 1927 (May 28, 1994). Looks at prion as an agent for infectious diseases.

Goldman, Lee, and J. Claude Bennett, eds. *Cecil Textbook of Medicine.* 21st ed. Philadelphia: W. B. Saunders, 2000. A classic medical reference text that covers prion diseases. Bibliography, index.

Groschup, Martin H., and Hans A. Kretzschmar, eds. *Prion Diseases: Diagnosis and Pathogenesis.* New York: Springer, 2000. Comprehensive collection of research on the pathogenesis of prion diseases in humans and other animals, pharmacology, epidemiology and diagnosis, and more. Illustrations (some color), bibliography.

Harris, David A., ed. *Prions: Molecular and Cellular Biology.* Portland, Oreg.: Horizon Scientific Press, 1999. Focuses on the cellular, biochemical, and genetic aspects of prion diseases. Illustrations, bibliography, index.

Klitzman, Robert. *The Trembling Mountain: A Personal Account of Kuru, Cannibals, and Mad Cow Disease.* New York: Plenum Trade, 1998. Autobiographical account of a study of kuru disease in Papua New Guinea. Illustrations, index.

Prusiner, Stanley B. "The Prion Diseases." *Scientific American* 272 (January, 1995). Discusses how meat consumption by humans can lead to prion diseases.

_____, ed. *Prion Biology and Diseases.* Cold Spring Harbor, N.Y.: Cold Spring Harbor Laboratory Press, 1999. Prusiner, who won the 1997 Nobel Prize in Medicine for his discovery of prions, and contributors provide an overview of research. Illustrations (some color), bibliography, index.

Rabenau, Holger F., Jindrich Cinatl, and Hans Wilhelm Doerr, eds. *Prions: A Challenge for Science, Medicine, and Public Health System.* New York: Karger, 2001. Focuses on the etiological, clinical, and diagnostic aspects of prion diseases, as well as epidemiology, disease management, and how prions might be inactivated. Illustrations, bibliography, index.

Ratzan, Scott C., ed. *The Mad Cow Crisis: Health and the Public Good.* New York: New York University Press, 1998. A look at the disease from scientific, historical, political, health, preventive, and management perspectives. Illustrations, bibliography, index.

Web Site of Interest

National Organization for Rare Disorders. http://www.rarediseases.org. Searchable site by type of disorder. Includes background information on Creutzfeldt-Jakob syndrome, a list of other names for the disorder, and a list of related organizations.

A three-dimensional image of the ras protein. (U.S. Department of Energy Human Genome Program, http://www.ornl.gov/hgmis)

Protein Structure

Field of study: Molecular genetics

Significance: *Proteins have three-dimensional structures that determine their functions, and slight changes in overall structure may significantly alter their activity. Correlation of protein structure and function can provide insights into cellular metabolism and its many interconnected processes. Because most diseases result from improper protein function, advances in this field could lead to effective molecular-based disease treatments.*

Key terms

AMINO ACID: the basic subunit of a protein; there are twenty commonly occurring amino acids, any of which may join together by chemical bonds to form a complex protein molecule

ENZYMES: proteins that are able to increase the rate of chemical reactions in cells without being altered in the process

HYDROGEN BOND: a weak bond that helps stabilize the folding of a protein

POLYPEPTIDE: a chain of amino acids joined by chemical bonds

R GROUP: a functional group that is part of an amino acid that gives each amino acid its unique properties

Protein Structure and Function

Proteins consist of strings of individual subunits called amino acids that are chemically bonded together with peptide bonds. Once amino acids are bonded, the resulting molecule is called a polypeptide. The properties and arrangement of the amino acids in the polypeptide cause it to fold into a specific shape or conformation that is required for proper protein function. Proteins have been called the "workhorses" of t0he cell because they perform most of the activities encoded in the genes of the cell. Proteins function by binding to other molecules, frequently to other proteins. The precise three-dimensional shape of a protein determines the specific molecules it will be able to bind to, and for many proteins binding is specific to just one other specific type of molecule.

In 1973, Christian B. Anfinsen performed experiments that showed that the three-dimensional structure of a protein is determined by the sequence of its amino acids. He used a protein called ribonuclease (RNase), an enzyme that degrades RNA in the cell. The ability of ribonuclease to degrade RNA is dependent upon its ability to fold into its proper three-dimensional shape. Anfinsen showed that if the enzyme was completely unfolded by heat and chemical treatment (at which time it would not function), it formed a linear chain of amino acids. Although there were 105 possible conformations that the enzyme could take upon refolding, it would refold into the single correct functional conformation upon removal of heat and chemicals. This established that the amino acid sequences of proteins, which are specified by the genes of the cell, carry all of the information necessary for proteins to fold into their proper three-dimensional shapes.

To understand protein conformation better, it is helpful to analyze the underlying levels of structure that determine the final three-dimensional shape. The primary structure of a polypeptide is the simplest level of structure and is, by definition, its amino acid sequence. Because primary structure of polypeptides ultimately determines all succeeding levels of structure, knowing the primary structure should theoretically allow scientists to predict the final three-dimensional structure. Building on a detailed knowledge of the structure of many proteins, scientists can now develop computer programs that are able to predict three-dimensional shape with some degree of accuracy, but much more research will be required to increase the accuracy of these methods.

Primary Protein Structure

There are twenty naturally occurring amino acids that are commonly found in proteins, and each of these has a common structure consisting of a nitrogen-containing amino group ($-NH_2$), a carboxyl group ($-COOH$), a hydrogen atom (H), and a unique functional group referred to as an R group, all bonded to a central carbon atom (known as the alpha carbon, or Cα) as shown in the following figure:

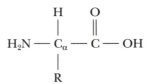

The uniqueness of each of the twenty amino acids is determined by the R group. This group may be as simple as a hydrogen atom (in the case of the amino acid glycine) or as complex as a ring-shaped structure (as found in the amino acid phenylalanine). It may be charged, either positively or negatively, or it may be uncharged.

Cells join amino acids together to form peptides (strings of up to ten amino acids), polypeptides (strings of ten to one hundred amino acids), or proteins (single or multiple polypeptides folded and oriented to one another so they are functional). The amino acids are joined together by covalent bonds, called peptide bonds (in the box in the following figure), between the carbon atom of the carboxyl group of one amino acid ($-COOH$) and the nitrogen atom of the amino group ($-NH_2$) of the next adjacent amino acid:

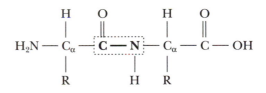

During the formation of the peptide bond, a molecule of water (H_2O) is lost (an -OH from the carboxyl group and an -H from the amino group), so this reaction is also called a dehydration synthesis. The result is a dipeptide (a peptide made of two amino acids joined by a peptide bond) that has a "backbone" of nitrogens and carbons (N—Cα—C—N—Cα—C) with other elements and R groups protruding from the backbone. An amino acid may be joined to the growing peptide chain by formation of a peptide bond between the carbon atom of the free carboxyl group (on the right of the preceding figure) and the nitrogen atom of the amino acid being added. The end of a polypeptide with an exposed carboxyl group is called the C-terminal end, and the end with an exposed amino group is called the N-terminal end.

The atoms and R groups that protrude from the backbone are capable of interacting with each other, and these interactions lead to higher-order secondary, tertiary, and quaternary structures.

Secondary Structure

The next level of structure is secondary structure, which involves the formation of hydrogen bonds between the oxygen atoms in carboxyl groups with the hydrogen atoms of amino groups from different parts of the polypeptide. Hydrogen bonds are weak bonds that form between atoms that have a very strong attraction for electrons (such as oxygen or nitrogen), and a hydrogen atom that is bound to another atom with a very strong attraction for electrons. Secondary structure does not involve the formation of bonds with R groups or atoms that are parts of R groups, but involves bonding just between amino and carboxyl groups that are in the peptide bonds making up the backbone of polypeptides.

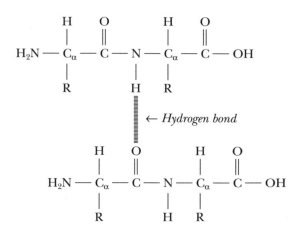

← *Hydrogen bond*

These hydrogen bonds between backbone molecules lead to the formation of two major types of structures: alpha helices and beta-pleated sheets. An alpha helix is a rigid structure shaped very much like a telephone cord; it spirals around as the oxygen of one amino acid of the chain forms a hydrogen bond with the hydrogen atom of an amino acid five amino acids away on the protein strand. The rigidity of the structure is caused by the large number of hydrogen bonds (individually weak but collectively strong) and the compactness of the helix that forms. Many alpha helices are found in proteins that function to maintain cell structure.

Beta sheets are formed by hydrogen bonding between amino acids in different regions (often very far apart on the linear strand) of a polypeptide. The shape of a beta-pleated sheet may be likened to the bellows of an accordion or a sheet of paper that has been folded multiple times to form pleats. Because of the large number of hydrogen bonds in them, beta sheets are also strong structures, and they form planar regions that are often found at the bottom of "pockets" inside proteins to which other molecules attach.

In addition to alpha helices or beta-pleated sheets, other regions of the protein may have no obvious secondary structure; these regions are said to have a "random coil" shape. It is the combinations of random coils, alpha helices, and beta sheets that form the secondary structure of the protein.

Tertiary Structure

The final level of protein shape (for a single polypeptide or simple protein) is called tertiary structure. Tertiary structure is caused by the numerous interactions of R groups on the amino acids and of the protein with its environment, which is usually aqueous (water based). Various R groups may either be attracted to and form bonds with each other, or they may be repelled from each other. For example, if an R group has an overall positive electrical charge, it will be attracted to R groups with a negative charge but repelled from other positively charged R groups. For a polypeptide with one hundred amino acids, if amino acid number 6 is negatively charged, it could be attracted to a positively charged amino acid at position 74, thus bringing two ends of the protein that are linearly distant into close proximity. Many of these attractions lead to the formation of hydrogen, ionic, or covalent bonds. For example, sulfur is contained in the R groups of a few of the amino acids, and sometimes a disulfide bond (a covalent bond) will be formed between two of these. It is the arrangement of disulfide bonds in hair proteins that gives hair its physical properties of curly versus straight. Hair permanent treatments actually break these disulfide bonds and then reform them when the hair is arranged as desired. Many other R groups in the protein will also be attracted to or repelled from each other, leading to an overall folded shape that is most stable. In addition, because most proteins exist in an aqueous environment in the cell, most proteins are folded such that their amino acids with hydrophilic R groups (R groups attracted to water) are on the outside, while their amino acids with hydrophobic R groups (R groups repelled from water) are tucked away in the interior of the protein.

Quaternary Structure

Many polypeptides are nonfunctional until they physically associate with another polypeptide, forming a functional unit made up of two or more subunits. Proteins of this type are said to have quaternary structure. Quaternary structure is caused by interactions between the R groups of amino acids of two different polypeptides. For example, hemoglobin, the oxygen-carrying protein found in red blood cells, functions as a tetramer, with four polypeptide subunits.

Because secondary, tertiary, and quaternary interactions are caused by the R groups of the specific amino acids, the folding is ultimately dictated by the amino acid sequence of the protein. Although there may be numerous possible final conformations that a polypeptide could take, it usually assumes only one of these, and this is the conformation that leads to proper protein function. Many polypeptides are capable of folding into their final conformation spontaneously. More complex ones may need the assistance of other proteins, called chaperones, to help in the folding process.

Impact and Applications

The function of a protein may be altered by changing its shape, because proper function is dependent on proper conformation. Many genetic defects are detrimental because they represent a mutation that results in a change in protein structure. Changes in protein conformation are also an integral part of metabolic control in cells. Normal cellular processes are controlled by "turning on" and "turning off" proteins at the appropriate time. A protein's activity may be altered by attaching a molecule or ion to that protein that results in a change of shape. Because the shape is caused by R group interactions, binding of a charged ion such as calcium to the protein will alter these interactions and thus alter the shape and function of the protein. One molecular "on/off" switch that is used frequently within a cell involves the attachment or removal of a phosphate group to or from a protein. Attachment of a phosphate will significantly alter the shape of the protein by repelling negatively charged amino acids and attracting positively charged amino acids, which will either activate the protein to perform its function (turn it on) or deactivate it (turn it off).

Cancer and diseases caused by bacterial or viral infections are often the result of nonfunctional proteins that have been produced with incorrect shapes or that cannot be turned on or off by a molecular switch. The effects may be minor or major, depending upon the protein, its function, and the severity of the structural deformity. Understanding how a normal protein is shaped and how it is altered in the disease process allows for the development of drugs that may block the disease. This may be accomplished by blocking or changing the effect of the protein of interest or by generating drugs or therapies that mimic the normal functioning of the protein. Thus, understanding protein structure is essential for understanding proper protein function and for developing molecular-based disease treatments.

—*Sarah Lea McGuire, updated by Bryan Ness*
See also: Central Dogma of Molecular Biology; DNA Repair; DNA Replication; DNA Structure and Function; Genetic Code; Genetic Code, Cracking of; Molecular Genetics;

Protein Synthesis; RNA Structure and Function; RNA Transcription and mRNA Processing; RNA World; Synthetic Genes.

Further Reading

Banaszak, Leonard J. *Foundations of Structural Biology.* San Diego: Academic Press, 2000. Focuses on three-dimensional visualization strategies for proteins and DNA segments. Illustrations, bibliography, index.

Brändén, Carl-Ivar, and John Tooze. *Introduction to Protein Structure.* 2d ed. New York: Garland, 1999. Covers research into the structure and logic of proteins. Illustrations (mostly color), bibliography, index.

Darnell, James, et al. *Molecular Cell Biology.* 4th ed. New York: W. H. Freedman, 2000. Provides both summary and detailed accounts of protein structure and the chemical bonds that lead to the various levels of structure. Illustrations, laser optical disc, bibliography, index.

Johnson, George B. *How Scientists Think: Twenty-one Experiments That Have Shaped Our Understanding of Genetics and Molecular Biology.* Dubuque, Iowa: Wm. C. Brown, 1996. Gives an excellent introductory account of Christian Anfinsen's experiments leading to the determination that primary sequence dictates protein shape. Illustrations, index.

McRee, Duncan Everett. *Practical Protein Crystallography.* 2d ed. San Diego: Academic Press, 1999. Introductory protein structure handbook. Illustrations (some color), bibliography, index.

Maddox, Brenda. *Rosalind Franklin: The Dark Lady of DNA.* New York: HarperCollins, 2002. Biography of the foundational but little recognized work of the physical chemist, whose photographs of DNA were critical in helping James Watson, Francis Crick, and Maurice Wilkins discover the double-helical structure of DNA, for which they won the Nobel Prize in 1962. Illustrations, bibliography, index.

Murphy, Kenneth P. *Protein Structure, Stability, and Folding.* Totowa, N.J.: Humana Press, 2001. Describe cutting-edge experimental and theoretical methodologies for investigating these proteins and protein folding.

Protein Synthesis

Field of study: Molecular genetics

Significance: *Cellular proteins can be grouped into two general categories: proteins with a structural function that contribute to the three-dimensional organization of a cell, and proteins with an enzymatic function that catalyze the biochemical reactions required for cell growth and function. Understanding the process by which proteins are synthesized provides insight into how a cell organizes itself and how defects in this process can lead to disease.*

Key terms

AMINO ACID: the basic subunit of a protein; there are twenty commonly occurring amino acids, any of which may join together by chemical bonds to form a complex protein molecule

PEPTIDE BOND: the chemical bond between amino acids in protein

POLYPEPTIDE: a linear molecule composed of amino acids joined together by peptide bonds; all proteins are functional polypeptides

RNA: ribonucleic acid, that molecule that acts as the messenger between genes in DNA and their protein product, directing the assembly of proteins; as an integral part of ribosomes, RNA is also involved in protein synthesis

TRANSLATION: the process of forming proteins according to instructions contained in an RNA molecule

The Flow of Information from Stored to Active Form

The cell can be viewed as a unit that assembles resources from the environment into biochemically functional molecules and organizes these molecules in three-dimensional space in a way that allows cellular growth and replication. In order to carry out this organizational process, a cell must have a biosynthetic means to assemble resources into useful molecules, and it must contain the information required to produce the biosynthetic and structural machinery. DNA serves as the stored form of this information, whereas protein is its active form. Although there are thousands of different proteins in cells, they either serve a structural role or are enzymes that catalyze the biosynthetic reactions of a cell. Following the discovery of the structure of DNA in 1953 by James Watson and Francis Crick, scientists began to study the process by which the information stored in this molecule is converted into protein.

Proteins are linear, functional molecules composed of a unique sequence of amino acids. Twenty different amino acids are used as the protein building blocks. Although the information for the amino acid sequence of each protein is present in DNA, protein is not synthesized directly from this source. Instead, RNA serves as the intermediate form from which proteins are synthesized. RNA plays three roles during protein synthesis. Messenger RNA (mRNA) contains the information for the amino acid sequence of a protein. Transfer RNAs (tRNAs) are small RNA molecules that serve as adapters that decipher the coded information present within an mRNA and bring the appropriate amino acid to the polypeptide as it is being synthesized. Ribosomal RNAs (rRNAs) act as the engine that carries out most of the steps during protein synthesis. Together with a specific set of proteins, rRNAs form ribosomes that bind the mRNA, serve as the platform for tRNAs to decode an mRNA, and catalyze the formation of peptide bonds between

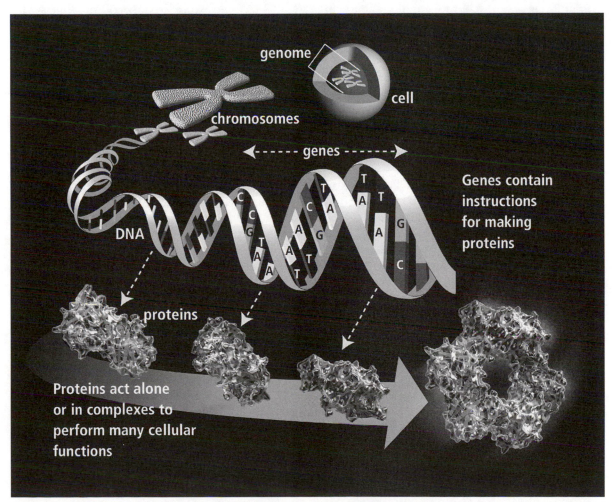

The fundamental steps in protein synthesis. (U.S. Department of Energy Human Genome Program, http://www.ornl.gov/hgmis)

amino acids. Each ribosome is composed of two subunits: a small (or 40s) and a large (or 60s) subunit, each of which has its own function. The "s" in 40s and 60s is an abbreviation for Svedberg units, which are a measure of how quickly a large molecule or complex molecular structure sediments (or sinks) to the bottom of a centrifuge tube while being centrifuged. The larger the number, the larger the molecule.

Like all RNA, mRNA is composed of just four types of nucleotides: adenine (A), guanine (G), cytosine (C), and uracil (U). Therefore, the information in an mRNA is contained in a linear sequence of nucleotides that is converted into a protein molecule composed of a linear sequence of amino acids. This process is referred to as "translation," since it converts the "language" of nucleotides that make up an mRNA into the "language" of amino acids that make up a protein. This is achieved by a three-letter genetic code in which each amino acid in a protein is specified by a three-nucleotide sequence in the mRNA called a codon. The four possible "letters" means that there are sixty-four possible three-letter "words." As there are only twenty amino acids used to make proteins, most amino acids are encoded by several different codons. For example, there are six different codons (UCU, UCC, UCA, UCG, AGU, and AGC) that specify the amino acid serine, whereas there is only one codon (AUG) that specifies the amino acid methionine. The mRNA, therefore, is simply a linear array of codons (that is, three-nucleotide "words" that are "read" by tRNAs together with ribosomes). The region within an mRNA containing this sequence of codons is called the coding region.

Before translation can occur in eukaryotic cells, mRNAs undergo processing steps at both ends to add features that will be necessary for translation (These processing steps do not occur in prokaryotic cells.) Nucleotides are structured such that they have two ends, a 5′ and a 3′ end, that are available to form chemical bonds with other nucleotides. Each nucleotide present in an mRNA has a 5′ to 3′ orientation that gives a directionality to the mRNA so that the RNA begins with a 5′ end and finishes in a 3′ end. The ribosome reads the coding region of an mRNA in a 5′ to 3′ direction. Following the synthesis of an mRNA from its DNA template, one guanine is added to the 5′ end of the mRNA in an inverted orientation and is the only nucleotide in the entire mRNA present in a 3′ to 5′ orientation. It is referred to as the cap. A long stretch of adenosine is added to the 3′ end of the mRNA to make what is called the poly-A tail.

Typically, mRNAs have a stretch of nucleotide sequence that lies between the cap and the coding region. This is referred to as the leader sequence and is not translated. Therefore, a signal is necessary to indicate where the coding region initiates. The codon AUG usually serves as this initiation codon; however, other AUG codons may be present in the coding region. Any one of three possible codons (UGA, UAG, or UAA) can serve as stop codons that signal the ribosome to terminate translation. Several accessory proteins assist ribosomes in binding mRNA and help carry out the required steps during translation.

The Translation Process: Initiation

Translation occurs in three phases: initiation, elongation, and termination. The function of the 40s ribosomal subunit is to bind to an mRNA and locate the correct AUG as the initiation codon. It does this by binding close to the cap at the 5′ end of the mRNA and scanning the nucleotide sequence in its 5′ to 3′ direction in search of the initiation codon. Marilyn Kozak identified a certain nucleotide sequence surrounding the initiator AUG of eukaryotic mRNAs that indicates to the ribosome that this AUG is the initiation codon. She found that the presence of an A or G three nucleotides prior to the AUG and a G in the position immediately following the AUG were critical in identifying the correct AUG as the initiation codon. This is referred to as the "sequence context" of the initiation codon. Therefore, as the 40s ribosomal subunit scans the leader sequence of an mRNA in a 5′ to 3′ direction, it searches for the first AUG in this context and may bypass other AUGs not in this context.

Nahum Sonenberg demonstrated that the scanning process by the 40s subunit can be impeded by the presence of stem-loop structures present in the leader sequence. These form

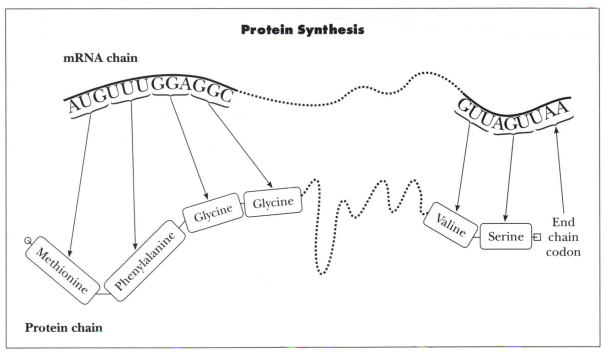

Protein Synthesis

mRNA chain

AUGUUUGGAGGC .... GUUAGUUAA

Methionine Phenylalanine Glycine Glycine Valine Serine End chain codon

Protein chain

Protein synthesis is directed by messenger RNA (mRNA). The order of the amino acids in the protein chain is controlled by the order of the bases in the mRNA chain. It takes a codon of three bases to specify one amino acid.

from base pairing between complementary nucleotides present in the leader sequence. Two nucleotides are said to be complementary when they join together by hydrogen bonds. For instance, the nucleotide (or base) A is complementary to U, and these two can form what is called a "base pair." Likewise, the nucleotides C and G are complementary. Several accessory proteins, called eukaryotic initiation factors (eIFs), aid the binding and scanning of 40s subunits. The first of these, eIF4F, is composed of three subunits called eIF4E, eIF4A, and eIF4G. The protein eIF4E is the subunit responsible for recognizing and binding to the cap of the mRNA. The eIF4A subunit of eIF4F, together with another factor called eIF4B, functions to remove the presence of stem-loop structures in the leader sequence through the disruption of the base pairing between nucleotides in the stem loop. The protein eIF4G is the large subunit of eIF4F, and it serves to interact with several other proteins, one of which is eIF3. It is this latter initiation factor that the 40s subunit first associates with during its initial binding to an mRNA.

Through the combined action of eIF4G and eIF3, the 40s subunit is bound to the mRNA, and through the action of eIF4A and eIF4B, the mRNA is prepared for 40s subunit scanning. As the cellular concentration of eIF4E is very low, mRNAs must compete for this protein. Those that do not compete well for eIF4E will not be translated efficiently. This represents one means by which a cell can regulate protein synthesis. One class of mRNA that competes poorly for eIF4E encodes growth-factor proteins. Growth factors are required in small amounts to stimulate cellular growth. Sonenberg has shown that the overproduction of eIF4E in animal cells leads to a reduction in the competition for this protein, and mRNAs such as growth-factor mRNAs that were previously poorly translated when the concentration of eIF4E was low are now translated at a higher rate when eIF4E is abundant. This in turn results in the overproduction of growth factors, which leads to uncontrolled growth, a characteristic typical of cancer cells.

A protein that specifically binds to the poly-A tail at the 3' end of an mRNA is called the poly-

A-binding protein (PABP). Discovered in the 1970's, the only function of this protein was thought to be to protect the mRNA from attack at its 3′ end by enzymes that degrade RNA. Daniel Gallie demonstrated another function for PABP by showing that the PABP-poly-A-tail complex was required for the function of the eIF4F-cap complex during translation initiation. The idea that a protein located at the 3′ end of an mRNA should participate in events occurring at the opposite end of an mRNA seemed strange initially. However, RNA is quite flexible and is rarely present in a straight, linear form in the cellular environment. Consequently, the poly-A tail can easily approach the cap at the 5′ end. Gallie showed that PABP interacts with eIF4G and eIF4B, two initiation factors that are closely associated with the cap, through protein-to-protein contacts. The consequence of this interaction is that the 3′ end of an mRNA is held in close physical proximity to its cap. The interaction between these proteins stabilizes their binding to the mRNA, which in turn promotes protein synthesis. Therefore, mRNAs can be thought of as adopting a circular form during translation that looks similar to a snake biting its own tail. This idea is now widely accepted by scientists.

One additional factor, called eIF2, is needed to bring the first tRNA to the 40s subunit. Along with the initiator tRNA (which decodes the AUG codon specifying the amino acid methionine), eIF2 aids the 40s subunit in identifying the AUG initiation. Once the 40s subunit has located the initiation codon, the 60s ribosomal subunit joins the 40s subunit to form the intact 80s ribosome. (Svedberg units are not additive; therefore, a 40s and 60s unit joined together do not make a 100s unit.) This marks the end of the initiation phase of translation.

The Translation Process: Elongation and Termination

During the elongation phase, tRNAs bind to the 80s ribosome as it passes over the codons of the mRNA, and the amino acids attached to the tRNAs are transferred to the growing polypeptide. Binding of the tRNAs to the ribosome is assisted by an accessory protein called eukary-

otic elongation factor 1 (eEF1). A codon is decoded by the appropriate tRNA through base pairing between the three nucleotides that make up the codon in the mRNA and three complementary nucleotides within a specific region (called the anticodon) within the tRNA. The tRNA binding sites in the 80s ribosome are located in the 60s subunit. The ribosome moves over the coding region one codon at a time, or in steps of three nucleotides, in a process referred to as "translocation." When the ribosome moves to the next codon to be decoded, the tRNA containing the appropriate anticodon will bind tightly in the open site in the 60s subunit (the A site). The tRNA that bound to the previous codon is present in a second site in the 60s subunit (the P site). Once a new tRNA has bound to the A site, the ribosomal RNA itself catalyzes the formation of a peptide bond between the growing polypeptide and the new amino acid. This results in the transfer of the polypeptide attached to the tRNA present in the P site to the amino acid on the tRNA present in the A site. A second elongation factor, eEF2, catalyzes the movement of the ribosome to the next codon to be decoded. This process is repeated one codon at a time until a stop codon is reached.

The termination phase of translation begins when the ribosome reaches one of the three termination or stop codons. These are also referred to as "nonsense" codons as the cell does not produce any tRNAs that can decode them. Accessory factors, called release factors, are also required to assist this stage of translation. They bind to the empty A site in which the stop codon is present, and this triggers the cleavage of the bond between the completed protein from the last tRNA in the P site, thereby releasing the protein. The ribosome then dissociates into its 40s and 60s subunits, the latter of which diffuses away from the mRNA. The close physical proximity of the cap and poly-A tail of an mRNA maintained by the interaction between PABP and the initiation factors (eIF4G and eIF4B) is thought to assist the recycling of the 40s subunit back to the 5′ end of the mRNA to participate in a subsequent round of translation.

Impact and Applications

The elucidation of the process and control of protein synthesis provides a ready means by which scientists can manipulate these processes in cells. In addition to infectious diseases, insufficient dietary protein represents one of the greatest challenges to world health. The majority of people living today are limited to obtaining their dietary protein solely through the consumption of plant matter. Knowledge of the process of protein synthesis may allow molecular biologists to increase the amount of protein in important crop species. Moreover, most plants contain an imbalance in the amino acids needed in the human diet that can lead to disease. For example, protein from corn is poor in the amino acid lysine, whereas the protein from soybeans is poor in methionine and cysteine. Molecular biologists may be able to correct this imbalance by changing the codons present in plant genes, thus improving this source of protein for those people who rely on it for life.

—*Daniel R. Gallie*

See also: Central Dogma of Molecular Biology; DNA Repair; DNA Replication; DNA Structure and Function; Genetic Code; Genetic Code, Cracking of; Molecular Genetics; Protein Structure; RNA Structure and Function; RNA Transcription and mRNA Processing; RNA World; Synthetic Genes.

Further Reading

Crick, Francis. "The Genetic Code III." *Scientific American* 215 (October, 1966). The co-discoverer of DNA's helical structure provides a good summary of the code specifying the amino acids.

Lake, James. "The Ribosome." *Scientific American* 245 (August, 1981). Summarizes information about the structure and function of ribosomes.

Lewin, Benjamin. *Genes VII*. New York: Oxford University Press, 2001. Details the translational process.

Rich, Alexander, and Sung Hou Kim. "The Three Dimensional Structure of Transfer RNA." *Scientific American* 238 (January, 1978). Presents a structural description of tRNA.

Proteomics

Fields of study: Molecular genetics; Techniques and methodologies

Significance: *The study of proteomics and its relationship to genomics currently focuses on the vast family of gene-regulating proteins. These polypeptides and their functions affect the expression of various genetically related diseases, such as Alzheimer's and cancer. By focusing on the interrelated groups of regulator functions, geneticists are learning the connections between structure, abundance within the cell, and how each protein relates to expression.*

Key terms

CHROMATOGRAPHY: a separation technique involving a mobile solvent and a stationary, adsorbent phase

MASS SPECTROSCOPY: a method of analyzing molecular structure in which sample molecules are ionized and the resulting fragmented particles are passed through electric and magnetic fields to a detector

PERIPHERAL PROTEINS: proteins of the chromosome that do not directly affect transcription

PROTEIN FOLDING STRUCTURE: the three-dimensional structure of proteins created by the folding of linked amino acids upon each other; this structure is held together by intermolecular forces, such as hydrogen bonds and ionic attractions

PROTEIN MARKER: a sequence of DNA that chemically attracts a particular regulatory protein sequence or structure

REGULATORS: proteins that control the transcription of a gene

SENILE PLAQUES: protein sections that are no longer functional and clutter the intercellular space of the brain, disrupting proper processes

TRANSCRIPTION: the process by which mRNA is formed using DNA as a template

TRANSLATION: the process of building a protein by bonding amino acids according to the mRNA marker present

What Is "Proteomics"?

Historically, much of the focus in genetic research has been on genes and completion of the Human Genome Project. More recently, the focus has shifted to a new and related topic, the proteome. Proteins are known to perform most of the important functions of cells. Therefore, proteomics is, essentially, the study of proteins in an organism and, most important, their function. There are many aspects to the understanding of protein function, including where a particular protein is located in the cell, what modifications occur during its activity, what ligands may bind to it, and its activity. Researchers are seeking to identify all the proteins made in a given cell, tissue, or organism and determine how those proteins interact with metobolites, with themselves, and with nucleic acids. By studying proteomics, scientists hope to uncover underlying causes of disease at the cellular level, invent better methods of diagnosis, and discover new, more efficient medicines for the treatment of disease.

Proteomics has moved to the forefront of molecular research, especially in the area of drug research. Neither the structure nor the function of a gene can be predicted from the DNA sequence alone. Although genes code for proteins, there is a large difference between the number of messenger (mRNA) molecules transcribed from DNA and the number of proteins in a cell. In addition, two hundred known modifications occur during the stages between transcription and post-translation, including phosphorylation, glycosylation, proteolytic processing, deamidation, sulfation, and nitration. Other factors that affect the expression of proteins include aging, stress, environmental forces, and drugs. In addition, changes to the sequence of amino acids may occur during or after translation.

Methods of Proteomic Research

In order to study the functions of a protein, it must be separated from other proteins or contaminants, purified, and structurally characterized. These are the major tasks facing researchers in the field.

In order to obtain a sufficient quantity of a particular protein for study, the coding plasmid can be injected into *Escherichia coli* bacteria and the cells will translate the protein multiple times. Alternatively, it must be extracted from biological tissues. The desired polypeptide must then be separated from cells or tissues that may contain thousands of unique proteins. This can be accomplished by homogenizing the tissue, extracting the proteins with solvents or by centrifugation, and further purifying the protein by various means, including high-pressure liquid chromatography (HPLC, separation by solubility differences) and two-dimensional (2-D) gel electrophoresis (separation of molecules by charge and molecular mass).

Structural characterization begins with establishing the order of linked amino acids in the protein. This can be accomplished by the classical techniques of using proteases to fragment the protein chemically and then analyzing the fragments by separation and spectroscopic analysis. The molecular mass of small polypeptides can be investigated by employing several techniques involving mass spectrometry (MS). Sequentially coupled mass spectrometers (the "tandem" MS/MS techniques) are being used to analyze the amino acid sequence and molecular masses of isoloated larger polypeptides. These MS/MS analyses are sometimes added to a separation method, such as HPLC, to analyze mixtures of polypeptides.

Historically, Linus Pauling used analytical data from X-ray diffraction (or crystallography) to determine the three-dimensional, helical structure of proteins. The method is still being used to investigate the structures of proteins and ligand-protein complexes. Such studies may lead to significant improvements in the design of medicinal drugs. One significant drawback to analyzing protein structure by X-ray diffraction, however, is that the method requires a significant quantity (approximately 1 milligram) of the protein. Transmission electron microscopy (TEM), which uses electron beams to produce images and diffraction patterns from extremely small samples or regions of a sample, is therefore often preferable. The TEM method may involve auxiliary techniques to analyze data, including enhancement of images by means of computer software.

Although such methods provide valuable information in analyzing the structure of proteins, they suffer from the loss of spatial information that occurs when tissues are homogenized, when the protein is obtained from a manufactured, bacterial environment, or when it is otherwise isolated. Matrix-assisted laser desorption/ionization time-of-flight (MALDI-TOF) mass spectrometry is a complementary method of analysis that does not yield structural information but provides protein profiles from intact tissue, allowing comparison of diseased versus normal tissue.

Large databases of mass spectroscopic data are being assembled to assist in future identification of know proteins. Further databases of proteome information include particular molecular masses, charges, and, in some cases, connections to the genes regulated or the parent gene of the peptide in question. Scientists hope to relate regulators and the complex web of peripheral proteins that affect the function of each gene.

Challenges and Limitations of Current Methods

The amount of data being obtained by proteomics research poses a problem in organizing and processing the information obtained on proteins. The Human Proteome Organization (HUPO) and the European Bioinformatics Institute (EBI) are two organizations whose purposes include the management and organization of proteomics information and databases, and the facilitation of the advancement of this scientific endeavor.

Analyzing MS data from proteins and relating the complex array of proteins within a single cell to the linear genetic material of DNA present challenges to researchers that they are tackling through computer algorithms, programs, and databases. The SWISS-PROT database, for example, is an annotated protein-sequence database maintained by the Swiss Bioinformatics Institute.

Other obstacles to relating proteins to parent genes include the loss of quaternary structure during separation and the presence of post-translation processing, which can alter the amino acid sequence to the extent that it be-

comes almost unrecognizable from the parent gene. A lack of protein amplification methods—techniques that would produce more copies of a protein to aid in study—requires sensitive analysis methods and increasingly strong detectors. Currently new methods are being developed, but the limit of study is as large as 1 nanometer.

Disease

Proteins often act as markers for disease. As researchers study proteins, they have found that disease may be characterized by some proteins that are being overproduced, not being produced at all, or being produced at inappropriate times. As the correlation of proteins to disease becomes clearer, better diagnostic tests and drugs are being explored. For example, Alzheimer's disease and Down syndrome are associated with a common protein fragment as the major extracellular protein component of senile plaques.

Researchers are investigating changes in protein expression in heart disease and heart failure, and several hundred cardiac proteins have already been identified. The study of proteomics in immunological diseases has revealed that there is a connection between the human neutrophil α-defensins (HNPs) and human immunodeficiency virus, HIV-1. HNPs are small, cysteine-rich, cationic antimicrobial proteins that are stored in the azurophilic granules of neutrophils and released during phagocytosis to kill ingested foreign microbes. To date, the three most abundant forms of the protein have been implicated in suppressing HIV-1 in vivo.

Similarly, cancer is being studied to find a roster of proteins that are present in cancerous cells but not in normal cells. A joint effort from the National Cancer Institute and the Food and Drug Administration is searching for the differences between cancerous and normal cells, and also for protein "markers."

Possible Future Directions

Although proteomics is a relatively new area of genetic research, the importance of the sugar coatings of proteins and cells is gaining attention, under the name "glycomics." This area of study has arisen because of the many roles of sugar coatings in important cellular functions, including the immunological recognition sites, barriers, and sites for attack by pathogens.

—*Audrey Krumbach, Kayla Williams, and Massimo D. Bezoari*

See also: Bioinformatics; Genomics; Human Genetics; Human Genome Project; Protein Structure; Protein Synthesis.

Further Reading

Liebler, David G. *Introduction to Proteomics: Tools for the New Biology.* Totowa, N.J.: Humana Press, 2001. Basics of protein and proteome analysis, key concepts of proteomics, workings of the analytical instrumentation, overview of software tools, and applications of protein and peptide separation techniques, mass spectrometry, and more.

Link, Andrew J., ed. *2-D Proteome Analysis Protocols.* Nashville, Tenn.: Vanderbilt University Medical Center. Practical proteomics, presenting techniques with step-by-step instructions for laboratory researchers. Fifty-five chapters prepared by more than seventy specialists.

Modern Drug Discovery (October, 2002). The entire issue is devoted to proteomics, with many interesting articles on methods of research, Web sites, and computer-assisted methods of data analysis.

Web Sites of Interest

Cambridge Healthtech Institute. http://www.genomicglossaries.com/content/proteomics.asp. This site provides a useful glossary of technical terms used in proteomics, as well as many links to related sites.

Human Proteomics Organization. http://www.hupo.org. HUPO works to consolidate regional proteome organizations into a worldwide group, conduct scientific and educational activities, and disseminate knowledge about both the human proteome and model organisms.

Pseudogenes

Field of study: Molecular genetics
Significance: *Pseudogenes are DNA sequences derived from partial copies, mutated complete copies of functional genes, or normal copies of a gene that has lost its control sequences and therefore cannot be transcribed. They may originate by gene duplication or retrotransposition. They are apparently nonfunctional regions of the genome that may evolve at a maximum rate, free from the evolutionary constraints of natural selection.*

Key terms

INTRONS: noncoding segments of DNA within a gene that are removed from pre-messenger RNA (pre-mRNA) as a part of the process of producing a mature mRNA

LONG INTERSPERSED SEQUENCES (LINES): long repeats of DNA sequences scattered throughout a genome

NEUTRAL THEORY OF MOLECULAR EVOLUTION: the theory that most DNA sequence evolution is a result of mutations that are neutral with respect to the fitness of the organism

RETROTRANSPOSON (RETROPOSON): a DNA sequence that is transcribed to RNA and reverse transcribed to a DNA copy able to insert itself at another location in the genome

REVERSE TRANSCRIPTASE: an enzyme, isolated from retroviruses, that synthesizes a DNA strand from an RNA template

SHORT INTERSPERSED SEQUENCES (SINES): short repeats of DNA sequences scattered throughout a genome

Definition and Origin

Pseudogenes are DNA sequences that resemble genes but are not correctly transcribed or translated to a functional polypeptide. If a functional gene is duplicated so that there are two nonhomologous copies of it in the genome, one of the copies can retain the code for the original polypeptide product, while the other is free from such constraints, since one copy of the gene is sufficient to produce the protein. Because mutations in one copy do not destroy the gene's function, they may be retained, and the unneeded copy can evolve more quickly. It may change to produce a different, functional polypeptide (and effectively become a new gene), or it may remain nonfunctional as a pseudogene. There are two types of pseudogenes, defined by how they were produced: nonprocessed and processed.

Nonprocessed Pseudogenes

Nonprocessed (or duplicated) pseudogenes arise when a portion of the original gene is duplicated, with portions necessary for proper functioning missing or altered or when the complete original gene is duplicated. They can be identified by the presence of introns and may have mutations in the promoter that prevent transcription or the correct removal of introns, or they may have other mutations (such as premature stop codons) within exons that result in translation of a nonfunctional polypeptide. A series of tandem duplications of a gene can result in clustered gene families, which can include expressed genes, expressed pseudogenes (which are transcribed but produce no functional polypeptide), and nonexpressed pseudogenes that are not transcribed. The alpha-globulin and beta-globulin clusters are examples of such gene families. Other examples of nonprocessed pseudogenes include members of the immunoglobulin (Ig) and major histocompatibility complex (MHC) gene families.

Processed Pseudogenes

Processed pseudogenes originate from transcribed RNA copies of genes that are copied back to DNA by the enzyme reverse transcriptase. Processed pseudogenes are usually integrated into the genome in a different location from the original gene. Reverse transcriptase is an enzyme produced by retroviruses, which have RNA genomes that are reverse transcribed to DNA when the viruses infect host cells. Retrotransposons, which are related to retroviruses, are DNA sequences that transpose or duplicate themselves by reverse transcription of a transcribed RNA copy of the sequence.

Often, retrotransposons will carry along a copy of the surrounding host DNA, resulting in the duplication of that sequence—a processed pseudogene. Because the introns of a gene are removed from the RNA transcript, processed pseudogenes are not exact copies of the original DNA sequence; the introns are missing. Copies of protein-coding genes copied by this mechanism are members of a type of repetitive DNA called LINES (for long interspersed sequences) and exist in multiple copies scattered around the genome, each up to several thousand base pairs in length. Short processed pseudogenes are members of another class of repetitive DNA called SINES (short interspersed sequences of up to several hundred base pairs in length) and result from the retrotransposon-mediated copying of tRNA or rRNA genes. SINES of this type are sometimes very abundant in genomes because they may have internal promoters, so that they are more easily transcribed, and therefore transposed. The most prominent of SINES are those that are members of the *Alu* family, which occur an average of once every six thousand base pairs in the human genome.

Pseudogenes and Neutral Evolution

In 2003, Japanese researchers reported the discovery of a mouse pseudogene that is involved in regulating the expression of its related "functional" gene. This discovery suggested that at least some pseudogenes may have important functions. Although pseudogenes are very commonly found across genomes, most do not appear to serve any function, and until further research uncovers more functional pseudogenes this assumption appears warranted. Their abundance can be explained by the tendency of duplicated sequences to be further copied. Retrotransposition increases the number of copies of processed pseudogenes, and gene duplication leading to unprocessed pseudogenes favors mechanisms that generate additional copies, leading to clustered gene families. Natural selection does not tend to eliminate these additional copies because their presence does not harm the organism as long as there is at least one functional copy of the original gene. In other words, pseudogenes are selectively neutral.

Because of their selective neutrality, pseudogenes are especially useful for estimating neutral mutation rates in genomes. The neutral theory of evolution predicts that, because of the constraints of selection, functional regions of the genome (such as the exons, or coding sequences, of genes) will evolve more slowly than less critical sequences, such as introns, or nonfunctional sequences like pseudogenes. The number of nucleotide differences between homologous sequences of related species can be used to calculate estimates of evolutionary rates, and such estimates support the neutral theory: the greatest rates of divergence occur within pseudogenes. Using comparisons from several pseudogenes, researchers can establish the baseline neutral mutation rate for a group of species.

—Stephen T. Kilpatrick
See also: Gene Families; Repetitive DNA.

Further Reading

Graur, Dan, and Wen-Hsiung Li. *Fundamentals of Molecular Evolution.* 2d ed. Sunderland, Mass.: Sinauer Associates, 1999. A detailed review of the topic, including the importance of pseudogenes in determining genomic rates of neutral evolution.

Lewin, Benjamin. *Genes VII.* New York: Oxford University Press, 2001. Provides several examples of clustered gene families that include pseudogenes, and describes the mechanism for the origin of processed pseudogenes.

Li, Wen-Hsiung, Takashi Gojobori, and Masatoshi Nei. "Pseudogenes as a Paradigm of Neutral Evolution." *Nature* 292, no. 5820 (1981): 237-239. Discusses the predictions of the neutral theory, how rates of nucleotide substitution may be calculated, and how data from pseudogenes support the neutral theory.

Pseudohermaphrodites

Field of study: Diseases and syndromes
Significance: *Pseudohermaphrodites are individuals born with either ambiguous genitalia or external genitalia that are the opposite of their chromosomal sex. These individuals need a thorough medical evaluation and appropriate medical intervention to help ensure a healthy, well-adjusted life.*

Key terms

AMBIGUOUS GENITALIA: external sexual organs that are not clearly male or female
GENOTYPE: the sum total of the genes present in an individual
GONADS: organs that produce reproductive cells and sex hormones, for example, testes in males and ovaries in females
KARYOTYPE: the number and kind of chromosomes present in every cell of the body (normal female karyotype is 46,XX and normal male karyotype is 46,XY)
PHENOTYPE: the physical appearance and physiological characteristics of an individual, which depends on the interaction of genotype and environment

Normal Fetal Development

Prior to nine weeks gestational age, a male and a female fetus have identical external genitalia (sexual organs) consisting of a phallus and labioscrotal folds. The phallus develops into a penis in males and a clitoris in females; labioscrotal folds become the scrotum in males and the labial folds in females. Early in development, the gonads can develop into either testes or ovaries. In a fetus with a normal male karyotype (46,XY), the primitive gonads become testes, which produce testosterone. Testosterone in turn causes enlargement of the primitive phallus into a penis. It is the presence of the Y chromosome, and in particular a small, sex-determining region of the Y chromosome termed the *SRY* locus, that drives the formation of the testes. The presence of the *SRY* locus appears to be essential for development of a normal male.

Pseudohermaphroditism

A true hermaphrodite is born with both ovarian and testicular tissue. A male pseudohermaphrodite has a 46,XY karyotype with either female genitalia or ambiguous genitalia (but only testicular tissue); a female pseudohermaphrodite has a 46,XX karyotype with either male genitalia or ambiguous genitalia (but only ovarian tissue). Ambiguous genitalia typically consist of a small, abnormally shaped, phalluslike structure, often with hypospadias (in which urine comes from the base of the penis instead of the tip) and abnormal development of the labioscrotal folds (not clearly a scrotum or labia). A vaginal opening may be present.

Most cases of pseudohermaphroditism result from abnormal exposure to increased or decreased amounts of sex hormones during embryonic development. The most common cause of female pseudohermaphroditism is exposure of a female fetus to increased levels of testosterone during the first half of pregnancy. Maternal use of anabolic steroids can cause this condition, but the most common genetic cause of increased testosterone exposure is congenital adrenal hyperplasia (CAH). CAH results from an abnormality in the enzymatic pathways of the fetus that make both cortisol (a stress hormone) and the sex steroids (such as testosterone). At several points in these pathways, there may be a nonfunctioning enzyme that results in too little production of cortisol and too much production of the sex steroids. This will result in partial masculinization of the external genitalia of a female embryo. Females with CAH are usually born with an enlarged clitoris (often mistakenly thought to be a penis) and partial fusion of the labia. Males can also have CAH, but the excess testosterone does not affect their genital development since a relatively high level of testosterone exposure is a normal part of their development.

The most common causes of male pseudohermaphroditism are abnormalities of testosterone production or abnormalities in the testosterone receptor at the cellular level. One example is a deficiency in 5-alpha-reductase, the enzyme that converts testosterone to dihydrotestosterone (DHT). When there is a deficiency of this enzyme, there will be a deficiency of DHT, which is the hormone primarily responsible for masculinization of external genitalia. A male who lacks DHT will have female-appearing external genitalia or ambiguous genitalia at birth. Often these individuals are reared as females, but at puberty they will masculinize because of greatly increased production of testosterone. These individuals may actually develop into nearly normal-appearing males. Abnormalities of the testosterone receptor can also result in a range of different conditions in affected males, from normal female appearance (a totally defective receptor) to ambiguous genitalia (partially defective receptor) in a 46,XY male. These individuals will not masculinize at puberty because no matter how much testosterone or DHT they produce, their bodies cannot respond to the hormones.

Both male and female pseudohermaphroditism can result from chromosomal abnormalities. The absence or dysfunction of the *SRY* locus produces an individual with normal female genitalia but a 46,XY karyotype. Individuals with a 46,XX karyotype who have the *SRY* locus transposed to one of their X chromosomes will have a normal male appearance.

Impact and Applications

Some forms of pseudohermaphroditism are life threatening, and so early diagnosis is imperative. Both males and females with CAH are at risk for sudden death caused by low cortisol levels and other hormone deficiencies. Early diagnosis is relatively easy in affected females since their genital abnormalities are noticeable at birth. Affected males are often not recognized until they have a life-threatening event, which usually occurs in the first two weeks of life. Treatment of CAH consists of appropriate hormone supplementation that, if instituted early in life, can help prevent serious problems. CAH is inherited in an autosomal recessive manner, so parents of an affected individual have a 25 percent chance of having another affected child with each pregnancy.

The sex of rearing of a child with ambiguous genitalia is usually determined by the child's type of pseudohermaphroditism. Typically, sex of rearing will be based on the chromosomal sex of the child. These children may need sex hormone supplementation or surgery to assist in developing gender-appropriate genitalia. Children with pseudohermaphroditism with normal-appearing genitalia at birth may not be recognized until puberty, when abnormal masculinization or feminization may occur.

These individuals need medical evaluation and karyotype determination to guide the proper medical treatment.

—*Patricia G. Wheeler*

See also: Biological Clocks; Gender Identity; Hermaphrodites; Homosexuality; Human Genetics; Metafemales; RNA Transcription and mRNA Processing; Steroid Hormones; Testicular Feminization Syndrome; X Chromosome Inactivation; XYY Syndrome.

Further Reading

Hunter, R. H. F. *Sex Determination, Differentiation, and Intersexuality in Placental Mammals.* New York: Cambridge University Press, 1995. Discusses the genetic determination of sex in mammals. Illustrations (some color), bibliography, index.

Meyer-Bahlburg, Heino. "Intersexuality and the Diagnosis of Gender Identity Disorder." *Archives of Sexual Behavior* 23 (1994). Addresses gender identity and its relation to pseudohermaphroditism.

Simpson, J. L. "Disorders of the Gonads, Genital Tract, and Genitalia." In *Emery and Rimoin's Principles and Practice of Medical Genetics,* edited by David L. Rimoin et al. 3 vols. New York: Churchill Livingstone, 2002. A detailed account of male and female pseudohermaphroditism. Illustrations, bibliography, index.

Speiser, Phyllis W., ed. *Congenital Adrenal Hyperplasia.* Philadelphia: W. B. Saunders, 2001. Contents address prenatal treatment, newborn screening, new treatments, surgery, gender, sexuality, cognitive function, and pregnancy outcomes. Illustrated.

Web Sites of Interest

Intersex Society of North America. http://www.isna.org. The society is "a public awareness, education, and advocacy organization which works to create a world free of shame, secrecy, and unwanted surgery for intersex people (individuals born with anatomy or physiology which differs from cultural ideals of male and female)." Includes links to information on such conditions as clitoromegaly, micropenis, hypospadias, ambiguous genitals, early genital surgery, adrenal hyperpla-

sia, Klinefelter syndrome, androgen insensitivity, and testicular feminization.

Johns Hopkins University, Division of Pediatric Endocrinology, Syndromes of Abnormal Sex Differentiation. http://www.hopkinsmedicine.org/pediatricendocrinology. Site provides a guide to the science and genetics of sex differentiation, including a glossary. Click on "patient resources."

National Organization for Rare Disorders (NORD). http://www.rarediseases.org. Offers information and articles about rare genetic conditions and diseases, including XYY syndrome, in several searchable databases.

Punctuated Equilibrium

Fields of study: Evolutionary biology; Population genetics

Significance: *Punctuated equilibrium is a model of evolutionary change in which new species originate abruptly and then exist through a long period of stasis. This model is important as an explanation of the stepwise pattern of species change seen in the fossil record.*

Key terms

ALLOPATRIC SPECIATION: a theory that suggests that small parts of a population may become genetically isolated and develop differences that would lead to the development of a new species

HETEROCHRONY: a change in the timing or rate of development of characters in an organism relative to those same events in its evolutionary ancestors

PHYLETIC GRADUALISM: the idea that evolutionary change proceeds by a progression of tiny changes, adding up to produce new species over immense periods of time

Evolutionary Patterns

Nineteenth century English naturalist Charles Darwin viewed the development of new species as occurring slowly by a shift of characters within populations, so that a gradual transition from one species to another took place. This is now generally referred to as

Stephen Jay Gould in his office on the Harvard campus in 1997. Together with Niles Eldredge, Gould developed the theory of punctuated equilibrium to explain gaps in the fossil record. (AP/Wide World Photos)

phyletic gradualism. A number of examples from the fossil record were put forward to support this view, particularly that of the horse, in which changes to the feet, jaws, and teeth seem to have progressed in one direction over a long period of time. Peter Sheldon in 1987 documented gradual change in eight lineages of trilobites over a three-million-year period in the Ordovician period of Wales. Despite these, and other, examples (some of which have been reinterpreted), it is clear that the fossil record more commonly shows a picture of populations that are stable through time but are separated by abrupt morphological breaks. This pattern was recognized by Darwin but was attributed by him to the sketchy and incomplete nature of the fossil record. So few animals become fossilized, and conditions for fossilization are so rare, that he felt only a fragmentary sampling of gradual transitions was present, giving the appearance of abrupt change.

One hundred years later, the incompleteness of the fossil record no longer seemed convincing as an explanation. In 1972, Niles Eldredge and Stephen Jay Gould published their theory of the evolutionary process, called by them "punctuated equilibrium." This model explains the lack of intermediates by suggesting that evolutionary change occurs only in short-lived bursts in which a new species arises abruptly from a parent species, often with relatively large morphological changes, and thereafter remains more or less stable until its extinction.

The Process of Punctuated Equilibrium

A number of explanations have been put forward to show how this process might take place. One of these, termed allopatric speciation, was first proposed by Ernst Mayr in 1963. He pointed out that a reproductive isolating mechanism is needed to provide a barrier to gene flow and that this could be provided by geographic isolation. Allopatric or geographical isolation could result when the normal range of a population of organisms is reduced or fragmented. Parts of the population become separated in peripheral isolates, and if the population is small, it may become modified rapidly by natural selection or genetic drift, particularly if it is adapting to a new environment. This type of process is commonly called the founder effect, because it is the characteristics of the small group of individuals that will overwhelmingly determine the possible characteristics of their descendants. As the initial members of the peripheral isolate may be few in number, it might take only a few generations for the population to have changed enough to become reproductively isolated from the parent population. In the fossil record, this will be seen as a period of stasis representing the parent population, followed by a rapid morphological change as the peripheral population is isolated from it and then replaces it, either competitively or because it has become extinct or has moved to follow a shifting habitat. Because this is thought to take place rapidly in small populations, fossilization potential is low, and unequivocal examples are not common in the fossil record. However, in 1981, Peter Williamson published a well-documented example from the Tertiary period of Lake Turkana in Kenya, which showed episodes of stasis and rapid change in populations of freshwater mollusks. The increases in evolutionary rate were apparently driven by severe environmental change that caused parts of the lake to dry up.

Punctuated changes may also have taken place because of heterochrony, which is a change in the rate of development or timing of appearance of ancestral characters. Paedomorphosis, for example, would result in the retention of juvenile characters in the adult, while its opposite, peramorphosis, would result in an adult morphologically more advanced than its ancestor. Rates of development could be affected by a mutation, perhaps resulting in the descendant growing for much longer than the ancestral form, thus producing a giant version. These changes would be essentially instantaneous and thus would show as abrupt changes of species in the fossil record.

Impact and Applications

The publication of the idea of punctuated equilibrium ignited a storm of controversy that still persists. It predicts that speciation can be very rapid, but more important, it is consistent with the prevalence of stasis over long periods of time so often observed in the fossil record. Species had long been viewed as flexible and responsive to the environment, but fossil species showed no change over long periods despite a changing environment. Biologists have thus had to review their ideas about the concept of species and the processes that operate on them. Species are now seen as real entities that have characteristics that are more than the sum of their component populations. Thus the tendency of a group to evolve rapidly or slowly may be intrinsic to the group as a whole and not dependent on the individuals that compose it. This debate has helped show that the fossil record can be important in detecting phenomena that are too large in scale for biologists to observe.

—David K. Elliott

See also: Artificial Selection; Consanguinity and Genetic Disease; Evolutionary Biology; Genetic Load; Hardy-Weinberg Law; Inbreeding and Assortative Mating; Molecular Clock Hypothesis; Natural Selection; Population Genetics; Speciation.

Further Reading

Eldredge, Niles. *Time Frames: The Rethinking of Darwinian Evolution and the Theory of Punctuated Equilibria.* New York: Simon & Schuster, 1985. The theory's coauthor explains how punctuated equilibria complements Darwin's thesis. Illustrations, bibliography, index.

Eldredge, Niles, and Stephen Jay Gould. "Punctuated Equilibria: An Alternative to Phyletic Gradualism." In *Models in Paleobiology*, edited by Thomas J. M. Schopf. San Francisco:

Freeman, Cooper, 1972. The 1972 paper that introduced the theory of punctuated equilibrium to the scientific community. Illustrations, bibliography.

Gould, Stephen Jay. "The Meaning of Punctuated Equilibria and Its Role in Validating a Hierarchical Approach to Macroevolution." In *Perspectives on Evolution*, edited by Roger Milkman. Sunderland, Mass.: Sinauer Associates, 1982. The founder of the theory expands on its implications for evolution.

Gould, Stephen Jay, and Niles Eldredge. "Punctuated Equilibrium: The Tempo and Mode of Evolution Reconsidered." *Paleobiology* 3 (1977). A follow-up to the original exposition of the theory.

Prothero, Donald R. *Bringing Fossils to Life: An Introduction to Paleobiology*. New York: McGraw-Hill, 2003. Includes an introductory discussion of punctuated equilibrium. Bibliography, index.

Somit, Albert, and Steven A. Peterson, eds. *The Dynamics of Evolution: The Punctuated Equilibrium Debate in the Natural and Social Sciences*. Ithaca, N.Y.: Cornell University Press, 1992. Provides an overview of the punctuated equilibrium debate, including updates by Gould and Eldredge. Bibliography, index.

Quantitative Inheritance

Field of study: Population genetics

Significance: *Quantitative inheritance involves metric traits. These traits are generally associated with adaptation, reproduction, yield, form, and function. They are thus of great importance to evolution, conservation biology, psychology, and especially to the improvement of agricultural organisms.*

Key terms

GENOTYPE: the genetic makeup of an organism at all loci that affect a quantitative trait

HERITABILITY: the proportion of phenotypic differences among individuals that are a result of genetic differences

METRIC TRAITS: traits controlled by multiple genes with small individual effects and continuously varying environmental effects, resulting in continuous variation in a population

PHENOTYPE: the observed expression of a genotype that results from the combined effects of the genotype and the environment to which the organism has been exposed

The Genetics Underlying Metric Traits

An understanding of the genetics affecting metric traits came with the unification of the Mendelian and biometrical schools of genetics early in the 1900's. The statistical relationships involved in inheritance of metric traits such as height of humans were well known in the late 1800's. Soon after that, Gregor Mendel's breakthrough on particulate inheritance, obtained from work utilizing traits such as colors and shapes of peas, was rediscovered. However, some traits did not follow Mendelian inheritance patterns. As an example, Francis Galton crossed pea plants having uniformly large seeds with those having uniformly small seeds. The seed size of the progeny was intermediate. However, when the progeny were mated among themselves, seed size formed a distribution from small to large with many intermediate sizes.

How could particulate genetic factors explain a continuous distribution? The solution was described early in the twentieth century when Swedish plant breeder Herman Nilsson-Ehle crossed red and white wheat. The resulting progeny were light red in color. When matings were made within the progeny, the resulting kernels of wheat ranged in color from white to red. He was able to categorize the wheat into five colors: red, intermediate red, light red, pink, and white. Intermediate colors occurred with greater frequency than extreme colors. Nilsson-Ehle deduced that particulate genetic factors (now known as alleles) were involved, with red wheat inheriting four red alleles, intermediate red inheriting three red alleles, light red inheriting two red alleles, pink inheriting one red allele, and white inheriting no red alleles. These results were consistent with Mendel's findings, except that two sets of factors (now known as loci) were controlling this trait rather than the single locus observed for the traits considered by Mendel. Further, these results could be generalized to account for additional inheritance patterns controlled by more than two loci. Quantitative inheritance was mathematically described by British statistician and geneticist Ronald A. Fisher.

Under many circumstances, the environment also modifies the expression of traits. A combination of many loci with individually small effects alone would produce a rough bell-shaped distribution for a quantitative trait. Environmental effects are continuous and are independent of genetic effects. Environmental effects blur the boundaries of the genetic categories and can make it difficult or impossible to identify the effects of individual loci for many quantitative traits. The distribution of phenotypes, reflecting combined genetic and environmental effects, is typically a smooth, bell-shaped curve.

Genetic and environmental effects jointly influence the value of most metric traits. The relative magnitudes of genetic and environmental effects are measured using heritability statistics. Although essentially equivalent, heritability has several practical definitions. One definition states that heritability is equal to the proportion of observed differences among organisms for a trait due to genetic differences. For example, if one-quarter of the differences among cows for the amount of milk they produce are caused by differences among their genotypes, the heritability of milk production is 25 percent. The remaining 75 percent of differences among the animals are attributed to environmental effects. An alternative definition is that heritability is equal to the proportion of differences among sets of parents that are passed on to their progeny. For example, if the average height of a pair of parents is 8 inches (20 centimeters) more than the mean of their population and the heritability of height is 50 percent, their progeny would be expected to average 4 inches (10 centimeters) taller than their peers in the population.

Fundamental Relationships of Quantitative Genetics

Two relationships are fundamental to the understanding and application of quantitative genetics. First, there is a tendency for likeness among related individuals. Although similarities of human stature and facial appearance within families are familiar to most people, similar relationships hold for such traits in all organisms. Correlation among relatives exists for such diverse traits as blood pressure, plant height, grain yield, and egg production. These correlations are caused by relatives sharing a portion of genes in common. The more closely the individuals are related, the greater the proportion of genes that are shared. Identical twins share all their genes, and full brothers and sisters or parent and offspring are expected to share one-half their genes. This relationship is commonly utilized in the improvement of agricultural organisms. Individuals are chosen to be parents based on the performance of their relatives. For example, bulls of dairy breeds are chosen to become widely used as sires based on the milk-producing ability of their sisters and daughters.

The second fundamental relationship is that, in organisms that do not normally self-fertilize, vigor is depressed in progeny that result from the mating of closely related individuals. This effect is known as inbreeding depression. It may be the basis of the social taboos regarding incestuous relationships in humans and for the dispersal systems for some other species of mammals such as wolves. Physiological barriers have evolved to prevent fertilization between close relatives in many species of plants. Some mechanisms function as an anatomical inhibitor to prevent union of pollen and ova from the same plant; in maize, for example, the male and female flower are widely separated on the plant. Indeed, in some species such as asparagus and holly trees, the sexes are separated in different individuals; thus all seeds must consequently result from cross-pollination. In other systems, cross-pollination is required for fertile seeds to result. The pollen must originate from a plant genetically different from the seed parent. These phenomena are known as self-incompatibility and are present in species such as broccoli, radishes, some clovers, and many fruit trees.

The corollary to inbreeding depression is hybrid vigor, a phenomenon of improved fitness that is often evident in progeny resulting from the mating of individuals less related than the average in a population. Hybrid vigor has been utilized in breeding programs to achieve remarkable productivity of hybrid seed corn as well as crossbred poultry and livestock. Hybrid vigor results in increased reproduction and efficiency of nutrient utilization. The mule, which results from mating a male donkey to a female horse, is a well-known example of a hybrid that has remarkable strength and hardiness compared to the parent species, but which is, unfortunately, sterile.

Quantitative Traits of Humans

Like other organisms, many traits of humans are quantitatively inherited. Psychological characteristics, intelligence quotient (IQ), and birth weight have been studied extensively. The heritability of IQ has been reported to be high. Other personality characteristics such as incidence of depression, introversion, and enthusiasm have been reported to be highly heritable. Musical ability is another characteristic under some degree of genetic control. These results have been consistent across replicated studies and are thus expected to be reliable; however, some caution must be exercised when considering the reliability of results from individual studies. Most studies of heritability in humans have involved likeness of twins reared together and apart. The difficulty in obtaining such data results in a relatively small sample size, at least relative to similar experiments in animals. An unfortunate response to studies of quantitative inheritance in humans was the eugenics movement.

Birth weight of humans is of interest because it is both under genetic control and subject to influence by well-known environmental factors, such as smoking by the mother. Birth weight is subject to stabilizing selection, in which individuals with intermediate values have the highest rates of survival. This results in genetic pressure to maintain the average birth weight at a relatively constant value.

Quantitative Characters in Agricultural Improvement

The ability to meet the demand for food by a growing world population is dependent upon continuously increasing agricultural productivity. Reserves of high-quality farmland have nearly all been brought into production, and a sustainable increase in the harvest of fish is likely impossible. Many countries that struggle to meet the food demands of their populations are too poor to increase agricultural yields through increased inputs of fertilizer and chemicals. Increased food production will, therefore, largely depend on genetic improvement of the organisms produced by farmers worldwide.

Most characteristics of economic value in agriculturally important organisms are quantitatively inherited. Traits such as grain yield, baking quality, milk and meat production, and efficiency of nutrient utilization are under the influence of many genes as well as the production environment. The task of breeders is not only to identify organisms with superior genetic characteristics but also to identify those breeds and varieties well adapted to the specific environmental conditions in which they will be produced. The type of dairy cattle that most efficiently produces milk under the normal production circumstances in the United States, which includes high health status, unlimited access to high-quality grain rations, and protection from extremes of heat and cold, may not be ideal under conditions in New Zealand in which cattle are required to compete with herdmates for high-quality pasture forage. Neither of these animals may be ideal under tropical conditions where extremely high temperatures, disease, and parasites are common.

Remarkable progress has been made in many important food crops. Grain yield has responded to improvement programs. Development of hybrid corn increased yield severalfold over the last few decades of the twentieth century. Development of improved varieties of small grains resulted in an increased ability of many developing countries to be self-sufficient in food production. Grain breeder Norman Borlaug won the Nobel Peace Prize in 1970 for his role in developing grain varieties that contributed to the Green Revolution.

Can breeders continue to make improvements in the genetic potential for crops, livestock, and fish to yield enough food to support a growing human population? Tools of biotechnology are expected to increase the rate at which breeders can make genetic change. Ultimately, the answer depends upon the genetic variation available in the global populations of food-producing organisms and their wild relatives. The potential for genetic improvement of some species has been relatively untapped. Domestication of fish for use in aquaculture and utilization of potential crop species such as amaranth are possible food reserves. Wheat, corn, and rice provide a large proportion of the calories supporting the world population. The yields of these three crop species have already benefited from many generations of selective breeding. For continued genetic improvement, it is critical that variation not be lost through the extinction of indigenous strains and wild relatives of important food-producing organisms.

Impact and Applications

Molecular genetics and biotechnology have also added new tools for analyzing the genetics of quantitative traits. In any organism that has had its genome adequately mapped, genetic markers can be used to determine the number of loci involved in a particular trait. In carefully constructed crosses geneticists look for statistical correlations between markers and the trait of interest. When a high correlation is found, the marker is said to represent a quantitative trait locus (QTL). Often a percentage effect for each QTL can be determined and because the location of markers is typically known, the potential location of the gene can also be inferred (that is, somewhere near the marker). A good understanding of the QTLs involved in the expression of a quantitative trait can help determine the best way to improve the organism.

Although QTLs are much easier to discover in organisms where controlled crosses are possible, studies have also been carried out in humans. In humans, geneticists must rely on whatever matings have happened, and due to ethical limitations, cannot set up specific crosses. Studies in humans have attempted to quantify the number of QTLs responsible for such things as IQ and various physical traits. One study even purported to show that homosexuality is genetically based. Although there is some support for such studies, much controversy surrounds them, and ethicists continue to worry that conclusions from such research will be used in a new wave of eugenics. In spite of the risk of misusing an improved understanding of human quantitative traits, human biology and medicine stand to benefit.

—*William R. Lamberson, updated by Bryan Ness*

See also: Artificial Selection; Biofertilizers; Consanguinity and Genetic Disease; Epistasis; Genetic Load; Hardy-Weinberg Law; Hybridization and Introgression; Inbreeding and Assortative Mating; Mendelian Genetics; Polygenic Inheritance; Population Genetics; Speciation; Twin Studies.

Further Reading

Falconer, D. S., and Trudy F. MacKay. *Introduction to Quantitative Genetics.* 4th ed. Reading, Mass.: Addison-Wesley, 1996. The standard text on the subject, outlining the genetics of differences in quantitative phenotypes and their applications to animal breeding, plant improvement, and evolution.

Race

Field of study: Human genetics and social issues

Significance: *Humans typically have been categorized into a small number of races based on common traits, ancestry, and geography. Knowledge of human genomic diversity has increased awareness of ambiguities associated with traditional racial groups. The sociopolitical consequences of using genetics to devalue certain races are profound, and based on the available data, are completely baseless.*

Key terms

EUGENICS: a movement concerned with the improvement of human genetic traits, predominantly by the regulation of mating

HUMAN GENOME DIVERSITY PROJECT: an extension of the Human Genome Project in which DNA of native people around the world is collected for study

POPULATION: a group of geographically localized, interbreeding individuals

RACE: a collection of geographically localized populations with well-defined genetic traits

History of Racial Classification

Efforts to classify humans into a number of distinct types date back at least to the ancient Greeks. Applying scientific principles to separate people into races has been a goal for more than two centuries. In 1758, the founder of biological classification, Swedish botanist Carolus Linnaeus, arranged humans into four principal races: *Americanus, Europeus, Asiaticus,* and *Afer.* Although geographic location was his primary organizing factor, Linnaeus also described the races according to subjective traits such as temperament. Despite his use of archaic criteria, Linnaeus did not give superior status to any of the races.

Johann Friedrich Blumenbach, a German naturalist and admirer of Linnaeus, developed a classification with lasting influence. Many of his contemporaries believed that different groups of humans arose separately in several regions of the world. Blumenbach, on the other hand, strongly believed in one form of human and believed that physical variations among races were chiefly caused by differences in environment. Therefore, his scheme sought to show a gradual change in bodily appearance, all deviating from an original type. Blumenbach maintained that the original forms, which he named "Caucasian," were those primarily of European ancestry. His final classification, published in 1795 in *On the Natural Variety of Mankind,* consisted of five races: Caucasian, Malay, Ethiopian, American, and Mongolian. Two races directly radiated from the Caucasians: the Malay and the American. The Malay (Pacific islanders) then generated the Ethiopian (Africans), while the American (from the New World) gave rise to the Mongolian (East Asians). The fifth race, the Malay, was added to Linnaeus's classification to show a step-by-step change from the original body type.

After Linnaeus and Blumenbach, many variations of their categories were formulated, chiefly by biologists and anthropologists. Classification "lumpers" combined people into only a few races (for example, black, white, and Asian). "Splitters" separated the traditional groups into many different races. One classification scheme divided all Europeans into Alpine, Nordic, and Mediterranean races. Others split Europeans into ten different races. No one scheme of racial classification came to be accepted throughout the scientific community.

Genetic Diversity Among Races

The genetic components of a population are produced by three primary factors: natural selection, nonadaptive genetic change, and mating between neighboring populations. The first two factors may result in differences between populations, and reproductive isolation, either voluntary or because of geographic isolation, perpetuates the distinctions. Natural selection refers to the persistence of genetic traits favorable in a specific environment. For example, a widely held assumption concerns skin color, primarily a result of the pigment melanin. Melanin offers some shielding from ultraviolet solar rays. According to this theory, people living in regions with concentrated ultraviolet exposure have increased melanin synthesis and, therefore, dark skin color conferring additional protection against skin cancer. Individuals with genes for increased melanin have enhanced survival rates and reproductive opportunities. The reproductive opportunities produce offspring that inherit those same genes for increased melanin. This process results in a higher percentage of the population with elevated melanin production genes. Therefore, genes coding for melanin production are favorable and persist in these environments.

The second factor contributing to the genetic makeup of a population is nonadaptive genetic change. This process involves random genetic mutations. Mutations are changes resulting in modified forms of the same gene. For example, certain genes are responsible for eye color. Individuals contain alternate forms of these genes, or alleles, which result in observed differences in eye color. Alleles resulting from nonadaptive genetic change may remain in the population because of their neutral nature. In other words, they are not harmful or beneficial. Because these traits are impartial to environmental influences, they may endure from generation to generation. Different populations will spontaneously produce, persist, and delete them. Genetic difference between populations caused by these random mutations and isolation is called genetic drift.

The third factor, mating between individuals from neighboring groups, tends to merge traits from several populations. This genetic mixing often results in offspring with blended characteristics and only moderate variations between adjacent groups.

Several studies have compared the overall genetic complement of various human populations. On average, any two people of the same or a different race diverge genetically by a mere 0.2 percent. It is estimated that only 0.012 percent contributes to traditional racial variations. Hence, most of the genetic dissimilarities between a person of African descent and a person of European descent are also different between two individuals with the same ancestry. The genes do not differ. It is the proportion of individuals expressing a specific allele of a gene that varies from population to population.

Upon closer examination, it was found that Africa is unequaled with respect to cumulative genetic diversity. If overall genetic distinctness is evaluated, numerous races are found in Africa, Khoisan Africans of southern Africa being the most distinct. According to one theory, the remainder of the human species (including Asians, Europeans, and aboriginal Australians) corresponds to only one other race.

Conflicts Concerning Definitions of Race

Linnaeus developed a scientific system of classification that is fundamentally still in use. This approach involves separating all organisms first into broad groups based on general characteristics. These large groups are broken down further into smaller and smaller groups, each subdivision containing individuals with more similarities. For example, humans are found within the large kingdom containing all types of animals. Animals are separated based on the formation of a backbone. Of those animals containing a backbone, humans are placed into a set with all mammals and then further cataloged with other primates. Each succeeding classification unit contains individuals more alike, since the characteristics used to define each subdivision are more specific. Eventually, all organisms are placed into a species category. Humans belong in the species *Homo sapiens*. By definition, a sexually reproducing species contains all individuals that can mate and produce fertile offspring. Race is analogous to a more specific unit, the subspecies, a fundamentally distinct subgroup within one species.

For a racial or subspecies classification scheme to be objective and biologically meaningful, researchers must decide carefully which heritable characteristics (passed to future generations genetically) will define, or separate, the races. Several principles are considered. First, the discriminating traits must be discrete. In other words, differences among races must be distinguishable, not continually changing by small degrees between populations. Second, everyone placed within a specific race must possess the selected trait's defining variant. Features used to describe a race must agree. This means that all of the selected characteristics are found consistently in each member. For example, if blue eyes and brown hair are chosen as defining characteristics, everyone designated as belonging to that race must share both of those characteristics. Individuals placed in other races should not exhibit this particular combination. The purpose of using these characteristics is to distinguish groups. Consequently, if traits are shared by members of two or more races, their defining value is poor. Third, individuals of the same race must have descended from a common ancestor, unique to those people. Many shared characteristics present in individuals of a race may be traced to that ancestor by heredity. Based on the preceding defining criteria (selection of discrete traits, agreement of traits, and common ancestry), pure representatives of each racial category should be detectable.

Many researchers maintain that traditional races do not conform to accepted scientific principles of subspecies classification. For example, the traits used to define traditional human races are rarely discrete. Skin color, a prominent characteristic employed, is not a well-defined trait. Approximately five genes influence skin color significantly, but fifty or so likely contribute. Pigmentation in humans results from a complex series of biochemical pathways regulated by amounts of enzymes (molecules that control chemical reactions) and enzyme inhibitors, along with environmental factors. Like most complex traits involving many genes, human skin color varies on a continuous gradation. From lightest to darkest, all intermediate pigmentations are represented. Color may vary widely even within the same family. The boundary between black and white is an arbitrary, human-made border, not one imposed by nature.

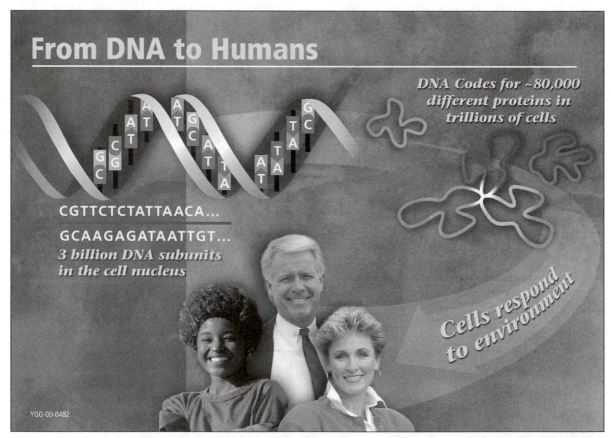

On average, any two people of the same or a different race diverge genetically by a mere 0.2 percent, and only 0.012 percent contribute to traditional racial variations. Allelic variations account for most of the superficial differences perceived as race. (U.S. Department of Energy Human Genome Program, http://www.ornl.gov/hgmis)

In addition, traditional defining racial characteristics, such as skin color and facial characteristics, are not found in all members of a race; they are not in agreement. For example, many Melanesians, indigenous to Pacific islands, have pigmentation as dark as any human but are not classified as "black." Another example concerns unclassifiable populations. For example, many individuals native to India have Caucasoid facial features and very dark skin, yet live in Asia. When traditional racial characteristics are examined closely, many groups are left with no conventional race. No "pure" genetic representatives of any traditional race exist.

Common ancestry, or evolutionary relationships, must also be considered. Genetic studies have shown that Africans do not belong to a single "black" heritage. In fact, several lineages are found in Africa. An even greater variance is found in African Americans. Besides a diverse African ancestry, it is estimated that, on average, 20 to 30 percent of African American heritage is European or Native American. Yet all black Americans are consolidated into one race.

The true diversity found in humans is not patterned according to accepted standards of the subspecies. Only at extreme geographical distances are notable differences found. However, "in-between" populations have always been in existence because of mating, and therefore gene flow, between neighboring groups. Consequently, human populations in close proximity have more genetic similarities than distant populations. It is the population itself that best illustrates the pattern of human diversity. Well-defined genetic borders between human populations are not observed, and racial boundaries in classification schemes are often formed arbitrarily.

Theories of Human and Racial Evolution

Advances in DNA technology have greatly aided researchers in their quest to reconstruct the history of *Homo sapiens* and its various subgroups. Analysis of human DNA has been performed on both nuclear and mitochondrial DNA. Mitochondria are organelles responsible for generating cellular energy. Each mitochondrion contains a single, circular DNA molecule accounting for approximately 0.048 percent of the entire genetic complement. In 1987, geneticist Rebecca L. Cann compared mitochondrial DNA from many populations: African, Asian, Caucasian, Australian, and New Guinean. Agreeing with other mitochondrial and nuclear DNA studies, the results indicated that Africans were the most genetically variable by a significant extent. The results suggested to Cann that Africa was the root of all humankind and that humans first arose there 100,000 to 200,000 years ago. Several lines of research, including DNA analysis of humanoid fossils, provide evidence for this theory.

Many scientists are using genetic markers to decipher the migrations that fashioned past and present human populations. For example, DNA comparisons revealed three Native American lineages. Some scientists believe one migration crossed the Bering Strait, most likely from Mongolia. Only after further migration throughout the Americas were the three American Indian lineages formed. Another theory states that three separate Asian migrations occurred, each bringing a different lineage. Another example is the South African Lemba community. DNA analysis gives credence to their claim as one of the lost tribes of Israel. Considering the cumulative evidence, many scientists regard a more correct depiction of human populations to be a roughly inverted version of Blumenbach's. Asians arose from Africans, and Europeans are Asian and North African hybrids. However, interpretations of DNA analyses are, almost inevitably, controversial. Multiple theories abound and are revised as additional research is performed.

Sociopolitical Implications

Race is often portrayed as a natural, biological division, the result of geographic isolation and adaptation to local environment. However, confusion between biological and cultural classification obscures perceptions of race. When individuals describe themselves as "black," "white," or "Hispanic," for example, they are usually describing cultural heredity as well as biological similarities. The relative importance of perceived cultural affiliations or genetics

Descendants of Sally Hemings, an African American slave of Thomas Jefferson who is known by DNA evidence to have had children by him, pose at Jefferson's home, Monticello, during a July, 2003, reunion. (AP/Wide World Photos)

varies depending on the circumstances. Examples illustrating the ambiguities are abundant. Nearly all people with African American ancestry are labeled black, even if they have a white parent. In addition, dark skin color designates one as belonging to the black race, including Africans and aboriginal Australians, who have no common genetic lineage. State laws, some on the books until the late 1960's, required a "Negro" designation for anyone with one-eighth black heritage (one black great-grandparent).

Unlike biological boundaries, cultural boundaries are sharp, repeatedly motivating discrimination, genocide, and war. The frequent use of biology to devalue certain races and excuse bigotry has profound implications for individuals and society. In the early and mid-twentieth century, the eugenics movement, advocating the genetic improvement of the human species, translated into laws against interracial marriage, sterilization programs, and mass murder. Harmful effects include accusations of deficiencies in intelligence or moral character based on traditional racial classification.

The frequent use of biology to devalue certain races and excuse bigotry has profound implications for individuals and society. Blumenbach selected Caucasians (who inhabit regions near the Caucasus Mountains, a Russian and Georgian mountain range) as the original form of humans because in his opinion they were the most beautiful. All other races deviated from this ideal and were, therefore, less beautiful. Despite Blumenbach's efforts not to demean other groups based on intelligence or moral character, the act of ranking in any form left an ill-fated legacy.

Many scientists are attempting to reconcile the negativities associated with racial studies. The Human Genome Diversity Project, a global undertaking, has requested that researchers collect and store DNA from indigenous populations around the world. These samples will be available to all qualified scientists. Results of the studies may include gene therapy treatments and greater success with organ transplantation. A more thorough understanding of the genetic diversity and unity in the species *Homo sapiens* will as a result be possible.

—*Stacie R. Chismark*

See also: Biological Determinism; Eugenics; Eugenics: Nazi Germany; Evolutionary Biology; Genetic Engineering: Social and Ethical Issues; Heredity and Environment; Intelligence; Miscegenation and Antimiscegenation Laws; Sociobiology; Sterilization Laws.

Further Reading

Cavalli-Sforza, Luigi L. *The Great Human Diasporas: A History of Diversity and Evolution.* Translated by Serah Thorne. Reading, Mass.: Addison-Wesley, 1995. Argues that humans around the world are more similar than different. Basing his discussion on a study of genetic data of fifteen populations from five continents, the author explains how human groups colonized the earth 100,000 years ago. Illustrations, maps, bibliography, index.

Cavalli-Sforza, Luigi L., et al. *The History and Geography of Human Genes.* Princeton, N.J.: Princeton University Press, 1996. Often referred to as a "genetic atlas," this volume contains fifty years of research comparing heritable traits, such as blood groups, from more than one thousand human populations. Illustrations, maps, bibliography, index.

Devlin, Bernie, et al. *Intelligence, Genes, and Success: Scientists Respond to "The Bell Curve."* New York: Springer, 1997. Presents a scientific and statistical reinterpretation of the arguments by Richard Herrnstein and Charles Murray, in their book *The Bell Curve* (see below), about the heritability of intelligence and about IQ and social success. Bibliography, index.

Fish, Jefferson M., ed. *Race and Intelligence: Separating Science from Myth.* Mahwah, N.J.: Lawrence Erlbaum, 2002. An interdisciplinary collection disputing race as a biological category and arguing that there is no general or single intelligence and that cognitive ability is shaped through education. Bibliography, index.

Fraser, Steven, ed. *The "Bell Curve" Wars: Race, Intelligence, and the Future of America.* New York: Basic Books, 1995. Brief, critical response to the book by Herrnstein and Murray by scholars from a variety of disciplines and backgrounds. Bibliography.

Gates, E. Nathaniel, ed. *The Concept of "Race" in Natural and Social Science.* New York: Garland, 1997. Argues that the concept of race, as a form of classification based on physical characteristics, was arbitrarily conceived during the Enlightenment and is without scientific merit. Illustrations, map, bibliography.

Gould, Stephen Jay. *The Mismeasure of Man.* Rev. ed. New York: W. W. Norton, 1996. Presents a historical commentary on racial categorization and a refutation of theories espousing a single measure of genetically fixed intelligence. Bibliography, index.

Graves, Joseph L., Jr. *The Emperor's New Clothes: Biological Theories of Race at the Millennium.* New Brunswick, N.J.: Rutgers University Press, 2001. Argues for a more scientific approach to debates about race, one that takes human genetic diversity into account. Illustrations, bibliography, index.

Herrnstein, Richard J., and Charles Murray. *The Bell Curve: Intelligence and Class Structure in America.* New York: Free Press, 1994. The authors maintain that IQ is a valid measure of intelligence, that intelligence is largely a product of genetic background, and that differences in intelligence among social classes play a major part in shaping American society. Illustrations, bibliography, index.

Kevles, Daniel J. *In the Name of Eugenics: Genetics and the Uses of Human Heredity.* Cambridge, Mass.: Harvard University Press, 1995. Discusses genetics both as a science and as a social and political perspective, and how the two often collide to muddy the boundaries of science and opinion.

Valencia, Richard R., and Lisa A. Suzuki. *Intelligence Testing and Minority Students: Foundations, Performance Factors, and Assessment Issues.* Thousand Oaks, Calif.: Sage, 2000. Historical and multicultural perspective on intelligence and its often assumed relation with socioeconomic status, home environment, test bias, and heredity. Illustrations, bibliography, index.

Web Sites of Interest

Genetics and Identity Project. http://www.bioethics.umn.edu/genetics_and_identity. Project looks at the ways genetic research affects racial, ethnic, and familial identities.

Human Genome Project Research Institute, Minorities, Race, and Genomics. http://www.ornl.gov/techresources/human_genome/elsi/minorities.html. Site provides information on race and genetic research, particularly on how that research affects minority communities.

National Academies Press, Evaluating Human Genetic Diversity. http://www.nap.edu. A free, downloadable book on human genetic diversity, which includes the chapter "Human Rights and Human Genetic-Variation Research."

Repetitive DNA

Field of study: Molecular genetics
Significance: *Eukaryotic nuclei contain repetitive DNA elements of different origin, which constitute between 20 and 90 percent of the genome depending on the species. The presence and type of repetitive DNA elements have provided insights into gene flow, forensic investigations, biomedicine, and genomic mapping.*

Key terms

NUCLEOTIDE: the basic unit of DNA, consisting of a five-carbon sugar, a nitrogen-containing base, and a phosphate group

POLYMORPHISM: the presence of many different alleles for a particular locus in individuals of the same species

RETROTRANSPOSITION: a subset of the replicative transposable elements that transpose through an RNA intermediate

TANDEM REPETITIVE DNA (TR-DNA): DNA are composed of repeating units that are oriented in "head-to-tail" arrays

VARIABLE NUMBER TANDEM REPEAT (VNTR): a type of DNA sequence in which a short sequence is repeated over and over; chromosomes from different individuals frequently have different numbers of the basic repeat, and if many of these variants are known, the sequence is termed a hypervariable

Types of Repetitive DNA

The nuclear genomes of eukaryotes are characterized by repetitive DNA elements consisting of nucleotide sequences that vary in length and base composition and that are localized to a particular region of the genome or dispersed throughout the genome (for example, on different chromosomes). Some repetitive DNA elements are found in the genome a few times, whereas others may be repeated millions or billions of times; thus, the percentage of the total genome represented by repetitive DNA varies widely among taxa.

There are two major classes of tandem repetitive DNAs (TR-DNAs): those that are localized to a particular region (or regions) of the genome and those that are dispersed throughout the genome. TR-DNAs are composed of repeating units that are oriented in "head-to-tail" arrays. The repetitive units of an array may include genes, promoters, and intergenic spacers or repeats of simple nucleotide sequences. For example, in the kangaroo rat the simple sequence AAG is repeated 2.4 billion times.

Localized TR-DNA is often composed of members of multigene families. For example, in humans there are 350 copies of the ribosomal RNA (rRNA) genes on five different chromosomes that occur as tandemly repeated arrays. Transfer RNA (tRNA) and immunoglobin genes represent other examples of multigene families that are tandemly repeated. However, most localized TR-DNA consists of simple, noncoding repetitive DNA sequences that often, but not always, can be found in heterochromatic or centromeric regions.

Dispersed TR-DNA sequences are scattered throughout the genome and can be divided into two major groups: short interspersed elements (SINEs) and long interspersed repeats (LINEs).

Origin and Evolution of Dispersed DNA Elements

SINEs are nonviral retropseudogenes that were derived from genes encoding small, untranslated RNAs (for example, tRNAs). The RNA transcript was reverse transcribed into DNA and then was inserted into the genome. In their current state, although they resemble the genes they were from which they derived, they no longer function properly. SINEs are also examples of transposable elements capable of "jumping" from one locus to another via an RNA intermediate.

The best-characterized SINEs in humans are highly repetitive *Alu* sequences, so named because they are cleaved multiple times by the endonuclease AluI, derived from the bacterium *Arthrobacter luteus*. Between 500,000 and 1 million *Alu* copies are scattered across the human genome, each approximately three hundred nucleotides in length. *Alu* sequences may constitute as much as 5 percent of the human genome.

LINEs are derived from a viral ancestor and are also capable of transposition. The most common LINE element in humans, constituting 5 percent of the human genome, is termed L1. There are about 200,000 copies of L1 in each diploid cell. Full-length, functional (that is, transpositionally competent) L1 elements are approximately 6 kilobase pairs (kb) in length, but most copies of L1 are truncated at the 5′ end and incapable of moving. Full-length L1 copies contain two protein-coding regions, or open reading frames (ORFs): ORF-1 and ORF-2. ORF-1 encodes an RNA-binding protein, and ORF-2 codes for reverse transcriptase.

Classification of Simple Tandem Repeats

Simple sequences that are tandemly repeated are classified into four major groups based on three characteristics: the number of nucleotides in the repetitive unit, the number of times the unit is repeated, and whether or not the element is localized or scattered across the genome. Satellite DNA is composed of basic units, ranging from two to hundreds of nucleotides in length, that are repeated more than one thousand times. Satellite DNA represents an example of a localized simple repeat that is typically found in centromeric regions. Units of between nine and one hundred nucleotides that are tandemly repeated ten to one hundred times and scattered throughout the genome are known as minisatellites. Microsatellites are also dispersed elements composed of short repeats of a basic unit one to six nucleotides in length that is tandemly repeated ten to one hundred times at each locus. The most common microsatellite loci in humans are dinucleotide arrays of $(CA)_N$. However, on average there is at least one tri- or tetranucleotide microsatellite locus per 10 kb of human genomic DNA. Finally, the basic unit of dispersed *Alu* sequences is one to five nucleotides in length, and this unit is repeated ten to forty times per locus.

Polymorphism at Loci Composed of Simple Tandem Repeats

For purposes of convenience, the four groups of simple tandem repeats discussed above (satellite DNA, minisatellites, microsatellites, and *Alu* sequences) are sometimes collectively referred to as variable number tandem repeats (VNTRs).

Separate VNTR loci are thought of as alleles; therefore, in humans each VNTR locus will be represented by two alleles, one paternal and the other maternally inherited. All VNTR loci exhibit high rates of mutation. For these reasons, VNTR loci are highly polymorphic, that is, there are a large number of alleles at any given locus. This polymorphism can be assayed using laboratory techniques such as polymerase chain reaction (PCR) or Southern blotting to examine the differences in the lengths of the alleles (repetitive elements) at a particular locus.

Length differences at VNTR loci arise as a result of mispairing of repeats during replication, mitosis, or meiosis theoretically resulting in the loss or gain of one to many of the repeat units. Empirical studies and computer-based modeling experiments have demonstrated that each mutation usually increases or decreases the number of repeated units of an allele in a "one-step" manner. In other words, most mutations result in the loss or gain of only one repeated unit.

The multiallelic variation that arises through variation in repeat copy number provides genetic markers useful for many different applications. For example, under conditions of random mating and because of high mutation rates at VNTR loci, most individuals within the human population are heterozygous at any selected VNTR locus. This observation directly led to the origin of DNA fingerprinting (or DNA profiling), which is now considered admissible forensic evidence in many judicial systems worldwide. Length variation of VNTRs creates a powerful tool for identity analysis (for example, paternity testing) and is routinely used by population geneticists to examine gene flow among populations. In the fields of genomics and biomedicine, VNTR loci are useful genetic landmarks for mapping the location of other genes of interest, that is, those with a particular function or others implicated in disease.

Are Interspersed Repeated Elements "Junk" DNA?

Repeated DNA elements were once believed to be "selfish" or "junk" DNA, concerned only with their own proliferation within the host cell's genome. Recent studies, however, reveal that repetitive elements interact with the genome with profound evolutionary consequences. For example, satellite DNA found near the centromere may play a role in assembling and fusing chromosomal microtubules during cell division. It is also now clear that transposable genetic elements such as SINEs, LINEs, and *Alu* sequences may have played a significant role in the evolution of particular proteins. For example, *Alu* elements flanking the primordial human growth hormone gene are responsible for the evolution of a relatively new member of the gene family, the chorionic somatomammotropin gene. Transposable repeated elements may have contributed substantially to the origin of new gene functions by initiating a copy of an existing gene (which, over time, can acquire a different function) or by creating "composite" genes composed of domains from two or more previously unrelated genes.

Transposable Elements and Human Disease

Retrotranspositions of LINEs and SINEs into coding or noncoding genomic DNAs represent major insertional mutations. The effects of such insertions vary but are usually deleterious, leading to debilitating human diseases. Among a growing list of diseases known in some cases to be caused by the insertion of LINEs or SINEs are Duchenne muscular dystrophy, Glanzmann thrombasthenia, hemophilia, hypercholesterolemia, neurofibromatosis, Sandhoff disease, and Tay-Sachs disease. Translocation of repeated sequences has also been demonstrated to "turn on" tumorogenic oncogenes (for example, one type of colon cancer).

Other studies have shown that "unstable" minisatellite, microsatellite, and *Alu* loci can also cause disease. In short, there seems to be a threshold number of repeats of the basic nucleotide unit that can be accommodated at a given locus. When this threshold is exceeded by overamplification of the basic repeated unit, serious diseases may arise. Among those diseases attributed to overamplification of tandem repeats of simple sequences are fragile X syndrome and Huntington's disease.

—*J. Craig Bailey*

See also: Aging; Anthrax; Chromosome Structure; Chromosome Walking and Jumping; DNA Fingerprinting; Gene Families; Genome Size; Genomics; Human Genetics; Model Organism: *Neurospora crassa*; Molecular Clock Hypothesis; Pseudogenes; RFLP Analysis; Telomeres.

Further Reading

Li, Wen-Hsiung. *Molecular Evolution.* Sunderland, Mass.: Sinauer Associates, 1997. Provides a basic introduction to the different types of variable tandem repeats, their uses in the biological sciences, and how they affect genome organization.

Maichele, A. J., N. J. Farwell, and J. S. Chamberlain. "A B2 Repeat Insertion Generates Alternate Structures of the Mouse Muscle Gamma-phosphorylase Kinase Gene." *Genomics* 16, no. 1 (1993): 139-149. An excellent example of how retrotransposition of repetitive DNA elements may alter the function of, or give rise to, new structural proteins.

Maraia, Richard J., ed. *The Impact of Short Interspersed Elements (SINEs) on the Host Genome.* Austin, Tex.: R. G. Landes, 1995. A comprehensive treatise on the origin, evolution, and functional roles that SINEs play in the biology of organisms and in biomedicine.

Restriction Enzymes

Fields of study: Genetic engineering and biotechnology; Molecular genetics

Significance: *Restriction enzymes are bacterial enzymes capable of cutting DNA molecules at specific nucleotide sequences. Discovery of these enzymes was a pivotal event in the development of genetic engineering technology, and they are routinely and widely used in molecular biology.*

Key terms

ENZYME: a molecule, usually a protein, that is used by cells to facilitate and speed up a chemical reaction

METHYLATION: the process of adding a methyl chemical group (one carbon atom and three hydrogen atoms) to a particular molecule, such as a DNA nucleotide

NUCLEASE: a type of enzyme that breaks down the sugar-phosphate backbone of nucleic acids such as DNA and RNA

NUCLEOTIDES: the building blocks of nucleic acids, composed of a sugar, a phosphate group, and nitrogen-containing bases

Discovery and Role of Restriction Enzymes in Bacteria

Nucleases are a broad class of enzymes that destroy nucleic acids by breaking the sugar-phosphate backbone of the molecule. Until 1970, the only known nucleases were those that destroyed nucleic acids nonspecifically—that is, in a random fashion. For this reason, these enzymes were of limited usefulness for working with nucleic acids such as DNA and RNA. In 1970, molecular biologist Hamilton Smith discovered a type of nuclease that could fragment DNA molecules in a specific and therefore pre-

dictable pattern. This nuclease, *Hin*dII, was the first restriction endonuclease or restriction enzyme. Smith was working with the bacterium *Haemophilus influenzae* (*H. influenzae*) when he discovered this enzyme, which was capable of destroying DNA from other bacterial species but not the DNA of *H. influenzae* itself. The term "restriction" refers to the apparent role these enzymes play in destroying the DNA of invading bacteriophages (bacterial viruses), while leaving the bacterial cell's own DNA untouched. A bacterium with such an enzyme was said to "restrict" the host range of the bacteriophage.

As more restriction enzymes from a wide variety of bacterial species were discovered in the 1970's, it became increasingly clear that these enzymes could be useful for creating and manipulating DNA fragments in unique ways. What was not clear, however, was how these enzymes were able to distinguish between bacteriophage DNA and the bacterial cell's own DNA. A chemical comparison between DNA that could and could not be fragmented revealed that the DNA molecules differed slightly at the restriction sites (the locations the enzyme recognized and cut). Nucleotides at the restriction site were found to have methyl (CH_3) groups attached to them, giving this phenomenon the name DNA methylation.

The conclusion was that the methylation somehow protected the DNA from attack, and this could account for Smith's observation that *H. influenzae* DNA was not destroyed by its own restriction enzyme; presumably the enzyme recognized a specific methylation pattern on the DNA molecule and left it alone. Foreign DNA (from another species, for example) would not have the correct methylation pattern, or it might not be methylated at all, and could therefore be fragmented by the restriction enzyme. Hence, restriction enzymes are now regarded as part of a simple yet effective bacterial defense mechanism to guard against foreign DNA, which can enter bacterial cells with relative ease.

Mechanism of Action

To begin the process of cleaving a DNA molecule, a restriction enzyme must first recognize the appropriate place on the molecule. The recognition site for most restriction enzymes involves a short, usually four- to six-nucleotide, palindromic sequence. A palindrome is a word or phrase that reads the same backward and forward, such as "Otto" or "madam"; in terms of DNA, a palindromic sequence is one that reads the same on each strand of DNA but in opposite directions. *Eco*RI (derived from the bacterium *Escherichia coli*) is an example of an enzyme that has a recognition site composed of nucleotides arranged in a palindromic sequence:

——GAATTC——
——CTTAAG——

If the top sequence is read from left to right or the bottom sequence is read from right to left, it is always GAATTC.

An additional consideration in the mechanism of restriction enzyme activity is the type of cut that is made. When a restriction enzyme cuts DNA, it is actually breaking the "backbone" of the molecule, consisting of a chain of sugar and phosphate molecules. This breakage occurs at a precise spot on each strand of the double-stranded DNA molecule. The newly created ends of the DNA fragments are then informally referred to as "sticky ends" or "blunt ends." These terms refer to whether single-stranded regions of DNA are generated by the cutting activity of the restriction enzyme. For example, the enzyme *Eco*RI is a "sticky end" cutter; when the cuts are made at the recognition site, the result is:

—GAATTC— → —G   AATTC—
—CTTAAG—   —CTTAA   G—

The break in the DNA backbone is made just after the G in each strand; this helps weaken the connections between the nucleotides in the middle of the site, and the DNA molecule splits into two fragments. The single-stranded regions, where the bases TTAA are not paired with their complements (AATT) on the other

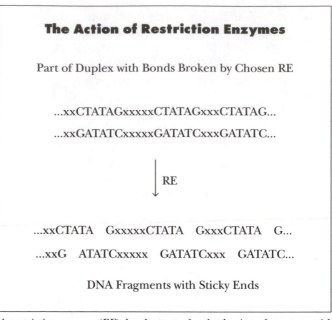

The Action of Restriction Enzymes

Part of Duplex with Bonds Broken by Chosen RE

...xxCTATAGxxxxxCTATAGxxxCTATAG...

...xxGATATCxxxxxGATATCxxxGATATC...

↓ RE

...xxCTATA   GxxxxxCTATA   GxxxCTATA   G...

...xxG   ATATCxxxxx   GATATCxxx   GATATC...

DNA Fragments with Sticky Ends

A restriction enzyme (RE) breaks part of a duplex into fragments with "sticky ends." Each x denotes an unspecified base in a nucleotide unit.

strand, are called overhangs; however, the bases in one overhang are still capable of pairing with the bases in the other overhang as they did before the DNA strands were cut. The ends of these fragments will readily stick to each other if brought close together (hence the name "sticky ends").

Enzymes that create blunt ends make a flush cut and do not leave any overhangs, as demonstrated by the cutting site of the enzyme *Alu*I:

——AGCT——   →   ——AG   CT——
——TCGA——   ——TC   GA——

Because of the lack of overhanging single-strand regions, these two DNA fragments will not readily rejoin. In practice, either type of restriction enzyme may be used, but enzymes that produce sticky ends are generally favored over blunt-end-cutting enzymes because of the ease with which the resulting fragments can be rejoined.

Impact and Applications

It is no exaggeration to say that the entire field of genetic engineering would have been impossible without the discovery and widespread use of restriction enzymes. On the most basic level, restriction enzymes allow scientists to create recombinant DNA molecules (hybrid molecules containing DNA from different sources, such as humans and bacteria). No matter what the source, DNA molecules can be cut with restriction enzymes to produce fragments that can then be rejoined in new combinations with DNA fragments from other molecules. This technology has led to advances such as the production of human insulin by bacterial cells such as *Escherichia coli.*

The DNA of most organisms is relatively large and complex; it is usually so large, in fact, that it becomes difficult to manipulate and study the DNA of some organisms, such as humans. Restriction enzymes provide a convenient way to cut large DNA molecules very specifically into smaller fragments that can then be used more easily in a variety of molecular genetics procedures.

Another area of genetic engineering that is possible because of restriction enzymes is the production of restriction maps. A restriction map is a diagram of a DNA molecule showing where particular restriction enzymes cut the molecule and the molecular sizes of fragments that are generated. The restriction sites can then be used as markers for further study of the DNA molecule and to help geneticists locate important genetic regions. Use of restriction enzymes has also revealed other interesting and useful markers of the human genome, called restriction fragment length polymorphisms (RFLP). RFLP refers to changes in the size of restriction fragments caused by mutations in the recognition site for a particular restriction enzyme. More specifically, the recognition site is mutated so that the restriction enzyme no longer cuts there; the result is one long fragment where, before the mutation, there would have been two shorter fragments. These changes in fragment length can then be used as markers for the region of DNA in question. Because they result from mutations in the DNA sequence, they are inherited from one generation to the next. Thus these mutations have been a valuable tool for molecular biologists in producing a map of human DNA and for those scientists involved in "fingerprinting" individuals by means of their DNA.

—*Randall K. Harris, updated by Bryan Ness*

See also: Bacterial Genetics and Cell Structure; Bioinformatics; Biopharmaceuticals; Blotting: Southern, Northern, and Western; Cloning; Cloning Vectors; DNA Fingerprinting; Forensic Genetics; Gender Identity; Genetic Engineering; Genetic Engineering:

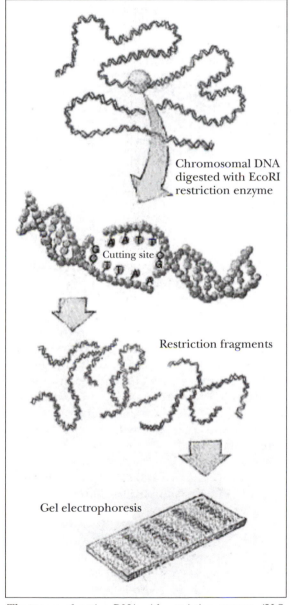

Chromosomal DNA digested with EcoRI restriction enzyme

Cutting site

Restriction fragments

Gel electrophoresis

The process of cutting DNA with restriction enzymes. (U.S. Department of Energy Human Genome Program, http://www.ornl .gov/hgmis)

Historical Development; Genetic Engineering: Social and Ethical Issues; Genomic Libraries; Model Organism: *Xenopus laevis*; Molecular Genetics; Population Genetics; RFLP Analysis; Shotgun Cloning; Synthetic Genes.

Further Reading

Drlica, Karl. *Understanding DNA and Gene Cloning: A Guide for the Curious.* 4th ed. New York: John Wiley & Sons, 2003. Provides basic information about restriction enzymes and their use in cloning. Illustrations, bibliography, index.

Lewin, Benjamin. *Genes VII.* Oxford, N.Y.: Oxford University Press, 2001. Provides a detailed yet highly readable explanation of restriction and methylation in bacteria. Illustrations, bibliography, index.

Watson, James D., et al. *Recombinant DNA.* New York: Scientific American Books, 1992. An excellent resource for the general reader wishing to understand the basics of genetic engineering. Full-color illustrations, diagrams, bibliography, index.

Reverse Transcriptase

Fields of study: Genetic engineering and biotechnology; Molecular genetics

Significance: *Retroviruses infect eukaryotic cells, using reverse transcriptases (RTs) to turn their RNA genomes to DNA that enables their host to use the DNA to make new virus particles. Retroviral DNA, often dormant for years before new virus particles are released, can be oncogenic, giving infected cells high incidences of cancer. Purified RTs are used to make RNA into DNA for biotechnology.*

Key terms

DEOXYRIBONUCLEOSIDE TRIPHOSPHATE (DNTP): one of four monomers (dATP, dCTP, dGTP, dTTP) incorporated into DNA

DNA POLYMERASE: an enzyme that catalyzes the formation of a DNA strand using a template DNA or RNA molecule as a guide

PRIMER: A short piece of single-stranded DNA that can hybridize to denatured DNA and provide a start point for extension by a DNA polymerase

PROOFREADING ACTIVITY: enzyme activity in DNA polymerase that fixes errors made in copying templates

RETROVIRUSES: viruses that possess RNA genomes with genetic information that flows from RNA to host DNA via reverse transcriptases

Genetic Information Flow and Retroviruses

The central dogma of molecular genetics states that information flow is from DNA to RNA to proteins. RNA polymerase transcribes RNA using a DNA template. For structural genes, the transcribed RNA is a messenger RNA (mRNA), which is used by ribosomes to produce a protein. To maintain and reproduce its DNA, an organism uses RNA to make DNA, via DNA polymerase. It was long believed by geneticists that there were no exceptions to the central dogma.

Some viruses, retroviruses, possess RNA genomes with genetic information flow from RNA to DNA (via reverse transcriptases), and back, before translation. Retroviruses have been isolated from cancers and cancer tissue cultures from birds, rodents, primates, and humans, and some retroviruses cause a high incidence of certain cancers. Flow of retroviral genetic information from RNA to DNA was proposed in 1964 by Howard Temin (1934-1994) for Rous sarcoma virus. Temin, along with David Baltimore, jointly received the Nobel Prize in Physiology or Medicine in 1975 for independently discovering the enzyme reverse transcriptase (RT).

Rous sarcoma virus causes tumors in birds. Temin's hypothesis was based on effects of nucleic acid synthesis inhibitors on replication of the virus. First, the process was inhibited by actinomycin D, an inhibitor of DNA-dependent RNA synthesis. Furthermore, DNA synthesis inhibition by cytosine arabinoside, early after infection, stopped viral replication. Therefore, a DNA intermediate seemed involved in viral replication. The expected process was termed reverse transcription because RNA becomes DNA instead of DNA becoming RNA.

Howard M. Temin. (© The Nobel Foundation)

RT Discovery and Properties

Retrovirus infection begins with injection of RT and single-stranded RNA into host cells. RT (an RNA-dependent DNA polymerase) causes biosynthesis of viral DNA using an RNA template from the retrovirus HIV (human immunodeficiency virus), the causative agent of acquired immunodeficiency syndrome (AIDS). RTs have been purified from many retroviruses. Avian, murine, and human RTs have been studied most. All have ribonuclease H (RNase H) activity on the same protein as polymerase activity. Ribonuclease H degrades RNA strands of DNA-RNA hybrids. A nuclease that degrades DNA is later involved in retrovirus DNA integration into host cell DNA. Most biochemical properties of purified RTs are common to them and other DNA polymerases. For example, all require the following for DNA synthesis: a primer on which synthesis begins, a template which is copied, and a supply of the four dNTPs.

An RT converts a retroviral single-stranded RNA genome to "integrated double-stranded DNA" as follows: First a hybrid (DNA-RNA) duplex is made from viral RNA, as an antiparallel DNA strand is produced. The RNA-directed DNA polymerase activity of RT is primed by host cell transfer RNA, which binds to the viral RNA. Then, the viral RNA strand is destroyed by RNase H, and the first DNA strand now becomes the template for synthesis of a second antiparallel DNA strand. Resultant duplex DNA is next integrated into a host cell chromosome, where it is immediately used to make virus particles or, alternatively, it takes up residence in the host cell's genome, remaining unused—often for years—until it is activated and causes cancer or production of new viruses.

Importance of Reverse Transcriptases

RTs can use almost any RNA template for DNA synthesis. Low RT template specificity allows RT to be used to make DNA copies of a

David Baltimore. (© The Nobel Foundation)

wide variety of RNAs in vitro. This has been very useful in molecular biology, especially in production of exact DNA copies of purified RNAs. Once the copies are made by RT, they can be cloned into bacterial expression vectors, where mass quantities of the gene product can be produced. It has also been shown that RT activity takes part in making telomeres (protective chromosome ends). Telomere formation and maintenance are essential cell processes, related to life span and deemed important to understanding cancer.

RTs are also important in treatment of acquired immunodeficiency syndrome (AIDS). The drugs most useful for AIDS treatment are RT inhibitors such as zidovudine, didanosine, zalcitabine, and stavudine. RTs are also associated with the difficulty in maintaining successful long-term AIDS treatment, due to rapid development of resistant HIV in individual AIDS patients. The resistance is postulated to be due to RT's lack of a proofreading component. Inadequate proofreading in sequential replication of HIV viral particles from generation to generation is believed to cause the rapid mutation of the viral genome.

—*Sanford S. Singer*

See also: cDNA Libraries; Central Dogma of Molecular Biology; Model Organism: *Chlamydomonas reinhardtii*; Pseudogenes; Repetitive DNA; RNA Isolation; RNA Structure and Function; RNA World; Shotgun Cloning.

Further Reading

Goff, S. P. "Retroviral Reverse Transcriptase: Synthesis, Structure, and Function." *JAIDS* 3 (1990): 817-831. A solid, well-illustrated paper.

Joklik, Wolfgang K., ed. *Microbiology: A Centenary Perspective*. Washington, D.C.: ASM Press, 1999. Thoroughly reviews important microbiology issues, including reverse transcriptase papers by Howard Temin and David Baltimore.

Litvack, Simon. *Retroviral Reverse Transcriptases*. Austin, Tex.: R. G. Landes, 1996. Describes discovery, biosynthesis, structure, inhibitors, and action mechanisms of reverse transcriptase. Illustrations and bibliographic references.

O'Connell, Joe, ed. *RT-PCR Protocols*. Totowa, N.J.: Humana Press, 2002. Collects several papers on the use of reverse transcription polymerase chain reaction in analysis of mRNA, quantitative methodologies, detection of RNA viruses, genetic analysis, and immunology. Tables, charts, index.

Shippen-Lentz, D., and E. H. Blackburn. "Functional Evidence for an RNA Template in Telomerase." *Science* 247 (1990): 546-552. Points to the RT activity in telomerases. Illustrated.

Varmus, Harold. "Retroviruses." *Science* 240 (1988): 1427-1435. The discoverer of oncogenes describes properties of different retroviruses, including the mechanism of reverse transcription.

RFLP Analysis

Field of study: Techniques and methodologies

Significance: *RFLP analysis was the first simple method available for distinguishing individuals based on DNA sequence differences. The conceptual basis for this technique is still widely used in genetics, although RFLP analysis has been largely supplanted by other, faster and more powerful techniques for the comparison of genetic differences.*

Key terms

GEL ELECTROPHORESIS: a method for separating DNA molecules by size by applying electric current to force DNA through a matrix of agarose, which inhibits the migration of larger DNA fragments more than small DNA fragments

RESTRICTION ENZYMES: proteins that recognize specific DNA sequences and then cut the DNA, normally at the same sequence recognized by the enzymes

SOUTHERN BLOTTING: a method for transferring DNA molecules from an agarose gel to a nylon membrane; once the DNA is on the membrane, it is incubated with a DNA containing an identifiable label and is then used to detect similar or identical DNA sequences on the membrane

The Procedure

Restriction fragment length polymorphism (RFLP) analysis is a method for distinguishing individuals and analyzing relatedness, based on genetic differences. RFLP analysis relies on small DNA sequence differences that lead to the loss or gain of restriction enzyme sites in a chromosome or to the change in size of a DNA fragment bracketed by restriction enzyme sites. These sequence differences lead to a different pattern of bands on a gel (reminiscent of a bar code) that varies from individual to individual.

RFLP analysis starts with the isolation of DNA. Typically, DNA isolation requires the use of detergents, protein denaturants, RNA degrading enzymes, and alcohol precipitation to separate the DNA from the other cellular components. This DNA could be isolated from a blood sample provided by an individual, from evidence left at the scene of a crime, or from other sources of cells or tissues.

The purified DNA is then digested with a molecular "scissors" called a restriction enzyme. Restriction enzymes recognize and cut precise sequences, typically six base pairs in length. If one base pair is changed in that recognition sequence, the enzyme will not cut the DNA at that point. On the other hand, if a sequence that is not recognized by a restriction enzyme is altered by mutation, so that it now is recognized, the DNA will be cleaved at that point. In other cases, the DNA sequences recognized by the restriction enzymes themselves are not changed, but the length of DNA between two restriction enzyme sites differs between individuals. These types of mutations occur with enough regularity that often even two closely related individuals will have some detectable differences in the sizes of DNA fragments produced from restriction enzyme digestion.

Once the DNA has been digested with a restriction enzyme, it is separated by size in an agarose gel. At this point, the DNA appears, to the eye, to be a smear of molecules of all sizes, and it is not generally possible to differentiate the DNAs from different individuals at this stage. The size-fractioned DNA is next transferred to a nylon membrane in a process called Southern blotting. The result of the transfer is that the location and arrangement of the DNA fragments in the gel is maintained on the membrane, but the DNA is now single-stranded (critical for the next step in the process) and much easier to handle.

The final step in the process is to detect specific DNA fragments on the membrane. This is done by using a DNA fragment that is labeled to act as a probe, to home in on and identify similar DNA sequences on the membrane. Before use, the probe is made single-stranded, so it can bind to the single-stranded DNA on the membrane. The probe DNA can be labeled with radioactivity, in which case it is detected using X-ray film. The probe DNA can also be labeled with molecules that are bound by proteins, and the proteins can then be detected either directly or indirectly.

In a case in which a restriction enzyme site has been added or removed, the probe is normally a DNA fragment that is found in only one location in the genome. In cases in which one is looking at the size of fragments bracketed by restriction enzyme sites, the probe DNA is normally a DNA molecule that is found in several sites in the genome, and the DNA fragments that are identified in this analysis are ones that tend to vary between individuals. In many cases, the probe DNA binds to regions of DNA that consist of variable number tandem repeats (VNTRs). The number of VNTRs tends to vary between different individuals and, consequently, these sequences are useful for identification.

Applications

One of the earliest uses of this technique in clinical medicine was in the prenatal diagnosis of sickle-cell disease. Previous work had shown that many individuals with the disease had a mutation in their DNA that eliminated a restriction enzyme site in a gene encoding a hemoglobin protein. This information was used to develop a diagnostic RFLP procedure. A section of the hemoglobin gene is used as a probe. The size of restriction enzyme fragments identified is different in individuals who have sickle-cell disease (and therefore have two mutant alleles) compared with individuals who carry either one mutant allele or have two unmutated hemoglobin alleles. This method allowed for the identification of affected fetuses using DNA from cells isolated from amniotic fluid (a much simpler and safer procedure than the previous method of diagnosis, which required isolating fetal red blood cells).

Another widely reported use of RFLP analysis has been in forensic science. RFLP methods have been critical in helping to identify criminals, and these methods have also helped exonerate innocent people. The first application of RFLP in forensic analysis was in the case of the murders of two young girls in England, in 1983 and 1986. Initially, a seventeen-year-old boy confessed to the murders. RFLP analysis, using DNA from the crime scene, indicated that he was not the murderer. After extensive investigation, including RFLP analysis of DNA from more than forty-five hundred men, a suspect was identified. Confronted with the evidence, the suspect pleaded guilty to both murders and was jailed for life. Since then, RFLP analysis has been used in thousands of criminal cases. Other forensic applications of RFLP include its use as evidence in court cases involving paternity determinations and its role in identifying the bodies of missing persons who otherwise could not be identified.

In addition to the clinical and forensic applications described above, RFLP analysis has been used in many subdisciplines of biology since the early 1980's. The applications of RFLP analysis range from the conservation of endangered species to the identification of strains of bacteria associated with disease outbreaks to basic research involving the classification of organisms.

Although RFLP analysis has been widely used since its inception, it is increasingly being displaced by polymerase chain reaction (PCR) methods, which typically are much faster and require much less DNA. RFLP analysis was, however, an important step in the introduction of modern DNA analysis into the biology laboratory and the courtroom. The guiding principle behind RFLP analysis—identifying individuals, strains, and species, based on DNA sequence differences—is still a part of more recently developed techniques.

—*Patrick G. Guilfoile*

See also: Blotting: Southern, Northern, and Western; Chromosome Theory of Heredity; DNA Fingerprinting; Gender Identity; Genetic Engineering; Genetic Testing; Model Organism: *Arabidopsis thaliana*; Paternity Tests; Polymerase Chain Reaction; Prenatal Diagnosis; Restriction Enzymes.

Further Reading

Chang, J. C., and Y. W. Kan. "A Sensitive New Prenatal Test for Sickle-Cell Anemia." *New England Journal of Medicine* 307 (1982): 30-32. A short and readable scientific paper describing one of the first applications of RFLP analysis to clinical medicine. The same issue of the journal also has another, somewhat more detailed, article on the same topic.

Guilfoile, P. *A Photographic Atlas for the Molecular Biology Laboratory.* Englewood, Colo.: Morton, 2000. An illustrated guide to molecular biology techniques, including a substantial illustrated section on RFLP analysis.

Jeffreys, A., V. Wilson, and S. L. Thein. "Individual-Specific Fingerprints of Human DNA." *Nature* 316 (1985): 76-79. A technical article that describes some of the background information that led to the use of RFLP analysis in forensic science.

Orkin, S. H., P. F. Little, H. H. Kazazian, Jr., and C. D. Boehm. "Improved Detection of the Sickle Mutation by DNA Analysis: Application to Prenatal Diagnosis." *New England Journal of Medicine* 307 (1982): 32-36.

RNA Isolation

Field of study: Molecular genetics

Significance: *All cells in an organism or population of organisms of the same species contain the same (or nearly the same) set of genes. Therefore, understanding which genes are expressed under different conditions is critical to answering many questions in biology, including how cells differentiate into tissues, how cells respond to different environments, and which genes are expressed in tumor cells. The starting point for answering those questions is RNA isolation.*

Key terms

cDNA library: a set of copies, or clones, of all or nearly all mRNA molecules produced by cells of an organism

complementary DNA (cDNA): also called copy DNA, DNA that copies RNA molecules, made using the enzyme reverse transcriptase

microarray analysis: a method, requiring isolated RNA, that allows simultaneous determination of which of thousands of genes are transcribed (expressed) in cells

reverse transcriptase polymerase chain reaction (RT-PCR): a technique, requiring isolated RNA, for quickly determining if a gene or a small set of genes are transcribed in a population of cells

RNA: ribonucleic acid, the macromolecule in the cell that acts as an intermediary between the genetic information stored as DNA and the manifestation of that genetic information as proteins

RNases: ribonucleases, or cellular enzymes that catalyze the breakdown of RNA

Cell Lysis

RNA isolation is a difficult proposition. RNA has a short life span in cells (as short as minutes in bacteria), and it is somewhat chemically unstable. In addition, enzymes that degrade RNA (RNases) are widespread in the environment, further complicating the task of separating intact RNA from other molecules in the cell.

The first step in RNA isolation is rapidly breaking open cells under conditions where RNA will not be degraded. One method involves freezing cells immediately in liquid nitrogen, then grinding the cells in liquid nitrogen in order to prevent any RNA degradation. Other methods involve lysing cells in the presence of strong protein denaturants so that any RNases present in the cell or the environment will be rapidly inactivated. The difficulty of the cell lysis step depends substantially on the type of cell involved. Bacterial and fungal cells are typically much more difficult to break open than cells from mammals. As a consequence, it is often more difficult to isolate intact RNA from bacteria and fungi.

Protein Denaturation and Further Purification

The next step in RNA isolation is to denature all proteins from the cell, to ensure that RNases will be inactive. In many cases, this is done at the same time as cell lysis. RNases are among the most resilient enzymes known, capable of being boiled or even autoclaved, yet retaining the ability to cleave RNA once they cool down. Consequently, the RNA next needs to be separated from RNases and other proteins to ensure that it will remain intact.

The separation of RNA from the rest of the macromolecules in the cell can be accomplished in a number of ways. One of the older methods for purifying RNA uses ultracentrifugation in very dense cesium chloride solutions. During high-speed centrifugation, these solutions create a gradient, with the greatest density at the bottom of the tube. RNA is the densest macromolecule in the cell, so it forms a pellet in the bottom of the ultracentrifuge tube. A more recently developed technique for RNA purification involves the use of columns that bind RNA but not other macromolecules. The columns are washed to remove impurities, such as DNA and proteins, and then the RNA is eluted from the column matrix. Another, more recently developed technique is based on the observation that, at an appropriate pH (level of acidity), RNA partitions into the water phase of a water-organic mixture. DNA and proteins either are retained at the boundary of the water-organic mixture or are dissolved in the organic phase.

Once the RNA is isolated, it needs to be handled carefully to ensure that it will not be degraded. Normally this involves re-suspending the RNA in purified water, adding an alcohol solution, and storing it at ¯70 or ¯80 degrees Celsius (¯94–¯112 degrees Fahrenheit). The purified RNA can then be used in a variety of techniques that help determine which genes are being transcribed in particular cells or tissues. These techniques include RT-PCR, northern hybridization, microarray analysis, and the construction of cDNA libraries.

Special RNA Isolation Procedures

In some cases, a geneticist wants to isolate only RNA from the cytoplasm of the cell, since

RNA from the nucleus may be more heterogeneous. In this case, cells are lysed using a gentle detergent that disrupts the cytoplasmic membrane, without disturbing the nuclear membrane. Centrifugation is used to separate the nuclei from the cytoplasm, and then the cytoplasmic RNA is further purified as described above.

For some procedures, such as RT-PCR, the RNA sometimes needs to be further purified to ensure that no contaminating DNA is present. In this case, the RNA sample may be treated with the enzyme DNase I, which destroys DNA but leaves RNA intact.

For other procedures, like cDNA library construction, the RNA is often purified to remove ribosomal RNA (rRNA), transfer RNA (tRNA), and other stable RNAs, since the majority of RNA in the cell (typically more than 90 percent) is rRNA and tRNA. In this case, the RNA solution is treated by incubating it with single-stranded DNA containing a chain of eighteen to twenty thymine nucleotides, either on a column or in solution. Messenger RNA (mRNA) from eukaryotes contains runs of twenty to two hundred adenine nucleotides that bind to the single-stranded DNA and allow the mRNA to be purified away from the stable RNAs.

Like most techniques in genetics, RNA isolation methods have improved greatly over the years. With advances in methods for studying gene expression such as microarray analysis, isolating intact RNA is a technique that is more critical than ever in the modern genetics laboratory.

—*Patrick G. Guilfoile*

See also: cDNA Libraries; DNA Isolation; DNA Structure and Function; Polymerase Chain Reaction; Reverse Transcriptase; RNA Structure and Function; RNA Transcription and mRNA Processing.

Further Reading

Ausubel, Fredrick, Roger Brent, Robert Kingston, David Moore, J. Seidman, and K. Struhl. *Current Protocols in Molecular Biology.* Hoboken, N.J.: John Wiley & Sons, 1998. A regularly updated compendium that includes RNA isolation protocols from several different laboratories.

Farrell, Robert. *RNA Methodologies.* 2d ed. San Diego, Calif.: Academic Press, 1998. Probably the definitive book on RNA techniques, including RNA isolation. Includes a substantial amount of background information as well as detailed protocols.

O'Connell, Joe, ed. *RT-PCR Protocols.* Totowa, N.J.: Humana Press, 2002. Collects several papers on the use of reverse transcription polymerase chain reaction in analysis of mRNA, quantitative methodologies, detection of RNA viruses, genetic analysis, and immunology. Tables, charts, index.

Sambrook, J., and D. W. Russel, eds. *Molecular Cloning: A Laboratory Manual.* 3d ed. Cold Spring Harbor, N.Y.: Cold Spring Harbor Laboratory Press, 2000. The latest edition of one of the most popular guides to molecular biology protocols. Includes several RNA isolation procedures.

RNA Structure and Function

Field of study: Molecular genetics

Significance: *Ribonucleic acid (RNA), a molecule that plays many roles in the storage and transmission of genetic information, exists in several forms, each with its own unique function. RNA acts as the messenger between genes in the DNA and their protein product, directing the assembly of proteins. RNA is also an integral part of ribosomes, the site of protein synthesis, and some RNAs have been shown to have catalytic properties. Understanding the structure and function of RNA is important to a fundamental knowledge of genetics; in addition, many developing medical therapies will undoubtedly utilize special RNAs to combat genetic diseases.*

Key terms

MESSENGER RNA: a type of RNA that carries genetic instructions, copied from genes in DNA, to the ribosome to be decoded during translation

RETROVIRUS: a special type of virus that carries its genetic information as RNA and converts it into DNA that integrates into the cells of the virus's host organism

RIBOSOMAL RNA: a type of RNA that forms a major part of the structure of the ribosome

RIBOSOMES: organelles that function in protein synthesis and are made up of a large and a small subunit composed of proteins and ribosomal RNA (rRNA) molecules

RIBOZYME: an RNA molecule that can function catalytically as an enzyme

TRANSCRIPTION: the synthesis of an RNA molecule directed by RNA polymerase using a DNA template

TRANSFER RNA: a form of RNA that acts to decode genetic information present in mRNA, carries a particular amino acid, and is vital to translation

TRANSLATION: the synthesis of a protein molecule directed by the ribosome using information provided by an mRNA

The Chemical Nature of RNA

Ribonucleic acid (RNA) is a complex biological molecule that is classified along with DNA as a nucleic acid. Chemically, RNA is a polymer (long chain) consisting of subunits called ribonucleotides linked together by phosphodiester bonds. Each ribonucleotide consists of three parts: the sugar ribose (a five-carbon simple sugar), a negatively charged phosphate group, and a nitrogen-containing base. There are four types of ribonucleotides, and the differences among them lie solely in which of four possible bases each contains. The four bases are adenine (A), guanine (G), cytosine (C), and uracil (U).

The structures of DNA and RNA are very similar, with the following differences. The sugar found in the nucleotide subunits of DNA is deoxyribose, which differs slightly from the ribose found in the ribonucleotides of RNA. In addition, while DNA nucleotides also contain four possible bases, there is no uracil in DNA; instead, DNA nucleotides contain a different base called thymine (T). Finally, while DNA exists as a double-stranded helix in nature, RNA is almost always single-stranded. Like DNA, a single RNA strand has a 5'-to-3' polarity. These numbers are based on which carbon atom is exposed at the end of the polymer, each of the carbon atoms being numbered around the sugar molecule.

The Folding of RNA Molecules

The function of an RNA molecule is determined by its nucleotide sequence, which represents information derived from DNA. This nucleotide sequence is called the primary structure of the molecule. Many RNAs also have an important secondary structure, a three-dimensional shape that is also important for the function of the molecule. The secondary structure is determined by hydrogen bonding between parts of the RNA molecule that are complementary. Complementary pairing is always between A and U ribonucleotides and C and G ribonucleotides. Hydrogen bonding results in double-stranded regions in the secondary structure.

Since RNA is single-stranded, it was recognized shortly after the discovery of some of its major roles that its capacity for folding is great and that this folding might play an important part in the functioning of the molecule. Base pairing often represents local interactions, and a common structural element is a "hairpin loop" or "stem loop." A hairpin loop is formed when two complementary regions are separated by a short stretch of bases so that when they fold back and pair, some bases are left unpaired, forming the loop. The net sum of these local interactions is referred to as the RNA's secondary structure and is usually important to an understanding of how the RNA works. All transfer RNAs (tRNAs), for example, are folded into a secondary structure that contains three stem loops and a fourth stem without a loop, a structure resembling a cloverleaf in two dimensions.

Finally, local structural elements may interact with other elements in long-range interactions, causing more complicated folding of the molecule. The full three-dimensional structure of a tRNA molecule from yeast was finally confirmed in 1978 by several groups independently, using X-ray diffraction. In this process, crystals of a molecule are bombarded with X rays, which causes them to scatter; an expert can tell by the pattern of scattering how the different atoms in the molecule are oriented with respect to one another. The cloverleaf arrangement of a tRNA undergoes further folding so that the entire molecule takes on a roughly *L-*

shaped appearance in three dimensions. An understanding of the three-dimensional shape of an RNA molecule is crucial to understanding its function. By the late 1990's, the three-dimensional structures of many tRNAs had been worked out, but it had proven difficult to do X-ray diffraction analyses on most other RNAs because of technical problems. More advanced computer programs and alternate structure-determining techniques are enabling research in this field to proceed.

Synthesis and Stability of RNA

RNA molecules of all types are continually being synthesized and degraded in a cell; even the longest-lasting ones exist for only a day or two. Shortly after the structure of DNA was established, it became clear that RNA was synthesized using a DNA molecule as a template, and the mechanism was worked out shortly thereafter. The entire process by which an RNA molecule is constructed using the information in DNA is called transcription. An enzyme called RNA polymerase is responsible for assembling the ribonucleotides of a new RNA complementary to a specific DNA segment (gene). Only one strand of the DNA is used as a template (the sense strand), and the ribonucleotides are initially arranged according to the base-pairing rules. A DNA sequence called the "promoter" is a site RNA polymerase can bind initially and allows the process of RNA synthesis to begin. At the appropriate starting site, RNA polymerase begins to assemble and connect the nucleotides according to the complementary pairing rules, such that for every A nucleotide in the DNA, RNA polymerase incorporates a U ribonucleotide into the RNA being assembled. The remaining pairing rules stipulate that a T in DNA denotes an A in RNA and that a C in DNA represents a G in RNA (and vise versa). This process continues until another sequence, called a "terminator," is reached. At this point, the RNA polymerase stops transcription, and a new RNA molecule is released.

Much attention is rightfully focused on transcription, since it controls the rate of synthesis of each RNA. It has become increasingly clear, however, that the amount of RNA in the cell at a given time is also strongly dependent on RNA stability (the rate at which it is degraded). Every cell contains several enzymes called ribonucleases (RNases) whose job it is to cut up RNA molecules into their ribonucleotides subunits. Some RNAs last only thirty seconds, while others may last up to a day or two. The signals regulating RNA degradation are being studied, and although much has been learned, many details remain unclear. It is important to remember that both the rates of synthesis (transcription) and degradation ultimately determine the amount of functional RNA in a cell at any given time.

Three Classes of RNA

While all RNAs are produced by transcription, several classes of RNA are created, and each has a unique function. By the late 1960's, three major classes of RNAs had been identified, and their respective roles in the process of protein synthesis had been identified. In general, protein synthesis refers to the assembly of a protein using information encoded in DNA, with RNA acting as an intermediary to carry information and assist in protein building. In 1956, Francis Crick, one of the scientists who had discovered the double-helical structure of DNA, referred to this information flow as the "central dogma," a term that continues to be used, although exceptions to it are now known.

A messenger RNA (mRNA) carries a complementary copy of the DNA instructions for building a particular protein. In eukaryotes it typically represents the information from a single gene and carries the information to a ribosome, the site of protein synthesis. The information must be decoded to make a protein. Nucleotides are read in groups of three (called codons). In addition, mRNAs contain signals that tell a ribosome where to start and stop translating.

Ribosomal RNA (rRNA) is part of the structure of the ribosome. Four different rRNAs interact with many proteins to form functional ribosomes that direct the events of protein synthesis. One of the rRNAs interacts with mRNA to orient it properly so translation can begin at the correct location. Another rRNA acts to facilitate the transfer of the growing polypeptide from one tRNA to another (peptidyl transferase activity).

Transfer RNA (tRNA) serves the vital role of decoding the genetic information. There are at least twenty and usually more than forty different tRNAs in a cell. On one side, tRNAs contain an "anticodon" loop, which can base-pair with mRNA codons according to their sequence and the base-pairing rules. On the other side, each contains an amino acid binding site, with the appropriate amino acid for its anticodon. In this way, tRNAs recognize the codons and supply the appropriate amino acids. The process continues until an entire new polypeptide has been constructed.

The attachment of the correct amino acids is facilitated by a group of enzymes called tRNA amino acyl synthetases. Each type of tRNA has a corresponding synthetase that facilitates the attachment of the correct amino acid to the amino acid binding site. The integrity of this process is crucial to translation; if only one tRNA is attached to an incorrect amino acid, the resulting proteins will likely be nonfunctional.

Split Genes and mRNA Processing in Eukaryotes

In bacterial genes, there is a colinearity between the segment of a DNA molecule that is transcribed and the resulting mRNA. In other words, the mRNA sequence is complementary to its template and is the same length, as would be expected. In the late 1970's, several groups of scientists made a seemingly bizarre discovery regarding mRNAs in eukaryotes (organisms whose cells contain a nucleus, including all living things that are not bacteria): The sequences of mRNAs isolated from eukaryotes were not collinear with the DNA from which they were transcribed. The coding regions of the corresponding DNA were interrupted by seemingly random sequences that served no apparent function. These "introns," as they came to be known, were apparently transcribed along with the coding regions (exons) but were somehow removed before the mRNA was translated. This completely unexpected observation led to further investigations that revealed that mRNA is extensively processed, or modified, after its transcription in eukaryotes.

After a eukaryotic mRNA is transcribed, it contains several to many introns and is referred to as immature, or a "pre-mRNA." Before it can become mature and functional, three major processing events must occur: splicing, the addition of a 5′ cap, and a "tail." The process of splicing is complex and occurs in the nucleus with the aid of "spliceosomes," large complexes of RNAs and proteins that identify intervening sequences and cut them out of the pre-mRNA. In addition, spliceosomes rejoin the exons to produce a complete, functional mRNA. Splicing must be extremely specific, since a mistake causing the removal of even one extra nucleotide could change the final protein, making it nonfunctional. During splicing, capping and the addition of a poly-A tail take place. A so-called cap, which consists of a modified G nucleotide, is added to the beginning (5′ end) of the pre-mRNA by an unconventional linkage. The cap appears to function by interacting with the ribosome, helping to orient the mature mRNA so that translation begins at the proper end. A tail, which consists of many A nucleotides (often two hundred or more), is attached to the 3′ end of the pre-mRNA. This so-called poly-A tail, which virtually all eukaryotic mRNAs contain, seems to be one factor in determining the relative stability of an mRNA. These important steps must be performed after transcription in eukaryotes to produce a functional mRNA.

Other Important Classes of RNA and Specialized Functions

The traditional roles of RNA in protein synthesis were originally considered its only roles. RNA in general, while considered an important molecule, was thought of as a "helper" in translation. This all began to change in 1982, when the molecular biologists Thomas Cech and Sidney Altman, working independently and with different systems, reported the existence of RNA molecules that had catalytic activity. This means that RNA molecules can function as enzymes; until this time, it was believed that all enzymes were protein molecules. The importance of these findings cannot be overstated, and Cech and Altman ultimately shared the 1989 Nobel Prize in Chemistry for the discovery of these RNA enzymes, or "ribozymes."

Both of these initial ribozymes catalyzed reactions that involved the cleavage of other RNA molecules—that is, they acted as nucleases. Subsequently, many ribozymes have been found in various organisms, from bacteria to humans. Some of them are able to catalyze different types of reactions, and there are new ones reported every year. Thus ribozymes are not a mere curiosity but play an integral role in the molecular machinery of many organisms. Their discovery also gave rise to the idea that at one point in evolutionary history, molecular systems composed solely of RNA, performing many roles, existed in an "RNA world."

At around the same time as these momentous discoveries, still other classes of RNAs were being discovered, each with its own specialized functions. In 1981, Jun-ichi Tomizawa discovered RNA interference (RNAi), the first example of what would become another major class of RNAs, the "antisense RNAs" or "interference RNAs." The RNAs in this group are complementary to a target molecule (usually an mRNA) and can bind to that target via complementary base pairing. RNAi binding usually plays a regulatory role, often acting to prevent translation of the relevant mRNA to modulate the expression of the protein for which it codes. Most of these antisense RNAs are encoded by the same gene as their target, but a group called the "transencoded antisense RNAs" actually have their own genes, which are separate and distinct from their target molecule's gene. This is especially significant because the complementarity between antisense RNA and the target is often not perfect, resulting in interesting interactions with unique structural features. The prototype of this class of RNAs, micF RNA, was discovered in 1983 by Masayori Inouye and subsequently characterized by Nicholas Delihas. An understanding of the binding of this special type of antisense RNA to its target will provide insights into RNA-RNA interactions that may be vital for use in genetic therapy. Research on RNAi molecules continues, and many new insights into genetic control after transcription have been gained.

Another major class of RNAs, the small nuclear RNAs (snRNAs), was also discovered in the early 1980's. Molecular biologist Joan Steitz was working on the autoimmune disease systemic lupus when she began to characterize the snRNAs. There are six different snRNAs, now called U1-U6 RNAs. These RNAs exist in the nucleus of eukaryotic cells and play a vital role in mRNA splicing. They associate with proteins in the spliceosome, forming so-called ribonucleoprotein complexes (snRNPs, pronounced "snurps"), and play a prominent role in detecting proper splice sites and directing the protein enzymes to cut and paste at the proper locations.

It has been known since the late 1950's that many viruses contain RNA, and not DNA, as their genetic material. This is another fascinating role for RNA. The viruses that cause influenza, polio, and a host of other diseases are RNA viruses. Of particular note are a class of RNA viruses known as retroviruses. Retroviruses, which include human immunodeficiency virus (HIV), the virus that causes acquired immunodeficiency syndrome (AIDS) in humans, use a special enzyme called reverse transcriptase to make a DNA copy of their RNA when they enter a cell. The DNA copy is inserted into the DNA of the host cell, where it is referred to as a "provirus," and never leaves. This discovery represents one of the exceptions to the central dogma. In the central dogma, RNA is always made from DNA, and retroviruses have reversed this flow of information. Clearly, understanding the structures and functions of the RNAs associated with these viruses will be important in attempting to create effective treatments for the diseases associated with them.

An additional role of RNA was noted during the elucidation of the mechanism of DNA replication. It was found that a small piece of RNA, called a "primer," must be laid down by the enzyme primase, an RNA polymerase, before DNA polymerase can begin. RNA primers are later removed and replaced with DNA. Also, it is worth mentioning that the universal energy-storing molecule of all cells, adenosine triphosphate (ATP), is in fact a version of the RNA nucleotide containing adenine (A).

Impact and Applications

The discovery of the many functions of RNA, especially its catalytic ability, has radically changed the understanding of the functioning of genetic and biological systems and has revolutionized the views of the scientific community regarding the origin of life. The key to understanding how RNA can perform all of its diverse functions lies in elucidating its many structures, since structure and function are inseparable. Much progress has been made in establishing the structures of hundreds of RNA molecules; several methods, including advanced computer programs, are making it easier to predict and analyze RNA structure. Three-dimensional modeling is much more difficult, and while the three-dimensional structures of several RNAs have been worked out, much work remains.

In terms of basic research and genetic engineering, the discovery of antisense RNAs and ribozymes has facilitated many procedures, providing insight at the molecular level of genetic processes that would have been difficult to obtain without this knowledge and the tools it has made available. Additionally, plants, bacteria, and animals have been genetically engineered to alter the expression of some of their genes, in many cases making use of the new RNA technology. An example is the genetically engineered tomato, which does not ripen until it is treated at the point of sale. This tomato was created by inserting an antisense RNA gene; when it is expressed, it inactivates the mRNA that codes for the enzyme involved in production of the ripening hormone.

Although success in human gene therapy has been limited, the usage of retroviruses to introduce ribozymes, antisense RNAs, or a combination of both into genetically defective cells offers great promise for the future in fighting a wide variety of diseases, from AIDS and cancer to cystic fibrosis and sickle-cell disease. One thing is clear: RNA will play an important role in increasing the understanding of genetics and in the revolution of gene therapy. RNA is one of the most structurally interesting and functionally diverse of all the biological molecules.

—*Matthew M. Schmidt, updated by Bryan Ness*
See also: Ancient DNA; Antisense RNA; Chromosome Structure; DNA Isolation; DNA Repair; DNA Replication; DNA Structure and Function; Genetic Code; Genetic Code, Cracking of; Molecular Genetics; Noncoding RNA Molecules; One Gene-One Enzyme Hypothesis; Protein Structure; Protein Synthesis; Repetitive DNA; RNA Isolation; RNA Transcription and mRNA Processing; RNA World.

Further Reading

Eckstein, Fritz, and David M. J. Lilley, eds. *Catalytic RNA*. New York: Springer, 1996. Offers a comprehensive overview of ribozyme diversity and function. Illustrations (some color), bibliography, index.

Erickson, Robert P., and Jonathan G. Izant, eds. *Gene Regulation: Biology of Antisense RNA and DNA*. New York: Raven Press, 1992. Provides both a comprehensive overview of natural antisense RNA function and prospects for its uses in gene therapy. Illustrations, bibliography, index.

Murray, James A. H., ed. *Antisense RNA and DNA*. New York: Wiley-Liss, 1992. Presents experimental approaches. Illustrations, bibliography, index.

Simons, Robert W., and Marianne Grunberg-Manago, eds. *RNA Structure and Function*. Cold Spring Harbor, N.Y.: Cold Spring Harbor Laboratory Press, 1997. An advanced text that takes a detailed look at the various structures of RNA, their relationships to function, and the techniques for determining RNA structure. Illustrations, bibliography, index.

Watson, James D., et al. *Molecular Biology of the Gene*. 5th ed. Menlo Park, Calif.: Benjamin Cummings, 2003. Discusses RNA structures and their relationship to function. Illustrations, bibliography, index.

RNA Transcription and mRNA Processing

Field of study: Molecular genetics
Significance: *Translation of messenger RNA molecules (mRNAs) occurs even while transcription is taking place in prokaryotes. In eukaryotes the process is much more complex, with transcription occurring in the nucleus, followed by multiple pro-*

cessing steps before a mature mRNA is ready to be translated. All of these extra steps are required for mRNAs to be transported out of the nucleus and for recognition by ribosomes in the cytoplasm.

Key terms

MESSENGER RNA (mRNA): the form of RNA that contains the coding instructions used to make a polypeptide by ribosomes

RNA POLYMERASE: the enzyme that transcribes RNA using a strand of DNA as a template

TRANSCRIPTION: the process that converts DNA code into a complementary strand of RNA (mRNA) containing code that can be interpreted by ribosomes

TRANSLATION: the process, mediated by ribosomes, in which the genetic code in an mRNA is used to produce a polypeptide, the ultimate product of structural genes

RNA Polymerase

Transcription is the process whereby the directions for making a protein are converted from DNA-based instructions to RNA-based instructions. This step is required in the process of expressing a gene as a polypeptide, because ribosomes, which assemble polypeptides, can read only RNA-based messages. Although transcription is complicated and involves dozens of enzymes and proteins, it is much simpler in prokaryotes than in eukaryotes. Because prokaryotes lack a nucleus, transcription and translation are linked processes both occurring in the cytoplasm. In eukaryotes, transcription and translation occur as completely separate processes, transcription occurring in the nucleus and translation occurring in the cytoplasm. (It is now known that some translation also occurs in the nucleus, but apparently only a small amount, probably less than 10 percent of the translation occurring in a cell.)

In eukaryotes there are three different types of RNA polymerase that transcribe RNA using a strand of DNA as a template (there is a single type of RNA polymerase in prokaryotes). Two of them, called RNA polymerase I (pol I) and RNA polymerase III (pol III), specialize in transcribing types of RNA that are functional products themselves, such as ribosomal RNA (rRNA) and transfer RNA (tRNA). These

RNAs are involved in translation. RNA polymerase II (pol II) transcribes RNA from structural genes, that is, genes that code for polypeptides. Pol II therefore is the primary RNA polymerase and the one that will be the focus of this article when discussing transcription in eukaryotes.

Transcription in Prokaryotes

The first step in transcription is for RNA polymerase to identify the location of a gene. In prokaryotes many genes are clustered together in functional groups called operons. For example, the lactose (*lac*) operon contains three genes, each coding for one of the enzymes needed to metabolize the sugar lactose. At the beginning of each operon are two control sequences, the operator and the promoter. The promoter is where RNA polymerase binds, in preparation for transcription. The operator is a control region that determines whether RNA polymerase will be able to bind to the promoter. The operator interacts with other proteins that determine when the associated operon should be expressed. They do this by either preventing RNA polymerase from binding to the promoter or by assisting it to bind.

RNA polymerase recognizes promoters by the specific base-pair sequences they contain. Assuming all conditions are correct, RNA polymerase binds to the promoter, along with another protein called the sigma factor (σ). The beginning of genes are detected with the aid of σ. Transcription begins at a leader sequence a little before the beginning of the first gene and continues until RNA polymerase reaches a termination signal. If the operon contains more than one gene, all of the genes are transcribed into a single long mRNA, each gene separated from its neighbors by a spacer region. The mRNA is put together by pairing ribonucleotides with their complementary nucleotides in the DNA template. In place of thymine (T), RNA uses uracil (U); otherwise the same bases are present in RNA and DNA, the others being adenine (A), guanine (G), and cytosine (C). The pairing relationships are as follows, the DNA base listed first in each pair: A-U, T-A, G-C, and C-G.

RNA polymerase catalyzes the joining of

ribonucleotides as they pair with the DNA template. Each mRNA is constructed beginning at the 5′ end (the phosphate end) and ending with the 3′ end (the hydroxyl end). Even while transcription is taking place, ribosomes begin binding to the mRNA to begin translation. As soon as RNA polymerase has completed transcribing the genes of an operon, it releases from the DNA and soon binds to another promoter to begin the process all over again.

Transcription in Eukaryotes

Transcription in eukaryotes differs from the process in prokaryotes in the following major ways: (1) genes are transcribed individually instead of in groups; (2) DNA is complexed with many proteins and is highly compacted, and therefore must be "unwound" to expose its promoters; (3) transcription occurs in a separate compartment (the nucleus) from translation, most of which occurs in the cytoplasm; and (4) initially transcription results in a pre-messenger RNA (pre-mRNA) molecule that must be processed before it emerges as a mature mRNA ready for translation. Additionally, mRNAs are much longer-lived in eukaryotes.

The first step in transcription is for RNA polymerase to find a gene that needs to be transcribed. Only genes occurring in regions of the DNA that have been unwound are prepared for potential transcription. RNA polymerase binds to an available promoter, which is located just before a gene and has a region in it called the TATA box (all promoters have the consensus sequence TATAAAA in them). RNA polymerase is unable to bind to the promoter without

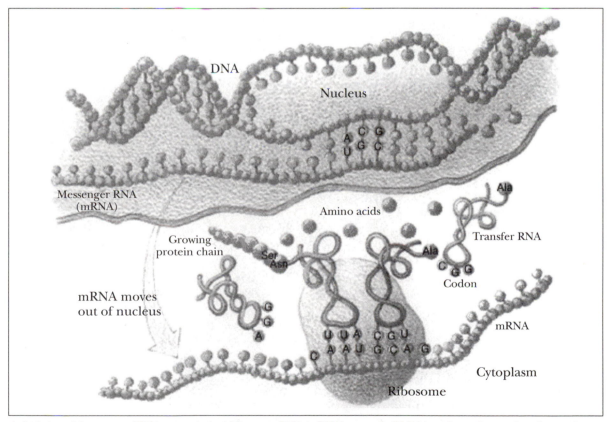

A depiction of the process of RNA transcription. Messenger RNA (mRNA) moves the DNA's template or instructions for protein synthesis (genetic code, or arrangement of bases) from the cell nucleus out into the cytoplasm, where it binds to a ribosome, the cell's "protein factory." Transfer RNA molecules then synthesize amino acids by linking to a codon on the mRNA and transferring the resulting amino acid to a growing chain of amino acids, the protein molecule. (U.S. Department of Energy Human Genome Program, http://www.ornl.gov/hgmis)

assistance from over a dozen other proteins, including a TATA-binding protein, several transcription factors, activators, and coactivators. There are other DNA sequences further upstream than the promoter that control transcription too, thus accounting for the fact that some genes are transcribed more readily, and therefore more often, than others.

Once RNA polymerase has bound to the promoter, it begins assembling an RNA molecule complementary to the DNA code in the gene. It starts by making a short leader sequence, then transcribes the gene, and finishes after transcribing a short trailer sequence. Transcription ends when RNA polymerase reaches a termination signal in the DNA. The initial product is a pre-mRNA molecule which is much longer than the mature mRNA will be.

mRNA Processing in Eukaryotes

Pre-mRNAs must be processed before they can leave the nucleus and be translated at a ribosome. Three separate series of reactions play a part in producing a mature mRNA: (1) intron removal and exon splicing, (2) 5′ capping, and (3) addition of a poly-A tail. Not all transcripts require all three modifications, but most do.

The reason pre-mRNAs are much longer than their respective mature mRNAs has to do with the structure of genes in the DNA. The coding sequences of almost all eukaryotic genes are interrupted with noncoding regions. The noncoding regions are called introns, because they represent "intervening" sequences, and the coding regions are called exons. For an mRNA to be mature it must have all the introns removed and all the exons spliced together into one unbroken message. Special RNA/protein complexes called small nuclear ribonucleoprotein particles, or snRNPs (pronounced as "snurps" by geneticists), carry out this process. The RNAs in the snRNPs are called small nuclear RNAs or snRNAs. Several snRNPs grouped together form a functional splicing unit called a spliceosome. Spliceosomes are able to recognize short signal sequences in pre-mRNA molecules that identify the boundaries of introns and exons. When a spliceosome has found an intron, it binds correctly, and through formation of a lariat-shaped structure, it cuts

the intron out and splices the exons that were on each side of the intron to each other. Genes may have just a few introns, or they may have a dozen or more. Why eukaryotes have introns at all is still an open question, as introns, in general, appear to have no function.

While intron removal and exon splicing are taking place, both ends of maturing mRNAs must also be modified. At the 5′ end (the end with an exposed phosphate) an enzyme adds a modified guanosine nucleotide called 7-methylguanosine. This special nucleotide is added so that ribosomes in the cytoplasm can recognize the correct end of mRNAs, and it probably also prevents the 5′ end of mRNAs from being degraded.

At the 3′ end of maturing mRNAs another enzyme, called polyadenylase, adds a string of adenine nucleotides. Polyadenylase actually recognizes a special signal in the trailer sequence, at which it cuts and then adds the adenines. The result is what is called a poly-A tail. Initially geneticists did not understand the function of poly-A tails, but now it appears that they protect mRNAs from enzymes in the cytoplasm that could break them down. Essentially, poly-A tails are the main reason mRNAs in eukaryotes survive so much longer than mRNAs in prokaryotes.

Once the modifications have been completed, mRNAs are ready to be exported from the nucleus and will now travel through nuclear pores and enter the cytoplasm, where awaiting ribosomes will translate them, using the RNA code to build polypeptides.

Transcription and Disease

Ordinarily transcription works like a well-oiled machine, and only the right genes are transcribed at the right time so that just the right amount of protein product is produced. Unfortunately, due to the great complexity of the system, problems can occur that lead to disease. It has been estimated that about 15 percent of all genetic diseases may be due to improper intron removal and exon splicing in pre-mRNA molecules. Improper gene expression accounts for many other diseases, including many types of cancer.

Beta-thalassemia, a genetic disorder causing

Cooley's anemia, is caused by a point mutation (a change in a single nucleotide) that changes a cutting and splicing signal. As a result, the mature mRNA has an extra piece of intron, making the mRNA longer and causing a reading frame shift. A reading frame shift causes everything from the mutation forward to be skewed, so that the code no longer codes for the correct amino acids. Additionally, as in the case of Cooley's anemia, a reading frame shift often introduces a premature stop codon. The gene involved codes for the beta chain of hemoglobin, the protein that carries oxygen in the blood, and this mutation results in a shortened polypeptide that does not function properly.

A single point mutation in a splicing site can have even more far-reaching consequences. In 2000, researchers in Italy discovered an individual who was genetically male (having one X and one Y chromosome) but was phenotypically female. She had no uterus or ovaries and only superficial external female anatomy, making her a pseudohermaphrodite. This condition can be caused by defects either in androgen production or in the androgen receptor. In this case, the defect was a simple point mutation in the androgen receptor gene that led to one intron being retained in the mature mRNA. Within the intron was a stop codon, which meant when the mRNA was translated, a shorter, nonfunctional polypeptide was formed. The subject did show a very small response to androgen, so apparently some of the pre-mRNAs were being cut and spliced correctly, but not enough to produce the normal male phenotype.

The same kinds of mutations as those discussed above can lead to cancer, but mutations that change the level of transcription of proto-oncogenes can also lead to cancer. Proto-oncogenes are normal genes involved in regulating the cell cycle, and when these genes are overexpressed they become oncogenes (cancer-causing genes). Overexpression of proto-oncogenes leads to overexpression of other genes, because many proto-oncogenes are transcription factors, signal proteins that interact with molecules controlling intracellular growth and growth factors released by cells to stimulate other cells to divide.

Overexpression can occur when there is a mutation in one of the control regions upstream from a gene. For example, a mutation in the promoter sequence could cause a transcription factor, and thus RNA polymerase, to bind more easily, leading to higher transcription rates. Other control regions, such as enhancer sequences, often far removed from the gene itself, may also affect transcription rates.

Anything that causes the transcription process to go awry will typically have far-reaching consequences. Geneticists are just beginning to understand some of the underlying errors behind a host of genetic diseases, and it should be no surprise that some of them involve how genes are transcribed. Knowing what the problem is, unfortunately, does not usually point to workable solutions. When the primary problem is an excessive rate of transcription, specially designed antisense RNA molecules (RNA molecules that are complementary to mRNA molecules) might be designed that will bind to the overexpressed mRNAs and disable them. This approach is still being tested. In the case of point mutations that derail the cutting and splicing process, the only solution may be gene therapy, a technique still not considered technically possible and not expected to be feasible for some time to come.

—Bryan Ness

See also: Ancient DNA; Antisense RNA; Cancer; Chromosome Structure; DNA Isolation; DNA Repair; DNA Replication; DNA Structure and Function; Genetic Code; Genetic Code, Cracking of; Molecular Genetics; Noncoding RNA Molecules; One Gene-One Enzyme Hypothesis; Protein Structure; Protein Synthesis; Pseudohermaphrodites; Repetitive DNA; RNA Isolation; RNA Structure and Function; RNA World.

Further Reading

Hampsey, Michael. "Molecular Genetics of the RNA Polymerase II General Transcriptional Machinery." *Microbiology and Molecular Biology Reviews* 62, no. 2 (1998): 465-503. Overview of the role of RNA polymerase II.

Latchman, David S. "Transcription-Factor Mutations and Disease." *The New England Journal of Medicine* 334 (1996): 28-33. A general overview of the kinds of diseases caused by mutations in transcription factor genes. Includes an overview of potential treatments.

Macfarlane, W. M. "Transcription." *Journal of Clinical Pathology: Molecular Pathology* 53 (2000): 1-7. A general introduction with an emphasis on transcription factors and their role in transcription.

Ptashne, Mark, and Alexander Gann. *Genes and Signals.* Cold Spring Harbor, N.Y.: Cold Spring Harbor Laboratory Press, 2002. A nice overview of transcription and related topics, readable by advanced high school students and undergraduates.

Shatkin, Aaron J., and James L. Manley. "The Ends of the Affair: Capping and Polyadenylation." *Nature Structural Biology* 7, no. 10 (2000): 838-842.

White, Robert J. *Gene Transcription: Mechanisms and Control.* Malden, Mass.: Blackwell, 2001. An in-depth look at all aspects, including regulation, of transcription. Aimed mostly at upper undergraduate students, but begins with the basics.

RNA World

Fields of study: Evolutionary biology; Molecular genetics

Significance: *The RNA world is a theoretical time in the early evolution of life, during which RNA molecules played important genetic and enzymatic roles that were later taken over by molecules of DNA and proteins. Ideas about RNA's ancient functions have led to new concepts of the origin of life and have important implications in the use of gene therapy to treat diseases.*

Key term

RIBOSOMAL RNA (rRNA): a type of RNA that forms a major part of the structure of the ribosome

RIBOSOME: an organelle that functions in protein synthesis, composed of a large and a small subunit composed of proteins and ribosomal RNA molecules

RIBOZYME: an RNA molecule that can function catalytically as an enzyme

The Central Dogma and the Modern Genetic World

Soon after the discovery of the double-helical structure of DNA in 1953 by James Watson and Francis Crick, Crick proposed an idea regarding information flow in cells that he called the "central dogma of molecular biology." Crick correctly predicted that in all cells, information flows from DNA to RNA to protein. DNA was known to be the genetic material, the "library" of genetic information, and it had been clear for some time that the enzymes that actually did the work of facilitating chemical reactions were invariably protein molecules. The discovery of three classes of RNA during the 1960's seemed to provide the link between the DNA instructions and the protein products.

In the modern genetic world, cells contain three classes of RNA that act as helpers in the synthesis of proteins from information stored in DNA, a process called translation. A messenger RNA (mRNA) is "transcribed" from a segment of DNA (a gene) that contains information about how to build a particular protein and carries that information to the cellular site of protein synthesis, the ribosome. Ribosomal RNAs (rRNAs) interacting with many proteins make up the ribosome, whose major job is to coordinate and facilitate the protein-building procedure. Transfer RNAs (tRNAs) act as decoding molecules, reading the mRNA information and correlating it with a specific amino acid. As the ribosome integrates the functions of all three types of RNA, polypeptides are built one amino acid at a time. These polypeptides, either singly or in aggregations, can then function as enzymes, ultimately determining the capabilities and properties of the cell in which they act.

While universally accepted, the central dogma led many scientists to question how this complex, integrated system came about. It seemed to be a classic "chicken and egg" dilemma: Proteins could not be built without instructions from DNA, but DNA could not replicate and maintain itself without help from protein enzymes. The two seemed mutually dependent upon each other in an inextricable way. An understanding of the origins of the modern genetic system seemed far away.

The Discovery of Ribozymes

In 1983, a discovery was made that seemed so radical it was initially rejected by most of the scientific community. Molecular biologists Thomas Cech and Sidney Altman, working independently and in different systems, announced the discovery of RNA molecules that possessed catalytic activity. This meant that RNA itself can function as an enzyme, obliterating the idea that only proteins could function catalytically.

Cech had been working with the protozoan *Tetrahymena*. In most organisms except bacteria, the coding portions of DNA genes (exons) are interrupted by noncoding sequences (introns), which are transcribed into mRNA but which must be removed before translation. Protein enzymes called nucleases are usually responsible for cutting out the introns and joining together the exons in a process called splicing. The molecule with which Cech was working was an rRNA that contained introns but could apparently remove them and rejoin the coding regions without any help. It was a self-splicing RNA molecule, which clearly indicated its enzymatic capability. Altman was working with the enzyme ribonuclease (RNase) P in bacteria, which is responsible for cutting mature tRNA molecules out of an immature RNA segment. RNase P thus also acts as a nuclease. It was known for some time that RNase P contains both a protein and an RNA constituent, but Altman was ultimately able to show that it was the RNA rather than the protein that actually catalyzed the reaction.

The importance of these findings cannot be overstated, and Cech and Altman ultimately shared the 1989 Nobel Prize in Chemistry for the discovery of these RNA enzymes, or ribozymes (joining the terms "ribonucleic acid" and "enzymes"). Subsequently, many ribozymes have been found in various organisms, from bacteria to humans. Some of them are able to catalyze different types of reactions, and new ones are periodically reported. Ribozymes have thus proven to be more than a mere curiosity, playing an integral role in the molecular machinery of many organisms.

At around the same time as these important discoveries, still other functions of RNA were being identified. While perhaps not as dramatic as the ribozymes, antisense RNAs, small nuclear RNAs, and a variety of others further proved the versatility of RNA. While understanding the roles of ribozymes and other unconventional RNAs is important to the understanding of genetic functioning in present-day organisms, these discoveries were more intriguing to many scientists interested in the origin and evolution of life. In a sense, the existence of ribozymes was a violation of the central dogma, which implied that information was ultimately utilized solely in the form of proteins. While the central dogma was not in danger of becoming obsolete, a clue had been found that might possibly allow a resolution, at least in theory, to questions about whether the DNA or the protein came first. The exciting answer: perhaps neither.

The RNA World Theory and the Origin of Life

Given that RNA is able to store genetic information (as it certainly does when it functions as mRNA) and the new discovery that it could function as an enzyme, there was no longer any need to invoke the presence of either DNA or protein as necessities in the first living system. The first living molecule would have to be able to replicate itself without any help, and just such an "RNA replicase" has been proposed as the molecule that eventually led to life as it is now known. Like the self-splicing intron of *Tetrahymena*, this theoretical ribozyme could have worked on itself, catalyzing its own replication. This RNA would therefore have functioned as both the genetic material and the replication enzyme, allowing it to make copies of itself without the need for DNA or proteins. Biologist Walter Gilbert coined the term "RNA world" for this interesting theoretical period dominated by RNA. Modern catalytic RNAs can be thought of as molecular fossils that remain from this period and provide clues about its nature.

How might this initial RNA have come into being in the first place? Biologist Aleksandr Oparin predicted in the late 1930's that if simple gases thought to be present in Earth's early atmosphere were subjected to the right conditions (energy in the form of lightning, for example), more complex organic molecules would be formed. His theory was first tested in 1953 and was resoundingly confirmed. A mixture of methane, ammonia, water vapor, and hydrogen gas was energized with high-voltage electricity, and the products were impressive: several amino acids and aldehydes, among other organic molecules. Subsequent experiments have been able to produce ribonucleotide bases. It seems reasonable, then, that nucleotides could have been present on the early Earth and that their random linkage could lead to the formation of an RNA chain.

After a while, RNA molecules would have found a way to synthesize proteins, which are able to act as more efficient and diverse enzymes than ribozymes by their very nature. Why are proteins better enzymes than ribozymes? Since RNA contains only four bases that are fundamentally similar in their chemical properties, the range of different configurations and functional capabilities is somewhat limited as opposed to proteins. Proteins are constructed of twenty different amino acids whose functional groups differ widely in terms of their chemical makeup and potential reactivity. It is logical to suppose, therefore, that proteins eventually took over most of the roles of RNA enzymes because they were simply better suited to doing so. Several of the original or efficient ribozymes would have been retained, and those are the ones that can be observed today.

How could a world composed strictly of RNAs, however, be able to begin protein synthesis? While it seems like a tall order, scientists have envisioned an early version of the ribosome that was composed exclusively of RNA. Biologist Harry Noller reported in the early 1990's that the activity of the modern ribosome that is responsible for catalyzing the formation of peptide bonds between amino acids is in fact carried out by rRNA. This so-called peptidyltransferase activity had always been attributed to one of the ribosomal proteins, and rRNA had been envisioned as playing a primarily structural role. Noller's discovery that the large ribosomal RNA is actually a ribozyme allows scientists to picture a ribosome working in roughly the same way that modern ones do,

without containing any proteins. As proteins began to be synthesized from the information in the template RNAs, they slowly began to assume some of the RNA roles and probably incorporated themselves into the ribosome to allow it to function more efficiently.

The transition to the modern world would not be complete without the introduction of DNA as the major form of the genetic material. RNA, while well suited to diverse roles, is actually a much less suitable genetic material than DNA for a complex organism (even one only as complex as a bacterium). The reason for this is that the slight chemical differences between the sugars contained in the nucleotides of RNA and DNA cause the RNA to be more reactive and much less chemically stable; this is good for a ribozyme but clearly bad if the genetic material is to last for any reasonable amount of time. Once DNA initially came into existence, therefore, it is likely that the relatively complex organisms of the time quickly adopted it as their genetic material; shortly thereafter, it became double-stranded, which facilitated its replication immensely. This left RNA, the originator of it all, relegated to the status it enjoys today; molecular fossils exist that uncover its former glory, but it functions mainly as a helper in protein synthesis.

This still leaves the question of how DNA evolved from RNA. At least two protein enzymes were probably necessary to allow this process to begin. The first, ribonucleoside diphosphate reductase, converts RNA nucleotides to DNA nucleotides by reducing the hydroxyl group located on the 2′ carbon of ribose. Perhaps more important, the enzyme reverse transcriptase would have been necessary to transcribe RNA genomes into corresponding DNA versions. Examples of both of these enzymes exist in the modern world.

Some concluding observations are in order to summarize the evidence that RNA and not DNA was very likely the first living molecule. No enzymatic activity has ever been attributed to DNA; in fact, the 2′ hydroxyl group that RNA possesses and DNA lacks is vital to RNA's ability to function as a ribozyme. Furthermore, ribose is synthesized much more easily than deoxyribose under laboratory conditions. All modern cells synthesize DNA nucleotides from RNA precursors, and many other players in the cellular machinery are RNA-related. Important examples include adenosine triphosphate (ATP), the universal cellular energy carrier, and a host of coenzymes such as nicotinamide adenine dinucleotide (NAD), derived from B vitamins and vital in energy metabolism.

Impact and Applications

The discovery of ribozymes and the other interesting classes of RNAs has dramatically altered the understanding of genetic processes at the molecular level and has provided compelling evidence in support of exciting new theories regarding the origin of life and cellular evolution. The RNA world theory, first advanced as a radical and unsupported hypothesis in the early 1970's, has gained almost universal acceptance by scientists. It is the solution to the evolutionary paradox that has plagued scientists since the discovery and understanding of the central dogma: Which came first, DNA or proteins? Since they are inextricably dependent upon each other in the modern world, the idea of the RNA world proposes that, rather than one giving rise to the other, they are both descended from RNA, that most ancient of genetic and catalytic molecules. Unfortunately, the RNA world model is not without its problems.

In the mid- to late 1990's, several studies on the stability of ribose, the sugar portion of ribonucleotides, showed that it breaks down relatively easily, even in neutral solutions. A study of the decay rate of ribonucleotides at different temperatures also caused some concern for the RNA world theory. Most current scenarios see life arising in relatively hot conditions, at least near boiling, and the instability of ribonucleotides at these temperatures would not allow for the development of any significant RNA molecules. Ribonucleotides are much more stable at 0 degrees Celsius (32 degrees Fahrenheit), but evidence for a low-temperature environment for the origin of life is limited. Consequently, some evolutionists are suggesting that the first biological entities might have relied on something other than RNA, and that the RNA world was a later development. Therefore, although the RNA world seems like a plausible

model, another model is now needed to establish the precursor to the RNA world.

Apart from origin-of-life concerns, the discoveries that led to the RNA world theory are beginning to have a more practical impact in the fields of industrial genetic engineering and medical gene therapy. The unique ability of ribozymes to find particular sequences and initiate cutting and pasting at desired locations makes them powerful tools. Impressive uses have already been found for these tools in theoretical molecular biology and in the genetic engineering of plants and bacteria. Most important to humans, however, are the implications for curing or treating genetically related disease using this powerful new RNA-based technology.

Gene therapy, in general, is based on the idea that any faulty, disease-causing gene can theoretically be replaced by a genetically engineered working replacement. While theoretically a somewhat simple idea, in practice it is technically very challenging. Retroviruses may be used to insert DNA into particular target cells, but the results are often not as expected; the new genes are difficult to control or may have adverse side effects. Molecular biologist Bruce Sullenger pioneered a new approach to gene therapy, which seeks to correct the genetic defect at the RNA level. A ribozyme can be engineered to seek out and replace damaged sequences before they are translated into defective proteins. Sullenger has shown that this so-called trans-splicing technique can work in nonhuman systems and, in 1996, began trials to test his procedure in humans.

Many human diseases could be corrected using gene therapy technology of this kind, from inherited defects such as sickle-cell disease to degenerative genetic problems such as cancer. Even pathogen-induced conditions such as acquired immunodeficiency syndrome (AIDS), caused by the human immunodeficiency virus (HIV), could be amenable to this approach. It is ironic and gratifying that an understanding of the ancient RNA world holds promise for helping scientists to solve some of the major problems in the modern world of DNA-based life.

—*Matthew M. Schmidt, updated by Bryan Ness*

See also: Ancient DNA; Antisense RNA; Chromosome Structure; DNA Isolation; DNA Repair; DNA Replication; DNA Structure and Function; Genetic Code; Genetic Code, Cracking of; Molecular Genetics; Noncoding RNA Molecules; One Gene-One Enzyme Hypothesis; Protein Structure; Protein Synthesis; Repetitive DNA; RNA Isolation; RNA Structure and Function; RNA Transcription and mRNA Processing.

Further Reading

De Duve, Christian. "The Beginnings of Life on Earth." *American Scientist* 83 (September/ October, 1995). Discusses several scenarios regarding RNA's possible involvement in life's origins.

Gesteland, Raymond F., Thomas R. Cech, and John F. Atkins, eds. *The RNA World: The Nature of Modern RNA Suggests a Prebiotic RNA.* 2d ed. Cold Spring Harbor, N.Y.: Cold Spring Harbor Laboratory Press, 1999. An advanced, detailed look at the theories behind the RNA world, the evidence for its existence, and the modern "fossils" that may be left from this historic biological period. Illustrated (some color).

Hart, Stephen. "RNA's Revising Machinery." *Bioscience* 46 (May, 1996). A discussion of the use of ribozymes in gene therapy.

Horgan, John. "The World According to RNA." *Scientific American* 189 (January, 1996). Summarizes the accumulated evidence that RNA molecules once served both as genetic and catalytic agents.

Miller, Stanley L. *From the Primitive Atmosphere to the Prebiotic Soup to the Pre-RNA World.* NASA CR-2076334007116722. Washington, D.C.: National Aeronautics and Space Administration, 1996. Miller discusses his famous 1953 experiment, whereby he produced organic amino acids from inorganic materials in laboratory, igniting a new understanding of how RNA and DNA work.

Watson, James D., et al. *The Molecular Biology of the Gene.* 5th ed. Menlo Park, Calif.: Benjamin Cummings, 2003. Includes a comprehensive discussion of all aspects of the RNA world.

Shotgun Cloning

Field of study: Genetic engineering and biotechnology

Significance: *Shotgun cloning is the random insertion of a large number of different DNA fragments into cloning vectors. A large number of different recombinant DNA molecules are generated, which are then introduced into host cells, often bacteria, and amplified. Because a large number of different recombinant DNAs are generated, there is a high likelihood one of the clones contains a fragment of DNA of interest.*

Key terms

CLONING VECTOR: a plasmid or virus into which foreign DNA can be inserted to amplify the number of copies of the foreign DNA

MARKER: a gene that encodes an easily detected product that is used to indicate that foreign DNA is in an organism

RECOMBINANT DNA: a novel DNA molecule formed by the joining of DNAs from different sources

RESTRICTION ENDONUCLEASE: an enzyme that recognizes a specific nucleotide sequence in a piece of DNA and causes cleavage of the DNA; often simply called a restriction enzyme

Recombinant DNA Cloning and Shotgun Cloning

Before the development of recombinant DNA cloning, it was very difficult to study DNA sequences. Cloning a DNA fragment allows a researcher to obtain large amounts of that specific DNA sequence to analyze without interference from the presence of other DNA sequences. There are many uses for a cloned DNA fragment. For example, a DNA fragment can be sequenced to determine the order of its nucleotides. This information can be used to determine the location of a gene and the amino acid sequence of the gene's protein product. Cloned pieces of DNA are also useful as DNA probes. Because DNA is made of two strands that are complementary to each other, a cloned piece of DNA can be used to probe for copies of the same or similar DNA sequences in

other samples. A cloned gene can also be inserted into an expression vector where it will produce the gene's protein product.

Shotgun cloning begins with the isolation of DNA from the organism of interest. In separate test tubes, the DNA to be cloned and the cloning vector DNA are digested (cut) with a restriction endonuclease that cuts the vector in just one location and the foreign DNA many times. Many restriction endonucleases create single-stranded ends that are complementary, so the end of any DNA molecule cut with that endonuclease can join to the end of any other DNA cut with the same endonuclease. When the digested vector and foreign DNA are mixed, they join randomly and are then sealed using DNA ligase, an enzyme that seals the small gap between two pieces of DNA. This creates recombinant DNA molecules composed of a copy of the vector and a random copy of foreign DNA. The recombinant DNA molecules are then introduced into host cells where the cloning vector can replicate each time the cell divides, which is approximately every twenty minutes in the case of *Escherichia coli*. The resulting collection of clones, each containing a potentially different fragment of foreign DNA, called a genomic library. If a large collection of clones is produced, it is likely that every part of the genome from which the DNA came will be represented somewhere in the genomic library.

The presence of the cloning vector in host cells is determined by selecting for a marker gene in the cloning vector. Most vectors have two marker genes, and often both are different antibiotic resistance genes. A common example is the plasmid pBR322, which has a tetracycline and an ampicillin resistance gene. A restriction endonuclease cuts once somewhere in the tetracycline resistance gene, and if a foreign DNA fragment becomes incorporated, the resulting recombinant plasmid will have a nonfunctional tetracycline resistance gene. A bacterial cell transformed with a recombinant plasmid will therefore be resistant to ampicillin, but will be sensitive to tetracycline. Many plasmids will not incorporate any foreign DNA and will be nonrecombinant. Cells that are transformed with a nonrecombinant plasmid will be resistant to both tetracycline and ampicillin. After the bacterial cells have been transformed, they are grown on a medium with ampicillin. The only cells that will survive will be those that have received a plasmid vector. To determine which cells have received a recombinant plasmid, the colonies are carefully transferred onto a new medium that has both ampicillin and tetracycline. On this medium, only cells with nonrecombinant plasmids will survive. Thus, colonies that grew on the first media, but not on the second, contain recombinant plasmids. Cells from these colonies are collected and grown, each in a separate tube, and these constitute a genomic library.

Once a genomic library has been produced, the DNA fragments contained in it can be screened and analyzed in various ways. Using the right techniques, specific genes can be found, which can then be used in future analyses and experiments.

Alternatives to Shotgun Cloning

In shotgun cloning, many different DNA fragments from an organism are cloned, and then the specific DNA clone of interest is identified. The number of clones can be reduced, making the search easier, if the DNA of interest is known to be in a restriction endonuclease fragment of a specific size. DNA can be size-selected before cloning using gel electrophoresis, in which an electric current carries DNA fragments through the pores or openings of an agarose gel. DNA migrates through the gel based on DNA fragment size, with the smaller fragments traveling more rapidly than the larger fragments. DNA of a specific size range can be isolated from the gel and then used for cloning. Finally, to clone a piece of DNA known to code for a protein, scientists can use an enzyme called reverse transcriptase to make DNA copies (called a complementary DNA or cDNA) of isolated messenger RNA (mRNA). The cDNA is then cloned, in a similar manner to that already discussed, to produce what is called a cDNA library. One of the advantages of this approach is that the number of clones is greatly reduced.

—*Susan J. Karcher, updated by Bryan Ness*

See also: cDNA Libraries; Cloning; Cloning Vectors; Genomic Libraries; Genomics; Human Genome Project; Restriction Enzymes.

Further Reading

Glick, Bernard R. *Molecular Biotechnology: Principles and Applications of Recombinant DNA.* Washington, D.C.: ASM Press, 2003. Covers the scientific principles of biotechnology and gives applications. Color illustrations.

Kreuzer, Helen, and Adrianne Massey. *Recombinant DNA and Biotechnology: A Guide for Students.* 2d ed. Washington, D.C.: ASM Press, 2001. For high school level. Gives introductory text and activities to learn the basics of molecular biology and biotechnology.

_____. *Recombinant DNA and Biotechnology: A Guide for Teachers.* 2d ed. Washington, D.C.: ASM Press, 2001. For high school and introductory college level. Presents a guide to biotechnolgy with history, applications, simple protocols, and exercises.

Micklos, David A, Greg A. Freyer, and David A. Crotty. *DNA Science: A First Course.* Cold Spring Harbor, N.Y.: Cold Spring Harbor Laboratory Press, 2003. Gives an introduction to molecular biology techniques for high school or beginning college students. Includes background text and laboratory procedures.

Watson, James D., et al. *Recombinant DNA.* New York: Scientific American Books, 1992. Provides a description of shotgun cloning as well as an overview of many other cloning methods. Full-color illustrations, diagrams, bibliography, index.

Sickle-Cell Disease

Field of study: Diseases and syndromes
Significance: *Sickle-cell disease is a treatable hereditary blood disease that occurs mainly among people of African, Caribbean, and Mediterranean descent, which has led to concerns, particularly in the United States, that it might be used as a surrogate for discrimination against particular racial groups. It is one of the most well documented exam-*

ples of an evolutionary process known as heterozygote advantage, an important means by which genetic variability is preserved.

Key terms

HEMOGLOBIN: a molecule made up of two alpha and two beta amino acid chains whose precise chemical and structural properties normally allow it to bind with oxygen in the lungs and transport it to other parts of the body

HETEROZYGOUS: having two different forms (alleles) of the same gene (locus), each inherited from a different parent

HOMOZYGOUS: having the same allele from both parents

Genetics and Early Research

Sickle-cell disease, also known as sickle-cell anemia, is a hereditary blood disease found primarily among people of African, Caribbean, and Mediterranean descent. Studies of the incidence of the disease in families led to recognition that the illness is manifested only in individuals who receive the sickle-cell allele from both parents. In most circumstances, individuals who inherit the sickle-cell allele from only one parent display no symptoms of the disease; however, they are carriers of the sickle-cell gene and may pass it on to their children.

In 1910, James B. Herrick, a Chicago physician, first described the characteristically "sickle" or bent appearance of the red blood cells after which the disease is named in blood taken from an anemic patient. In the mid-1930's, Linus Pauling, working with graduate student Charles Coryell, demonstrated that hemoglobin undergoes a dramatic structural change as it combines and releases oxygen. Upon learning that red blood cells from sickle-cell disease patients only assume their characteristic form in the oxygen-deprived venous blood system, Pauling proposed in 1949 that sickle-cell disease was the result of a change in

Linus Pauling (right), with George Beadle, c. 1952. (California Institute of Technology)

the normal amino acid sequence of hemoglobin that interferes with its binding properties.

Three years later, while working with another graduate student, Harvey Itano, Pauling isolated normal hemoglobin and sickle-cell hemoglobin from an individual with anemia using a technique known as electrophoresis. They conducted this investigation by loading hemoglobin onto a paper medium and subjecting it to an electrical current, the presumption being that if the two molecules differed in overall electrical charge, one would migrate along the path of the current faster than the other. In this way, Pauling and Itano established that normal and sickle-cell hemoglobins differ in their respective electrical charges, and people who are heterozygous for the sickle-cell gene have hemoglobin of both types.

In the mid-1950's, Vernon Ingram approached the problem using a more sophisticated version of Pauling's procedure. Ingram first treated hemoglobin of the two types with an enzyme (trypsin) that broke the complex hemoglobin molecules into smaller polypeptides and then used electrophoretic techniques on the resulting polypeptides to determine precisely where in their respective amino acid sequences the two hemoglobins differed from one another. Ingram was able to show that normal and sickle-cell hemoglobin differ by only a single amino acid out of a total of more than three hundred: Where the normal hemoglobin gene codes for glutamic acid in the sixth position of the beta-globin, the sickle-cell gene substitutes another amino acid (usually lysine). Ingram's work provided proof of Pauling's earlier proposal, making sickle-cell disease the very first example of a genetic disease being traced to its precise origin at the molecular level.

Physiological Basis, Symptoms, and Treatments

This substitution of lysine for glutamic acid in the beta chain of the hemoglobin molecule has a profound effect on its biological properties under conditions of oxygen deprivation. Hemoglobin coded for by the sickle-cell gene causes the beta chains of the hemoglobin to stick to one another as long, rigid rods and consequently deforms the normally smooth, donut-shaped appearance of the red blood cell to a characteristic sickle shape that prevents it from squeezing through tiny blood capillaries.

Symptoms of the disease appear about six months after birth, when the last of fetal hemoglobin, a type of hemoglobin that increases the oxygen supply of blood, leaves the infant's body. The severity of the illness varies widely among individuals. Some develop severe anemia as deformed red blood cells are removed more rapidly from the bloodstream (an average "life" of seventeen versus forty days). They may also experience periodic bouts of severe pain ("pain crises"), strokes, and blindness, all thought to be the direct result of sickled cells clogging blood vessels and thereby depriving tissues of oxygen. Heterozygous carriers of the gene normally display no symptoms, although some have been known to become ill under extreme circumstances, such as high altitudes.

A great deal of progress has been made in the diagnosis and treatment of sickle-cell disease. This includes both a variety of pain management therapies and the use of antibiotics such as penicillin to prevent infections. Although there is no cure, several promising experimental therapies for this disease are under investigation, including the use of bone marrow transplants (transplants of the mast cells that give rise to red blood cells from people not having the disease) and

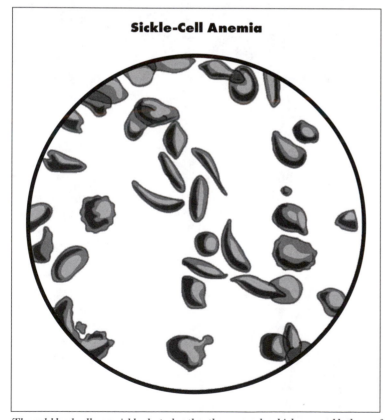

Sickle-Cell Anemia

The red blood cells are sickle-shaped rather than round, which causes blockage of capillaries. (Hans & Cassidy, Inc.)

hydroxyurea, a chemical thought to stimulate the production of fetal hemoglobin.

Attention to and funding for research on sickle-cell disease has increased since World War II, although misinformation about the disease persists. Many have raised concerns that carriers of the disease are discriminated against, both by potential employers and insurance companies. Several organizations were established in the United States in the early 1970's to promote education, treatment, and research for the disease, including Howard University's Center for Sickle-Cell Disease, founded by Ronald B. Scott in 1972. Today, forty U.S. states, the District of Columbia, Puerto Rico, and the Virgin Islands, were screening newborns for the sickle-cell trait.

Evolutionary Significance

In most cases, hereditary diseases with such negative consequences as those associated with sickle-cell disease are kept at low frequencies in populations by natural selection; that is, individuals who carry genes for hereditary diseases are less likely to survive and reproduce than those who carry the normal form of the gene. The continued presence of defective genes in a population therefore reflects the action of chance mutations. Yet the sickle-cell gene is much more common than one would expect if its frequency in a population was caused by mutation alone.

In some areas that are associated with a high incidence of malaria, such as the equatorial belt of Africa, some tribes have been found to have frequencies of the sickle-cell gene as high as 40 percent. This curious correlation between high frequencies of the sickle-cell gene and areas where malaria is common led Anthony C. Alison to suggest, in 1953, that the sickle-cell gene provides an advantage in such environments. Malaria is a deadly, mosquito-borne disease caused by a microscopic parasite, *Plasmodium vivax*, which uses human red blood cells as hosts for part of its life cycle. People who have normal hemoglobin are vulnerable to the disease, and people who are homozygous for the sickle-cell gene in malaria-infested regions die quite early in life because of anemia and other complications. However, when the red cells of people who are heterozygous for the sickle-cell gene are invaded by the malarial parasite, the red cells adhere to blood vessel walls, become deoxygenated, and assume the sickled shape, prompting both their destruction and that of their parasitic invader. This provides the heterozygous carrier with a natural resistance to malaria and explains the relatively high frequency of the sickle-cell gene in such environments. Sickle-cell disease thus represents a particularly well-documented example of a selective process known as heterozygote advantage, in which individuals heterozygous for a given gene have a greater probability of surviving or reproducing than either homozygote. This is an important phenomenon from an evolutionary standpoint because it provides a mechanism by which genetic diversity in a population may be preserved.

—David Wijss Rudge

See also: Amniocentesis and Chorionic Villus Sampling; Biopharmaceuticals; Genetic Engineering; Genetic Screening; Genetic Testing; Hardy-Weinberg Law; Incomplete Dominance; Mutation and Mutagenesis; RFLP Analysis; Shotgun Cloning.

Further Reading

Allison, Anthony. "Sickle Cells and Evolution." *Scientific American* 202 (August, 1956). An early discussion of sickle-cell disease as an evolutionary phenomenon.

Anionwu, Elizabeth N., and Karl Atkin. *The Politics of Sickle Cell and Thalassaemia.* Philadelphia: Open University Press, 2001. Discussion centers on the experiences of patients and their families. Bibliography, index.

Ingram, Vernon M. "How Do Genes Act?" *Scientific American* 204 (January, 1958). Summarizes author's research on the amino acid structure of hemoglobin.

Serjeant, Graham R., and Beryl E. Serjeant. *Sickle Cell Disease.* 3d ed. New York: Oxford University Press, 2001. Discusses the biology of sickle-cell disease and the diseases's management. Illustrations, maps, bibliography, index.

Steinberg, Martin H., et al., eds. *Disorders of Hemoglobin: Genetics, Pathophysiology, and Clini-*

cal Management. Foreword by H. Franklin Bunn. New York: Cambridge University Press, 2001. Covers the diseases' molecular and genetic bases, their epidemiology and genetic selection, and their diagnoses and treatments. Illustrations (some color), bibliography, index.

Tapper, Melbourne. *In the Blood: Sickle Cell Anemia and the Politics of Race.* Philadelphia: University of Pennsylvania Press, 1999. Explores anthropological, genetic, medical, and political texts, and other discourses on race, to discuss how the disease has come to be known as a "black" disease since its medical identification in 1910. Index.

Wailoo, Keith. *Dying in the City of the Blues: Sickle Cell Anemia and the Politics of Race and Health.* Chapel Hill: University of North Carolina Press, 2001. Examines medical literature, patient accounts, black newspapers, blues lyrics, and other popular sources, and discusses how individuals made sense of and lived with the disease, even before its medical and scientific recognition. Illustrations, bibliography, index.

Web Sites of Interest

Dolan DNA Learning Center, Your Genes Your Health. http://www.ygyh.org. Sponsored by the Cold Spring Harbor Laboratory, this site, a component of the DNA Interactive Web site, offers information on more than a dozen inherited diseases and syndromes, including sickle-cell disease.

Sickle Cell Information Center. http://www.scinfo.org. This site provides news and research updates, worldwide resource links, and an interactive link for children.

Signal Transduction

Field of study: Molecular genetics
Significance: *Signal transduction consists of all of the molecular events that occur between the arrival of a signaling molecule at a target cell and its response. A significant proportion of the genome in animals consists of genes involved in cell signal-* *ing. The protein products of these genes allow cells to communicate with each other in order to coordinate their metabolism, movements, and reproduction. Failure of cells to communicate properly can lead to cancer, defects in embryological development, and many other disorders.*

Key terms

CELL CYCLE: the orderly sequence of events by which a cell grows, duplicates its chromosomal DNA, and partitions the DNA into two new cells

CELL SIGNALING: communication between cells that occurs most commonly when one cell releases a specific "signaling" molecule that is received by another cell

RECEPTORS: molecules in target cells that bind specifically to a particular signaling molecule

TARGET CELL: the cell that receives and responds to a signaling molecule

Signal Transduction Pathways

Signal transduction can occur by a number of different, often complex, sequences of molecular events called signal transduction pathways, which result in several kinds of target cell response, including the turning on of genes, the activation of metabolic pathways, and effects on the cell cycle. Among the signaling molecules found in higher organisms are hormones, local mediators that produce local physiological effects, growth factors that act locally to promote growth, and survival factors that act locally to repress cell suicide (apoptosis). Growth factors and survival factors are particularly important during embryological development, when they orchestrate the changes in cell types, positions, and numbers that give rise to the new organism.

Types of Receptors

Most signal transduction pathways begin with the binding of signaling molecules to specific receptors in target cells. Signaling molecules are often referred to as receptor ligands. The binding of the ligand to its receptor initiates a signal transduction pathway. A cell can respond to a particular signaling molecule only if it possesses a receptor for it.

Receptors are protein molecules. There are two categories of them, based on location in the cell: receptors that are intracellular and receptors that are anchored in the cell's surface membrane. The membrane-anchored receptors can be further divided based on the steps of the signal transduction pathway that they initiate: receptors that bind to and activate GTP-binding proteins (G proteins), receptors that are enzymes, and receptors that are ion channels. Receptors that are channels bind neurotransmitters or hormones and increase or decrease the flow of specific ions into the cell, leading to a physiological response by the cell. These receptors generally do not have a direct effect on gene expression (although changes in a cell's calcium ion concentrations can influence gene expression). Each of the other receptor types stands at the head of a signal transduction pathway that is characteristic for each receptor type and can lead to gene expression. In what follows, some of the more common transduction pathways that can lead to gene expression are described.

Intracellular Receptors

Intracellular receptors include the receptors for lipid-soluble hormones such as steroid hormones. Some of these receptors are in the cell's cytoplasm and some are in the nucleus. Hormone molecules enter the cell by first diffusing across the membrane and then binding to the receptor. Before the hormones enter the cell, the receptors are attached to "chaperone" proteins, which hold the receptor in a configuration that allows hormone binding but prevents it from binding to DNA. Hormone molecules displace these chaperone molecules, enabling the receptor to bind to DNA. If the receptor is a cytoplasmic receptor, the hormone-receptor complex is first transported into the nucleus, where it binds to a specific DNA nucleotide sequence called a hormone response element (HRE) that is part of the promoter of certain genes. In most cases the receptors bind as dimers; that is, two hormone-receptor complexes bind to the same HRE. The receptor-hormone complex functions as a transcription factor, promoting transcription of the gene and production of a protein that the cell was

not previously producing. The hormone hydrocortisone, for example, triggers the synthesis of the enzymes aminotransferase and tryptophan oxygenase. A single hormone such as hydrocortisone can turn on synthesis of two or more proteins if each of the genes for the proteins contains an HRE. In some cases, when hormone-receptor complexes bind to an HRE, they suppress transcription rather than promote it.

G Protein-Binding Receptors

Many hormones, growth factors, and other signaling molecules bind to membrane receptors that can associate with and activate heterotrimeric G proteins when a signaling molecule is bound to the receptor. Heterotrimeric G proteins are a family of proteins that are present on the cytoplasmic surface of the cell membrane. Many cell types in the body contain one or more of these family members, and different cell types contain different ones. All heterotrimeric G proteins are made up of three subunits: the alpha, beta, and gamma subunits. The alpha subunit has a binding site for GTP or GDP (hence the name G proteins) and is the principal part of the protein that differs from one heterotrimeric G protein family member to another. When the receptor is empty (no signal molecule attached), these G proteins have GDP bound to the alpha subunit and the G protein is not bound to the receptor.

However, when a signaling molecule binds to the receptor, the cytoplasmic domain of the receptors changes shape so that it now binds to the G protein. In binding to the receptor, the G protein also changes shape, causing GDP to leave and GTP to bind instead. Simultaneously, the alpha subunit detaches from the beta-gamma subunit and both the alpha subunit and the beta-gamma subunit detach from the receptor. The alpha subunit or the beta-gamma subunit (depending upon the particular G-protein family member involved and the cell type) then activates (or with some G-protein family members, inhibits) one of several enzymes, most commonly adenylate cyclase or phospholipase C. Alternatively, they can open or close a membrane ion channel, altering the electrical properties of the cell; for example, potassium ion channels in heart muscle cells can be opened by G proteins in response to the neurotransmitter acetylcholine.

In cases where adenylate cyclase or phospholipase C is activated, these enzymes catalyze reactions that produce molecules called second messengers, which, through a series of steps, activate proteins that lead to a physiological response (such as contraction of smooth muscle), a biochemical response (such as glycogen synthesis) or a genetic response (such as activating a gene).

Activation of adenylate cyclase causes it to catalyze the conversion of adenosine triphosphate (ATP) to the second messenger cyclic adenosine monophosphate (cAMP), which in turn activates a protein called protein kinase A which, in some cells, moves into the nucleus and phosphorylates and activates transcription factors such as CREB (CRE-binding protein). CREB binds to a specific DNA sequence in the promoter of certain genes called the CRE (cAMP-response element), as well as to other transcription factors, to activate transcription of the gene. In other cells, protein kinase A activates enzymes or other proteins involved in physiological or metabolic responses.

Activation of phospholipase C catalyzes the breakdown of a glycolipid component of the cell membrane called phosphatidylinositol bisphosphate (PIP2) into two second messengers, inositol triphosphate (IP3) and diacylglycerol (DAG). DAG activates a protein called protein kinase C (PK-C), which in turn activates other proteins, leading to various cell responses, including, in certain cells of the immune system, activation of transcription factors which turn on genes involved in the body's immune response to infection. IP3 causes the release of calcium ions stored in the endoplasmic reticulum. These ions bind and activate the protein calmodulin, which activates a variety of proteins, leading in most cases to a physiological response in the cell.

Catalytic Receptors

Catalytic receptors are receptors that function as enzymes, catalyzing specific reactions in the cell. The part of the receptor that is in the cytoplasm (the cytoplasmic domain) has catalytic capability. Binding of a signaling molecule to the external domain of the receptor activates the catalytic activity of the cytoplasmic domain. There are several kinds of catalytic receptors based on the type of reaction they catalyze; these include receptor tyrosine phosphatases, receptor guanylate cyclases, receptor serine/threonine kinases, and receptor tyrosine kinases. Receptor tyrosine kinases (RTKs) are the most common of these.

RTKs are the receptors for many growth factors and at least one hormone. For example, they are the receptors for fibroblast growth factor (FGF), epidermal growth factor (EGF), platelet-derived growth factor (PDGF), nerve growth factor (NGF), and insulin. RTKs play a role in regulating many fundamental processes, such as cell metabolism, the cell cycle, cell proliferation, cell migration, and embryonic development. In most cases, when a ligand binds to this type of receptor, a conformational (shape) change occurs in the receptor so that it binds to another identical receptor-ligand complex to produce a double or dimeric receptor. The dimeric receptor then catalyzes a cytoplasmic reaction in which several tyrosine amino acids in the cytoplasmic domain of the receptor itself are phosphorylated. The phosphorylated tyrosines then function as docking sites for several other proteins, each of which can initiate one of the many branches of the RTK signal transduction pathway, leading to the various cell responses. One of the major branches of the RTK pathway that in many cases results in gene expression begins with the binding of the G protein ras (ras is not one of the trimeric G proteins discussed above) to the activated RTK receptor via adapter proteins. Binding of ras to the adapter proteins activates it by allowing it to bind GTP instead of GDP. Activated ras then phosphorylates the enzyme MEK, which phosphorylates and activates an enzyme of the MAP kinase family. In cases where this enzyme is MAP kinase itself, the enzyme dimerizes, moves into the nucleus, and activates genes, usually many genes, by phosphorylating and activating their transcription factors

Signal Transduction and the Cell Cycle

The biochemical machinery that produces the cell cycle consists of several cyclins whose concentrations rise and fall throughout the cell cycle. Cyclins activate cyclin-dependent kinases (cdk's), which activate the proteins that carry out the events of each stage of the cell cycle. In higher organisms, control of the cell cycle is carried out primarily by growth factors. In the absence of growth factors, many cells will stop at a point in the cell cycle known as the G_1 checkpoint and cease dividing. The cell cycle is started when the cells are exposed to a growth factor. For example, some growth factors start cell division by binding to a membrane receptor and initiating the RTK/MAP kinase signal transduction pathway. The activated transcription factor that results from this pathway activates a gene called *myc*. The protein that is produced from this gene is itself a transcription factor, which activates the cyclin D gene, which produces cyclin D, an important component of the cell cycle biochemical machinery. Cyclin D activates cyclin-dependent kinase 4 (cdk4), which drives the cell into the G_1 phase of the cell cycle. cdk4 also causes an inhibiting molecule called pRB to be removed from a transcription factor for the cyclin E gene. Cyclin E is then produced and activates cyclin-dependent kinase 2 (cdk2), which drives the cell into the S phase of the cell cycle, during which chromosomal DNA is replicated, leading to cell division by mitosis.

Signal Transduction and Cancer

Cancer is caused primarily by uncontrolled cell proliferation. Since many signal transduction pathways lead to cell proliferation, it is not surprising that defects in these pathways can lead to cancer. For example, as described above, many growth factors promote cell proliferation by activating the RTK/MAP kinase signal transduction pathway. In that pathway a series of proteins is activated (ras, MAP kinase, and so on). If a mutation occurred in the gene for one of these, ras for example, such that the mutant ras protein is always activated rather than being activated only when it binds to the receptor, then the cell would always be dividing and cancerous growth could result. Another example would be if the gene for pRB that binds to and inhibits the cyclin E transcription factor were mutated such that the pRB could never bind to the transcription factor; then the cell would divide continuously. Mutations in both ras and pRB are in fact known to cause cancer in humans.

—Robert Chandler

See also: Burkitt's Lymphoma; Cancer; Cell Cycle, The; Cell Division; DNA Replication; Gene Regulation: Bacteria; Model Organism: *Saccharomyces cerevisiae*; Oncogenes; One Gene-One Enzyme Hypothesis; Steroid Hormones; Tumor-Suppressor Genes.

Further Reading

Alberts, Bray, et al. *Molecular Biology of the Cell.* New York: Garland, 2002. A condensed version of one of the standard textbooks in the field of cell biology. One chapter contains essential, basic information about signal transduction.

Cell 103, no. 2 (October 13, 2000): 181-320. A special issue of the journal *Cell* devoted entirely to the topic of cell signaling. The cited pages contain three "minireviews" and eleven reviews of the relatively recent primary literature. A good entry into the primary literature.

Gomperts, Kramer, et al. *Signal Transduction.* San Diego, Calif.: Academic Press, 2002. Provides comprehensive yet readable coverage of signal transduction; contains excellent illustrations and citations of other literature.

Hoch, James A., and Thomas J. Silhavy, eds. *Two-Component Signal Transduction.* Washington, D.C.: ASM Press, 1995. Written for microbiologists working in the areas of gene expression, pathogenesis, and bacterial metabolism, covers the molecular and cellular biology of a wide variety of two-component signal transduction systems in bacteria. Illustrated.

Lodish, Harvey, et al. *Molecular Cell Biology.* New York: W. H. Freeman, 2000. One of the standard textbooks in the field of cell biology. Chapter 20 provides detailed information on signal transduction.

Smallpox

Fields of study: Diseases and syndromes;
Viral genetics

Significance: *Smallpox is a poxvirus disease of humans existing in two forms; the more virulent and frequently lethal form is* Variola major, *and a milder form is* Variola minor. *Smallpox is very contagious, requiring strict quarantine measures and aggressive vaccination programs to contain and eradicate outbreaks. Although smallpox was eradicated globally in 1977, at least two research stocks exist, and there is concern that clandestine-held stocks of the virus may be used as weapons of bioterrorism.*

Key terms

BIOTERRORISM: the use of living organisms as instruments or weapons of terror, such as the deliberate introduction of smallpox, anthrax, or other diseases into civilian populations

POXVIRUS: any of the family of viruses that produces pustules on the surface of the skin

TRANSMISSIBILITY: the rate at which a disease spreads from primary to secondary cases

Definitions

Smallpox is a member of the *Poxviridae* family of viruses, which are the largest and most complex of all known viruses. Poxviruses are named for the characteristic rash or pox lesions that occur during most infections. The poxviruses include a number of familiar diseases such as smallpox, cowpox, rabbitpox, sheeppox, and fowlpox. Two subfamilies of poxviruses are recognized based on their hosts. The orthopoxvirus subfamily comprises viruses that affect vertebrates and includes smallpox; the poxviruses of the subfamily parapoxviruses infect invertebrates, primarily insects. There are two types of variola, the poxvirus that causes smallpox: *Variola major* causes the more virulent and lethal form of smallpox in humans, and *V. minor* causes a milder form of smallpox. Both varieties infect only humans and monkeys. Other names or synonyms for smallpox include alastrim, amaas, Kaffir mil pox, West Indian modified smallpox, and para-smallpox.

History and Symptoms

Historically one of the most devastating and lethal of all human diseases, smallpox is named for the small pustules that occur as a rash over the skin of the victim. Smallpox symptoms include a rash that spreads over the entire body, high fever, chills, aches and pains, and vomiting. The most lethal form, black or hemorrhagic smallpox, results in death within two to six days. The fatality rate varies with health and previous exposure of the local population but ranges from 30 to 90 percent.

Humans have had a long and unfortunate history of association with smallpox. The disease apparently originated in India and spread westward into the Middle East and Northern Africa several thousand years ago. An Egyptian mummy of the Twentieth Dynasty shows the characteristic scarring associated with smallpox. Warriors returning from the Crusades brought the disease back with them. In the following centuries smallpox became endemic throughout much of Europe and became a rite of passage for much of the population—those who contracted smallpox and survived were marked by its scars throughout life. In time, the population built up a partial immunity to the disease. Smallpox was carried by Europeans to the New World and to Australia during the Age of Exploration. It was spread to the immunologically defenseless Amerindians of North America and Aboriginals of Australia with devastating effect and may have contributed to the ease of European settlement following the decimation of tribal peoples in both areas, as it caused widespread death and devastation among the indigenous populations and was at least partly responsible for the depopulation of natives in the newly discovered lands. Before its eradication, smallpox was endemic throughout the world, with major centers of the disease in Africa, Asia, and the Middle East.

Genetics of Smallpox

The poxviruses are the largest and most complex of all the viruses that have so far been identified in animals. The variola virus that causes smallpox has a brick-shaped outer envelope and a dumbbell-shaped core that contains the smallpox genome. The smallpox genome is composed of linear, double-stranded DNA containing more than two hundred genes. Chemically, the smallpox virion consists of 90 percent protein, 3 percent DNA, and 5 percent lipid. The DNA genome codes for several hundred polypeptides, including several transcriptases responsible for replication of the virus within the cells of the host.

Replication of smallpox begins when the virus attaches to the surface of a host cell. After binding to receptors on the plasma membrane of the host cell, the host cell passes the virus into the cytoplasm by endocytosis. Once inside the cell, the virus becomes trapped in a lysosome vesicle in the cytoplasm. The first step in removing its viral coat probably occurs at this stage, as host cell enzymes dissolve the viral envelope. The viral core, containing the DNA, then exits the lysosome and enters the cytoplasm, where the viral genome can be expressed. One of the first steps involves the production of enzymes that degrade the proteins of the viral core, which releases the naked viral DNA into the cytoplasm. Additional transcription takes place, initially producing structural proteins and enzymes, including DNA polymerase, which promotes the replication of the viral DNA. Finally, the late messenger RNA (mRNA) is transcribed, producing additional structural proteins and assembly enzymes that complete virion construction. During viral replication, most host-cell protein synthesis is blocked, because transport of host-cell mRNA molecules though the nuclear envelope into the cytoplasm is prevented.

Newly completed virons exit the host cell through microvilli on the cell surface or fuse with the cell membrane, after which they exit the cell by the process of exocytosis. Once in the tissue, fluids, and bloodstream, the newly

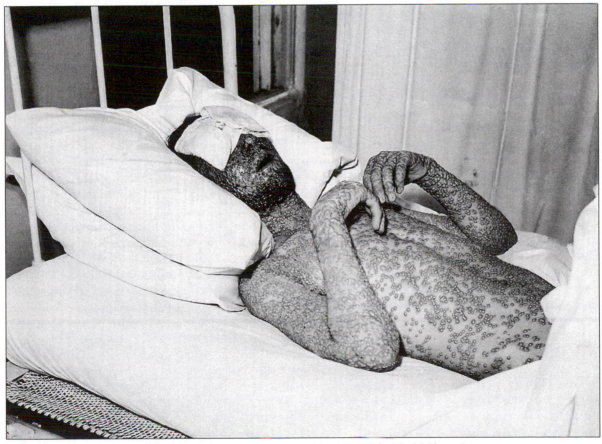

A man with advanced smallpox in 1941. (AP/Wide World Photos)

released and highly infectious viral particles can invade and replicate in other host cells.

Transmission

Smallpox is transmitted from one human to another, either by direct contact or via droplets released into the air during sneezing and coughing fits. The virus does not live long outside the human body and does not reproduce outside the human body. No natural animal carriers of variola other than monkeys, which are also susceptible, are known for the smallpox disease. In extremely rare cases smallpox is transmitted by carriers that are themselves immune to the disease but can transmit the disease to others. Still, only a few droplets settling on another person are sufficient to transmit smallpox. Because of the virulence and mode of transmission, public health regulations specify decontamination procedures. Living quarters, bedding, clothes, and other articles of infected persons must be thoroughly cleansed by heat or with formaldehyde, or destroyed altogether.

Pathogenesis and Symptoms

Infection occurs when the variola virus enters the respiratory mucosa of the nasal or pharyngeal region of the upper respiratory tract of humans. Apparently, only a few viral particles are needed to produce an infection. After a few hours or a few days, the virus migrates to and invades cells in the lymph nodes of the nasopharyngeal region, where it enters the cells, following which rapid reproduction occurs. After a few days, it enters the bloodstream, a condition called viremia. At this time symptoms of smallpox appear. The virus spreads into lymph nodes, spleen, and bone marrow, where reproduction continues rapidly. By the eighth day of infection, the virus is contained in white blood cells or leukocytes, which transmit it to the small blood vessels in the dermis of the skin as well as in the mucosa that lines the mouth and pharynx.

Following an incubation period of about two weeks (the range is between seven and seventeen days), symptoms appear, including high fever, headache, nausea, malaise, and often backache. Accompanying these symptoms is a rash that begins in the mouth and spreads across the face, forearms, trunk, and legs. The rash is first confined to a reddish or purplish swelling of the blood vessels but soon becomes pustular as little round nodules appear on the surface of the body. If the patient recovers, the pustules crust over and the resultant scabs eventually split, which causes scarring of the face.

Death occurs within a few days following the appearance of the rash, most commonly from toxemia caused by variola antigens and various immune complexes circulating in the blood. In some cases the disease is followed by encephalitis. Smallpox fatalities typically occur because of complications such as pneumonia, septicemia, and nephritis (kidney failure). Survivors often suffer from general scarring, ulcers, scarring of the cornea leading to blindness, and skin abscesses. Treatment of survivors with chemotherapy has reduced the severity of many of these complications.

The considerably less virulent form of smallpox, *Variola minor*, produces a much less severe illness characterized by fever, chills, and a milder rash. The same conditions are sometimes seen in patients who have previously been vaccinated or even as a response to vaccinations.

Treatment and Control

Despite decades of research, there is no specific treatment for smallpox other than bed rest and application of antibiotics to prevent secondary infections. Therefore, only prevention of spread by quarantine of infected persons prevents epidemics. Immediate recognition of the disease remains the strongest control measure, followed by vaccination of all health care personnel and others that may come in contact with infected persons.

Widespread and aggressive inoculation programs conducted during the first half of the twentieth century eradicated smallpox from most regions of the world, including North America, Eurasia, and Oceania, largely as a result of the success of the vaccination process originally developed by Edward Jenner. By 1967 smallpox was found only in thirty-three countries and had an annual infection rate of

10 million to 15 million cases. In that year the World Health Organization (WHO) initiated a campaign to eliminate smallpox completely as a human disease, concentrating in Africa, India, and Indonesia. The last case of smallpox in Asia was reported in Bangladesh in 1975 and the last known smallpox victim was recorded in Somalia in October, 1977. Eradication was considered accomplished by 1979. The cost of the eradication campaign was $150 million.

Most researchers conclude that the effective eradication of smallpox was made possible for several reasons: (1) smallpox cases could be quickly and positively identified, (2) there are no natural carriers that serve as disease reservoirs, (3) humans were the only carriers, (4) individuals who survived did not continue to harbor the virus, and (5) the smallpox vaccine proved highly effective.

Recombinant DNA Technology and Vaccinia Viruses

Vaccinia viruses can absorb comparatively large amounts of foreign DNA without losing their ability to replicate, giving rise to the idea that they may provide a vehicle for providing immunity for other viral diseases of humans. One of several ongoing investigations involves insertion of 22-25 kilobase pairs into vaccinia. Experiments using this technique have produced vaccinia strains that encode surface proteins (antigens) of a number of important viruses, including influenza, hepatitis B, and herpesvirus. One possible outcome of these recombinant DNA experiments is the production of vaccinia strains that can serve as vaccines for several viral diseases simultaneously.

Smallpox as a Bioterrorism Weapon

Since its official eradication in 1979, only two stocks of smallpox officially remain; one stock is held at the Centers for Disease Control in Atlanta, Georgia, and the other is kept at VECTOR, Novosibirsk, in central Russia. However, there remains the possibility that clandestine stocks still exist, and these stocks may serve as potential bioterrorism weapons, either to be used against military or civilian populations or to be mounted as international threats. The use of smallpox as a bioterrorism weapon would be classed as an international crime, but prevention of its use is difficult unless all existing stocks can be identified and destroyed.

Smallpox is a potential bioterrorism weapon because of its transmissibility, its known lethality, and the general lack of immunity of much of the global population. Because of its bioterrorism potential, research is now centered on rapid identification methods that enable the early detection of smallpox as well as aggressive vaccination programs for individuals most at risk, who have been identified as health care workers. In addition, smallpox vaccinations were reinstated in 2002 for some U.S. military personnel and some health care workers, essentially those considered at highest risk. The vaccine is made from live but weakened vaccinia virus that is pricked into the skin. The characteristic blister scabs over within three weeks. During this time it is possible to transmit the virus to other parts of the body and to other people. Reactions to the vaccine range from a mild soreness around the vaccination site to more severe effects that may include brain inflammation and a rare and progressive bacterial inflammation called vaccinia that is sometimes fatal. For these reasons, mass vaccinations of the general public have been discouraged.

See also: Anthrax; Bacterial Resistance and Super Bacteria; Biological Weapons; Emerging Diseases; Gene Regulation: Viruses; Hereditary Diseases; Viral Genetics.

—Dwight G. Smith

Further Reading

Anderson, R. M., and R. M. May. *Infectious Diseases of Humans: Dynamics and Control.* Oxford, England: Oxford University Press, 1992. Smallpox and other major diseases of humans are described and discussed.

Brooks, G. F., J. S. Butel, and S. A. Morse. *Medical Microbiology.* 21st ed. Stamford, Conn.: Appleton and Lange, 1998. Includes a summary of biological and medical properties of the virus that causes smallpox.

Fenner, F., D. A. Henderson, I. Arita, Z. Jezek, and I. D. Ladnyi. *Smallpox and Its Eradication.* Geneva, Switzerland: World Health Organization Report, 1988. The detailed story of the eradication of smallpox as a disease of humans from the global scale. Some of this report is technical, but the effort to eradicate smallpox is thoroughly described.

Miller, Judith, Stephen Engelberg, and William Broad. *Germs: Biological Weapons and America's Secret War.* New York: Simon & Schuster, 2001. This book, written by three *New York Times* reporters, explores the ideas and actions of scientists and politicians involved in the past, present, and future of germ warfare. Forty-two pages of notes and a select bibliography.

Preston, R. *The Demon in the Freezer: A True Story.* New York: Random House. 2002. This book, available in both print and audio, explores the use of smallpox stocks for research and evaluates the potential of genetically engineered smallpox as a weapon of mass destruction.

U.S. Department of Defense. *Twenty-first Century Bioterrorism and Germ Weapons: U.S. Army Field Manual for the Treatment of Biological Warfare Agent Casualties (Anthrax, Smallpox, Plague, Viral Fevers, Toxins, Delivery Methods, Detection, Symptoms, Treatment, Equipment).* Washington, D.C.: U.S. Department of Defense Manual, 2002. Available to the public, this is the standard reference manual for members of the Armed Forces Medical Services.

World Health Organization. *Future Research on Smallpox Virus Recommended.* Geneva, Switzerland: World Health Organization Press, 1999. This press release emphasizes the need for smallpox research in light of its potential use as a weapon in the bioterrorism arsenal.

Web Sites of Interest

National Organization for Rare Disorders. http://www.rarediseases.org. Searchable site by type of disorder. Includes background information on smallpox and a list of related resources.

Centers for Disease Control. http://www.bt.cdc.gov/agent/smallpox/index.asp. The CDC's Web page on smallpox includes information on the disease and posts the latest on smallpox vaccines.

Sociobiology

Fields of study: History of genetics; Human genetics and social issues; Population genetics

Significance: *Sociobiology attempts to explain social interactions among members of animal species from an evolutionary perspective. The application of the principles of sociobiology to human social behavior initiated severe criticism and accusations of racism and sexism.*

Key terms

ALTRUISM: the capacity of one individual to behave in a way that benefits another individual of the same species at some cost to the actor

EUSOCIALITY: an extreme form of altruism and kin selection in which most members of the society do not reproduce but rather feed and protect their relatives

KIN SELECTION: a special type of altruistic behavior in which the benefactor is related to the actor

RECIPROCAL ALTRUISM: a type of altruism in which the benefactor may be expected to return the favor of the actor

SOCIETY: a group of individuals of the same species in which members interact in relatively complex ways

History

Sociobiology is best known from the works of Edward O. Wilson, especially his 1975 book *Sociobiology: The New Synthesis.* This work both synthesized the concepts of the field and initiated the controversy over the application of sociobiological ideas to humans. However, the concepts and methods of sociobiology did not start with Wilson; they can be traced to Charles Darwin and others who studied the influence of genetics and evolution on behavior. Sociobiologists attempt to explain the genetics and evolution of social activity of all types, ranging from flocking in birds and herd formation in mammals to more complex social systems such as eusociality. "The new synthesis" attempted to apply genetics, population biology, and evolutionary theory to the study of social systems.

When sociobiological concepts were applied to human sociality, many scientists, especially social scientists, feared a return to scientific theories of racial and gender superiority. They rebelled vigorously against such ideas. Wilson was vilified by many of these scientists, and some observers assert that the term "sociobiology" generated such negative responses that scientists who studied in the field began using other names for it. At least one scientific journal dropped the word "sociobiology" from its title, perhaps in response to its negative connotations. However, the study of sociobiological phenomena existed in the social branches of animal behavior and ethology long before the term was coined. Despite the criticism, research has continued under the name sociobiology as well as other names, such as "behavioral ecology."

Sociobiology and the Understanding of Altruism

Sociobiologists have contributed to the understanding of a number of aspects of social behavior, such as altruism. Illogical in the face of evolutionary theory, apparently altruistic acts can be observed in humans and other animal groups. Darwinian evolution holds that the organism that leaves the largest number of mature offspring will have the greatest influence on the characteristics of the next generation. Under this assumption, altruism should disappear from the population as each individual seeks to maximize its own offspring production. If an individual assists another, it uses energy, time, and material it might have used for its own survival and reproduction and simultaneously contributes energy, time, and material to the survival and reproductive effort of the recipient. As a result, more members of the next generation should be like the assisted organism than like the altruistic one. Should this continue generation after generation, altruism would decrease in the population and selfishness would increase. Yet biologists have cataloged a number of altruistic behaviors.

When a prairie dog "barks," thus warning others of the presence of a hawk, the prairie dog draws the hawk's attention. Should it not just slip into its burrow, out of the hawk's reach?

When a reproductively mature acorn woodpecker stays with its parents to help raise the next generation, the woodpecker is bypassing its own reproduction for one or more years. Should it not leave home and attempt to set up its own nest and hatch its own young? Eusocial species, such as honeybees and naked mole rats, actually have many members who never reproduce; they work their entire lives to support and protect a single queen, several reproductive males, and their offspring. It would seem that all these altruistic situations should produce a decrease in the number of members of the next generation carrying altruistic genes in favor of more members with "selfish" genes.

Sociobiologists have reinterpreted some of these apparently altruistic acts as camouflaged selfishness. The barking prairie dog, for example, may be notifying the hawk that it sees the predator, that it is close to its burrow and cannot be caught; therefore, the hawk would be better off hunting someone else. Perhaps the young acorn woodpecker learns enough from the years of helping to make its fewer reproductive years more successful than its total reproductive success without the training period.

It is difficult, however, to explain the worker honeybee this way. The worker bee never gets an opportunity to reproduce. Sociobiologists explain this and other phenomena by invoking kin selection. Since the worker bees are closely related to the queen (as sisters or daughters), to reproductive males, and to other workers they help feed and protect, they share a large number of genes with them. If they help raise enough brothers and sisters (especially males and queens) to more than make up for the offspring they do not produce themselves, they will actually increase the proportion of individuals similar to themselves more than if they "selfishly" reproduced.

The prairie dog's behavior might be explained this way as well. The organisms the prairie dog is warning are primarily relatives. By warning them, the prairie dog helps preserve copies of its own genes in its relatives. If the cost of the behavior (an occasional barking prairie dog being captured by a hawk because the warning call drew the hawk's attention) is more than compensated for by the number of

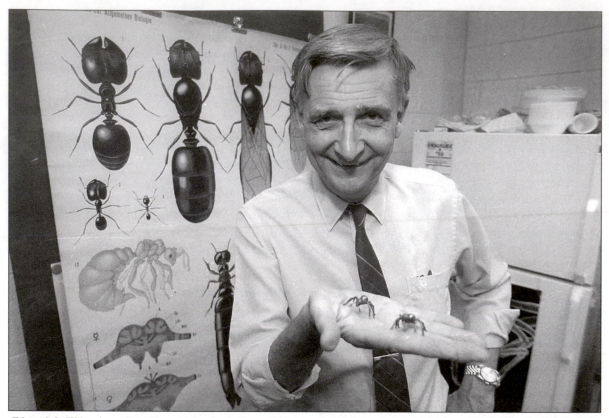

Edward O. Wilson's studies of insect behavior influenced his controversial theories of sociobiology. (AP/Wide World Photos)

relatives saved from the hawk by the warning, kin selection will preserve the behavior. The helper acorn woodpecker's behavior may be explained in similar ways, not as an altruistic act but as a selfish act to favor copies of the helper's genes in its relatives. Another explanation of altruism set forth by sociobiologists is reciprocity or reciprocal altruism: If the prairie dog is sometimes warned by others and returns the favor by calling out a warning when it sees a predator, the prairie dog town will be safer for all prairie dogs.

Opposition to the Application of Sociobiology to Humans

Wilson's new synthesis attempted to incorporate biology, genetics, population biology, and evolution into the study and explanation of social behavior. When the analyses turned to human sociality, critics feared that they would lead back to the sexist, racist, and determinist viewpoints of the early twentieth century. The argument over the relative importance of heredity or environment (nature or nurture) in determining individual success had been more or less decided in favor of the environment, at least by social scientists. Poor people were not poor because they were inherently inferior but because the environment they lived in did not give them an equal chance. Black, Hispanic, and other minority people were not inordinately represented among the poor because they were genetically inferior but because their environment kept them from using their genetic capabilities.

Sociobiologists entered the fray squarely on the side of an appreciable contribution from genetic and evolutionary factors. Few, if any, said that the environment was unimportant in the molding of racial, gender, and individual characteristics; rather, sociobiologists claimed that the genetic and evolutionary history of human individuals and groups played an important role in determining their capabilities, just as they do in other animals. Few, if any, claimed that this meant that one race, gender, or group was superior to another. However, many (if not all) sociobiologists were accused of promoting racist, sexist, and determinist ideas with their application of sociobiological concepts to humans.

Extremists on both sides of the question have confused the issues. Such extremists range from opponents of sociobiological ideas who minimize genetic or evolutionary influence on the human cultural condition to sociobiologists who minimize the role of environmental influences. In at least some minds, extremists in the sociobiological camp have done as much damage to sociobiology as its most ardent opponents. Sociobiology (by that or another name) will continue to contribute to the understanding of the social systems of animals and humans. The biological, genetic, and evolutionary bases of human social systems must be studied. The knowledge obtained may prove to be as enlightening as has sociobiology's contribution to the understanding of social systems in other animals.

—Carl W. Hoagstrom

See also: Aggression; Alcoholism; Altruism; Behavior; Biological Clocks; Biological Determinism; Criminality; Developmental Genetics; Eugenics; Gender Identity; Genetic Engineering: Medical Applications; Genetic Engineering: Social and Ethical Issues; Genetic Screening; Genetic Testing; Genetic Testing: Ethical and Economic Issues; Heredity and Environment; Homosexuality; Human Genetics; Inbreeding and Assortative Mating; Intelligence; Klinefelter Syndrome; Knockout Genetics and Knockout Mice; Miscegenation and Antimiscegenation Laws; Natural Selection; Twin Studies; XYY Syndrome.

Further Reading

Alcock, John. *The Triumph of Sociobiology.* Reprint. New York: Oxford University Press, 2003. Reviews the history of the controversies and debates surrounding Wilson's ideas on sociobiology. Illustrations, bibliography, index.

Blackmore, Susan J. *The Meme Machine.* Foreword by Richard Dawkins. New York: Oxford University Press, 1999. Argues that human behavior and cultural production, such as habits and the making of songs, ideas, and objects, are memetic. That is, they replicate in the form of memes, as do genes, within and between populations. Also argues that memes serve as the foundation of culture. Bibliography, index.

Cartwright, John. *Evolution and Human Behavior: Darwinian Perspectives on Human Nature.* Cambridge, Mass.: MIT Press, 2000. Offers an overview of the key theoretical principles of human sociobiology and evolutionary psychology and shows how they illuminate the ways humans think and behave. Argues that humans think, feel, and act in ways that once enhanced the reproductive success of our ancestors.

Cronk, Lee. *That Complex Whole: Culture and the Evolution of Human Behavior.* Boulder, Colo.: Westview Press, 1999. Discusses the links between behavioral and social scientists, who do not have a basic understanding of the import of culture on human behavior, and anthropologists, who in turn lack a complete understanding of evolutionary biology. Bibliography, index.

Cziko, Gary. *The Things We Do: Using the Lessons of Bernard and Darwin to Understand the What, How, and Why of Our Behavior.* Cambridge: MIT Press, 2000. Contrary to the Newton-inspired idea that humans react to the environment, Cziko argues that humans are less passive and reactive and more active beings, acting on their environments in order to shape their perceptions of the world. Illustrations, bibliography, index.

Segerstråle, Ullica. *Defenders of the Truth: The Battle for Science in the Sociobiology Debate and Beyond.* New York: Oxford University Press, 2000. Addresses Wilson's *Sociobiology* and the ensuing debates on determinism versus free will, nature versus nurture, adaptationism versus environmentalism, and others. Bibliography, index.

Van der Dennen, Johan M. G., David Smillie, and Daniel R. Wilson, eds. *The Darwinian Heritage and Sociobiology.* Westport, Conn.: Praeger, 1999. Interdisciplinary approach to Darwin's influence on sociobiology, and discussions of sociobiological perspectives on war and other forms of conflict, marital relations, and utopia. Illustrations, bibliography, index.

Wilson, Edward O. "The Biological Basis of Morality." *Atlantic Monthly*, April, 1998. Wilson argues that ethical and moral reasoning comes not from outside human nature, as if God-given, but from human nature itself in an ever-changing world.

_____. *On Human Nature.* Cambridge, Mass.: Harvard University Press, 1978. A look at the significance of biology and genetics on how we understand human behaviors, including aggression, sex, and altruism and the institution of religion.

_____. *Sociobiology: The New Synthesis.* Cambridge, Mass.: Belknap Press of Harvard University Press, 1975. The text that brings together Wilson's theories on the genetic, biological, and evolutionary basis of social systems.

Web Site of Interest

The Open Directory Project, Sociobiology. http://dmoz.org/science/biology/socio biology. Comprehensive list of sites devoted to sociobiology, including links to sites covering the science of sociobiology.

Speciation

Field of study: Population genetics

Significance: *Speciation, the biological formation of new species, has produced the wide variety of living things on earth. Although speciation can be caused by other forces or events, natural selection is considered the primary mechanism promoting speciation.*

Key terms

ALLOPATRIC SPECIATION: the genetic divergence of populations caused by separation from each other by a geographic barrier such as a mountain range or an ocean

POPULATION: a group of organisms of the same species in the same place at the same time and thus potentially able to mate; populations are the basic unit of speciation

REPRODUCTIVE ISOLATING MECHANISM: a characteristic that prevents an individual of one species from interbreeding (hybridizing) with a member of another species

SPECIES: a class of organisms with common attributes; individuals are usually able to produce fertile offspring only when mating with members of their own species

SYMPATRIC SPECIATION: the genetic divergence of populations that are not separated geographically

Species Concepts

Before the time of Charles Darwin, physical appearance was the only criterion for classifying an organism. This "typological species concept" was associated with the idea that species never change (fixity of species). This way of defining a species causes problems when males and females of the same species look different (as with peacocks and peahens) or when there are several different color patterns among members of a species (as with many insects). Variability within species, whether it is a visible part of their anatomy, an invisible component of their biochemistry, or another characteristic such as behavior, is an important element in understanding how species evolve.

The "biological species concept" uses reproduction to define a species. It states that a species is composed of individuals that can mate and produce fertile offspring in nature. This concept cannot be used to classify organisms such as bacteria, which do not reproduce sexually. It also cannot be used to classify dead specimens or fossils. This definition emphasizes the uniqueness of each individual (variability) in sexually reproducing species. For example, in the human species (*Homo sapiens*), there are variations in body build, hair color and texture, ability to digest milk sugar (lactose), and many other anatomical, biochemical, and behavioral characteristics. All of these variations are the result of genetic mutations, or changes in genes.

According to evolutionary scientist Ernst Mayr, to a "population thinker," variation is reality and type is an abstraction or average; to a "typological thinker," variation is an illusion and type is the reality. Typological thinking is similar to typecasting or stereotyping, and it cannot explain the actual variability seen in species, just as stereotyping does not recognize the variability seen in people. Additional definitions, such as the "evolutionary species concept," include the continuity of a species' genes through time or other factors not addressed by the biological species concept.

Isolation and Divergence of Populations

Species are composed of unique individuals that are nevertheless similar enough to be able to mate and produce fertile offspring. However, individuals of a species are infrequently in close enough proximity to be able to choose a mate from all opposite-sex members of the same species. Groups of individuals of the same species that are at least potential mates because of proximity are called populations.

The basic type of speciation in most sexually reproducing organisms is believed to be "allopatric," in which geographic isolation (separation) of the species into two or more populations is followed by accumulation of differences (divergence) between the populations that eventually prevent them from interbreeding. These differences are caused primarily by natural selection of characteristics advantageous to populations in different environments. If both populations were in identical environments after geographic isolation, they would be much less likely to diverge or evolve into new species.

Another type of speciation is "sympatric," in which populations are not separated geographically, but reproduction between them cannot occur (reproductive isolation) for some other reason. For example, one population may evolve a mutation that makes the fertilized egg (zygote) resulting from interbreeding with the other population incapable of surviving. Another possibility is a mutation that changes where or when individuals are active so that members of the different populations never encounter one another.

Darwin thought that divergence, and thus speciation, occurred gradually by the slow accumulation of many small adaptations "se-

lected" by the environment. More recently, it has been recognized that a very small population, or even a "founder" individual, may be the genetic basis of a new species that evolves more rapidly. This process, called genetic drift, is essentially random. For example, which member of an insect species is blown to an island by a storm is not determined by genetic differences from other members of the species but by a random event (in this case, the weather). This individual (or small number of individuals) is highly unlikely to contain all of the genetic diversity of the entire species. Thus the new population begins with genetic differences that may be enhanced by its new environment. Speciation proceeds according to the allopatric model, but faster. However, extinction of the new population may also occur.

Plants are able to form new species by hybridization (crossbreeding) more often than are animals. When plants hybridize, postmating incompatibility between the chromosomes of the parents and the offspring may immediately create a new, fertile species rather than a sterile hybrid, as in animals such as the mule. A frequent method of speciation in plants is polyploidy, in which two or more complete sets of chromosomes end up in the offspring. (Usually, one complete set is made up of half of each parent's chromosomes.)

Many species reproduce asexually (without the exchange of genes between individuals that defines sexual reproduction). These include bacteria and some plants, fish, salamanders, insects, rotifers, worms, and other animals. In spite of the fact that reproductive isolation has no meaning in these organisms, they are species whose chromosomes and genes differ from those of their close relatives.

Impact and Applications

Environmentalists and scientists recognize that the biodiversity created by speciation is essential to the functioning of the earth's life-support systems for humans as well as other species. Some practical benefits of biodiversity include medicines, natural air and water purification, air conditioning, and food.

The impact of understanding the genetic basis of evolving species cannot be underestimated. Artificial selection (in which humans decide which individuals of a species survive and reproduce) of plants has produced better food crops (for example, modern corn from teosinte) and alleviated hunger in developing nations by creating new varieties of existing species (for example, rice). Hybridization of animals has resulted in mules and beefaloes for the farm (both of which are sterile hybrids rather than species). Artificial selection of domesticated animals has produced the many breeds of horses, dogs, and cats (each of which is still technically one species). Genetic engineering promises to create crops that resist pests, withstand frost or drought, and contain more nutrients. Finally, understanding the genetics of the evolving human species has broad implications for curing disease and avoiding birth defects.

—*Barbara J. Abraham*

See also: Artificial Selection; Evolutionary Biology; Hardy-Weinberg Law; Hybridization and Introgression; Lateral Gene Transfer; Natural Selection; Polyploidy; Population Genetics; Punctuated Equilibrium.

Further Reading

Crow, Tim J., ed. *The Speciation of Modern Homo Sapiens.* Oxford, England: Oxford University Press, 2002. Chapters cover sexual selection, the question of whether or not *Homo sapiens* speciate on the Y chromosome, and what the Y chromosome might reveal about the origin of humans. Illustrations, bibliography.

Giddings, L. V., Kenneth Y. Kaneshiro, and Wyatt W. Anderson, eds. *Speciation, and the Founder Principle.* General principles of speciation among both plants and animals, with emphasis on the founder principle, covered by seventeen intertionally known geneticists.

Mayr, Ernst. *One Long Argument: Charles Darwin and the Genesis of Modern Evolutionary Thought.* Cambridge, Mass.: Harvard University Press, 1991. Includes a chapter ("How Species Originate") that points out that Darwin's explanation of speciation was limited by his lack of understanding of the origin of genetic variation (mutation and recombination). Illustrations, bibliography, index.

Stem Cells

Fields of study: Cellular biology; Human genetics and social issues

Significance: *Stem cells, which can be manipulated to create unlimited amounts of specialized tissue, may be used to treat a variety of diseases and injuries that have destroyed a patient's cells, tissues, or organs. Stem cells could also be used to gain a better understanding of how genetics works in the early stages of cell development and may play a role in the testing and development of drugs.*

Key terms

ADULT STEM CELL: an undifferentiated cell found among differentiated cells in a tissue or organ of an adult organism

BLASTOCYST: a preimplantation embryo consisting of a hollow ball of two layers of cells

CELL DIFFERENTIATION: the process whereby a precursor cell produces progeny that are capable of expressing a different set of genes

EMBRYONIC STEM CELL: an undifferentiated cell derived from the inner cell mass of a blastocyst

MULTIPOTENCY: the ability of cells to form progeny that can differentiate into one of the different types of cells that form the living organism

PLURIPOTENCY: the ability of a cell to give rise to all the differentiated cell types in an embryo

TOTIPOTENCY: the ability of a single cell to express the full genome in the cells to which it gives rise by cell division

Types of Stem Cells

Stem cells are defined by their ability to renew themselves, their lack of differentiation, and their ability to diversify into other cell types. There are three major classes of stem cells: totipotent, pluripotent, and multipotent. Totipotent cells can differentiate to become all of the cells that make up an embryo, all of the extraembryonic tissues, and all of the post-embryonic tissues and organs. Pluripotent cells have the potential to become almost all of the tissues found in an embryo but are not capable of giving rise to supporting cells and tissues.

Multipotent cells are specialized stem cells capable of giving rise to one class of cells.

A fertilized egg, or zygote, is totipotent. The zygote first divides into two cells about one day after fertilization and becomes an embryo. The embryonic cells remain totipotent for about four days after fertilization. At that point, the embryo consists of about eight cells. As the cells of the embryo continue to divide, they form a hollow sphere. The approximately fifty to one hundred cells on the inner side of the sphere are pluripotent and will continue developing to form the embryo, while the cells on the outer surface will give rise to the extraembryonic tissues, such as the placenta and the umbilical cord.

Multipotent stem cells are found in a variety of tissues in adult mammals and are sometimes referred to as adult stem cells. They are specialized stem cells that are committed to giving rise to cells that have a particular function. Identities of some multipotent stem cells have been confirmed. Hematopoietic stem cells give rise to all the types of blood cells. Mesenchymal stem cells in the bone marrow give rise to a variety of cell types: bone cells, cartilage cells, fat cells, and other kinds of connective tissue cells such as those in tendons. Neural stem cells in the brain give rise to its three major cell types: nerve cells (neurons) and two categories of nonneuronal cells, astrocytes and oligodendrocytes. Skin stem cells occur in the basal layer of the epidermis and at the base of hair follicles. The epidermal stem cells give rise to keratinocytes, which migrate to the surface of the skin and form a protective layer. The follicular stem cells can give rise to both the hair follicle and the epidermis.

Stem cells in adult mammalian tissues are rare and difficult to isolate. There is considerable debate concerning the plasticity of stem cells in adults. Plasticity is the ability of multipotent cells to exhibit pluripotency, such as the capacity of hematopoietic stem cells to differentiate into neurons.

Behavior in Cell Culture

During the 1980's researchers first established in vitro culture conditions that allowed embryonic stem cells to divide without differentiating. Embryonic stem cells are relatively easy to grow in culture but appear to be genetically unstable; mice cloned from embryonic stem cells by nuclear transfer suffered many genetic defects as a result of the genetic instability of the embryonic stem cells. As embryonic stem cells divide in culture, they lose the tags that tell an imprinted gene to be either turned on or turned off during development. Researchers have found that even clones made from sister stem cells show differences in their gene expression. However, these genetic changes, while having defined roles in fetal development, may have little significance in therapeutic uses, because the genes involved do not serve a critical role in adult differentiated cells.

Embryonic stem cells in laboratory bottles, 2001. (AP/Wide World Photos)

Paraplegic protesters in Washington, D.C., staged this mock "hanging in the balance" in April, 2002, to dramatize the need for embryonic stem cell research and to urge Congress to allow it. Embryonic stem cell research was subsequently limited to an existing number of cell lines already being investigated. (AP/Wide World Photos)

Unlike embryonic stem cells, adult stem cells do not divide prolifically in culture. When these stem cells do divide in culture, their division is unlike that of most cells. Generally, when a cell divides in culture, the two daughter cells produced are identical in appearance as well as in patterns of gene expression. However, when stem cells divide in culture, at least one of the daughter cells retains its stem cell culture while the other daughter cell is frequently a transit cell destined to produce a terminally differentiated lineage. The genes expressed in a stem cell and a transit cell are significantly different. Therefore a culture of adult stem cells may become heterogeneous in a short time.

Potential Therapeutic Issues

Although stem cells have significant use as models for early embryonic development, an-

other major research thrust has been for therapeutic uses. Stem cell therapy has been limited almost exclusively to multipotent stem cells obtained from umbilical cord blood, bone marrow, or peripheral blood. These stem cells are most commonly used to assist in hematopoietic (blood) and immune system recovery following high-dose chemotherapy or radiation therapy for malignant and nonmalignant diseases such as leukemia and certain immune and genetic disorders. For stem cell transplants to succeed, the donated stem cells must repopulate or engraft the recipient's bone marrow, where they will provide a new source of essential blood and immune system cells.

In addition to the uses of stem cells in cancer treatment, the isolation and characterization of stem cells and in-depth study of their molecular and cellular biology may help scientists understand why cancer cells, which have certain properties of stem cells, survive despite very aggressive treatments. Once the cancer cell's ability to renew itself is understood, scientists can develop strategies for circumventing this property.

Research efforts are under way to improve and expand the use of stem cells in treating and potentially curing human diseases. Possible therapeutic uses of stem cells include treatment of autoimmune diseases such as muscular dystrophy, multiple sclerosis, and rheumatoid arthritis; repair of tissues damaged during stroke, spinal cord injury, or myocardial infarction; treatment of neurodegenerative diseases such as amyotrophic lateral sclerosis (ALS, commonly called Lou Gehrig's disease) and numerous neurological conditions such as Parkinson's, Huntington's, and Alzheimer's diseases; and replacement of insulin-secreting cells in diabetics.

Stem cells may also find use in the field of gene therapy, where a gene that provides a missing or necessary protein is introduced into an organ for a therapeutic effect. One of the most difficult problems in gene therapy studies has been the loss of expression (or insufficient expression) following introduction of the gene into more differentiated cells. Introduction of the gene into stem cells to achieve sufficient long-term expression would be a major advance. In addition, the stem cell is clearly a more versatile target cell for gene therapy, since it can be manipulated to become theoretically any tissue. A single gene transfer into a pluripotent stem cell could enable scientists to generate stem cells for blood, skin, liver, or even brain targets.

Ethical Issues Concerning Use

Stem cell research, particularly embryonic stem cell research, has unleashed a storm of controversy. One primary controversy surrounding the use of embryonic stem cells is based on the belief by opponents that a fertil-

A group of pro-life demonstrators urge a ban on embryonic stem cell research, seeing such investigations as tantamount to baby killing. (AP/Wide World Photos)

ized egg is fundamentally a human being with rights and interests that need to be protected. Those who oppose stem cell research do not want fetuses and fertilized eggs used for research purposes. Others accept the special status of an embryo as a potential human being yet argue that the respect due to the embryo increases as it develops and that this respect, in the early stages in particular, may properly be weighed against the potential benefits arising from the proposed research.

Another ethical issue concerns the method by which embryonic stem cells are obtained. Embryonic stem cells are isolated from two sources: surplus embryos produced by in vitro fertilization and embryos produced by somatic cell nuclear transfer (SCNT), often referred to as therapeutic cloning. In SCNT, genetic material from a cell in an adult's body is fused with an enucleated egg cell. With the right conditions, this new cell can then develop into an embryo from which stem cells could be harvested. Opponents argue that therapeutic cloning is the first step on the slippery slope to reproductive cloning, the use of SCNT to create a new adult organism. Proponents maintain that producing stem cells by SCNT using genetic material from the patient will eliminate the possibility of rejection when the resulting stem cells are returned to the patient.

Legal Status

On August 9, 2001, President George W. Bush announced that federal funds could be used to support research using the sixty human embryonic stem cell lines that had been derived before that date. However, there were no restrictions placed on the types of research that could be conducted on mouse embryonic stem cell lines and no federal law or policy prohibiting the private sector from isolating stem cells from human embryos. Several states have introduced legislation to encourage research on stem cells taken from human embryos.

As of March, 2003, neither reproductive cloning nor therapeutic cloning was forbidden by law in the United States. Congress was debating competing legislation; one bill proposed to ban both types of cloning, while an alternative proposal would ban only reproductive cloning.

A number of states already have laws that ban human cloning for reproductive purposes, while a small number of states forbid cloning of embryos for stem cells as well.

—Lisa M. Sardinia

See also: Aging; Alzheimer's Disease; Autoimmune Disorders; Bioethics; Cancer; Cell Culture: Animal Cells; Cell Culture: Plant Cells; Cell Cycle, The; Cell Division; Cloning; Cloning: Ethical Issues; Cloning Vectors; Cystic Fibrosis; Developmental Genetics; Eugenics; Eugenics: Nazi Germany; Gene Therapy; Gene Therapy: Ethical and Economic Issues; Genetic Engineering: Medical Applications; Huntington's Disease; In Vitro Fertilization and Embryo Transfer; Infertility; Knockout Genetics and Knockout Mice; Model Organism: *Mus musculus*; Organ Transplants and HLA Genes; Totipotency; Transgenic Organisms.

Further Reading

Holland, Suzanne, Karen Lebacqz, and Laurie Zoloth, eds. *The Human Embryonic Stem Cell Debate: Science, Ethics, and Public Policy (Basic Bioethics)*. Cambridge, Mass.: MIT Press, 2001. A collection of twenty essays organized into four sections: basic science and history of stem cell research, ethics, religious perspectives, and public policy.

Kaji, Eugene H., and Jeffrey M. Leiden. "Gene and Stem Cell Therapies." *Journal of the American Medical Association* 285, no. 5 (2001): 545-550. An overview of stem cells from a clinical viewpoint. Includes discussion of the feasibility of stem cell therapy, future research, and ethical issues.

Kiessling, Ann, and Scott C. Anderson. *Human Embryonic Stem Cells: An Introduction to the Science and Therapeutic Potential*. Boston: Jones and Bartlett, 2003. In the context of the social debate and public policy of the George W. Bush administration, addresses the various stem cell research from the perspectives of many disciplines, from cell biology, embryology, and endocrinology to transplantation medicine.

Marshak, Daniel R., Richard L. Gardner, and David Gottlieb, eds. *Stem Cell Biology*. Cold Spring Harbor, N.Y.: Cold Spring Harbor

Laboratory Press, 2002. Contains papers on early embryonic development, cell cycle controls, embryonal carcinoma cells as embryonic stem cells, stem cells of human adult bone marrow, intestinal epithelial stem cells, and much more, designed for researchers new to the field of stem cell biology.

Rao, Mahendra S., ed. *Stem Cells and CNS Development*. Totowa, N.J.: Humana Press, 2001. Collection of papers on neural stem cells, including multipotent cells in both embryos and adults, transplant therapy, drug and gene discovery, and much more. Designed for scientists.

Web Site of Interest

National Institutes of Health, Stem Cell Information. http://stemcells.nih.gov. Government site covering stem cell basics, the science of stem cell research, and links to related resources.

Sterilization Laws

Field of study: Human genetics and social issues

Significance: *Forced sterilization for eugenic reasons became legal throughout much of the United States and many parts of the world during the first half of the twentieth century. Though sterilization is an ineffective mechanism for changing the genetic makeup of a population, sterilization laws remain in effect in many states in the United States and other countries throughout the world.*

Key terms

NEGATIVE EUGENICS: the effort to improve the human species by discouraging or eliminating reproduction among those deemed to be socially or physically unfit

POSITIVE EUGENICS: the effort to encourage more prolific breeding among "gifted" individuals

STERILIZATION: an operation to make reproduction impossible; in tubal ligation, doctors sever the Fallopian tubes so that a woman cannot conceive a child

The Eugenics Movement and Sterilization Laws

The founder of the eugenics movement is considered to be Sir Francis Galton, who carried out extensive genetic studies of human traits. He thought that the human race would be improved by encouraging humans with desirable traits (such as intelligence, good character, and musical ability) to have more children than those people with less desirable traits (positive eugenics). With the development of Mendelian genetics shortly after the beginning of the twentieth century, research on improving the genetic quality of plants and animals was in full swing. Success with plants and domestic animals made it inevitable that interest would develop in applying those principles to the improvement of human beings. As some human traits became known to be under the control of single genes, some geneticists began to claim that all sorts of traits (including many behavioral traits and even social characteristics and preferences) were under the control of a single gene with little regard for the possible impact of environmental factors.

The Eugenics Record Office at Cold Springs Harbor, New York, was set up by Charles Davenport to gather and collate information on human traits. The eugenics movement became a powerful political force that led to the creation and implementation of laws restricting immigration and regulating reproduction. Some geneticists and politicians reasoned that since mental retardation and other "undesirable" behavioral and physical traits were affected by genes, society had an obligation and a moral right to restrict the reproduction of individuals with "bad genes" (negative genetics).

The state of Indiana passed the first sterilization law in 1907, which permitted the involuntary sterilization of inmates in state institutions. Inmates included not only "imbeciles," "idiots," and others with varying degrees of mental retardation (described as "feeble-minded") but also people who were committed for behavioral problems such as criminality, swearing, and slovenliness. By 1911, similar laws had been passed in six states, and, by the end of the 1920's, twenty-four states had similar sterilization laws. Although not necessarily strictly en-

In 1927, the U.S. Supreme Court, in its Buck v. Bell *decision, supported the eugenic principle that states could use involuntary sterilization to eliminate genetic defects from the population. The result was the sterilization of more than sixty thousand mainly young people deemed to be weak, "feebleminded," or otherwise genetically inferior. Two sterilized residents of Lynchburg, Virginia, where many such sterilizations occurred, observe a historical marker that commemorates the tragic decision.* (AP/Wide World Photos)

forced, twenty-two states currently have sterilization laws on the books.

The U.S. Supreme Court, in its 1927 *Buck v. Bill* decision, supported the eugenic principle that states could use involuntary sterilization to eliminate genetic defects from the population. The vote of the Court was eight to one. The court's reasoning went as follows:

> We have seen more than once that the public welfare may call upon the best citizens for their lives. It would be strange if it could not call upon those who already sap the strength of the state for these lesser sacrifices, often not felt to be such by those concerned, in order to prevent our being swamped with incompetence. It is better for all the world, if instead of waiting to execute degenerate offspring for crime, or to let them starve for their imbecility, society can prevent those who are manifestly unfit from continuing their kind. The principle that sustains compulsory vaccination is broad enough to cover cutting the Fallopian tubes.

Ironically, the sterilization laws of the United States and Canada served as models for the eugenics movement in Nazi Germany in its program to ensure so-called racial purity and superiority.

Impact and Applications

Two problems associated with eugenics are the subjective nature of deciding which traits are desirable and determining who should decide. These concerns aside, the question of whether there is a sound scientific basis for the desire to manipulate the human gene pool remains. Does the sterilization of individuals who are mentally retarded or who have some other mental or physical defect improve the human genetic composition? Involuntary sterilization of affected individuals would quickly reduce the incidence of dominant genetic traits. Individuals who were homozygous for recessive traits would also be eliminated. However, most harmful recessive genes are carried by individuals who appear normal and, therefore, would not be "obvious" for sterilization purposes. These "normal" people would continue to pass the "bad" gene on to the next generation, and a certain number of affected people would again be born. It would take an extraordinary number of generations to significantly reduce the frequency of harmful genes.

Although the number of involuntary sterilizations in the United States is now minimal, the impact sterilization laws had on the population through 1960 was far-reaching, as nearly sixty thousand people were sterilized. Other countries also had laws that allowed forced sterilizations, with many programs continuing into the 1970's. The province of Alberta, Canada, sterilized three thousand people before its law was repealed. Another sixty thousand were sterilized in Sweden. The story of sterilization and "euthanasia" in Germany needs no retelling. With the ability to decipher the human genome and implement improved genetic testing procedures, a danger exists that new programs of eugenics and involuntary sterilization might once again emerge.

—*Donald J. Nash*

See also: Criminality; Eugenics; Eugenics: Nazi Germany; Hardy-Weinberg Law; Miscegenation and Antimiscegenation Laws; Prion Diseases: Kuru and Creutzfeldt-Jakob Syndrome; Race.

Further Reading

Campbell, Annily. *Childfree and Sterilized: Women's Decisions and Medical Responses.* New York: Cassell, 1999. Explores the lives of twenty-three women who chose sterilization over bearing children and the prejudices and stereotypes they faced from the medical profession, which often deemed sterilization pathological. Bibliography, index.

Gallagher, Nancy L. *Breeding Better Vermonters: The Eugenics Program in the Green Mountain State.* Hanover, N.H.: University Press of New England, 2000. A biologist looks at the science of eugenics and the social, ethnic, and religious tensions brought about by the Eugenics Survey of Vermont, an organization in existence from 1925 to 1936.

Kevles, Daniel J. *In the Name of Eugenics: Genetics and the Uses of Human Heredity.* Cambridge, Mass.: Harvard University Press, 1995. A comprehensive introduction to the history of the eugenics movement and the development of sterilization laws. Discusses genetics both as a science and as a social and political perspective, and how the two often collide to muddy the boundaries of science and opinion.

Web Sites of Interest

Cold Spring Harbor Laboratory, Image Archive on the American Eugenics Movemement. http://www.eugenicsarchive.org/eugenics. Comprehensive and extensively illustrated site that covers the eugenics movement in the United States, including sterilization laws.

University of Vermont, Vermont Eugenics: A Documentary History. http://www.uvm.edu/~eugenics/sterilizationdl.html. A listing of original documents related to sterilization and eugenics in the United States, including a statement from the American Eugenics Society (1926) and related newspaper articles.

Steroid Hormones

Fields of study: Developmental genetics; Molecular genetics

Significance: *Steroid hormones—hormones containing a steroid ring derived from cholesterol—are important for many processes that control sex determination, reproduction, behavior, and metabolism. Mutations in the genes that produce or*

regulate the action of specific steroid hormones may lead to infertility, sterility, sex determination, osteoporosis, autoimmune diseases, heart abnormalities, and breast, uterine, and prostate cancer.

Key terms

ANABOLIC STEROIDS: drugs derived from androgens and used to enhance performance in sports

ANDROGENS: steroid hormones that cause masculinization

ESTROGENS: steroid hormones that produce female characteristics

GLUCOCORTICOIDS: steroid hormones that respond to stress and maintain sugar, salt, and body fluid levels

HORMONES: chemical messengers produced by endocrine glands and secreted into the blood

MINERALOCORTICOIDS: a group of steroid hormones important for maintenance of salt and water balance

PROGESTINS: steroid hormones important for pregnancy and breast development

TESTOSTERONE: the principal androgen, produced by the testes and responsible for male secondary sexual characteristics

Steroid Hormone Characteristics and Function

Steroid hormones represent a group of hormones that all contain a characteristic "steroid" ring structure. This steroid ring is derived from cholesterol, and cholesterol is the starting material for the production of different steroid hormones. Steroid hormones, like other types of hormones, are secreted by endocrine glands into the bloodstream and travel throughout the body before having an effect. All steroid hormones, although specific for the regulation of certain genes, function in a similar manner. Because steroid hormones are derived from cholesterol, they have the unique ability to diffuse through a cell's outer plasma membrane. Inside the cell, the steroid hormone binds to its specific receptor in the cytoplasm. Upon binding, the newly formed hormone-receptor complex relocates to the nucleus. In the nucleus, the hormone-receptor complex binds to the DNA in the promoter region of certain

genes at specific nucleotide sequences termed hormone-responsive elements. The binding of the hormone-receptor complex to hormone-responsive elements causes the increased production of transcription and protein production in most cases. In some instances, binding to a specific hormone-responsive element will stop the production of proteins that are usually made in the absence of the hormone.

There are two types (sex steroid and adrenal steroid) and five classes of steroid hormones. The sex steroid hormones include the androgens, estrogens, and progestins and are produced by the male testes (androgens) and female ovaries. Adrenal steroid hormones include glucocorticoids and mineralocorticoids and are produced by the adrenal glands.

Sex Steroid Hormones

Sex steroid hormone genes are responsible for determining the sex and development of males and females. Androgens are a group of steroid hormones that cause masculinization. The principal androgen is testosterone, which is produced by the testes and is responsible for male secondary sexual characteristics (growth of facial and pubic hair, deepening of voice, sperm production). Estrogens are sex steroid hormones produced in the ovaries and cause femininization. In addition, estrogens control calcium content in the bones, modulate other hormones produced in the ovary, modify sexual behavior, regulate growth of secondary sex characteristics (menstrual periods, breast development, pubic hair) and are essential for pregnancy to occur. The most potent estrogen is 17-beta estradiol. Progestins, including progesterone, are also sex steroid hormones. Progesterone is important for proper breast development and normal and healthy pregnancies; it functions in the mother to alter endometrial cells so the embryo can implant. The loss of progesterone at the end of a pregnancy aids in the beginning of uterine contractions.

Anabolic steroids are drugs derived from the male steroid hormone testosterone and were developed in the late 1930's to treat hypogonadism in men, a condition that results in insufficient testosterone production by the testes. During this same period, scientists discovered that anabolic steroids also increased the muscle mass in animals. These findings led to the use of anabolic steroids by bodybuilders, weight-lifters, and other athletes to increase muscle mass and enhance performance. Anabolic steroid use can seriously affect the long-term health of an individual and in women results in masculinization.

Adrenal Steroid Hormones

Adrenal steroid hormones are secreted from the adrenal cortex and are important for many bodily functions, including response to stress, maintenance of blood sugar levels, fluid balance, and electrolytes. The glucocorticoids represent one class of adrenal steroid hormone. The most important, cortisol, performs critically important functions; it helps to maintain blood pressure and can decrease the response of the body's immune system. Cortisol can also elevate blood sugar levels and helps to control the amount of water in the body. Elevated cortisol helps the body respond to stress. The glucocorticoids cortisone and hydrocortisone are used as anti-inflammatory drugs to control itching, swelling, pain, and other inflammatory reactions. Prednisone and prednisolone, also members of the glucocorticoid class of hormones, are the broadest anti-inflammatory and immunosuppressive medications available.

The second class of adrenal steroid hormones is the mineralocorticoids, including aldosterone, which helps maintain salt and water balance and increases blood pressure. Aldosterone is crucial for retaining sodium in the kidney, salivary glands, sweat glands, and colon.

Genetic Defects Affecting Sex Steroid Hormones

Defects in the genes involved in the production of sex steroid hormones can have serious consequences. Mutations in the androgen receptor, the receptor for testosterone, result in testicular feminization syndrome. In this syndrome, the individual has the genes of a male (XY) but develops, behaves, and appears female. Other gene defects in androgen biosynthesis often result in sterility. Genetic defects in estrogen receptors or estradiol biosynthesis lead to infertility. Reduced levels of estradiol have also been linked to bone loss (osteoporosis) and infertility, whereas excessive levels are associated with an increased risk of breast and uterine cancer. Similarly, genetic mutations in the progesterone production pathway or the progesterone receptor are associated with infertility. In addition, bone loss is one of the most serious results of progesterone deficiency, made worse by inappropriate diet and lack of exercise.

Genetic Defects Affecting Adrenal Steroid Hormones

Genetic abnormalities in adrenal steroid hormone biosynthesis are known to cause hypertension in some cases of congenital adrenal hyperplasia (CAH). In people with this condition, hypertension usually accompanies a characteristic phenotype with abnormal sexual differentiation. CAH is a family of autosomal recessive disorders of adrenal steroidogenesis. Each disorder has a specific pattern of hormonal abnormalities resulting from a deficiency of one of the enzymes necessary for cortisol synthesis. The most common form of CAH is 21-hydroxylase deficiency; however, in all forms, cortisol production is impaired, which results in an increase in adrenocorticotropin and the overproduction of androgen steroids.

There are two major forms of 21-hydroxylase deficiency. Classic CAH deficiency results in masculinized girls that are born with genital ambiguity and may possess both female and male genitalia. Nonclassic 21-hydroxylase deficiency does not produce ambiguous genitalia in female infants but may result in premature puberty, short stature, menstrual irregularities or lack of a menstrual cycle, and infertility. Familial glucocorticoid deficiency (FGD) is an extremely rare, genetic autosomal recessive condition in which a part of the adrenal glands are destroyed. These changes result in very low levels of cortisol. Although this disease is easily treatable if recognized, when left untreated it is often fatal or can lead to severe mental disability.

Recently, the genetic basis of four forms of severe hypertension transmitted on an autosomal basis has been determined. All of these conditions are characterized by salt-sensitive increases in blood pressure, indicating an increased mineralocorticoid effect. The four disorders—aldosteronism, mineralocorticoid excess syndrome, activating mutation of the mineralocorticoid receptor, and Liddle syndrome—are a consequence of either abnormal biosynthesis, abnormal metabolism, or abnormal action of steroid hormones and the development of hypertension. Adrenal insufficiency is known as Addison's disease and causes death within two weeks unless treated. Classical Addison's disease results from a loss of both cortisol and aldosterone secretion as a result of the near total or total destruction of both adrenal glands.

—*Thomas L. Brown*

See also: Aggression; Allergies; Autoimmune Disorders; Behavior; Cancer; Gender Identity; Heart Disease; Hermaphrodites; Human Genetics; Metafemales; Pseudohermaphrodites; Testicular Feminization Syndrome; X Chromosome Inactivation; XYY Syndrome.

Further Reading

Ethier, Stephen P., ed. *Endocrine Oncology.* Humana Press, 2000. Experts provide chapters on cancers of the breast, prostate, endometrium, and ovary from cellular and molecular perspectives, including the way that steroid hormones function in both normal processes and pathogenesis.

Freedman, Leonard P., and M. Karin, eds. *Molecular Biology of Steroid and Nuclear Hormone Receptors.* Boston: Birkhauser, 1999. A molecular perspective on steroid functions in both normal and cancerous cells.

Khan, Sohaib A., and George M. Stancel, eds. *Protooncogenes and Growth Factors in Steroid Hormone Induced Growth and Differentiation.* Boca Raton, Fla.: CRC Press, 1994. Experts from cancer centers discuss the roles of steroid hormones in cancer from the perspectives of biochemistry, physiology, development, genetics, endocrinology, and other disciplines.

Moudgil, V. K., ed. *Steroid Hormone Receptors: Basic and Clinical Aspects.* Boston: Birkhauser, 1994. A scientific researcher examines the structural and functional alterations in steroid hormone receptors induced by phosphorylation, and hormonal and antihormonal ligands.

National Institutes of Health. *Steroid Abuse and Addiction.* NIH 00-3721. Bethesda, Md.: Author, 2000. This pamphlet outines the dangers of steroid use for unapproved purposes such as bodybuilding.

Tilly, J., J. F. Strauss III, and M. Tenniswood, eds. *Cell Death in Reproductive Physiology.* New York: Springer, 1997. Describes the selective death of steroid-producing tissues.

Wynn, Ralph M., and W. Jollie, eds. *Biology of the Uterus.* 2d rev. ed. Boston: Kluwer Academic, 1989. Reviews the basic biology of pregnancy and the role of sex steroid hormones in pregancy.

Swine Flu

Fields of study: Diseases and syndromes; Viral genetics

Significance: *The swine flu outbreak of 1918 was the most lethal worldwide epidemic known to humankind. Constant genetic changes in the influenza virus that caused the disease keep alive the potential for such another pandemic to occur in the future.*

Key terms

ANTIGENIC DRIFT: minor changes in the H and N proteins of the influenza virus

ANTIGENIC SHIFT: the acquisition by a strain of influenza virus of a different H and/or N protein

HEMAGLUTTININ (H): a protein necessary for entry of the influenza virus into a host cell

NEURAMINIDASE (N): a protein necessary for exit of the influenza virus from an infected cell

PANDEMIC: a worldwide outbreak of a particular disease

Virus Structure and Replication

Swine flu is a respiratory disease of humans caused by the H1N1 subtype of influenza A virus. Various types of influenza A virus can be found in humans, birds, swine, and other animals. Human disease is spread most commonly person to person and, rarely, from animal to person. Although some antiviral drugs are available, treatment involves mainly supportive therapy. Vaccination prevents disease, but the genetic nature of the virus requires that vaccinations must be given annually to be effective. Advances in genetic technology are helping scientists to understand why the swine flu virus caused such a devastating epidemic in 1918 and to develop treatment and prevention strategies that will be effective against future potential epidemics of swine flu.

The swine flu virus is composed of eight segments of RNA surrounded by a lipid envelope. Embedded in the envelope are two proteins essential for viral replication, known as hemagluttinin (H) and neuraminidase (N). The infectious cycle begins when hemagluttinin binds to the surface of the host cell. Next, the viral envelope fuses with the host cell mem-

brane and the RNA is released. Inside the host cell viral RNA is replicated and new proteins are synthesized. Newly assembled virus particles bud through the host cell membrane to acquire their envelopes. Neuraminidase keeps the new virus particles from sticking to the dying host cell so they are free to infect surrounding cells. The cycle continues until either the host is dead or the host's immune system stops the spread of the virus. The majority of the host's immune response to infection is directed against the H and N proteins.

Type A influenza viruses are continuously changing in the amino acid sequence of their H and N proteins by a process called antigenic drift. These changes occur because mistakes are made during the replication of the viral RNA that codes for these proteins. Changes that result in H and N proteins that are no longer recognized by the host's immune system al-

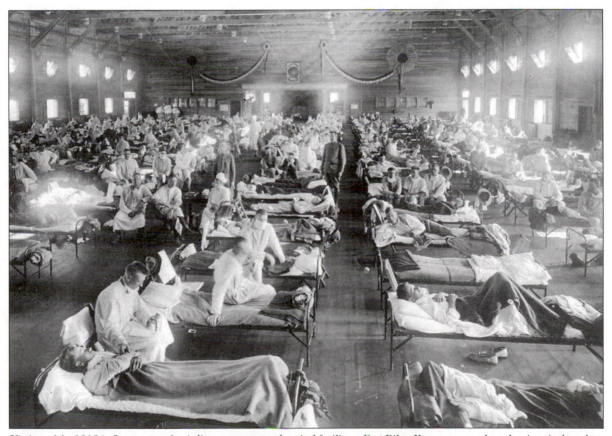

Victims of the 1918 influenza pandemic line an emergency hospital facility at Fort Riley, Kansas, near where the virus is thought to have originated. The pandemic caused at least 20 million deaths worldwide. (AP/Wide World Photos)

low that strain of virus to become prevalent over the strains of virus that are recognized. Antigenic drift is the reason that influenza vaccines are effective only for the year in which they are made. By the following year, the prevalent viruses have changed enough that the population is no longer immune to them.

Type A influenza viruses also change abruptly on a more sporadic basis due to antigenic shift. One type of antigenic shift, called reassortment, occurs when two different strains of influenza virus infect the same host cell. Because the genome is segmented, it is possible for RNA segments from one virus to get mixed up with RNA segments from the second virus when the new virus particles are made. Thus, a new virus containing genes from both viruses can arise. The influenza pandemics of so-called Asian flu in 1957 and Hong Kong flu in 1968 were a result of reassortment. A second type of antigenic shift occurs when an animal influenza virus jumps directly into the human population as occurred in the 1997 avian flu and the 1999 A(H9N2) outbreaks.

History

In 1918, an epidemic of swine flu killed more than 500,000 people in the United States and between 20 million and 50 million people worldwide—more that any other disease in such a short period of time in the history of humankind. After the influenza virus was isolated in 1933, scientists used blood tests to determine that a type A(H1N1) virus had caused the pandemic. In 1976, a second outbreak of A(H1N1) influenza was discovered in the United States in both humans and pigs (hence the name swine flu). The United States mobilized a massive vaccination program, but the predicted epidemic never followed.

Advances in genetic technology have enabled scientists to study RNA from the actual virus that caused the 1918 epidemic. Influenza genes were recovered from samples of the lung tissue of three victims by reverse transcription and polymerase chain reaction. A few genes have been sequenced and compared to known sequences of viral RNA from more recent outbreaks of influenza. It is not yet clear, however, what made the 1918 strain of virus so deadly.

Vaccinations

Constant genetic changes in influenza viruses dictate the development of new vaccines every year. The Global Influenza Surveillance Network, an arm of the World Health Organization (WHO), monitors viruses circulating in humans and identifies new strains, recommending annually a vaccine that targets the three most prevalent strains in circulation. Current research looks for vaccines that would be effective against all strains of influenza, so that new vaccines would not need to be developed each year or at least could be produced more quickly in case of a pandemic.

Future

Influenza experts agree that another pandemic is likely to happen. If a new strain of flu virus appears after antigenic shift against which the human population has no immunity, and that strain can cause illness and spread easily from person to person, an influenza pandemic can occur. Continuous global surveillance of influenza outbreaks, accompanied by full exchanges of information by national governments and their health agencies, is the key to identifying and preventing another pandemic. Advances in genetic technology will help solve the mystery of the 1918 swine flu and make improvements in vaccines and antiviral drugs that could help minimize an epidemic if another one occurs.

—Vicki J. Isola

See also: Antibodies; Bacterial Genetics and Cell Structure; Bacterial Resistance and Super Bacteria; Down Syndrome; Emerging Diseases; Gene Regulation: Viruses; Human Genome Project; Organ Transplants and HLA Genes; Restriction Enzymes; RNA Structure and Function; Smallpox.

Further Reading

Kolata, Gina B. *Flu: The Story of the Great Influenza Pandemic of 1918 and the Search for the Virus That Caused It.* New York: Farrar, Straus and Giroux, 1999. This story is important in understanding the current strategies in surveillance, prevention, and treatment of influenza.
Laver, W. Graeme, et al. "Disarming Flu Vi-

ruses." *Scientific American* 280, no. 1 (1999): 78-87. Describes the viral replication cycle, vaccine strategies, and virulence theories. Illustrated by helpful diagrams.

Taubenberger, J. K., et al. "Initial Genetic Characterization of the 1918 'Spanish' Influenza Virus." *Science* 275 (1997): 1793-1796. Describes the historic use of genetic technology to sequence the virus from the RNA of a victim of the 1918 pandemic.

Web Site of Interest

National Center for Infectious Diseases. http://www.cdc.gov/ncidod/diseases/flu/fluvirus.htm. This page on influenza provides basic information on the various forms of the virus, its effects, and treatments.

Synthetic Antibodies

Field of study: Immunogenetics
Significance: *Synthetic antibodies are artificially produced replacements for natural human antibodies. They are used to treat a variety of illnesses and promise to be an important part of medical technology in the future.*

Key terms

ANTIBODY: a protein molecule that binds to a substance in order to remove, destroy, or deactivate it

ANTIGEN: the substance to which an antibody binds

B CELLS: white blood cells that produce antibodies

MONOCLONAL ANTIBODIES: identical antibodies produced by identical B cells

The Development of Antibody Therapy

Natural antibodies are protein molecules produced by white blood cells known as B cells in response to the presence of foreign substances. A specific antibody binds to a specific substance, known as an antigen, in a way that renders it harmless or allows it to be removed from the body or destroyed. A person will produce antibodies naturally upon exposure to harmless versions of an antigen, a process

known as active immunization. Active immunization was the first form of antibody therapy to be developed and is used to prevent diseases such as measles and polio.

The oldest method of producing therapeutic antibodies outside the human body is known as passive immunization. This process involves exposing an animal to an antigen so that it develops antibodies to it. The antibodies are separated from the animal's blood and administered to a patient. Passive immunization is used to treat diseases such as rabies and diphtheria. A disadvantage of antibodies derived from animal blood is the possibility that the patient may develop an allergic reaction. Because the animal's antibodies are foreign substances, the patient's own antibodies may treat them as antigens, leading to fever, rash, itching, joint pain, swollen tissues, and other symptoms. Antibodies derived from human blood are much less likely to cause allergic reactions than antibodies from the blood of other animals. This led researchers to seek a way to develop synthetic human antibodies.

A major breakthrough in the search for synthetic antibodies was made in 1975 by Cesar Milstein and Georges Köhler. They developed a technique that allowed them to produce a specific antibody outside the body of a living animal. This method involved exposing an animal to an antigen, causing it to produce antibodies. Instead of obtaining the antibodies from the animal's blood, they obtained B cells from the animal's spleen. These cells are then combined with abnormal B cells known as myeloma cells. Unlike normal B cells, myeloma cells can reproduce identical copies of themselves an unlimited number of times. The normal B cells and the myeloma cells fuse to form cells known as hybridoma cells. Hybridoma cells are able to reproduce an unlimited number of times and are able to produce the same antibodies as the B cells. Those hybridoma cells that produce the desired antibody are separated from the others and allowed to reproduce. The antibodies produced this way are known as monoclonal antibodies.

Because human B cells do not normally form stable hybridoma cells with myeloma cells, B cells from mice are usually used. Because mouse antibodies are not identical to human antibodies, they may be treated as antigens by the patient's own antibodies, leading to allergic reactions. During the 1980's and 1990's, researchers began to develop methods of producing synthetic antibodies that were similar or identical to human antibodies. An antibody consists of a variable region, which binds to the antigen, and a constant region. The risk of allergic reactions can be reduced by combining variable regions derived from mouse hybridoma cells with constant regions from human cells. The risk can be further reduced by identifying the exact sites on the mouse variable region that are necessary for binding and integrating these sites into human variable regions. This method produces synthetic antibodies that are very similar to human antibodies.

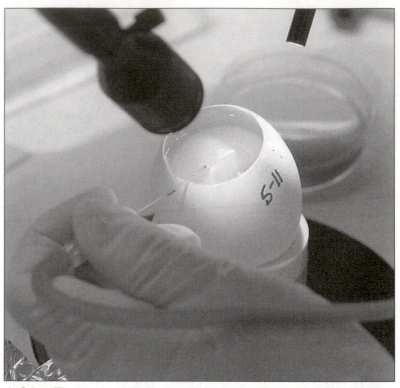

At Origen Therapeutics in Burlingame, California, a technician injects a chicken embryo with stem cells of another chicken embryo to which human antibodies have been added in order to make additional antibodies for pharmaceutical use. (AP/Wide World Photos)

Other methods exist to produce synthetic antibodies that are identical to human antibodies. A species of virus known as the Epstein-Barr virus can be used to change human B cells in such a way that they will fuse with myeloma cells to form stable hybridoma cells that produce human antibodies. Another method involves using genetic engineering to produce mice with B cells that produce human antibodies rather than mouse antibodies. One of the most promising techniques involves creating a "library" of synthetic human antibodies. This is done by using the polymerase chain reaction (PCR) to produce multiple copies of the genetic material within B cells. This genetic material contains the information that results in the production of proteins that come together to form antibodies. By causing these proteins to be produced and allowing them to combine at random, researchers are able to produce millions of different antibodies. The antibodies are then tested to detect those that bind to selected antigens.

Impact and Applications

Some synthetic antibodies are used to help prevent the rejection of transplanted organs. An antibody that binds to the heart drug digoxin can be used to treat overdoses of that drug. Antibodies attached to radioactive isotopes are used in certain diagnostic procedures. Synthetic antibodies have also been used in patients undergoing a heart procedure known as a percutaneous transluminal coronary angioplasty (PTCA). The use of a particular synthetic antibody has been shown to reduce the risk of having one of the blood vessels that supply blood to the heart shut down during or after a PTCA. Researchers also hope to develop synthetic antibodies to treat acquired immunodeficiency syndrome (AIDS) and septic shock, a syndrome caused by toxic substances released by certain bacteria.

The most active area of research involving synthetic antibodies in the 1990's was in the treatment of cancer. On November 26, 1997, the U.S. Food and Drug Administration approved a synthetic antibody for use in non-Hodgkin's lymphoma, a cancer of the white blood cells. It was the first synthetic antibody approved for use in cancer therapy.

—Rose Secrest

See also: Allergies; Anthrax; Antibodies; Autoimmune Disorders; Biopharmaceuticals; Blotting: Southern, Northern, and Western; Burkitt's Lymphoma; Cancer; Central Dogma of Molecular Biology; Cloning; Diabetes; Diphtheria; Genetic Engineering: Historical Development; Genetic Engineering: Industrial Applications; Hybridomas and Monoclonal Antibodies; Immunogenetics; Molecular Genetics; Multiple Alleles; Oncogenes; Organ Transplants and HLA Genes; Prion Diseases: Kuru and Creutzfeldt-Jakob Syndrome; Transgenic Organisms.

Further Reading

Coghlan, Andy. "A Second Chance for Antibodies." *New Scientist* 129 (February 9, 1991). An early discussion of the history and future of antibody therapy.

Kontermann, Roland, and Stefan Dübel, eds. *Antibody Engineering.* New York: Springer, 2001. A detailed look at basic methods, protocols for analysis, and recent and developing technologies. Illustrations, bibliography, index.

Mayforth, Ruth D. *Designing Antibodies.* San Diego: Academic Press, 1993. Methods of synthetic antibody production are described in detail. Illustrations, bibliography, index.

Synthetic Genes

Field of study: Genetic engineering and biotechnology

Significance: *Synthetic genes have been shown to function in biological organisms. Scientists hope that it will prove possible to restore normal function in diseased humans, animals, and plants by replacing defective natural genes with appropriately modified synthetic genes.*

Key terms

RESTRICTION ENZYME: an enzyme that cleaves, or cuts, DNA at specific sites with sequences recognized by the enzyme; also called restriction endonucleases

REVERSE TRANSCRIPTION: the synthesis of DNA from RNA

A Brief History

In 1871, Swiss physician Johann Friedrich Miescher reported that the chief constituent of the cell nucleus was nucleoprotein, or nuclein. Later it was established that the nuclei of bacteria contained little or no protein, so the hereditary material was named nucleic acid. At the end of the nineteenth century, German biochemist Albrecht Kossel identified the four nitrogenous bases: the purines adenine (A) and guanine (G) and the pyrimidines cytosine (C) and uracil (U). In the 1920's, Phoebus A. Levene and others indicated the existence of two kinds of nucleic acid: ribonucleic acid (RNA) and deoxyribonucleic acid (DNA); the latter contains thymine (T) instead of uracil.

The chemical identity of genes began to unfold in 1928, when Frederick Griffith discovered the phenomenon of genetic transformation. Oswald Avery, Colin MacLeod, and Maclyn McCarty (in 1944) and Alfred Hershey and Martha Chase (in 1952) demonstrated that DNA was the hereditary material. Following the elucidation of the structure of DNA in 1953 by James Watson and Francis Crick, pioneering efforts by several scientists led to the eventual synthesis of a gene. The successful enzymatic synthesis of DNA in vitro (in the test tube) in 1956, by Arthur Kornberg and colleagues, and that of RNA by Marianne Grunberg-Manago and Severo Ochoa also contributed to the development of synthetic genes. In 1961, Marshall Nirenberg and Heinrich Matthaei synthesized polyphenylalanine chains using a synthetic messenger RNA (mRNA). In 1965, Robert W. Holley and colleagues determined the complete sequence of alanine transfer RNA (tRNA) isolated from yeast. The interpretation of the genetic code by several groups of scientists throughout the 1960's was also clearly important.

In 1970, Har Gobind Khorana, along with twelve associates, synthesized the first gene: the gene for an alanine tRNA in yeast. There were no automatic DNA synthesizers available then. In 1976, Khorana's group synthesized the tyrosine suppressor tRNA gene of *Escherichia coli* (*E. coli*). The *lac* operator gene (twenty-one nucleotides long) was also synthesized, introduced into *E. coli*, and demonstrated to be functional. It took ten years to synthesize the first gene; by the mid-1990's, gene machines could synthesize a gene in hours.

Gene Synthesis

Protein engineering is possible by making targeted changes in a DNA sequence to produce a different product (protein) polypeptide with different properties, such as stress tolerance. The process of targeting a specific change in the nucleotide sequence (site-directed mutagenesis) allows the correlation of gene structure with protein function. Rapid sequencing with modern capillary DNA sequencers facilitates determination of the order of nucleotides that make up a gene in a matter of hours.

Once the sequence of a gene is known, it can be synthesized from nucleotides using gene machines. A gene machine is simply a chemical synthesizer made up of tubes, valves, and pumps that bonds nucleotides together in the right order under the direction of a computer. An intelligent person with a minimum of training can produce synthetic genes. A gene may be isolated from an organism using restriction enzymes (any of the several enzymes found in bacteria that serve to chop up the DNA of invading viruses), or it may be made on a gene machine. For example, the chymosin gene (an enzyme used in cheese making) in calves can be synthesized from its known nucleotide sequence instead of isolating it from calf DNA using restriction enzymes. Alternatively, chymosin mRNA can be obtained from calf stomach cells, which can be transformed into DNA through reverse transcription.

New or modified genes may be manufactured to obtain a desired product. Gene synthesis, coupled with automated rapid sequencing and protein analysis, has yielded remarkable dividends in medicine and agriculture. Genetic engineers are designing new proteins from scratch to learn more about protein function and architecture. With synthetic genes, the process of mutagenesis can be explored in greater depth. It is possible to produce various alterations at will in the nucleotide sequence of a gene and observe their effects on protein function. Such studies carry the potential to unravel many biochemical and genetic pathways that could be the key to a better understanding of health and disease.

—Manjit S. Kang

See also: Biopharmaceuticals; Cell Culture: Plant Cells; Cloning; Cloning Vectors; DNA Sequencing Technology; Gene Therapy; Genetic Engineering; Protein Synthesis; Restriction Enzymes; Reverse Transcriptase; Synthetic Antibodies.

Further Reading

Aldridge, Susan. *The Thread of Life: The Story of Genes and Genetic Engineering.* New York: Cambridge University Press, 1996. Provides a guide to DNA and genetic engineering.

Henry, Robert J. *Practical Applications of Plant Molecular Biology.* New York: Chapman & Hall, 1997. Gives protocols for important plant molecular biology techniques. Illustrations, bibliography, index.

Tay-Sachs Disease

Field of study: Diseases and syndromes

Significance: *Tay-Sachs disease (TSD) is a lethal disease inherited as an autosomal recessive disorder. Affected children are normal at birth, and symptoms are usually noticed by six months of age, after which they progressively worsen; the child usually dies at or before four years of age. There is no cure for this severe disorder of the nervous system, but an understanding of the genetic nature of the disorder has led to effective population screening, prenatal diagnosis, and genetic counseling.*

Key terms

GENETIC SCREENING: the testing of individuals for a disease-causing gene

HEXOSAMINIDASE A (HEX A): a lysosomal enzyme, the absence of which leads to Tay-Sachs disease

LYSOSOME: an organelle or structure in the cytoplasm of a cell that contains enzymes involved in the breakdown of metabolic products

PRENATAL DIAGNOSIS: the identification of a gene or disease in an embryo or fetus

Symptoms of Tay-Sachs Disease

Tay-Sachs disease (TSD) is an inherited birth defect that is named after Warren Tay, an English ophthalmologist, and Bernard Sachs, an American neurologist, who first described the disorder. TSD is one of the lysosomal storage disorders, as are Hurler's syndrome, Hunter's syndrome, Gaucher disease, and Fabry disease. Lysosomes are organelles found in the cytoplasm of cells and contain many enzymes that digest the cell's food and waste. TSD is caused by the lack of the enzyme hexosaminidase A (Hex A), which facilitates the breakdown of fatty substances and gangliosides in the brain and nerve cells. When Hex A is sufficiently lacking, as in TSD, gangliosides accumulate in the body and eventually lead to the destruction of the nervous system.

Children with TSD appear normal at birth and up to six months of age. During this time, they may show an exaggerated startle response to sound. Shortly after six months, more obvious symptoms appear. The child may show poor head control and an involuntary back-and-forth movement of the eyes. Also distinctive of TSD is a "cherry red spot" on the retina of the eye, first described by Tay, that usually appears after one year of age as atrophy of the optic nerve head occurs. The symptoms are progressive, and the child loses all the motor and mental skills developed to that point. Convulsions, increased motor tone, and blindness develop as the disease progresses. The buildup of storage material in the brain causes the head to enlarge, and brain weight may be 50 percent greater than normal at the time of death. There is no cure for TSD, and death usually occurs between two and four years of age, with the most common cause of death being pneumonia.

There are several forms of Tay-Sachs disease in addition to the classical, or infant, form already described. There is a juvenile form in which similar symptoms appear between two and five years of age, with death occurring around age fifteen. A chronic form of TSD has symptoms beginning at age five that are far milder than those of the infant and juvenile forms. Late-onset Tay-Sachs disease (LOTS) is a rare form in which there is some residual Hex A activity so that symptoms appear later in life and the disease progresses much more slowly.

Genetics of Tay-Sachs Disease

All forms of TSD are inherited as autosomal recessive disorders. One of the interesting features of TSD, as is true of some other genetic disorders, is its variation across ethnic groups. The Ashkenazi Jewish population, ancestors of most of the Jewish people in the United States, is a group of Jews of Eastern European descent. This group has a high incidence of TSD, about 1 in 3,600. Approximately one in thirty Ashkenazi Jews is a heterozygote (a person who carries one copy of the gene but does not show symptoms), compared to a figure of perhaps one in three hundred for the rest of the world's population. It is possible to screen the population and identify heterozygous individuals by means of a blood plasma assay that detects differences in Hex A activity. If two people are carriers of the gene, they have a one-quarter chance of having a child with TSD. If one or both individuals are not carriers, they can be reassured that their child will not have TSD. If both people are carriers, once pregnancy ensues, prenatal diagnosis can determine whether the developing fetus is affected. In cases of a positive diagnosis, couples can be counseled regarding therapeutic pregnancy termination.

Impact and Applications

Although much has been learned about the genetics of the Tay-Sachs gene and the protein deficiency that causes the disease, there is still no cure. Nevertheless, TSD provides an excellent example of how the medical community can assist a susceptible population in confronting an incurable genetic disease. The effective screening of populations at risk for TSD and prenatal detection of fetuses with TSD have served to dramatically reduce the overall incidence of this terrible disease.

—Donald J. Nash

See also: Genetic Counseling; Genetic Screening; Genetic Testing; Genetic Testing: Ethical and Economic Issues; Hereditary Diseases; In Vitro Fertilization and Embryo Transfer; Inborn Errors of Metabolism; Penetrance; Prenatal Diagnosis; Repetitive DNA.

Further Reading

Bach G., J. Tomczak, N. Risch, and J. Ekstein. "Tay-Sachs Screening in the Jewish Ashkenazi Population: DNA Testing Is the Preferred Procedure." *American Journal of Medical Genetics* 99 (February 15, 2001). Argues for using DNA testing as the most cost-effective and efficient way to screen for TSD.

Desnick, Robert J., and Michael M. Kaback, eds. *Tay-Sachs Disease.* San Diego: Academic, 2001. Detailed analysis of Tay-Sachs. Illustrations, bibliography, index.

National Tay-Sachs and Allied Diseases Association. *A Genetics Primer for Understanding Tay-Sachs and the Allied Diseases.* Brookline, Mass.: Author, 1995. An introductory overview of Tay-Sachs disease.

Parker, James N., and Philip M. Parker, eds. *The Official Parent's Sourcebook on Tay-Sachs Disease: A Revised and Updated Directory for the Internet Age.* San Diego: Icon Press, 2002. Topics include the essentials on Tay-Sachs disease, parents' rights, and insurance.

Zallen, Doris Teichler. *Does It Run in the Family? A Consumer's Guide to DNA Testing for Genetic Disorders.* New Brunswick, N.J.: Rutgers University Press, 1997. Covers the applications and social implications of testing for genetic disorders. Bibliography, index.

Web Site of Interest

Dolan DNA Learning Center, Your Genes Your Health. http://www.ygyh.org. Sponsored by the Cold Spring Harbor Laboratory, this site, a component of the DNA Interactive Web site, offers information on more than a dozen inherited diseases and syndromes, including Tay-Sachs disease.

Telomeres

Field of study: Cellular biology
Significance: *Telomeres, the ends of the arms of chromosomes of eukaryotes, become shorter as organisms age. They are thought to act biologically to slow chromosome shortening, which can lead to cell death caused by the loss of genes and may be related to aging and diseases such as cancer.*

Key terms

EUKARYOTE: a unicellular or multicellular organism with cells that contain a membrane-bound nucleus, multiple chromosomes, and membrane-bound organelles

PROKARYOTE: a unicellular organism with a single chromosome and lacking a nucleus or any other membrane-bound organelles

Eukaryotic Chromosomes and Telomeres

The DNA of bacteria and other related simple organisms (prokaryotes) consists of one double-stranded DNA molecule. Structurally and functionally, the prokaryotic chromosome contains one copy of most genes as well as DNA regions that control expression of these genes. Prokaryotic gene expression depends primarily upon a cell's moment-to-moment needs. An entire prokaryotic chromosome, its genome, usually encodes about one thousand genes.

The genomes of eukaryotes are much more complex and may include 100,000 or more genes. The number of chromosomes in different types of eukaryotes can range from just a few to several hundred. Each of these huge DNA molecules is linear rather than the circular molecule of the type seen in prokaryotes. In addition, many individual segments of the DNA of eukaryotes exist in multiple copies. For example, about 10 percent of the DNA of a eukaryote consists of "very highly repetitive segments" (VRS's), units that are less than ten deoxyribonucleotides long that are repeated up to several million times per cell. DNA segments that are several hundred deoxyribonucleotide units long represent about 20 to 25 percent of the DNA. They are repeated one thousand times or more per cell. The rest of the eukaryote DNA (from 65 to 70 percent of the total) consists of larger segments repeated once or a few times, the genes, and the DNA regions that control the expression of the genes.

Much of the repetitive DNA, called satellite DNA, does not seem to be involved in coding for proteins or RNAs involved in making proteins. Telomeres are part of this DNA and consist of pieces of DNA that are several thousand deoxyribonucleotide units long, found at both chromosome ends. They are believed to act to stabilize the ends of chromosomes and protect them from exonuclease enzymes that degrade DNA from the ends. Researchers have concluded this for two reasons. First, the enzymes that make two chromosomes every time a cell reproduces are unable to operate at the chromosome ends. Hence, the repeated reproduction of a eukaryote cell and its DNA will lead to the creation of shorter and shorter chromosomes, a process that can cause cell death when essential genes are lost. Second, as organisms age, the telomeres of their cells become shorter and shorter.

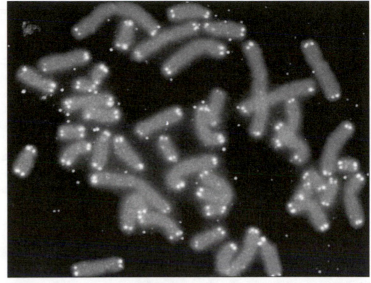

Telomeres appear as the lightened tips of the chromosomes shown here. (Robert Moyzis, University of California, Irvine, CA; U.S. Department of Energy Human Genome Program, http://www.ornl.gov/hgmis)

Telomerase Enzymes

When chromosomes are replicated in preparation for cell division, the internal segments are replicated by a complex process involving the enzymes primase and DNA polymerase. Primase lays down a small segment of RNA on the template strand of DNA, and DNA polymerase uses the primer to start replication. Making the end of a linear chromosome is a problem, however, because primers cannot consistently be produced at the very ends of the chromosomes. Consequently, with each cell division a small portion of the ends of newly replicated chromosomes is single-stranded and is trimmed off by exonucleases. This problem is solved by enzymes known as telomerases, which add telomeres to eukaryote chromosomes. Each telomerase contains a nucleic acid component (RNA) about 150 ribonucleotides long. This is equivalent to 1.5 copies of the appropriate repeat in the DNA telomere to be made. The enzyme uses this piece of RNA to make the desired DNA strand of the telomere. How the telomerase in any given species identifies the correct length of telomere repeat for a specific chromosome is not clearly understood, nor is the exact mechanism by which the DNA strand is made.

Telomerase activity can be lost in certain strains of simple eukaryotes, such as protozoa. When this happens to a given cell line, each cell division leads to the additional shortening of its telomeres. This process continues for a fixed number of cell divisions; it then ends with the death of the telomerase-deficient cell line.

A related observation has been made in humans. It has been shown that when human fibroblasts are grown in tissue culture, telomere length is longest when cells are obtained from young individuals. They are shorter in cells taken from the middle-aged, and very short in cells taken from the aged. Similar observations have been made with the fibroblasts from other higher eukaryotes as well as with other human cell types. In contrast, the process of telomere shortening does not happen when germ-cell lines—which in the whole organism produce sperm and ova—are grown in tissue culture. This suggests a basis for differences in longevity of the germ cells and the somatic cells that make up other human tissues.

Impact and Applications

The discovery and study of telomeres and telomerases produced new insights into DNA synthesis, the number of times a cell can reproduce, and the aging process. The circular DNA of bacteria (which are prokaryotes) allows them

Telomere Length in Clones

Dolly the sheep, the first mammal to be cloned from adult cells, was born on July 5, 1996. While Dolly ushered in a new era of mammalian cloning, her tenure as the cloning community's lovable mascot was, quite literally, short-lived. Dolly was euthanized on February 14, 2003, after being diagnosed with a progressive lung disease; she had already been suffering from debilitating arthritis. While Dolly's health problems could have resulted from "natural causes," both ailments are more characteristic of much older sheep. Sheep normally live to an age of about twelve, Dolly was only half that.

Dolly's early demise was actually foreshadowed in 1999, when the group which cloned her reported that Dolly's telomeres were shorter than expected for a sheep of her age. Dolly's telomeres were about the length one would expect if her cells had been six years old on the day she was born (Dolly was cloned from a six-year-old ewe). Since telomere length acts as a "molecular clock" that determines the age of a cell, researchers had hoped that this clock would somehow be "reset" upon transfer of an adult nucleus to a host ovum.

While clearly not the case for Dolly, this resetting of telomere length has been demonstrated in cloned cows. In 2000, Robert Lanza and colleagues reported that cloned cows had longer-than-normal telomeres. Will these "super cows" be able to live appreciably longer than normal cows? Only time will tell, since cows have a normal life span of about twenty years. What accounts for the difference between Dolly and these cloned cows? Subsequent research has shown that the type of cell used in the cloning process may be an important factor. In 2002, Norikazu Miyashita and colleagues reported that cows cloned from mammary gland cells (like Dolly) had shorter-than-normal telomeres, clones obtained from skin fibroblasts (connective tissue precursors, those used by Lanza) had longer-than-normal telomeres, and clones obtained from muscle cells showed no significant differences in telomere length.

The telomere length of clones may also be species-specific. Teruhiko Wakayama and colleagues cloned mice sequentially for six generations but saw no difference in telomere length in any of the clones produced. Mice, however, are known to have extremely long telomeres to begin with; also, unlike the somatic cells of cows or sheep, many of the somatic cells of mice are known to express the telomerase enzyme.

More research is necessary to understand exactly why certain animal clones are produced with shortened telomeres and others are not. Currently, our lack of knowledge on the subject remains one of the more compelling reasons not to attempt to clone a human at this time. A human clone produced with unusually short telomeres may, like Dolly, meet with an untimely death. In fact, patients with a human genetic disease called Hutchinson-Gilford progeria have skin fibroblasts with greatly reduced telomere lengths; persons affected with this disease live to an average age of about thirteen years.

—*James S. Godde*

to undergo many more cycles of reproduction than the somatic cells of the eukaryotes. The linear eukaryote chromosome may have evolved because such DNA molecules were too large to survive as circular molecules given their rigidity and fragility. In addition, the observation of telomere shortening in simple and complex eukaryotes raises the fascinating possibility that the life spans of organisms may be related to the conservation of telomeres associated with the replication of these structures by telomerases.

The role of telomere length in longevity is uncertain, but apparently significant. Cells grown in cell culture typically divide only a predictable number of times, and once this limit is reached they can no longer divide. At the same time, telomere length shortens with each division. Sometimes cells in culture will go through what is called a "crisis," after which they become "immortalized" and are able to divide an indefinite number of times. Immortal cells also actively express telomerases and maintain constant telomere lengths. Cancer cells typically exhibit these same characteristics. A better understanding of telomeres and telomerase expression might provide insights into aging and cancer, leading to a potential cure for cancer and age-related diseases.

—*Sanford S. Singer*

See also: Aging; Animal Cloning; Chromosome Mutation; Chromosome Structure; Cloning Vectors; DNA Replication; Molecular Genetics; Reverse Transcriptase.

Further Reading

Abstracts of Papers Presented at the 2001 Meeting on Telomeres and Telomerase. Arranged by Elizabeth H. Blackburn, Titia De Lange, and Carol Grieder. Cold Spring Harbor, N.Y.: Cold Spring Harbor Laboratory Press, 2001. Synopses of research and studies on telomeres and telomerase. Bibliography, index.

Blackburn, Elizabeth H. "Telomeres, Telomerase, and Cancer." *Scientific American*, February, 1996. Provides background on telomeres, telomerases, and their potential importance in carcinogenesis.

Blackburn, Elizabeth H., and Carol W. Greider, eds. *Telomeres.* Cold Spring Harbor, N.Y.: Cold Spring Harbor Laboratory Press, 1995. Covers the discovery, synthesis, and potential effects of telomeres on normal life, aging, neoplasms, and other pathologies. Illustrations, bibliography, index.

Double, John A., and Michael J. Thompson, eds. *Telomeres and Telomerase: Methods and Protocols.* Totowa, N.J.: Humana Press, 2002. A laboratory guide for exploring the world of the telomerase. Illustrations, bibliography, index.

Kipling, David, ed. *The Telomere.* New York: Oxford University Press, 1995. Describes telomeres, telomerases, relationships to cancer, and other aspects of potential telomere action.

Krupp, Guido, and Reza Parwaresch, eds. *Telomerases, Telomeres, and Cancer.* New York: Kluwer Academic/Plenum, 2003. Considers the way telomeres function in cancer. Illustrations, bibliography, index.

Lewis, Ricki. "Telomere Tales." *Bioscience* 48, no. 12 (December, 1998). An overview of some of the molecular research that supports the telomere shortening model of cellular aging.

Testicular Feminization Syndrome

Field of study: Diseases and syndromes

Significance: *The sex of a baby is usually determined at conception by the sex chromosomes, but other genetic events can alter the outcome. One such condition is testicular feminization syndrome, which causes a child with male chromosomes to be born with feminized genitals. Information gained from the study of this and similar conditions is being used to challenge the validity of sex-determination tests for athletes.*

Key terms

ANDROGEN RECEPTORS: molecules in the cytoplasm of cells that join with circulating male hormones

ANDROGENS: hormones that promote male body characteristics

DIFFERENTIATION: the process of changing from an unspecialized condition to a final specialized one

PHENOTYPE: the expressed characteristics, both physical and physiological, of an individual

SEX DETERMINATION: events that cause an embryo to become male or female

Development of Testicular Feminization Syndrome

Introductory biology courses teach that a fertilized egg that receives two X chromosomes at conception will be a girl, whereas a fertilized egg that receives an X and a Y chromosome will become a boy. However, other factors can also affect the development of a person's gender. Gender development in mammals begins at conception with the establishment of chromosomal sex (the presence of XX or XY chromosomes). Even twelve weeks into development, male and female embryos have the same external appearance. Internal structures for both sexes are also similar. However, the machinery has been set in motion to cause the external genitals to become male or female, with corresponding internal structures of the appropriate sex. The baby is usually born with the proper phenotype to match its chromosomal sex. However, development of the sex organs is controlled by several genes. This leaves a great deal of room for developmental errors to occur.

The primary gene involved in sex determination is carried on the Y chromosome. It is responsible for converting the early unisex gonads into testes. Once formed, the testes then produce the balance of androgen and estrogen that pushes development in the direction of the male phenotype. In the absence of this gene, the undetermined gonads become ovaries, and the female phenotype emerges. Therefore, the main cause of sex determination is not XX or XY chromosomes, but rather the presence or absence of the gene that promotes testis differentiation.

In order for the male hormones to have an influence on the development of the internal and external reproductive structures, the cells of those structures must receive a signal that they are part of a male animal. The androgens produced by the testes are capable of entering a cell through the cell membrane. Inside the cell, the androgens attach to specific protein receptor molecules (androgen receptors). Attachment causes the receptors to move from the cytoplasm into the nucleus of the cell. Once in the nucleus, the receptor-steroid complexes bind to DNA near genes that are designed to respond to the presence of these hormones. The binding event is part of the process that turns on specific genes—in this case, the genes that direct the process of building male genitals from the unisex embryonic structures as well as those that suppress the embryonic female uterus and tubes present in the embryo's abdomen.

In cases of testicular feminization, androgen receptors are missing from male cells. This is the result of a recessive allele located on the X chromosome. Because normal males have only one X, the presence of a recessive allele on that X will result in no production of the androgen receptor in that individual. The developing embryo is producing androgen in the testes; without the receptor molecules, however, the cells of the genitals are unable to sense the androgen and respond to it. For this reason, the disorder is sometimes known by an alternate name: androgen-insensitivity syndrome. The cells of the genitals are still capable of responding to estrogen from the testes. As a result, the genitals become feminized: labia and clitoris instead of a scrotum and penis, and a short, blind vagina. To the obstetrician and parents, the baby appears to be a perfect little girl. An internal examination would show the presence of testes rather than ovaries and the lack of a uterus and Fallopian tubes, but there would normally be no reason for such an examination.

Impact and Applications

Several events may lead to the diagnosis of this condition. The attempted descent of the testes into a nonexistent scrotum will cause pain that may be mistaken for the pain of a hernia; the presence of testes in the apparent girl will be discovered when the child undergoes repair surgery. In other cases, the child may seek medical help in the mid-teen years because she does not menstruate. Exploratory surgery would then reveal the presence of testes and the absence of a uterus. As a general rule, the testes are left in the abdomen until after puberty because they are needed as a source of estrogen to promote the secondary sex characteristics, such as breast development. Without this estrogen, the girl would remain childlike in body form. After puberty, the testes are usually removed because they have a tendency to become cancerous.

As a result of its phenotypic sex, an infant with testicular feminization is normally raised as a girl whose only problem is an inability to bear children. If the girl has athletic ability, however, other problems may arise. Since 1966, female Olympic athletes have had to submit to a test for the presence of the correct chromosomal sex. In the past, this has meant microscopic examination of cheek cells to count X chromosomes. In 1992, this technique was replaced by a test for the Y chromosome. Individuals who fail the "sex test," including those with testicular feminization syndrome, cannot compete against other women. Proponents argue that androgens aid muscle development, and the extra testosterone produced by the testes of a normal male would provide an unfair physical advantage. However, because people with testicular feminization syndrome are lacking androgen receptors, their muscle development would be unaffected by the extra androgen produced by the testes, and thus they would not be any stronger than well-conditioned women.

—*Nancy N. Shontz*

See also: Fragile X Syndrome; Gender Identity; Hereditary Diseases; Hermaphrodites; Klinefelter Syndrome; Metafemales; Pseudohermaphrodites; Steroid Hormones; XYY Syndrome.

Further Reading

Goodall, J. "Helping a Child to Understand Her Own Testicular Feminisation." *Lancet* 337 (January 5, 1991). Discusses how communicating with children in stages about their testicular feminization helps them cope emotionally.

Lemonick, Michael. "Genetic Tests Under Fire." *Time* 139 (February 24, 1992). The syndrome and its relationship to athletes and athletic performance.

Mange, Elaine Johansen, and Arthur P. Mange. *Basic Human Genetics.* 2d ed. Sunderland, Mass.: Sinauer Associates, 1999. Provides a detailed discussion of testicular feminization syndrome. Illustrations (some color), maps, laser optical disc, bibliography, index.

Web Sites of Interest

Intersex Society of North America. http://www.isna.org. The society is "a public awareness, education, and advocacy organization which works to create a world free of shame, secrecy, and unwanted surgery for intersex people (individuals born with anatomy or physiology which differs from cultural ideals of male and female)." Includes links to information on such conditions as clitoromegaly, micropenis, hypospadias, ambiguous genitals, early genital surgery, adrenal hyperplasia, Klinefelter syndrome, androgen insensitivity, and testicular feminization.

Johns Hopkins University, Division of Pediatric Endocrinology, Syndromes of Abnormal Sex Differentiation. http://www.hopkinsmedicine.org/pediatricendocrinology. Site provides a guide to the science and genetics of sex differentiation, including a glossary. Click on "patient resources."

Thalidomide and Other Teratogens

Field of study: Diseases and syndromes

Significance: *Teratogenesis is the development of defects in the embryo or fetus caused by exposure to chemicals, radiation, or other environmental conditions. Thalidomide, a sedative whose ingestion by pregnant women led to the birth of abnormal babies in the late 1950's and early 1960's, is one of the more publicized examples of a chemical teratogen.*

Key terms

CONGENITAL DEFECT: a defect or disorder that occurs during prenatal development

PEROMELIA: the congenital absence or malformation of the extremities caused by abnormal development of the limb bud from about the fourth to the eighth week after conception; the ingestion of thalidomide by pregnant women can cause this disorder in fetuses

Teratogenesis and Its Causes

Teratogenesis is the development of structural or functional abnormalities in an embryo or fetus due to the presence of a toxic chemical or other environmental factor. The term is derived from the Greek words *teras* (monster) and *genesis* (birth). The phenomenon is usually attributed to exposure of the mother to some causative agent during the early stages of pregnancy. These may include chemicals, excessive radiation exposure, viral infections, or drugs.

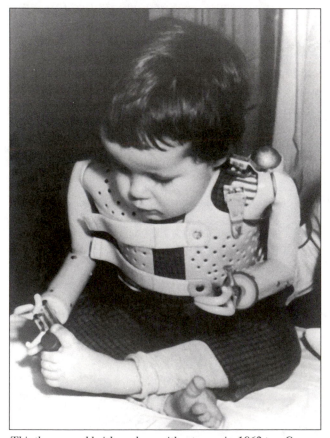

This three-year-old girl was born without arms in 1962 to a German mother who had taken thalidomide during her pregnancy. (AP/Wide World Photos)

Approximately 3 percent of the developmental abnormalities are attributed to drugs. Drugs that are taken by the father may be teratogenic only if they damage the chromosomes of a spermatozoan that then joins with the egg to form a zygote.

For many centuries, the impression that malformed babies were conceived as a result of the intercourse between humans and devils or animals dominated society. Seventeenth century English physiologist William Harvey attributed teratogenesis to embryonic development. In the nineteenth century, the French brothers Étienne and Isidore Geoffroy Saint-Hilaire outlined a systematic study on the science of teratology. In the United States, the importance of teratogens was first widely covered during the 1940's, when scientists discovered that pregnant women who were affected by German measles (rubella) often gave birth to babies that had one or more birth defects. In the 1940's and 1950's, the consumption of diethylstilbestrol (DES) before the ninth week of gestation to prevent miscarriage was found to produce cancer in the developing fetus. Animal studies have also shown that defective offspring result from the use of hallucinogens such as lysergic acid diethylamide (LSD).

A broader definition of teratogenesis may include other minor birth defects that are more likely to be genetically linked, such as clubfoot, cleft lip, and cleft palate. These defects can often be treated in a much more effective way than those caused by toxic substances. Clubfoot, for example, which can be detected by the unusual twisted position of one or both feet, may be treated with surgery and physical therapy within the first month after birth. Brachydactyly (short digits) in rabbits has been linked to a recessive gene that causes a local breakdown of the circulation in the developing bud of the embryo, which is followed by necrosis (tissue death) and healing. In more extreme cases of agenesis, such as limb absence, a fold of amnion (embryonic membrane) was found to cause strangulation of the limb. Agenesis has been observed with organs such as kidneys, bladders, testicles, ovaries, thyroids, and lungs. Other genetic teratogenic malformations include anencephaly (absence of brain at birth), microcephaly (small-size head), hydrocephaly (large-size head caused by accumulation of large amounts of fluids), spina bifida (failure of the spine to close over the spinal cord), cleft palate (lack of fusion in the ventral laminae), and hermaphrodism (presence of both male and female sexual organs).

Thalidomide and Its Impact

Thalidomide resembles glutethimide in its sedative action. Laboratory studies of the late 1950's and early 1960's had shown thalidomide to be a safe sedative for pregnant women. As early as 1958, the West German government made the medicine available without prescription. Other Western European countries followed, with the medicine available only upon physician's prescription. It took several years for the human population to provide the evidence that laboratory animals could not. German physician Widukind Lenz established the role of thalidomide in a series of congenital defects. He proved that administration of the drug during the first twelve weeks of the mother's pregnancy led to the development of phocomelia, a condition characterized by peromelia (the congenital absence or malformation of the extremities caused by the abnormal formation and development of the limb bud from about the fourth to the eighth week after conception), absence or malformation of the external ear, fusion defects of the eye, and absence of the normal openings of the gastrointestinal system of the body.

The United States escaped the thalidomide tragedy to a great extent because of the efforts of Frances O. Kelsey, M.D., of the U.S. Food and Drug Administration (FDA). She had serious doubts about the drug's safety and was instrumental in banning the approval of thalidomide for marketing in the United States. Other scientists such as Helen Brooke Taussig, a pioneer of pediatric cardiology and one of the physicians who outlined the surgery on babies with the Fallot (blue baby) syndrome, played a key role in preventing the approval of thalidomide by the FDA. It is estimated that about

seven thousand births were affected by the ingestion of thalidomide.

The thalidomide incident made all scientists more skeptical about the final approval of any type of medicine, especially those likely to be used during pregnancy. The trend intensified the fight against any chemicals that might affect the fetus during the first trimester, when it is particularly vulnerable to teratogens. Alcohol and tobacco drew many headlines in the media in the 1990's. Both have been shown to create congenital problems in mental development and learning abilities. At the same time, regulation of new FDA-approved medicine became much stricter, and efforts to study the long-term effects of various pharmaceuticals increased. Surprisingly, thalidomide itself has been used successfully in leprosy cases and, in conjunction with cyclosporine, to treat cases of the immune reaction that appears in many bone-marrow transplant patients. There is also a movement to use thalidomide in the treatment of acquired immunodeficiency syndrome (AIDS).

In addition to drugs, many other agents can affect fetal development. Essentially, any factor with the potential to cause DNA mutations has a high probability of being teratogenic. Consequently, early in pregnancy, women are advised to limit their exposure to a variety of potential teratogens, such as excess radiation, toxic chemicals, tobacco, alcohol, and other drugs. Prevention might even include work reassignment to limit or eliminate the woman's normal exposure to teratogens. Unfortunately, teratogenesis can occur early in the pregnancy, before the woman is even aware that she is pregnant. Prevention by avoidance is therefore essential.

—*Soraya Ghayourmanesh, updated by Bryan Ness*

See also: Congenital Defects; Prenatal Diagnosis.

Further Reading

Holmes, L. B. "Teratogen-Induced Limb Defects." *American Journal of Medical Genetics* 112 (October 15, 2002). Discusses limb defects, a common effect of human teratogens.

Stephens, Trent D., and Rock Brynner. *Dark Remedy: The Impact of Thalidomide and Its Revival as a Vital Medicine.* Cambridge, Mass.:

Perseus, 2001. Surveys the history of the birth defects epidemic from the 1960's through today, and discusses the search for an alternative to thalidomide that retains its curative effects. Bibliography, index.

Web Site of Interest

Teratology Society. http://www.teratology.org. The Teratology Society is a multidisciplinary scientific society founded in 1960, the members of which study the causes and biological processes leading to abnormal development and birth defects at the fundamental and clinical level, and appropriate measures for prevention.

Totipotency

Field of study: Cellular biology

Significance: *Totipotency is the ability of a living cell to express all of its genes to regenerate a whole new individual. Totipotent cells from plants have been used in tissue-culture techniques to produce improved plant materials that are pathogen-free and disease-resistant. Totipotent cells from animals are now being used to clone mammals, although ethical questions remain over whether cloning a human should be done.*

Key terms

MULTIPOTENT CELL: a stem cell capable of forming multiple differentiated tissues

PARTHENOGENESIS: asexual reproduction from a single egg without fertilization by sperm

PLURIPOTENT CELL: a stem cell that forms all types of differentiated tissues

UNIPOTENT CELL: a stem cell that forms only one differentiated tissue

Egg and Sperm Cells

In plants and animals, a whole organism is sexually reproduced from a zygote, a product of fusion between egg and sperm. Zygotes are totipotent. A zygote in the seed of a plant or in the uterus of a mammal divides by mitosis and has the potential to produce more cells, called embryonic cells, before developing into an adult individual. During embryonic cell divi-

sion, the cells begin to differentiate. Once differentiated, these specialized cells still possess all the genetic materials inherited from the zygote. Differentiated cells express or use some of their genes (not all) to produce their own specific proteins. For example, epidermal cells in human beings produce fibrous proteins called keratin to protect the skin, and red blood cells produce hemoglobin to help transport oxygen. Due to the differences in gene expression, differentiated cells have their own distinct structures and functions, and some differentiated cells are totipotent.

A whole organism can be asexually reproduced from a single egg without the sperm by a process called parthenogenesis. This occurs naturally in some insects, snakes, lizards, and amphibians, as well as in some plants. In this type of reproduction, the haploid chromosomes within an unfertilized egg duplicate, and the embryo develops as if the egg had been fertilized. The pseudo-fertilized eggs are totipotent and generate all female individuals. The females can reproduce under favorable environmental conditions without waiting for a mate. Like in vitro fertilization, parthenogenesis is used as a technique to create an embryo in the laboratory. Chromosomal duplication is induced in the egg cell to reproduce female individuals. However, no parthenogenic mammals had been developed.

It may become possible to produce males through a process called androgenesis. In the laboratory, the haploid chromosomes from one sperm may be induced to duplicate. As in animal cloning, the duplicated chromosomes, which are diploid, can be implanted into an enucleated egg cell (a cell from which the nucleus has been removed). Although androgenesis holds some promise, so far it has not produced normal embryos.

Cell Differentiation

Cell differentiation is a process whereby genetically identical cells become different or specialized for their specific functions. During differentiation, enzymes and other polypeptides, including other large molecules, are synthesized. Ribosomes and other cell structures are assembled. Differentiated cells express only

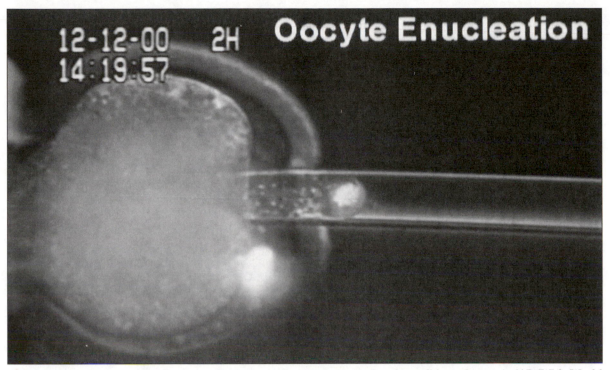

DNA is removed by pipette suction from a mammalian egg cell to prepare an enucleated egg cell for androgenesis. (AP/Wide World Photos)

some of their genes to make enzymes and other proteins.

Tissue differentiation is usually triggered by mitosis, followed by cytokinesis. Then differentiation occurs in the daughter cells. Often the two daughter cells have different structures and functions, but both retain the same genes. For example, the epidermal cell mitotically divides to produce one large and one small cell on the root surface; the large one maintains the role of epidermal cell as a root covering, whereas the small one becomes the root hair.

Totipotent Cells in Plant-Tissue Culture

Cuttings of plants and tissue-culture techniques have proven that many plant cells are totipotent. Tissue culture, however, helps to identify what specific type of cell is totipotent, because the technique uses a very small piece of known tissue. For example, if pith tissues from tobacco (*Nicotiana tobaccum*), soybean (*Glycine max*), and other dicot stems are cut off and cultured aseptically on an agar medium with proper nutrients and hormones, a clump of unspecialized and loosely arranged cells, called a callus, is formed. Each cell from the callus begins to divide and differentiate, forming a multicellular embryoid. One test tube can accommodate thousands of cells, and each embryoid has the potential to become a complete plantlet. Plantlets can be transplanted into the soil to develop into adult plants.

The phloem tissues from the roots of carrots (*Daucus* species) also exhibit totipotency. Cells in pollen grains of tobacco are totipotent, and they produce haploid plants. Using meristem tissues of shoot and root tips, the cells regenerate new plants that are free of viruses, bacteria, and fungi. Pathogen elimination is possible because vascular tissues (xylem and phloem), in which viruses move, do not reach the root or shoot apex. The protoplasts (cells without cell walls) from mesophyll cells of the leaf regenerate new plants.

Plant Hormones

Totipotency of plant cells is enhanced by the presence of hormones, such as auxins and cytokinins, in the culture media. Addition of auxins influences the expression of genes and causes physiological and morphological changes in plants. Addition of cytokinins promotes cell division, cytokinesis, and organ formation. If these are present in the proper ratio, callus from many plant species can be made to develop into an entire new plant. If the cytokinin-to-auxin ratio is high, cells in the callus divide and give rise to the development of buds, stems, and leaves. If the cytokinin-to-auxin ratio is low, root formation is favored. Totipotency of some plant cells is promoted by the addition of coconut water to the culture media—an indication that coconut water has the right proportion of cytokinin and auxin to regenerate an entire plant.

Importance of Totipotency in Plants

Clonal propagation of plants using tissue culturing is used commercially to mass-produce numerous ornamentals, vegetables, and forest trees. A major use of pathogen-free plants is for the storage of germ plasm and for transport of plant materials into different countries. It is also posssible to generate plants with desirable traits, such as resistance to herbicides and environmental stressors or tolerance of soil salinity, soil acidity, and heavy-metal toxicity. It is easier to select resistant or tolerant plants from a thousand cells than from a thousand plants.

Somatic Cells in Animal Cloning

Animals are more difficult to reproduce asexually than plants are. Somatic cells of animals become totitopotent when used as donor cells in cloning. The first successful animal cloned was a frog, *Xenopus laevis*. This cloning involved the use of a nucleus from the intestinal epithelial cells of a tadpole and an egg cell from a mature frog. In the laboratory, the nucleus from the egg cell was removed (enucleated) by micropipette. The tadpole's nucleus (the donor cell) was inserted into the enucleated frog's egg cell. The nuclei-injected egg cell underwent a series of embryonic developmental stages, including the blastula stage, developing into tadpoles that later died before becoming adults.

Cloning of Dolly the sheep (*Ovis* species) used the mammary cell of a six-year-old ewe as the donor cell. It was injected into the enucleated sheep's egg cell. Cloning a mammal requires a surrogate mother. The blastula stage of embryo was developed in vitro and was implanted into a surrogate mother. After five months, a lamb was born. The lamb was genetically identical to the sheep from which the mammary cell was taken. Today, cloning has been done by scientists to produce other animals, including cattle, pigs, monkeys, cats, and dogs. Cloning a human seems possible, but there are so many ethical and moral questions whether it should be done or not.

Stem Cells in Animal Cloning

Stem cells exhibit totipotency because they can generate new types of tissues. Some sources of stem cells are the blastocyst (the immature embryo), the fetus, the placenta, bone marrow, blood, skeletal muscle, and brain. Because there is no proof yet whether a single embryonic stem cell has the ability to regenerate into a complete individual, stem cells are generally only partially totipotent. A unipotent stem cell can form only one differentiated tissue. A multipotent stem cell can form multiple differentiated tissues. For example, stem cells from blood can form platelets, white blood cells, or red blood cells. The stem cells from skeletal muscle can form smooth muscle, cardiac muscle, bone, or cartilage. A pluripotent stem cell from embryo, brain, or bone marrow has the ability to develop all types of differentiated tissues of the body. For example, brain stem cells can be turned into all tissue types, including brain, muscles, blood cells, and nerves.

—*Domingo M. Jariel*

See also: Cell Culture: Plant Cells; Stem Cells.

Further Reading

Prentice, David A. *Stem Cells and Cloning.* New York: Benjamin Cummings, 2003. Discusses partial totipotency and differentiation of stem cells, including illustrations showing the sources of stem cells and how embryos can be developed by cloning techniques.

Russell, Peter J. *Genetics.* San Francisco, Calif.:

Benjamin Cummings, 2002. Discusses totipotency of some plants and animals. Includes illustrations on how cells from plants are used in tissue-culture techniques and how cells from animals are used in animal cloning.

Smith, Roberta H. *Plant Tissue Culture Techniques and Experiments*. New York: Academic Press, 2000. Discusses the totipotency of different types of cells in regenerating new plants in vitro using culture growth media. Includes an illustration of explant preparation and discusses media preparation and transplantation of regenerated plants.

Transgenic Organisms

Field of study: Genetic engineering and biotechnology

Significance: *Implanting genes from one organism into the genome of another enables scientists to study basic genetic mechanisms and inherited diseases and to create plants and animals with traits that are beneficial to humans.*

Key terms

GENOME: the complete genetic material carried by an individual

PLASMID: a circular piece of bacterial DNA that is often used as a vector

TRANSFORMATION: integration of foreign DNA into a cell

TRANSGENE: the foreign gene incorporated into a cell's DNA during transformation

VECTOR: a carrier molecule that introduces foreign genetic materials into a cell

Engineering Organisms

Domestication and selective breeding of animals and plants began before recorded history. In fact, historians propose, the shaping of organisms to fit human needs contributed to the rise of settled, complex culture. Until late in the twentieth century, farmers and scientists could breed novel strains only from closely related species or subspecies because the DNA had to be compatible in order to produce offspring that in turn were fertile.

In the late 1970's and early 1980's, molecular biologists learned how to surpass the limitations of selective breeding. They invented procedures for combining the DNA of species as distantly related as plants and animals. Organisms produced by such means are termed transgenic. This branch of genetic engineering made it possible to design novel organisms for genetic and biochemical research and for medical, agricultural, and ecological innovations. Commercial use of transgenic organisms also created worldwide controversy because of their potential threat to human health and the environment.

Transgenesis is much like gene therapy in that both transform cells for a specific purpose. However, whereas gene therapy targets only certain cells in order to cure a defect in them, transgenesis seeks to produce an entirely modified organism by incorporating the transgene into all the cells of the mature organism and changing the genome. This is done by transforming not only the somatic (body) cells of the host organism but also the germ cells, so that when the organism reproduces, the transgene will pass to the next generation. Transgenes perform their alterations by blocking the function of a host gene, by replacing the host gene with one that codes for a variant protein, or by introducing an additional gene.

Transgenic Animals

In 1978, yeast cells were the first to be transformed by insertion of foreign DNA, followed by mouse cells in 1979. Mouse embryos were transformed in 1980, which later led to the development of a "supermouse" that grew much larger than ordinary mice because it had received the gene for human growth hormone. Most of these transformations came after microinjection of DNA directly into cells. Later, scientists were able to deliver foreign genes into hosts by several other methods: incorporating them into retroviruses and then infecting target cells; electroinfusion, whereby an electric current passed the foreign DNA through the relatively flimsy animal cell wall; biolistics, a means of mechanically shooting a DNA bullet into cells; and conveying the DNA into an ovum aboard sperm. Two methods, developed at first for mice, are particularly successful in growing genetically modified animals after transformation. The first entails injecting transformed embryonic stem cells into a blastocyst (an early spherical form of an embryo). In the second, the DNA is inserted into the pronucleus of a freshly fertilized egg. The blastocyst or egg is then implanted into a foster mother for gestation.

The first complex transgenic animals were intended for genetic research. After disabling a specific gene, scientists could study its effect on the appearance, metabolic processes, and health of the mature animal. By 2003 thousands of genes had been tested. Also, research with mice transformed with human DNA enabled scientists to identify genes associated with breast and prostate cancers, cystic fibrosis, Alzheimer's disease, and severe combined im-

munodeficiency disorder (SCID). In 2001 the first transgenic primate, a rhesus monkey, was born, potentially supplying a research model genetically much more similar to humans than mice are.

Beginning in the late 1990's, transgenic animals were developed for production of proteins that can be used in pharmaceutical drugs to treat human disease. Accordingly, they have become known as "pharm animals." Lactating transgenic mice make tissue plasminogen activator in their milk. Similarly, transgenic sheep supply blood coagulation factor IX and alpha[1]-antitrypsin, transgenic pigs produce human hemoglobin, and transgenic cows make human lactoferrin. Scientists have also developed transgenic pigs that may supply tissue and organs for transplantation into humans without tissue rejection.

Transgenic Plants

Plant cells present greater difficulties for transformation because their cells walls are sturdier than animal cell walls. Microinjection and biolistics are possible but tricky and slow. A breakthrough for plant transgenesis came in 1983, when three separate teams of scientists used plasmids as vectors (carrier molecules) to infect plants with foreign DNA. The achievement came about because of research into plant tumors caused by crown gall disease. The pathogen, the soil bacterium *Agrobacterium tumefaciens*, caused the disease by ferrying bits of its own DNA into the genome of plants via plasmids, circular bits of extranuclear DNA. Scientists found that they could take the same plasmid, cut out bits of its DNA with enzymes and insert transgenes, and then use the altered plasmids as vectors to transform plants. Subsequently, scientists discovered that liposomes can be vectors. A liposome is a tiny ball of lipids that binds readily to a cell wall, opens a passage, and delivers any DNA that has been put inside it.

A great variety of transgenic plants have been designed for agriculture to produce genetically modified (GM) foods. The first to be marketed was a strain of tomato that ripened slowly so that it gained flavor by staying longer on the vine and remained ripe longer on super-

These two rhesus monkeys were born from cloned embryos in 1996. (AP/Wide World Photos)

market shelves. This Flavr Savr tomato was not a commercial success, however. Corn, cotton, soybeans, potatoes, and papayas received a gene from the bacterium *Bacillus thuringiensis* (*Bt*) that enables them to make a caterpillar-killing toxin; these are frequently referred to a *Bt* crops. Other crops have been made resistant to herbicides so that weeds can be easily killed without harming the food plants. Similarly, some transgenic crops tolerate salty or aluminum-rich soil, have less impact on the land because they require less water or tillage, or produce a high yield.

Like transgenic animals, some transgenic crops promise to deliver pharmaceuticals at lower costs and more conveniently than factory-made drugs. GM bananas and potatoes contain vaccines for protection against diarrheal diseases, such as cholera, and hepatitis B. In 2000, scientists reported invention of rice and wheat strains that produce anti-cancer antibodies. Golden rice, a transgenic strain that contains vitamin A, was developed to ward off blindness from vitamin A deficiency, which is a problem in countries that subsist largely on rice. Another strain has elevated iron levels to combat anemia. In a bid to reduce the health risk from smoking, a tobacco company developed a strain free of nicotine.

The Debate over Transgenesis

Transgenic organisms offer great benefits to humankind: deeper understanding of the genetic component in disease and aids in diagnosis; new, cheaper, more easily produced drugs; and crops that could help alleviate the growing hunger in the world. Yet during the 1990's protests against transgenesis began that are as contentious as any since the controversy over the pesticide DDT during the 1960's.

Some opponents object to the very fact that organisms are modified strictly for human benefit. They find such manipulations of life's essential code blasphemous and arrogant, or at the very least unethical and reckless. Furthermore, animal rights groups regard the production of transgenic pharm and research animals cruel and in violation of the natural rights of other species.

The greater portion of opponents, however, are concerned with specific dangers that transgenic organism may pose. Many consumers, most noticeably those in Europe, worry that GM foods contain hidden health risks. After transgenes were found to escape from crops and become part of wild plants, environmentalists proposed that there could be unforeseen and harmful ecological consequences, especially in the destruction of natural species and reduction of biodiversity.

Even those who welcome the creation of transgenic animals and plants are concerned about the legal and social effects. Principally, because biotechnology corporations can patent transgenic organisms, they potentially have great influence on agribusiness, perhaps to the detriment of small farmers and consumers.

—*Roger Smith*

See also: Antibodies; Biopesticides; Biopharmaceuticals; Genetic Engineering; Genetic Engineering: Agricultural Applications; Genetic Engineering: Medical Applications; Genetic Engineering: Risks; Genetic Engineering: Social and Ethical Issues; Genetically Modified (GM) Foods; Genomics; Human Growth Hormone; Hybridization and Introgression; Knockout Genetics and Knockout Mice; Lateral Gene Transfer; Model Organism: *Drosophila melanogaster*; Model Organism: *Mus musculus*; Model Organism: *Xenopus laevis*; Molecular Genetics; Viroids and Virusoids.

Further Reading

Brown, Kathryn, Karen Hopkin, and Sasha Nemecek. "GM Foods: Are They Safe?" *Scientific American* 284, no. 4 (2001): 52-57. Describes the risks and benefits in growing genetically modified foods and human consumption of them. Accompanied by graphics and tables that summarize and clarify technical matters.

Lurquin, Paul F. *The Green Phoenix: A History of Genetically Modified Plants.* New York: Columbia University Press, 2001. Written by a pioneer in the field of transgenic plants, this technically detailed but readable book requires a basic familiarity with microbiology and genetics. The author discusses the ecological and ethical controversies with insight and balance.

Nicholl, Desmond S. T. *An Introduction to Genetic Engineering.* 2d ed. New York: Cambridge University Press, 2002. A thorough, lucidly structured survey of the techniques and applications of genetic engineering. One chapter is devoted to transgenic plants and animals.

Velander, William, et al. "Transgenic Livestock as Drug Factories." *Scientific American* 276 (January, 1997). Explains how genetic engineering methods have resulted in the production of "pharm" animals whose milk contains large amounts of medicinal proteins.

Winston, Mark L. *Travels in the Genetically Modified Zone.* Cambridge, Mass.: Harvard University Press, 2002. A popular account of the agribusiness, government oversight, and science of genetically modified plants and animals. The science is explained cursorily for general readers.

Web Sites of Interest

Oak Ridge National Laboratory. Transgenic and Targeted Mutant Animal Database. http://www.ornl.gov/TechResources/Trans/hmepg.html. A searchable professional database about lines of genetically modified animals, methods used to create them and descriptions of the modified DNA, the expression of transgenes, and how transgenes are named.

TBASE: The Transgenic/Targeted Mutation Database, Jackson Laboratory, Bar Harbor, Maine. http://tbase.jax.org. Database of information about transgenic animals generated worldwide, searchable by species, technique, DNA construct, phenotype, laboratory. Features the "Knockout Model of the Month"—a discussion of new animal models—and a glossary.

Transgenic Crops: An Introduction and Resource Guide http://www.colostate.edu/programs/lifesciences/transgeniccrops. This richly illustrated site provides information about the history of plant breeding, the making of transgenic plants, government regulations, and risks and concerns. Also available in Spanish.

University of Michigan. Transgenic Animal Model Core. http://www.med.umich.edu/

tamc. A professional Web site for researchers seeking a host animal to test transgenes. However, it contains much useful general information about transgenics (especially transgenic rats), vectors, and laboratory procedures. With links and a photo gallery.

Transposable Elements

Fields of study: Bacterial genetics; Molecular genetics

Significance: *Transposable elements are discrete DNA sequences that have evolved the means to move (transpose) within the chromosomes. Transposition results in mutation and potentially large-scale genome rearrangements. Transposable elements contribute to the problem of multiple antibiotic resistance by mobilizing the genes of pathogenic bacteria for antibiotic resistance.*

Key terms

COMPOSITE TRANSPOSON: a transposable element that contains genes other than those required for transposition

RESISTANCE PLASMID (R PLASMID): a small, circular DNA molecule that replicates independently of the bacterial host chromosome and encodes a gene for antibiotic resistance

SELFISH DNA: a DNA sequence that has no apparent purpose for the host that spreads by forming additional copies of itself within the genome

TRANSPOSASE: an enzyme encoded by a transposable element that initiates transposition by cutting specifically at the ends of the element and randomly at the site of insertion

Jumping Genes

Transposable elements are DNA sequences that are capable of moving from one chromosomal location to another in the same cell. In some senses, transposable elements have been likened to intracellular viruses. The first genetic evidence for transposable elements was described by Barbara McClintock in the 1940's. She was studying the genetics of the pigmentation of maize (corn) kernels and realized that

Barbara McClintock

Though best known for her research on mobile genetic elements (for which she won the first unshared Nobel Prize in Physiology or Medicine awarded to a woman), Barbara McClintock's contributions to the field of genetics were many. McClintock's career in genetics spanned the development of the field itself. While an undergraduate at Cornell University (from which she earned her bachelor of science degree in 1923), she was invited to participate in a graduate course in genetics, then a fledgling discipline. She continued at Cornell as a graduate student (earning her Ph.D. in 1927), combining her interests in the microscopic internal structure of the cell (cytology) with the transmission of heritable traits (genetics).

McClintock's keen observational skills and her holistic approach to science allowed her to make significant advancements. It had only recently been established that the chromosomes (visible under the microscope) were the carriers of Gregor Mendel's "factors," or genes. McClintock used information gleaned from characteristics of the corn plant, *Zea mays* (maize), in conjunction with changes in its chromosomes to elucidate many aspects of genetic control. Her first major contribution was the identification and naming of the maize chromosomes. Shortly after, using cytological markers on the chromosomes, McClintock and graduate student Harriet Creighton demonstrated the correlation between patterns of inheritance and chromosomal crossover—the exchange of material between chromosomes. Breeding experiments focusing on "linkage groups" allowed McClintock to associate each of corn's ten chromosomes with the genes they carry.

At that point, Lewis Stadler, who was studying the mutagenic effects of X rays, sent some irradiated corn to McClintock. McClintock demonstrated that the resultant broken chromosomes can fuse into a ring, and she then hypothesized the existence of the telomere, a protective stabilizing structure at the end of the chromosome. At the University of Missouri, McClintock observed the ability of such broken chromosomes to go through a series of breakages and fusions (the breakage-fusion-bridge cycle) and identified concomitant chromosomal inversions and deletions. She ultimately discovered that certain genes could transfer from cell to cell and between chromosomes, thereby influencing the color patterns in the leaves and kernels of corn.

In 1941, McClintock moved to Cold Spring Harbor, New York, where she would remain. Doing much of her work before the discovery of the double helical structure of DNA, she rejected the simplistic one-way flow from DNA to RNA to protein outlined in the central dogma of molecular biology, seeking instead an explanation for the spatial and temporal variation in gene expression needed to link genetics to developmental and evolutionary change. Her own work showed that both the location and direction of genetic material, as well as the presence of other "controlling elements," had important effects on the expression of the gene.

In addition to this better-known work, McClintock identified the chromosomes of the bread mold *Neurospora* and described its meiotic cycle. She also headed a study aimed at conserving indigenous corn varieties in the Americas. In recognition of her place as one of the most distinguished scientists of the twentieth century, *The Barbara McClintock Papers* are available through the National Library of Medicine through its "Profiles in Science" Web site.

—*Lee Anne Martínez*

Barbara McClintock. (© The Nobel Foundation)

the patterns of inheritance were not following Mendelian laws. Furthermore, she surmised that insertion and excision of genetic material were responsible for the genetic patterns she observed. McClintock was recognized for this pioneering work with a Nobel Prize in Physiology or Medicine in 1983. It was not until the 1960's that the jumping genes that McClintock postulated were isolated and characterized. The first transposable elements to be well characterized were found in the bacteria *Escherichia coli* but have subsequently been found in the cells of many bacteria, plants, and animals.

Transposable elements are discrete DNA sequences that encode a transposase, an enzyme that catalyzes transposition. Transposition refers to the movement within a genome. The borders of the transposable element are defined by specific DNA sequences; often the sequences at either end of the transposable element are inverted repeats of one another. The transposase enzyme cuts the DNA sequences at the ends of the transposable element to initiate transposition and cuts the DNA at the insertion site. The site for insertion of the transposable element is not specific. Therefore, transposition results in random insertion into chromosomes and often results in mutation and genome rearrangement. In many organisms, transposition accounts for a significant fraction of all mutation. Although the details of the mechanism may vary, there are two basic mechanisms of transposition: conservative and replicative. In conservative transposition, the transposable element is excised from its original site and inserted at another. In replicative transposition, a copy of the transposable element is made and is inserted in a new location. The original transposable element remains at its initial site.

A subset of the replicative transposable elements includes the retrotransposons. These elements transpose through an RNA intermediate. Interestingly, their DNA sequence and organization are similar to those of retroviruses. It is likely that either retroviruses evolved from retrotransposons by gaining the genes to produce the proteins for a viral coat or retrotransposons evolved from retroviruses that lost the genes for a viral coat. This is one of the reasons that transposons are likened to viruses. Viruses can be thought of as transposons that gained the genes for a protein coat and thus the ability to leave one cell and infect others; conversely, transposons can be thought of as intracellular viruses.

Genetic Change and Selfish DNA

Transposition is a significant cause of mutation for many organisms. When McClintock studied the genetic patterns of maize kernel pigmentation, she saw the results of insertion and excision of transposable elements into and out of the pigment genes. Subsequently, it has been well established that mutations in many organisms are the result of insertion of transposable elements into and around genes. Transposition sometimes results in deletion mutations as well. Occasionally the transposase will cut at one end of the transposable element but skip the other end, cutting the DNA further downstream. This can result in a deletion of the DNA between the end of the transposable element and the cut site.

In addition to these direct results, it is believed that transposable elements may be responsible for large-scale rearrangements of chromosomes. Genetic recombination, the exchange of genetic information resulting in new combinations of DNA sequences, depends upon DNA sequence homology. Normally, recombination does not occur between nonhomologous chromosomes or between two parts of the same chromosome. However, transposition can create small regions of homology (the transposable element itself) spread throughout the chromosomes. Recombination occurring between homologous transposable elements can create deletions, inversions, and other large-scale rearrangements of chromosomes.

Scientists often take advantage of transposable elements to construct mutant organisms for study. The random nature of insertion ensures that many different genes can be mutated, the relatively large insertion makes it likely that there will be a complete loss of gene function, and the site of insertion is easy to locate to identify the mutated region.

Biologists often think of natural selection as working at the level of the organism. DNA sequences that confer a selective advantage to the organism are increased in number as a result of the increased reproductive success of the organisms that possess those sequences. It has been said that organisms are simply DNA's means of producing more DNA. In 1980, however, W. Ford Doolittle, Carmen Sapienza, Leslie Orgel, and Francis Crick elaborated on another kind of selection that occurs among DNA sequences within a cell. In this selection, DNA sequences are competing with each other to be replicated. DNA sequences that spread by forming additional copies of themselves will increase relative to other DNA sequences. There is selection for discrete DNA sequences to evolve the means to propagate themselves. One of the key points is that this selection does not work at the level of the organism's phenotype. There may be no advantage for the organism to have these DNA sequences. In fact, it may be that there is a slight disadvantage to having many of these DNA sequences. For this reason, DNA sequences that are selected because of their tendency to make additional copies of themselves are referred to as "selfish" DNA. Transposable elements are often cited as examples of selfish DNA.

Composite Transposons and Antibiotic Resistance

Some transposable elements have genes unrelated to the transposition process located between the inverted, repeat DNA sequences that define the ends of the element. These are referred to as composite transposons. Very frequently, bacterial composite transposons contain a gene that encodes resistance to antibiotics. The consequence is that the antibiotic resistance gene is mobilized: It will jump along with the rest of the transposable element to new DNA sites. Composite transposons may be generated when two of the same type of transposable elements end up near each other and flanking an antibiotic resistance gene. If mutations occurred to change the sequences at the "inside ends" of the transposable elements, the transposase would then only recognize and cut at the two "outside end" sequences to cause everything in between to be part of a new composite transposon.

Resistance to antibiotics is a growing public health problem that threatens to undo much of the progress that the antibiotic revolution made against infectious disease. Transposition of composite transposons is part of the problem. Transposition can occur between any two sites within the same cell, including between the chromosome and plasmid DNA. Plasmids are small, circular DNA molecules that replicate independently of the bacterial host chromosome. Resistance plasmids (R plasmids) are created when composite transposons carrying an antibiotic resistance gene insert into a plasmid. What makes this particularly serious is that some plasmids encode fertility factors (genes that promote the transfer of the plasmid from one bacteria to another). This provides a mechanism for rapid and widespread antibiotic resistance whenever antibiotics are used. The great selective pressure exerted by antibiotic use results in the spread of R plasmids throughout the bacterial population. This, in turn, increases the opportunities for composite transposon insertion into R plasmids to create multiple drug-resistant R plasmids. The first report of multiple antibiotic resistance caused by R plasmids was in Japan in 1957 when strains of *Shigella dysenteriae*, which causes dysentery, became resistant to four common antibiotics all at once. Some R plasmids encode resistance for up to eight different antibiotics, which often makes treatment of bacterial infection difficult. Furthermore, some plasmids are able to cause genetic transfer between bacterial species, limiting the usefulness of many antibiotics.

—*Craig S. Laufer*

See also: Antisense RNA; Archaea; Bacterial Resistance and Super Bacteria; Immunogenetics; Lateral Gene Transfer; Model Organism: *Escherichia coli*; Molecular Genetics; Mutation and Mutagenesis; Plasmids; Repetitive DNA.

Further Reading

Capy, Pierre, et al. *Dynamics and Evolution of Transposable Elements.* New York: Chapman & Hall, 1998. Addresses the structure of the transposable elements, heterochromatin, host phylogenies, the origin and coevolution of retroviruses, evolutionary links between telomeres and transposable elements, population genetics models of transposable elements, and more. Illustrations, maps, bibliography.

Keller, Evelyn Fox. *A Feeling for the Organism: The Life and Work of Barbara McClintock*. 10th anniversary ed. New York: W. H. Freeman, 1993. A now-classic look at McClintock and her pathbreaking work.

McClintock, Barbara. *The Discovery and Characterization of Transposable Elements: The Collected Papers of Barbara McClintock*. New York: Garland, 1987. Presents the papers on transposable elements that McClintock wrote between 1938 and 1984. Illustrations, bibliography.

McDonald, John F., ed. *Transposable Elements and Genome Evolution*. London: Kluwer Academic, 2000. Includes the seminal papers presented in 1992 at the University of Georgia during a meeting of molecular, population, and evolutionary geneticists to discuss the relevance of their research to the role played by transposable elements in evolution. Illustrations, bibliography, index.

Tumor-Suppressor Genes

Field of study: Molecular genetics
Significance: *Molecular analysis of tumor-suppressor genes has provided important information on mechanisms of cell cycle regulation and patterns of growth control in normal dividing cells and cancer cells. Tumor-suppressor genes represent cell cycle control genes that inhibit cell division and initiate cell death processes in abnormal cells. Mutations in these genes have been identified in many types of human cancer and play a critical role in the genetic destabilization and loss of growth control characteristic of malignancy.*

Key terms

CELL CYCLE: a highly regulated series of events critical to the initiation of cell division processes
ONCOGENES: a mutated or improperly expressed gene that can cause cancer; the nor-

mal form of an oncogene, called a protooncogene, is involved in regulating the cell cycle
p53 GENE: a tumor-suppressor gene, implicated in many types of cancer

Discovery of Tumor-Suppressor Genes

The existence of genes that play critical roles in cell cycle regulation by inhibiting cell division was predicted by several lines of evidence. In vitro studies involving the fusion of normal and cancer cell lines were often observed to result in suppression of the malignant phenotype, suggesting that normal cells contained inhibitors that could reprogram the abnormal growth behavior in the cancer cell lines. In addition, studies by Alfred Knudsen on inherited and noninherited forms of retinoblastoma, a childhood cancer associated with tumor formation in the eye, suggested that the inactivation of recessive genes as a consequence of mutation could result in the loss of function of inhibitory gene products critical to cell division control. With the advent of molecular methods of genetic analysis, the gene whose inactivation is responsible for retinoblastoma was identified and designated *Rb*.

Additional tumor-suppressor genes were identified by studies of DNA tumor viruses whose cancer-causing properties were found to result, in part, from the ability of specific viral gene products to inactivate host cell inhibitory gene products involved in cell cycle regulation. By inactivating these host cell proteins, the tumor virus removes the constraints on viral and cellular proliferation. The most important cellular gene product to be identified in this way is the p53 protein, named after its molecular weight. Genetic studies of human malignancies have implicated mutations in the *p53* gene in up to 75 percent of tumors of diverse tissue origin, including an inherited disorder called Li-Fraumeni syndrome, associated with many types of cancer. In addition, studies of other rare inherited malignancies have led to the identification of many other recessive genes whose inactivation contributes to oncogenic or cancer-causing mechanisms. Included in this list are the *BRCA1* and *BRCA2* genes in breast cancer, the *NF1* gene in neurofibromatosis, the *p16* gene in melanoma, and the *APC* gene in colorectal carcinoma. Each of these genes has also been implicated in nonhereditary cancers.

The Properties of Tumor-Suppressor Genes

Molecular analyses of the genetic and biochemical properties of tumor-suppressor genes have suggested that these gene products play critical but distinct roles in regulating processes involved in cellular proliferation. The *Rb* gene product represents a prototype tumor-suppressor gene that blocks progression of the cell cycle and cell division by binding to transcription factors in its active form. In order for cell division to occur in response to growth factor stimulation, elements of the signal cascade inactivate *Rb*-mediated inhibition by a mechanism involving the addition of phosphate to the molecule, a reaction called phosphorylation. Loss of *Rb* function as a consequence of mutation removes the brakes on this form of inhibitory control; the cell division machinery proceeds regardless of appropriate initiation by growth factors or other stimuli.

The *p53* tumor-suppressor gene product is a DNA-binding protein that regulates the expression of specific genes in response to genetic damage or other abnormal events that may occur during cell cycle progression. In response to p53 activation, the cell may arrest the process of cell division (by indirectly blocking Rb inactivation) to repair genetic damage before proceeding further along the cell cycle; alternatively, if the damage is too great, the *p53* gene product may initiate a process of cell death called apoptosis. The loss of p53 activity in the cell as a consequence of mutation results in genetic destabilization and the failure of cell death mechanisms to eliminate damaged cells from the body; both events appear to be critical to late-stage oncogenic mechanisms.

Impact and Applications

The discovery of tumor-suppressor genes has revealed the existence of inhibitory mechanisms critical to the regulation of cellular proliferation. Mutations that destroy the functional activities of these gene products cause the loss of growth control characteristic of cancer cells. Taken together, research on the patterns of oncogene activation and the loss of tumor-suppressor gene function in many types of human malignancy suggest a general model of oncogenesis. Molecular analyses of many tumors show multiple genetic alterations involving both oncogenes and tumor-suppressor genes, suggesting that oncogenesis (development of cancer) requires unregulated stimulation of cellular proliferation pathways along with a loss of inhibitory activities that operate at cell cycle checkpoints.

With respect to clinical applications, restoration of *p53* tumor-suppressor gene function by gene therapy appears to result in tumor regression in some experimental systems; however, much more work needs to be done in this area to achieve clinical relevance. More important, research on the mechanism of action of standard chemotherapeutic drugs suggests that cytotoxicity may be caused by p53-induced cell death; the absence of functional p53 in many tumors may account for their resistance to chemotherapy. Promising research suggests that it may be possible to elicit cell death in tumor cells lacking functional *p53* gene product in response to chemotherapy. The clinical significance of activating these p53-independent cell death mechanisms may be extraordinary.

—Sarah Crawford Martinelli

See also: Aging; Breast Cancer; Cancer; Cell Cycle, The; Cell Division; DNA Repair; Human Genetics; Human Genome Project; Model Organism: *Mus musculus*; Oncogenes.

Further Reading

Ehrlich, Melanie, ed. *DNA Alterations in Cancer: Genetic and Epigenetic Changes.* Natick, Mass.: Eaton, 2000. A comprehensive overview of the numerous and varied genetic alterations leading to the development and progression of cancer. Topics include oncogenes, tumor-suppressor genes, cancer predisposition, DNA repair, and epigenetic alteration such as methylation.

Fisher, David E., ed. *Tumor Suppressor Genes in Human Cancer.* Totowa, N.J.: Humana Press, 2001. Covers models used in suppressor gene studies, cancer drugs, and gene descriptions. Illustrated (some color).

Habib, Nagy A. *Cancer Gene Therapy: Past Achievements and Future Challenges.* New York: Kluwer Academic/Plenum, 2000. Reviews forty-one preclinical and clinical studies in cancer gene therapy, organized into sections on the vectors available to carry genes into tumors, cell cycle control, apoptosis, tumor-suppressor genes, antisense and ribozymes, immunomodulation, suicidal genes, angiogenesis control, and matrix metalloproteinase.

Lattime, Edmund C., and Stanton L. Gerson, eds. *Gene Therapy of Cancer: Translational Approaches from Preclinical Studies to Clinical Implementation.* 2d ed. San Diego: Academic Press, 2002. Provides a comprehensive review of the basis and approaches involved in the gene therapy of cancer.

Maruta, Hiroshi, ed. *Tumor-Suppressing Viruses, Genes, and Drugs: Innovative Cancer Therapy Approaches.* San Diego: Academic, 2002. An international field of experts address potential alternative cancer treatments, such as viral and drug therapy and gene therapy with tumor-suppressor genes.

Oliff, Alan, et al. "New Molecular Targets for Cancer Therapy." *Scientific American* 275 (September, 1996). Summarizes novel genetic approaches to cancer treatment.

Ruddon, Raymond. *Cancer Biology.* 3d ed. New York: Oxford University Press, 1995. A good general text.

Web Site of Interest

American Cancer Society. http://www.cancer.org. Site has searchable information on tumor suppressor genes.

Turner Syndrome

Field of study: Diseases and syndromes

Significance: *Turner syndrome is one of the most common genetic problems in women, affecting 1 out of every 2,000 to 2,500 women born. Short stature, infertility, and incomplete sexual development are the characteristics of this condition.*

Key terms

ESTROGENS: hormones or chemicals that stimulate the development of female sexual characteristics and control the female reproductive cycle

GROWTH HORMONE: a chemical that plays a key role in promoting growth in body size

KARYOTYPE: a laboratory analysis that confirms the diagnosis of Turner syndrome by documenting the absence or abnormality of one of the two X chromosomes normally found in women

SEX CHROMOSOMES: the chromosomes that control sexual determination during the development of males and females; females have two X chromosomes and males have one X and one Y chromosome

SYNDROME: a set of features or symptoms often occurring together and believed to stem from the same cause

Discovery of Turner Syndrome

Henry H. Turner, an eminent clinical endocrinologist, is credited with first describing Turner syndrome. In 1938, he published an article describing seven patients, ranging in age from fifteen to twenty-three years, who exhibited short stature, a lack of sexual development, arms that turned out slightly at the elbows, webbing of the neck, and low posterior hairline. He did not know what caused this condition. In 1959, C. E. Ford discovered that a chromosomal abnormality involving the sex chromosomes caused Turner syndrome. He found that most girls with Turner syndrome did not have all or part of one of their X chromosomes and argued that this missing genetic material accounted for the physical findings associated with the condition.

Turner syndrome begins at conception. The disorder results from an error during meiosis in the production of one of the parents' sex cells, although the exact cause remains unknown. Girls suspected of having Turner syndrome, usually because of their short stature, usually undergo chromosomal analysis. A simple blood test and laboratory analysis called a karyotype are done to document the existence of an abnormality.

Shortness is the most common characteristic of Turner syndrome. The incidence of short stature among women with Turner syndrome is virtually 100 percent. Women who have this condition are, on average, 4 feet 8 inches (1.4 meters) tall. The cause of the failure to grow is unclear. However, growth-promoting therapy with growth hormones has become standard. Most women with the syndrome also experience ovarian failure. Since the ovaries normally produce estrogen, women with Turner syndrome lack this essential hormone. This deficit results in infertility and incomplete sexual development. Cardiovascular disorders are the single source of increased mortality in women with this condition. High blood pressure is common.

Other physical features often associated with Turner syndrome include puffy hands and feet at birth, a webbed neck, prominent ears, a small jaw, short fingers, a low hairline at the back of the neck, and soft fingernails that turn up at the ends. Some women with Turner syndrome have a tendency to become overweight. Many women will exhibit only a few of these distinctive features, and some may not show any of them. This condition does not affect general intelligence. Girls with Turner syndrome follow a typical female developmental pattern with unambiguous female gender identification. However, another possible symptom is poor spatial perception abilities. For example, women with this condition may have difficulty driving, recognizing subtle social clues, and solving nonverbal mathematics problems; they may also suffer from clumsiness and attention-deficit disorder.

Treatments and Therapies

No treatment is available to correct the chromosome abnormality that causes this condition. However, injections of human growth hormone can restore most of the growth deficit. Unless they undergo hormone replacement therapy, girls with Turner syndrome will not menstruate or develop breasts and pubic hair. In addition to estrogen replacement therapy, women with Turner syndrome are often advised to take calcium and exercise regularly. Although infertility cannot be altered, pregnancy may be made possible through in vitro fertilization (fertilizing a woman's egg with sperm outside the body) and embryo transfer (moving the fertilized egg into a woman's uterus). Individuals with Turner syndrome can be healthy, happy, and productive members of society.

Nevertheless, because of its relative rarity, a woman with Turner syndrome may never meet another individual with this condition and may suffer from self-consciousness, embarrassment, and poor self-esteem. The attitudes of parents, siblings, and relatives are important in helping develop a strong sense of identity and self-worth. The Turner Syndrome Society of the United States is a key source of information and support groups. Advances in chromosomal analysis have proved helpful in the diagnosis and management of Turner syndrome. In addition, new developments in hormonal therapy for short stature and ovarian failure, combined with advances in in vitro fertilization, have significantly improved the potential for growth, sexual development, and fertility for afflicted individuals.

—*Fred Buchstein*

See also: Amniocentesis and Chorionic Villus Sampling; Dwarfism; Hereditary Diseases; Infertility; Klinefelter Syndrome; Mutation and Mutagenesis; Nondisjunction and Aneuploidy; X Chromosome Inactivation; XYY Syndrome.

Further Reading

Albertsson-Wikland, Kerstin, and Michael B. Ranke, eds. *Turner Syndrome in a Life Span Perspective: Research and Clinical Aspects.* New York: Elsevier, 1995. Focuses on molecular genetic evaluation, prenatal diagnosis, growth, medical and psychosocial management, pediatric and adult care, estrogen substitution therapy, bone mineralization, hearing, and more. Illustrations, bibliography, index.

Broman, Sarah H., and Jordan Grafman, eds. *Atypical Cognitive Deficits in Developmental Disorders: Implications for Brain Function.* Hillsdale, N.J.: Lawrence Erlbaum, 1994. Includes multidisciplinary research about Turner syndrome. Illustrations, bibliography, index.

Rieser, Patricia A., and Marsha Davenport. *Turner Syndrome: A Guide for Families.* Houston, Tex.: Turner Syndrome Society, 1992. Describes the syndrome's causes and features, and how families can help girls with the condition cope with the physical, social, and emotional concerns.

Rosenfeld, Ron G. *Turner Syndrome: A Guide for Physicians.* 2d ed. Houston, Tex.: Turner Syndrome Society, 1992. Definition, etiology, and management of Turner syndrome.

Saenger, Paul, and Anna-Maria Pasquino, eds. *Optimizing Health Care for Turner Patients in the Twenty-first Century.* New York: Elsevier, 2000. Covers the management of health care from infancy through the adult years. Discusses molecular genetics, prenatal diagnosis, cognitive function, reproduction, and more. Illustrations, index.

Web Sites of Interest

Intersex Society of North America. http://www .isna.org. The society is "a public awareness, education, and advocacy organization which works to create a world free of shame, secrecy, and unwanted surgery for intersex people (individuals born with anatomy or physiology which differs from cultural ideals of male and female)." Includes links to information on such conditions as clitoromegaly, micropenis, hypospadias, ambiguous genitals, early genital surgery, adrenal hyperplasia, Klinefelter syndrome, androgen insensitivity, and testicular feminization.

Johns Hopkins University, Division of Pediatric Endocrinology, Syndromes of Abnormal Sex Differentiation. http://www.hopkins medicine.org/pediatricendocrinology. A guide to the science and genetics of sex differentiation, including a glossary. Click on "patient resources."

National Institute of Child Health and Human Development. http://turners.nichd.nih.gov. Site provides information on the genetic and clinical features of the syndrome.

The Turner Syndrome Society of the United States. http://www.turner-syndrome-us.org. The main national support organization, offering resources and information.

Twin Studies

Field of study: Techniques and methodologies

Significance: *Studies of twins are widely considered to be the best way to determine the relative contributions of genetic and environmental factors to the development of human physical and psychological characteristics.*

Key terms

DIZYGOTIC: developed from two separate zygotes; fraternal twins are dizygotic because they develop from two separate fertilized ova (eggs)

MONOZYGOTIC: developed from a single zygote; identical twins are monozygotic because they develop from a single fertilized ovum that splits in two

ZYGOSITY: the degree to which two individuals are genetically similar

ZYGOTE: a cell formed from the union of a sperm and an ovum

The Origin of Twin Studies

Sir Francis Galton, an early pioneer in the science of genetics and a founder of the theory of eugenics, conducted some of the earliest systematic studies of human twins in the 1870's. Galton recognized the difficulty of identifying the extent to which human traits are biologically inherited and the extent to which traits are produced by diet, upbringing, education, and other environmental influences. Borrowing a phrase from William Shakespeare, Galton called this the "nature vs. nurture" problem. Galton reasoned that he could attempt to find an answer to this problem by comparing similarities among people who obviously shared a great deal of biological inheritance, with similarities among people sharing less biological inheritance. Twins offered the clearest example of people who shared common biological backgrounds.

Galton contacted all of the twins he knew and asked them to supply him with the names of other twins. He obtained information on ninety-four sets of twins. Of these, thirty-five sets were very similar, people who would today be called identical twins. These thirty-five pairs reported that people often had difficulty telling them apart. Using questionnaires and interviews, Galton compared the thirty-five identical pairs with the other twins. He found that the identical twins were much more similar to one another in habits, interests, and personalities, as well as in appearance. They were even much more alike in physical health and susceptibility to illness. The one area in which all individuals seemed to differ markedly was in handwriting.

Modern Twin Studies

Since Galton's time, researchers have discovered how biological inheritance occurs, and this has made possible an understanding of why twins are similar. It has also enabled researchers to make more sophisticated use of twins in studies that address various aspects of the nature vs. nurture problem. Parents pass their physical traits to their children by means of genes in chromosomes. Each chromosome carries two genes (called alleles) for every hereditary trait. One allele comes from the father and one comes from the mother. Any set of full brothers and sisters will share many of the same alleles, since all of their genes come from the same parents. However, brothers and sisters usually also differ substantially; each zygote (ovum, or egg, fertilized by a sperm cell) will combine alleles from the father and the mother in a unique manner, so different zygotes will develop into unique individuals. Even when two fertilized eggs are present at the same time, as in the case of dizygotic or fraternal twins, the two will have different combinations of genes from the mother and the father.

Identical twins are an exception to the rule of unique combinations of genes. Identical twins develop from a single zygote, a cell created by one union of egg and sperm. There-

In the background, the co-director of the Twins Reared Apart project, Nancy Segal, with twins Mark Newman and Gerald Levey, separated at birth. (AP/Wide World Photos)

fore, monozygotic twins (from one zygote) will normally have the same genetic makeup. Differences between genetic twins, researchers argue, must therefore be produced by environmental factors following birth.

The ideal way to conduct twin studies is to compare monozygotic twins who have been reared apart from each other in vastly different types of families or environments. This is rarely possible, however, because the number of twins separated at birth and adopted is relatively small. For this reason, researchers in most twin studies use fraternal twins as a comparison group, since the major difference between monozygotic and dizygotic twins is that the former are genetically identical. Statistical similarities among monozygotic twins that are not found among dizygotic twins are therefore believed to be caused by genetic inheritance.

Researchers use several types of data on twins to estimate the extent to which human characteristics are the consequence of genetics. One of the main sources for twin studies is the Minnesota Twin Registry. In the 1990's, this registry consisted of about 10,500 twins in Minnesota. They were found in Minnesota birth records from the years 1936 through 1955, and they were located and recruited by mail between 1985 and 1990. A second major source of twin studies is the Virginia Twin Registry. This is a register of twins constructed from a systematic review of public birth records in the Commonwealth of Virginia. A few other states also maintain records of twins. Some other organizations, such as the American Association of Retired Persons (AARP), keep records of twins who volunteer to participate and make these records available to researchers.

Zygosity, or degree of genetic similarity between twins, is usually measured by survey questions about physical similarity and by how often other people mistake one twin for the other. In some cases, zygosity may be determined more rigorously through analysis of DNA samples.

Problems with Twin Studies

Although twin studies are one of the best available means for studying genetic influences in human beings, there are a number of problems with this approach. Although twin studies

assume that monozygotic twins are biologically identical, some critics have claimed that there are reasons to question this assumption. Even though these twins tend to show greater uniformity than other people, developmental differences may emerge even in the womb after the splitting of the zygote.

Twins who show a great physical similarity may also be subject to environmental similarities so that traits believed to be caused by genetics may, in fact, be a result of upbringing. Some parents, for example, dress twins in matching clothing. Even when twins grow up in separate homes without being in contact with each other, their appearances and mannerisms may evoke the same kinds of responses from others. Physical attractiveness, height, and other characteristics often affect how individuals are treated by others so that the biologically based resemblances of twins can lead to common experiences.

Finally, critics of twin studies point out that twins constitute a special group of people and that it may be difficult to apply findings from twin studies to the population at large. Some studies have indicated that intelligence quotient (IQ) scores of twins, on average, are about five points below IQ scores in the general population, and twins may differ from the general population in other respects. It is conceivable that genetics plays a more prominent role in twins than in most other people.

Impact and Applications

Twin studies have provided evidence that a substantial amount of human character and behavior may be genetically determined. In 1976, psychologists John C. Loehlin and Robert C. Nichols published their analyses of the backgrounds and performances of 850 sets of twins who took the 1962 National Merit Scholarship test. Results showed that identical twins showed greater similarities than fraternal twins in abilities, personalities, opinions, and ambitions. A careful examination of backgrounds indicated that these similarities could not be explained by the similar treatment received by identical twins during upbringing.

Later twin studies continued to provide evidence that genes shape many areas of human life. Monozygotic twins tend to resemble each other in probabilities of developing mental illnesses, such as schizophrenia and depression, suggesting that these psychological problems are partly genetic in origin. A 1996 study published in the *Journal of Personality and Social Psychology* used a sample from the Minnesota Twin Registry to establish that identical twins are similar in probabilities of divorce. A 1997 study in the *American Journal of Psychiatry* indicated that there is even a great resemblance between twins in intensity of religious faith. Twin studies have offered evidence that homosexual or heterosexual orientation may be partly a genetic matter, although researcher Scott L. Hershberger has found that the genetic inheritance of sexual orientation may be greater among women than among men.

—*Carl L. Bankston III*

See also: Aging; Animal Cloning; Behavior; Cloning: Ethical Issues; Diabetes; DNA Fingerprinting; Gender Identity; Genetic Testing; Genetics in Television and Films; Heredity and Environment; Homosexuality; Intelligence; Prenatal Diagnosis; Quantitative Inheritance.

Further Reading

Hershberger, Scott L. "A Twin Registry Study of Male and Female Sexual Orientation." *Journal of Sex Research* 34, no. 2 (1997). Discusses the results of a Minnesota study of twins, homosexual orientation, sibling environment, and the genetic influence on sexuality.

Loehlin, John C., and Robert C. Nichols. *Heredity, Environment, and Personality: A Study of 850 Sets of Twins.* Austin: University of Texas Press, 1976. One of the most influential books on modern twin studies. Graphs, bibliography, index.

Piontelli, Alessandra. *Twins: From Fetus to Child.* New York: Routledge, 2002. A longitudinal study of the everyday lives of thirty pairs of twins, from life in the womb to age three. Illustrations, bibliography, index.

Spector, Tim D., Harold Snieder, and Alex J. MacGregor, eds. *Advances in Twin and Sib-Pair Analysis.* New York: Oxford University Press, 2000. Discusses background and context of twin and sib-pairs analysis, and epidemiological and biostatistical perspectives on the study of complex diseases. Illustrations, bibliography, index.

Steen, R. Grant. *DNA and Destiny: Nurture and Nature in Human Behavior.* New York: Plenum, 1996. A medical researcher and popular science writer summarizes evidence regarding the relative contributions of genetics and environment in shaping human personalities. Bibliography, index.

Wright, Lawrence. *Twins: And What They Tell Us About Who We Are.* New York: John Wiley & Sons, 1997. An overview of the use of twin studies in behavioral genetics, written for general readers. Bibliography, index.

Web Site of Interest

Minnesota Twin Family Study. http://www.psych.umn.edu/psylabs/mtfs. Site of an ongoing research study into the genetic and environmental factors of psychological development. Includes the discussion, "What's Special About Twins to Science?"

Viral Genetics

Field of study: Viral genetics
Significance: *The composition and structures of virus genomes are more varied than any identified in the entire bacterial, botanical, or animal kingdoms. Unlike the genomes of all other cells, which are composed of DNA, virus genomes may contain their genetic information encoded in either DNA or RNA. Viruses cannot replicate on their own but must instead use the reproductive machinery of host cells to reproduce themselves.*

Key terms

CAPSID: the protective protein coating of a virus particle

RIBOSOME: a cytoplasmic organelle that serves as the site for amino acid incorporation during the synthesis of protein

VIRIONS: mature infectious virus particles

What Is a Virus?

Viruses are submicroscopic, obligate intracellular parasites. This definition differentiates viruses from all other groups of living organisms. There exists more biological diversity within viruses than in all other known lifeforms combined. This is the result of viruses successfully parasitizing all known groups of living organisms. Viruses have evolved in parallel with other species by capturing and using genes from infected host cells for functions that they require to produce their progeny, to enhance their escape for their host's cells and immune system, and to survive the intracellular and extracellular environment. At the molecular level, the composition and structures of virus genomes are more varied than any others identified in the entire bacterial, botanical, or animal kingdoms. Unlike the genomes of all other cells composed of DNA, virus genomes may contain their genetic information encoded in either DNA or RNA. The nucleic acid comprising a virus genome may be single-stranded or double-stranded and may occur in a linear, circular, or segmented configuration.

The Need for a Host

It must be understood that virus particles themselves do not grow or undergo division. Virus particles are produced from the assembling of pre-formed components, whereas other agents actually grow from an increase in the integrated sum of their components and reproduce by division. The reason is that viruses lack the genetic information that encodes the apparatus necessary for the generation of metabolic energy or for protein synthesis (ribosomes). The most critical interaction between a virus and a host cell is the need of the virus for the host's cellular apparatus for nucleic acid and for the synthesis of proteins. No known virus has the biochemical or genetic potential to generate the energy necessary for producing all biological processes. Viruses depend totally on a host cell for this function.

Viruses are therefore not living in the traditional sense, but they nevertheless function as living things; they do replicate their own genes. Inside a host cell, viruses are "alive," whereas outside the host they are merely a complex assemblage of metabolically inert chemicals—basically a protein shell. Therefore, while viruses have no inner metabolism and cannot reproduce on their own, they carry with them the means necessary to get into other cells and then use those cells' own reproductive machinery to make copies of themselves. Viruses thrive at the host cell's expense.

Replication

The sole goal of a virus is to replicate its genetic information. The type of host cell infected by a virus has a direct effect on the process of replication. For viruses of prokaryotes (bacteria, primarily), reproduction reflects the physical simplicity of the host cell. For viruses with eukaryotic host cells (plants and animals), reproduction is more complex. The coding capacity of the genome forces the virus to choose a reproductive strategy. The strategy might involve near-total reliance on the host cell, resulting in a compact genome encoded for only a few essential proteins (+), or could involve a large, complex virus genome encoded with nearly all the information necessary for replication, relying on the host cell only for energy and ribosomes. Those viruses with an RNA genome plus messenger RNAs (mRNAs) have no need to enter the nucleus of their host cell, although during replication many often do. DNA genome viruses mostly replicate in the host cell's nucleus, where host DNA is replicated and the biochemical apparatus required for this process is located. Some DNA viruses (poxviruses) have evolved to contain the biochemical capacity to replicate in their host's cytoplasm, with a minimal need for the host cell's other functions.

Virus replication involves several stages carried out by all types of viruses, including the onset of infection, replication, and release of mature virions from an infected host cell. The stages can be defined in eight basic steps: attachment, penetration, uncoating, replication, gene expression, assembly, maturation, and release.

The first stage, attachment, occurs when a virus interacts with a host cell and attaches itself—binds with a virus-attachment protein (anitreceptor)—to a cellular receptor molecule in the cell membrane. The receptor may be a protein or a carbohydrate residue. Some complex viruses, such as herpesviruses, use more than one receptor and therefore have alternate routes of cellular invasion.

Shortly after attachment the target cell is penetrated. Cell penetration is usually an energy-dependent process, and the cell must be metabolically active for penetration to occur. The virus bound to the cellular receptor molecule is translocated across the cell membrane by the receptor and is engulfed by the cell's cytoplasm.

Uncoating occurs after penetration and results in the com-

plete or partial removal of the virus capsid and the exposure of the virus genome as a nucleoprotein complex. This protein complex can be a simple RNA genome or can be highly complex, as in the case of a retrovirus containing a diploid RNA genome responsible for converting a virus RNA genome into a DNA provirus.

How a virus replicates and the resulting expression of its genes depends on the nature of its genetic materials. Control of gene expression is a vital element of virus replication. Viruses use the biochemical apparatus of their

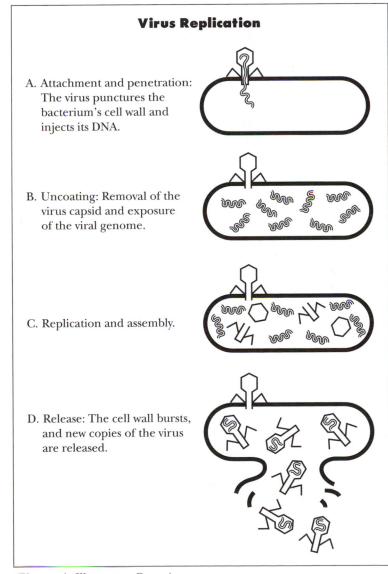

Virus Replication

A. Attachment and penetration: The virus punctures the bacterium's cell wall and injects its DNA.

B. Uncoating: Removal of the virus capsid and exposure of the viral genome.

C. Replication and assembly.

D. Release: The cell wall bursts, and new copies of the virus are released.

(Electronic Illustrators Group)

infected host cells to express their genetic information as proteins and do this by using the appropriate biochemical language recognized by the host cell. Viruses include double-stranded DNA viruses such as papoviruses, poxviruses, and herpesviruses; single-stranded sense DNA viruses such as parvoviruses; double-stranded RNA reoviruses; single-stranded sense RNA viruses such as flaviviruses, togaviruses, and claiciviruses; single-stranded antisense RNA such as filoviruses and bunyaviruses; single-stranded sense RNA with DNA intermediate retroviruses; and double-stranded DNA with RNA intermediate-like hepadnaviruses.

During assembly, the basic structure of the virus particle is formed. Virus proteins anchor themselves to the cellular membrane, and, as virus proteins and genome molecules reach a critical concentration, assembly begins. The result is that a genome is stuffed into a completed protein shell. The process of maturation prepares the virus particle for infecting subsequent cells and usually involves the cleavage of proteins to form matured products or conformational structural changes.

For most viruses, release is a simple matter of breaking open the infected cell and exiting. The breakage normally occurs through a physical interaction of proteins against the inner surface of the host cell membrane. A virus may also exit a cell by budding. Budding involves the creation of a lipoprotein envelope around the virion prior to the virion's being extruded out through the cell membrane.

—*Randall L. Milstein*

See also: DNA Structure and Function; Gene Regulation: Viruses; Genetic Engineering; Hybridomas and Monoclonal Antibodies; Oncogenes; Organ Transplants and HLA Genes; RNA Structure and Function; Viroids and Virusoids.

Further Reading

Becker, Yechiel, and Gholamreza Darai, eds. *Molecular Evolution of Viruses: Past and Present.* Boston: Kluwer Academic, 2000. Detailed research of the evolution of viruses by acquisition of cellular RNA and DNA, and how virus genes evade the host immune response.

Domingo, Esteban, Robert Webster, and John Holland, eds. *Origin and Evolution of Viruses.* New York: Academic Press, 1999. An interdisciplinary reference book consisting of chapters authored by leading researchers in the fields of RNA and DNA viruses; deals with the simplest, as well as the most complex, viral genomes known.

Holland, J. J., ed. *Current Topics in Microbiology and Immunology: Genetic Diversity of RNA Viruses.* New York: Springer-Verlag, 1992. Detailed collection of papers concerning the genetic and biological variabilities of RNA viruses, replicase error frequencies, the role of environmental selection pressures in the evolution of RNA populations, and the emergence of drug resistant virus genomes.

Zimmer, Carl. *Parasite Rex: Inside the Bizarre World of Nature's Most Dangerous Creatures.* New York: Free Press, 2000. A fascinating, general publication on parasites and their effects on their host organisms, with many references to the historical ramifications of viruses on biological evolution and the development of human culture.

Viroids and Virusoids

Field of study: Viral genetics

Significance: *Viroids are naked strands of RNA, 270 to 380 nucleotides long, that are circular and do not code for any proteins. However, some viroids are catalytic RNAs (ribozymes), able to cleave and ligate themselves. In spite of their simplicity, they are able to cause disease in susceptible plants, many of them economically important. Virusoids are similar to viroids, except that they require a helper virus to infect a plant and reproduce.*

Key terms

RNA POLYMERASE: an enzyme that catalyzes the joining of ribonucleotides to make RNA using DNA or another RNA strand as a template

RNASE: an enzyme that catalyzes the cutting of an RNA molecule

General Characteristics of Viroids and Virusoids

Viroids, and some virusoids, are circular, single-stranded RNA molecules, which normally appear as rods but when denatured by heating appear as closed circles. The rod-shaped structure is formed by extensive base pairing within the RNA molecule, and the secondary structure is divided into five structural domains. One domain is called the pathogenicity (P) domain, because differences among variant strains of the same species of viroid seem to correlate with differences in pathogenicity. Virusoids may also comprise linear RNA or, rarely, double-stranded circular RNA.

The difference between viroids and virusoids is in their mode of transmission. Viroids have no protective covering of any kind and are no more than the RNA that makes up their genetic material. They depend on breaks in a plant's epidermis or can travel with pollen or ovules to gain entry. Virusoids, also known as satellite RNAs, are packaged in the protein coat of other plant viruses, referred to as helpers, and are therefore dependent on the other virus.

Viroids are typically divided into two groups based on the nature of their RNA molecule. Group A is the smallest group, and their RNA has the ability to self-cleave. These include the avocado sunblotch and peach latent mosaic viroids. Group B contains all the other viroids, and their RNA is not capable of self-cleavage. Species in group B include the potato spindle tuber, coconut cadang, tomato plant macho, and citrus bent leaf viroids.

Virusoids are less well studied than viroids and, although more diverse, are most similar to group B viroids in that they cannot self-cleave. Examples include the tomato black ring virus viroid, the peanut stunt virus viroid, and the tobacco ringspot virus viroid. Because so little is known about virusoids, the remainder of this article will focus on viroids.

Viroid Pathogenesis

If infected leaves are homogenized in a blender and passed through an "ultrafilter" fine enough to exclude bacteria, the infection is easily transmitted to another plant by painting some of the filtrate on a leaf. Even billion-fold dilutions of the filtrate retain the ability to cause infection, suggesting that it is being replicated. RNase destroys infectivity, suggesting that the genetic material (RNA) is exposed to the medium, unlike viruses, which have a protective protein coat. When isolated from other cell components, an absorbance spectrum shows that viroids are pure nucleic acid, lacking a protein coat.

Although viroids are structurally simple and do not code for any proteins, they still cause disease. Although the molecular mechanisms of viroid pathogenesis are unknown, it is clear that the pathogenesis domain (P domain) is primarily responsible.

Changes in the sequence of nucleotides in the P domain have been correlated with pathogenicity. Some research suggests that the pathogenicity of a viroid strain is related to the resistance of the P domain to heat denaturation, with stability of this region being inversely related to severity. However, some evidence suggests that this may not be entirely true. In a series of nucleotide substitutions introduced by researchers into the P region of an intermediate strain (that is, intermediate in pathogenicity) of potato spindle tuber viroid (PSTVd), four showed viroid infectivity and pathogenicity that were the same as those of a previously reported severe strain of PSTVd. Altogether, eight different mutant strains were analyzed, and resistance to denaturation and PSTVd pathogenicity were not correlated in all cases.

Research is under way to understand how viroids move from cell to cell and traverse the cytoplasm to the nucleus, where many viroids replicate. There is evidence that a possible interaction might involve viroid RNA activating an RNA-activated protein kinase in response to a nucleotide sequence similar to that of the normal RNA activator. Protein kinases are integral to intracellular signaling pathways that control many aspects of cell metabolism. Once researchers understand the signals that viroids use to get around, it may be possible to devise treatments against them. A better understanding of the process may also shed light on normal biochemical communication pathways in plant cells.

Viroid Replication

Viroids replicate by a rolling circle mechanism, a method also used by some viruses. The original strand is referred to as the "+ strand," and complementary copies of it are called "− strands." Type A and B viroids replicate slightly differently. In type A viroids, the circular + strand is replicated by RNA-dependent RNA polymerase to form several linear copies of the RNA − strand connected end to end. Site-specific self-cleavage produces individual − strands later circularized by a host RNA ligase. Each − strand is finally copied by the RNA polymerase to make several linear copies of + strand RNA. Cleavage of this last strand makes individual RNA + strands, which are then circularized. Self-cleavage in viroids represents one of the cases in which RNA acts as an enzyme. The RNA forms a "hammerhead" structure that enzymatically cleaves the longer RNAs at just the right sites.

Replication of type B viroids is apparently mediated by normal host DNA-dependent RNA polymerase, which mistakes the viroid RNA for DNA. The overall process is similar to what happens with type A viroids, except that the − strand is not cleaved but instead is copied directly, yielding a + strand that is cleaved by host RNase to form individual copies that are ligated to become circular.

Economic Impact of Viroids

Genetically engineered plants in the future might make proteins that would essentially confer immunity by preventing viroids from entering the nucleus. With no access to the nucleus, a viroid would be incapable of replicating, effectively preventing the damage normally associated with viroid infection. Currently, no such transgenic plants exist, and viroids can reduce agricultural productivity if outbreaks are not checked quickly. The typical treatment is simply to destroy the affected plants, as there is no cure.

Although predominantly negative, viroids may have some potentially positive benefits. They have already been used in unique ways to study plant genetics, and they may provide insights into how plant proteins and nucleic acids move in and out of cell nuclei. It may also be possible to harness the benefits of viroid infection for certain agricultural applications, such as dwarfing citrus trees. Considerably more will need to be learned about viroids before they can be adequately controlled or used for human benefit.

—*Bryan Ness*

See also: DNA Structure and Function; Gene Regulation: Viruses; Genetic Engineering; Hybridomas and Monoclonal Antibodies; Oncogenes; Organ Transplants and HLA Genes; RNA Structure and Function; Viral Genetics.

Further Reading

Diener, T. O., R. A. Owens, and R. W. Hammond. "Viroids, the Smallest and Simplest Agents of Infectious Disease: How Do They Make Plants Sick?" *Intervirology* 35 (1993): 186-195. A review of the process whereby viroids cause pathology in plants.

Hammond, R. W. "Analysis of the Virulence Modulating Region of Potato Spindle Tuber Viroid (PSTVd) by Site-Directed Mutagenesis." *Virology* 187 (1992): 654-662. Report of experimental results on the potato spindle tuber viroid.

Lee, R. F., R. H. Brlansky, and L. W. Timmer. *Florida Citrus Pest Management Guide: Exocortis, Cachexia, and Other Viroids.* Gainesville: Plant Pathology Department; Citrus REC, Lake Alfred, Florida; Cooperative Extension Service, Institute of Food and Agricultural Sciences, University of Florida, 2003. A technical review of viroids as a cause of citrus disease.

Owens, R. A., W. Chen, Y Hu, and Y-H. Hsu. "Suppression of Potato Spindle Tuber Viroid Replication and Symptom Expression by Mutations, Which Stabilize the Pathogenicity Domain." *Virology* 208, no. 2 (1995): 554-564. Aimed at researchers.

Wassenegger, M., S. Heimes, L. Reidel, and H. L. Sanger. "RNA-Directed *De Novo* Methylation of Genomic Sequences in Plants." *Cell* 76 (1994): 567-576. Report on an experiment involving viroids that demonstrated the possibility that a mechanism of de novo methylation of genes might exist that can be targeted in a sequence-specific manner by their own mRNA.

X Chromosome Inactivation

Field of study: Developmental genetics
Significance: *Normal females have two X chromosomes, and normal males have one X chromosome. In order to compensate for the potential problem of doubling of gene products in females, one X chromosome is randomly inactivated in each cell.*

Key terms

BARR BODY: a highly condensed and inactivated X chromosome visible in female cells as a darkly staining spot in a prepared microscope slide

DOSAGE COMPENSATION: an equalization of gene products that can occur whenever there are more or fewer genes for specific traits than normal

MOSAIC: an individual possessing cells with more than one type of genetic constitution

SEX CHROMOSOMES: the X and Y chromosomes; females possess two X chromosomes, while males possess one X and one Y chromosome

The History of X Chromosome Inactivation

In 1961, Mary Lyon hypothesized that gene products were found in equal amounts in males and females because one of the X chromosomes in females became inactivated early in development. This hypothesis became known as the Lyon hypothesis, and the process became known as Lyonization, or X chromosome inactivation. Prior to this explanation, it was recognized that females had two X chromosomes and males had only one X chromosome, yet the proteins encoded by genes on the X chromosomes were found in equal amounts in females and males because of dosage compensation.

The principles of inheritance dictate that individuals receive half of their chromosomes from their fathers and the other half from their mothers at conception. Therefore, a female possesses two different X chromosomes (one from each parent). In addition to hypothesizing the inactivation of one X chromosome in each cell, the Lyon hypothesis also implies that the event occurs randomly. In any individual,

approximately one-half of the paternal X chromosomes and one-half of the maternal X chromosomes are inactivated. Thus, females display a mosaic condition since half of their cells express the X chromosome genes inherited from the father and half of their cells express the X chromosome genes inherited from the mother. In fact, this situation can be seen in individuals who inherit an allele for a different form of a protein from each parent: Some cells express one parent's protein form, while other cells express the other parent's protein form.

Prior to Lyon's hypothesis, it was known that a densely staining material could be seen in cells from females that was absent in cells from males. This material was termed a "Barr body," after Murray Barr. Later, it was shown that Barr bodies were synonymous with the inactivated X chromosome. Other observations led scientists to understand that the number of Barr bodies in a cell always represented one less than the number of X chromosomes in the cell. For example, one Barr body indicated the presence of two X chromosomes, and two Barr bodies indicated the presence of three X chromosomes.

Clinical Significance

The significance of Barr bodies became apparent with the observation that females lacking one Barr body or possessing more than one Barr body developed an abnormal appearance. Particularly intriguing were females with Turner syndrome. These females possess only one X chromosome per cell, a condition that is not analogous to normal females, who possess only one functional X chromosome per cell as a result of inactivation. The difference in the development of a Turner syndrome female and a normal female lies in the fact that both X chromosomes are active in normal females during the first few days of development. After this period, inactivation occurs randomly in each cell, as hypothesized by Lyon. In cases in which inactivation is not random, individuals may have a variety of developmental problems. Therefore, there is apparently a critical need for both X chromosomes to be active in females in early development for normal development to occur.

It is equally important that there not be more than two X chromosomes present during this early development. Females possessing three X chromosomes, and therefore two Barr bodies, are sometimes called superfemales or metafemales because of a tendency to be taller than average. These females are also two to ten times more likely to suffer from mild to moderate mental retardation.

The same phenomenon has been observed in males who possess Barr bodies. Barr bodies are not normally present in males because they have only one X chromosome. The presence of Barr bodies indicates the existence of an extra X chromosome that has become inactive. Just as in females, extra X chromosomes are also expressed in early development, and abnormal amounts of gene products result in abnormal physical characteristics and mental retardation. Males with Klinefelter syndrome have two X chromosomes and a Y chromosome. In cases in which males have more than two X chromosomes, the effects are even more remarkable.

Mechanism of X Inactivation

While it has been apparent since the 1960's that X inactivation is required for normal female development, the mechanism has been elusive. Only with the development of techniques to study the molecular events of the cell and its chromosomes has progress been made in understanding the process of inactivation. One process involved in turning off a gene (thus "shutting down" the process of transcription) is the alteration of one of the molecules of DNA known as cytosine. By adding a methyl group to the cytosine, the gene cannot produce the RNA necessary to make a protein. It is thought that this methyl group blocks the proteins that normally bind to the DNA so that transcription cannot occur. When methyl groups are removed from cytosines, the block is removed and transcription begins. This is a common means of regulating transcription of genes. Methylation is significantly higher in the inactivated X chromosome than in the activated X chromosome. As the genes on the chromosome become inactive, the chromosome condenses into the tightly packed mass

known as the Barr body. However, the process of methylation alone cannot entirely account for inactivation.

A region on the X chromosome called the X inactivation center (XIC) is considered the control center for X inactivation. In this region is a gene called the X inactivation specific transcripts (*XIST*) gene. At the time of its discovery, this gene was the only gene known to be functional in an inactivated chromosome. It produces an RNA that remains inside the nucleus.

Evidence in humans supports the hypothesis that the *XIST* gene is turned on and begins to make its RNA when the egg is fertilized. Studies with mice have shown that RNA is produced, at first, in low levels and from both X chromosomes. It has been shown in mice, but not humans, that prior to inactivation, Xist (lower-cased when referring to mouse genes) RNA is localized at the XIC site only, thus suggesting a potential role prior to actual inactivation of the chromosome. At this point, one X chromosome will begin to increase its production of XIST RNA; shortly thereafter, XIST RNA transcription from the other X chromosome ceases. It is not clear how XIST RNA initiates the process of inactivation and condensing of the inactive chromosome, but XIST RNA binds along the entire length of the inactive X chromosome in females. These results suggest that inactivation spreads from the XIC region toward the end of the chromosome and that XIST RNA is required to maintain an inactive state. If a mouse's *Xist* gene is mutated and cannot produce its RNA, inactivation of that X chromosome is blocked. Other studies have suggested that a product from a nonsex chromosome may interact with the XIC region, causing it to remain active. As expected, but not explained, the *XIST* gene is repressed, or expresses XIST RNA at only very low levels, in males with only one X chromosome.

No difference has been detected between maternally and paternally expressed *XIST* genes in humans. This has led scientists to suspect that *XIST* gene RNA may not be responsible for determining which X chromosome becomes inactivated. It is also not clear how the cell knows how many X chromosomes are present. The search for other candidates for these roles is under way. Finally, there are a few genes besides the *XIST* gene that are also active on the inactive X chromosome. How they escape the inactivation process and why this is necessary are also questions that must be resolved.

—*Linda R. Adkison*

See also: Fragile X Syndrome; Gender Identity; Hermaphrodites; Infertility; Klinefelter Syndrome; Metafemales; Pseudohermaphrodites; Testicular Feminization Syndrome; Turner Syndrome; XYY Syndrome.

Further Reading

Erbe, Richard W. "Single-Active-X Principle." *Scientific American Medicine* 2, section 9:IV (1995). Reviews the significance of gene dosage compensation in humans.

Latham, Keith E. "X Chromosome Imprinting and Inactivation in the Early Mammalian Embryo." *Trends in Genetics*, April, 1996. Discussion of observations on embryos with sex chromosomes from only one parent.

"X in a Cage." *Discover* 15 (March, 1994). Summarizes a mechanism for X chromosome inactivation.

Web Sites of Interest

Intersex Society of North America. http://www .isna.org. The society is "a public awareness, education, and advocacy organization which works to create a world free of shame, secrecy, and unwanted surgery for intersex people (individuals born with anatomy or physiology which differs from cultural ideals of male and female)." Includes links to information on such conditions as clitoromegaly, micropenis, hypospadias, ambiguous genitals, early genital surgery, adrenal hyperplasia, Klinefelter syndrome, androgen insensitivity, and testicular feminization.

Johns Hopkins University, Division of Pediatric Endocrinology, Syndromes of Abnormal Sex Differentiation. http://www.hopkins medicine.org/pediatricendocrinology. Site provides a guide to the science and genetics of sex differentiation, including a glossary. Click on "patient resources."

Xenotransplants

Field of study: Genetic engineering and biotechnology

Significance: *Xenotransplants are transplants of organs or cellular tissue between different species of animals, such as between pigs and humans. Although initial research in xenotransplantation focused primarily on transplanting organs such as hearts and kidneys, molecular biologists have also become interested in transplanting small amounts of cellular tissue or genetic material as part of therapeutic treatments. In addition, molecular biologists believe it may be possible to manipulate the DNA of animals in such a way to make their organs less prone to rejection if used in humans.*

Key terms

ANTI-REJECTION MEDICATION: drugs developed to counteract the natural immune system's reaction to transplanted organs

REJECTION: refusal of a patient's body to accept a transplanted organ

History

The idea of xenotransplants is actually quite old. During the eighteenth century, for example, transfusions of sheep's blood were believed to be therapeutic for certain human illnesses. Scientists also speculated about using animal organs to replace failing human ones. Until the mid-twentieth century, however, the human body's immune system prevented any successful xenotransplants from taking place. As the science of organ transplants between humans progressed, researchers became increasingly interested in experimenting with using animals as donors. Organ transplantation became an accepted medical treatment in humans, but there would never be enough donor organs available to treat every patient who could benefit from the procedure: As the demand for transplant surgery grew, the pool of available donor organs shrank in relation. As a result, one of the ethical dilemmas inherent in human organ transplant is that, almost always, in order for one person to receive a transplant, another person must die. Bone marrow and kidney transplants are among the few excep-

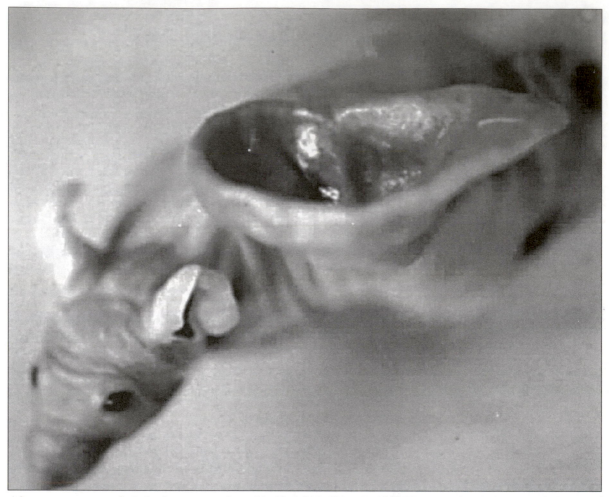

A human ear grows on the body of a mouse, engineered in Shanghai, China, in 1997. Human cells were used to grow the ear and then were inserted into the mouse body. (AP/Wide World Photos)

tions; the donor usually can donate bone marrow or one kidney and still survive. Nevertheless, for most organs the dilemma remains.

Researchers then suggested that organs could be harvested from compatible animals, eliminating both the need to wait for a compatible human donor and the shortage of usable organs. This is one form of xenotransplant, and it could eliminate the shortage of donor organs.

Early research focused on potential donor animals that were similar to humans, that is, primates such as baboons and chimpanzees. Perhaps the most publicized example of such a xenotransplant was the 1984 Baby Fay case, in which doctors in California transplanted a ba-

boon heart into a newborn human infant who otherwise had no chance for survival. The infant survived for several days before succumbing to complications. Researchers quickly learned that primates were not ideal candidates for organ donation; their organs were small in comparison to those of humans. Chimpanzee kidneys, for example, are too small to perform adequately in an adult human.

Researchers then turned their attention to pigs as possible donors. Swine make ideal donor candidates because they are physically large enough to have organs that can sustain humans, have a short gestation cycle, produce large litters of offspring, and, because they are routinely raised for meat production, are viewed as expendable by the general public. By using a combination of selective breeding and genetic manipulation, researchers hope to develop swine whose organs will be less susceptible to rejection by the human body. Pig heart valves are already routinely used in humans, with an estimated sixty thousand implanted annually. The use of larger organs has been less successful; transplant recipients experienced hyper-acute rejection. That is, their bodies immediately reacted to the foreign tissue by shutting off the flow of blood to it.

Ethical and Medical Concerns

Xenotransplantation presents a number of ethical and medical dilemmas. One major concern is the possibility that a virus, harmless to the donor animal, is transmitted to the human host and then proves fatal. Scientists worry that a potentially deadly disease epidemic could re-

Animal rights advocates and other activists protest the use of animals as "spare parts" factories for human organs in this demonstration in Munich, Germany, in November, 2000. (AP/Wide World Photos)

sult from using organs or other tissue from either swine or primates. While many researchers are confident that careful screening of donor animals would eliminate or minimize such risks, critics remain convinced that it is possible a virus could lie dormant and undetected in animals, causing problems only after the transplants occurred. Baboons, for example, carry a virus that has the potential to cause cancer in humans.

In addition to the medical issues raised, many bioethicists question the morality of using animals as a source of "spare parts" for humans. They are particularly troubled by the idea of possibly genetically altering a species such as swine in order to make their organs more compatible with human hosts. Proponents of xenotransplants counter these arguments by noting that humans have selectively bred animals for various purposes for thousands of years to eliminate certain characteristics while enhancing others. In addition, animals such as swine are already routinely slaughtered for human consumption.

Finally, there is the problem of human perceptions. While many people support the idea of xenotransplants on the genetic or cellular level, they are less enthusiastic about possible organ transplants. That is, while a majority of people surveyed said they would have no problem accepting a xenotransplant if it were part of gene therapy, far fewer were interested in possibly receiving a pig's heart if the need arose. If researchers do achieve successful xenotransplants using such organs, however, public perceptions could change. It is easy to question a medical procedure when it is still theoretical; it becomes much more difficult to do so after it becomes a reality.

—*Nancy Farm Männikkö*

See also: Animal Cloning; Cancer; Cloning; Cloning: Ethical Issues; Gene Therapy: Ethical and Economic Issues; Genetic Engineering: Historical Development; Heart Disease; Immunogenetics; In Vitro Fertilization and Embryo Transfer; Model Organism: *Mus musculus*; Model Organism: *Xenopus laevis*; Organ Transplants and HLA Genes; Stem Cells; Synthetic Antibodies; Totipotency; Transgenic Organisms

Further Reading

Bassett, Pamela. *Emerging Markets in Tissue Engineering: Angiogenesis, Soft and Hard Tissue Regeneration, Xenotransplant, Wound Healing, Biomaterials, and Cell Therapy.* Southborough, Mass.: D & MD Reports, 1999. Looks at xenotransplants from an economic perspective and discusses the potential for growth for biomedical firms entering the field.

Bloom, E. T., et al. "Xenotransplantation: The Potential and the Challenges." *Critical Care Nurse* 19 (April 1999): 76-83. Looks at the potential impact of xenotransplantation on patient care as well as the effects it might have on nursing responsibilities.

Cruz, J., et al. "Ethical Challenges of Xenotransplantation." *Transplant Proceedings* 32 (December, 2000): 2687. Members of a medical transplant team discuss questions about potential moral contradictions in using animal organs in human patients.

Daar, A. S. "Xenotransplants: Proceed with Caution." *Nature* 392 (March 5, 1998): 11. Sounds a warning about some of the potential risks involved in xenotransplants.

Persson, M. D., et al. "Xenotransplantation Public Perceptions: Rather Cells than Organs." *Xenotransplantation* 10 (2003): 72-79. Describes a public opinion survey done to gauge the public's willingness to support xenotransplant research and implementation.

"U.S. Decides Close Tabs Must Be Kept on Xenotransplants." *Nature* 405 (June 8, 2000): 606-607. Discussion of federal regulations regarding xenotransplant research.

XYY Syndrome

Field of study: Diseases and syndromes
Significance: *XYY syndrome occurs at fertilization and represents one of several human sex chromosome abnormalities. The resulting XYY male bears an extra Y chromosome that is associated with tall stature and possible intelligence and behavioral problems.*

Key terms

ANEUPLOIDY: possession of one or a few more or less than the normal number of chromosomes

GENETIC SCREENING: a medical technique that uses either fetal or adult cells to directly view the chromosomes of an individual to detect abnormalities in number or structure

MEIOSIS: cell division that produces sperm and egg cells having half the original number of chromosomes

NONDISJUNCTION: abnormal separation of chromosome pairs or duplicates during cell division, resulting in one daughter cell receiving an extra chromosome (or chromosomes) and the other daughter cell receiving a complementary number, less than normal

SEX CHROMOSOMES: the X and Y chromosomes, which determine the gender of an individual

Causes and Effects of XYY Syndrome

All normal human cells contain forty-six chromosomes consisting of twenty-three pairs; one member of each pair is contributed by the female parent and one by the male. Of these forty-six chromosomes, two chromosomes, designated X and Y, are known as the sex chromosome pair. Individuals with an XX pair are female, while those with an XY pair are male. Unlike the other twenty-two chromosome pairs, the X and Y chromosomes are strikingly different from each other in both size and function. While the Y chromosome is primarily concerned with maleness, the X chromosome contains information important to both genders.

During formation of sperm and eggs in the testes and ovaries, respectively, a unique form of nuclear division, known as meiosis (or reductional division), occurs during cell division that halves the chromosome number from forty-six to twenty-three. Sperm and eggs are thus carrying only one member of each pair of chromosomes, and the original number will be restored during fertilization. Because females only have the XX pair, their eggs can only have an X chromosome, while the male, having the XY pair, produces sperm bearing an X or a Y chromosome.

A common genetic error during sperm or

egg production is known as nondisjunction, which is the improper division of chromosomes between the daughter cells. Nondisjunction in the production of either gamete can result, at fertilization, in embryos without the normal forty-six chromosomes. XYY syndrome is one of several of these aneuploid conditions that involve the sex chromosome pair. While Klinefelter syndrome (an XXY male) and Turner syndrome (an X female) are more widely studied and recognized genetic diseases, the XYY male occurs with a frequency of 1 in 1,000 male births in the United States. Caused by a YY-bearing sperm fertilizing a normal X-bearing egg, the XYY embryo develops along a seemingly normal route and, unlike most other sex chromosome diseases, is not apparent at birth. In fact, identification of this disorder requires genetic testing or screening and is often discovered accidentally as a consequence of results from another genetic test. The only physical clue is unusually tall stature; otherwise, an affected male will be normal in appearance. The XYY male is also fertile, unlike those with aneuploidies involving other combinations of sex chromosomes, which usually result in sterility.

Behavioral and Research Implications

Interest in the association between aggression and the Y chromosome began in the years following World War II. Both psychologists and geneticists began intensive scrutiny of the genes that were located on the male sex chromosome. Men with multiple copies of the Y chromosome thus became the subjects of much of this research. Genetic links to violent, aggressive, and even criminal behavior were found, although many argued that below-average intelligence played a greater role. Many males with XYY syndrome do perform lower than average on standard intelligence tests and have a greater incidence of behavioral problems. The majority, however, lead normal lives and are indistinguishable from XY males.

The controversy surrounding this research began with a study at Harvard University that began in the early 1960's and ended in 1973 because of pressure from both public and scientific communities. The researchers screened all boys born at a Boston hospital, identifying those with sex chromosomal abnormalities. Because the parents of XYY boys were told of their children's genetic makeup and the possibility of lower intelligence and bad behavior, critics claimed that the researchers had biased the parents against their sons, causing the parents to treat the children differently. The environment would thus play a greater role than genetics in their behavior. Subsequent research has shown that the original hypothesis is at least partially accurate. There is a disproportionately large number of XYY males in prison populations, and they are usually of subaverage intelligence compared to other prisoners. It must be emphasized, however, that the majority of XYY males show neither low intelligence nor criminal behavior.

Scientists, doctors, geneticists, and psychologists now agree that the extra Y chromosome does cause above normal height, reading and math difficulties, and, in some cases, severe acne, but the explanation of the high prevalence of XYY men in prison populations has changed its focus from genes to environment. Large body size during childhood, adolescence, and early adulthood will no doubt cause people to treat these individuals differently, and they may in turn have learned to use their size defensively. Aggressive behavior, coupled with academic difficulties, may lead to further problems. Clearly, however, the majority of XYY males do well. The issue would be much easier to resolve if a YY or Y male existed, but because lack of an X chromosome results in spontaneous miscarriage, no YY or Y male embryo could ever survive.

—*Connie Rizzo*

See also: Aggression; Behavior; Criminality; Intelligence; Klinefelter Syndrome; Metafemales; Nondisjunction and Aneuploidy; Steroid Hormones; Testicular Feminization Syndrome; X Chromosome Inactivation.

Further Reading

Mader, Sylvia S. *Human Reproductive Biology*. 3d ed. New York: McGraw-Hill, 2000. Provides an excellent introduction to cell division, genetics, and sex from fertilization through birth. Illustrations, bibliography, index.
Tamarin, Robert H. *Principles of Genetics*. Bos-

ton: McGraw-Hill, 2002. A well-written reference text on genetics with complete discussions on aneuploidy, the sex chromosomes, genes, and abnormalities; it also includes a thorough reading list. Illustrations, maps (some color), bibliography, index.

Web Sites of Interest

Intersex Society of North America. http://www .isna.org. The society is "a public awareness, education, and advocacy organization which works to create a world free of shame, secrecy, and unwanted surgery for intersex people (individuals born with anatomy or physiology which differs from cultural ideals of male and female)." Includes links to information on such conditions as clitoromegaly, micropenis, hypospadias, ambiguous genitals, early genital surgery, adrenal hyperplasia, Klinefelter syndrome, androgen insensitivity, and testicular feminization.

Johns Hopkins University, Division of Pediatric Endocrinology, Syndromes of Abnormal Sex Differentiation. http://www.hopkins medicine.org/pediatricendocrinology. Site provides a guide to the science and genetics of sex differentiation, including a glossary. Click on "patient resources."

Biographical Dictionary of Important Geneticists

Altman, Sidney (1939-   ): Won the 1989 Nobel Prize in Chemistry, with Thomas R. Cech. Working independently Altman and Cech discovered that RNA, like proteins, can act as a catalyst; moreover, they found that when ribosomal RNA participates in translation of messenger RNA (mRNA) and the synthesis of polypeptides, it acts as a catalyst in some steps.

Anfinsen, Christian B. (1916-1995): Won the 1972 Nobel Prize in Chemistry. Anfinsen, studying the three-dimensional structure of the enzyme ribonuclease, proved that its conformation was determined by the sequence of its amino acids and that to construct a complete enzyme molecule no separate structural information was passed on from the DNA in the cell's nucleus.

Arber, Werner (1929-   ): First to isolate enzymes that modify DNA and enzymes that cut DNA at specific sites. Such restriction enzymes were critical in the developing field of molecular biology. Arber was awarded the 1978 Nobel Prize in Physiology or Medicine.

Aristotle (384-322 B.C.E.): Greek philosopher and scientist. Aristotle's *De Generatione* was devoted in part to his theories on heredity. Aristotle believed the semen of the male contributes a form-giving principle (*eidos*), while the menstrual blood of the female is shaped by the *eidos*. The philosophy implied it was the father only who supplied form to the offspring.

Auerbach, Charlotte (1899-1994): German-born geneticist who fled to England following the rise of the Nazi Party. Demonstrated that the mutations produced by mustard gases and other chemicals in *Drosophila* (fruit flies) were similar to those induced by X rays, suggesting a common mechanism.

Avery, Oswald Theodore (1877-1955): Immunologist and biologist who determined DNA to be the genetic material of cells. Avery's early work involved classification of the pneumococci, the common cause of pneumonia in the elderly. In 1944, he reported that the genetic information in these bacteria is DNA.

Bailey, Catherine (1921-   ): By applying methods of selective breeding, developed new varieties of fruits.

Baltimore, David (1938-   ): Along with Howard Temin, Baltimore isolated the enzyme RNA-directed DNA polymerase (reverse transcriptase), demonstrating the mechanism by which RNA tumor viruses can integrate their genetic material into the cell chromosome. Baltimore was awarded the 1975 Nobel Prize in Physiology or Medicine.

Barr, Murray Llewellyn (1908-1995): Canadian geneticist who discovered the existence of the Barr (Barr's) body, an inactive X chromosome found in cells from a female. The existence or absence of the body has been used in determining the sex of the individual from whom the cell originated.

Bateson, William (1861-1926): Plant and animal geneticist who popularized the earlier work of Gregor Mendel. In his classic *Mendel's Principles of Heredity* (1909), Bateson introduced much of the modern terminology used in the field of genetics. Bateson suggested the term "genetics" (from the Greek word meaning "descent") to apply to the field of the study of heredity.

Beadle, George Wells (1903-1989): Beadle's studies of the bread mold *Neurospora* demonstrated that the function of a gene is to encode an enzyme. Beadle and Edward Tatum were awarded the 1958 Nobel Prize in Physiology or Medicine for their one gene-one enzyme hypothesis.

Beckwith, Jonathan R. (1935-   ): Determined role of specific genes in regulating bacterial cell division. During the 1960's, he was among the first to isolate a specific gene. Beckwith is also known as a social activist in his arguments for the use of science for improvement of society.

Bell, Julia (1879-1979): British geneticist who applied statistical analysis in understanding hereditary medical disorders of the nervous system and limbs.

Benacerraf, Baruj (1920-   ): Won the 1980 Nobel Prize in Physiology or Medicine, with Jean Dausset and George D. Snell. Benacerraf, Dausset, and Snell each explained the genetic components of the major histocompatibility complex (MHC), the key to a person's immune system, and how the system produces antibodies to such a wide variety of foreign molecules and pathogens, such as viruses, fungi, and bacteria.

Berg, Paul (1926-   ): Developed DNA recombination techniques for insertion of genes in chromosomes. The techniques became an important procedure in understanding gene function and for the field of genetic engineering. Berg was awarded the 1980 Nobel Prize in Chemistry.

Bishop, John Michael (1936-   ): Determined that oncogenes, genetic information initially isolated from RNA tumor viruses, actually originate in normal host cells. Bishop was awarded the 1989 Nobel Prize in Physiology or Medicine for his discovery.

Bluhm, Agnes (1862-1943): German physician whose controversial theories on improvement of the "Race" through eugenics and fertility selection provided a basis for Nazi race theories. Among other aspects of her theory was the use of enforced sterilization.

Boring, Alice Middleton (1883-1955): Confirmed existing theories of the chromosomal basis of heredity. Her professional career consisted primarily in serving as a biology teacher to students in China between the world wars.

Borlaug, Norman (1914-   ): Won the 1970 Nobel Prize in Peace. Borlaug was a key figure in the Green Revolution in agriculture. Working as a geneticist and plant physiologist in a joint Mexican-American program, he developed strains of high-yield, short-strawed, disease-resistant wheat. His goal was to increase crop production and alleviate world hunger.

Botstein, David (1942-   ): Developed methods of localized mutagenesis for understanding the relationship between the structure and function of proteins. His development

of a linkage map involving human genes contributed to the progress of the Human Genome Project.

Boyer, Herbert W. (1936-   ): His isolation of restriction enzymes that produced a staggered cut on the DNA allowed for creation of so-called "sticky ends," which allowed DNA from different sources or species to be spliced together.

Brenner, Sydney (1927-   ): Molecular geneticist whose observations of mutations in nematodes (long, unsegmented worms) helped in understanding the design of the nervous system. Brenner was among the first to clone specific genes. He was awarded the Nobel Prize in Physiology or Medicine in 2002.

Brown, Michael S. (1941-   ): By studying the role of cell receptors in uptake of lipids from the blood, Brown discovered the genetic defect in humans associated with abnormally high levels of cholesterol. He was awarded the 1985 Nobel Prize in Physiology or Medicine.

Burnet, Frank Macfarlane (1899-1985): Proposed a theory of clonal selection to explain regulation of the immune response. Burnet was awarded the 1960 Nobel Prize in Physiology or Medicine.

Cairns, Hugh John (1922-   ): British virologist whose investigations of rates and mechanisms of DNA replication helped to lay the groundwork in studying the replication process.

Carroll, Christiane Mendrez (1937-1978): French geneticist and paleontologist, most noted for her taxonomic interpretations of early reptiles.

Cech, Thomas R. (1947-   ): Won the 1989 Nobel Prize in Chemistry, with Sidney Altman. Working independently, Cech and Altman discovered that RNA, like proteins, can act as a catalyst; moreover, Cech found that when ribosomal RNA participates in translation of mRNA and the synthesis of polypeptides, it acts as a catalyst in some steps.

Chargaff, Erwin (1905-2002): Determined that the DNA composition in a cell is characteristic of that particular organism. His discovery of base ratios, in which the concentration of adenine is equal to that of thymine, and guanine to that of cytosine, provided an important clue to the structure of DNA.

Cohen, Stanley N. (1935-): Developed the techniques for transfer of DNA between species, a major factor in the process of genetic engineering.

Collins, Francis Sellers (1950-): In 1989, Collins identified the gene that, when mutated, results in the genetic disease cystic fibrosis. Collins was instrumental in the identification of a number of genes associated with genetic diseases.

Correns, Carl Erich (1864-1935): German botanist who confirmed Gregor Mendel's laws through his own work on the garden pea. Correns was one of several geneticists who rediscovered Mendel's work in the early 1900's.

Crick, Francis Harry Compton (1916-): Along with James Watson, Crick determined the double-helix structure of DNA. Crick was awarded the Nobel Prize in Physiology or Medicine in 1962.

Darlington, Cyril (1903-1981): British geneticist who demonstrated changes in chromosomal patterns which occur during meiosis, leading to an understanding of chromosomal distribution during the process. He also described a role played by crossing over, or genetic exchange, in changes of patterns.

Darwin, Charles Robert (1809-1882): Naturalist whose theory of evolution established natural selection as the basis for descent with modification, more commonly referred to as evolution. His classic work on the subject, *On the Origin of Species by Means of Natural Selection* (1859), based on his five-year voyage during the 1830's on the British ship HMS *Beagle*, summarized the studies and observations that initially led to the theory. Darwin's pangenesis theory, first noted in *The Variation of Animals and Plants Under Domestication* (1868), later became the basis for the concept of the gene.

Darwin, Erasmus (1731-1802): British physician, inventor, and writer. In his classic *Zoonomia*, he advanced a theory of the role of the environment on genetic changes in organisms. A similar theory was later developed by Jean-Baptiste Lamarck. Darwin was the grandfather of both Charles Darwin and Francis Galton.

Dausset, Jean (1916-): Won the 1980 Nobel Prize in Physiology or Medicine, with Baruj Benacerraf and George D. Snell. Dausset, Benacerraf, and Snell each explained the genetic components of the major histocompatibility complex (MHC), the key to a person's immune system, and how the system produces antibodies to such a wide variety of foreign molecules and pathogens, such as viruses, fungi, and bacteria.

Delbrück, Max (1906-1981): A leading figure in the application of genetics to bacteriophage research, and later, with *Phycomyces*, a fungal organism. His bacteriophage course, taught for decades at Cold Spring Harbor, New York, provided training for a generation of biologists. He was awarded the 1969 Nobel Prize in Physiology or Medicine.

Demerec, Milislav (1895-1966): Croatian-born geneticist who was among the scientists who brought the United States to the forefront of genetics research. Demerec's experiments, based on the genetics of corn, addressed the question of what a gene represents. His work with bacteria included the determination of mechanisms of antibiotic resistance, as well as the existence of operons, closely linked genes which are coordinately regulated. Demerec was director of the biological laboratories in Cold Spring Harbor, New York, for many years among the most important sites of genetic research.

De Vries, Hugo (1848-1935): Dutch botanist whose hypothesis of intracellular pangenesis postulated the existence of pangenes, factors which determined characteristics of a species. De Vries established the concept of mutation as a basis for variation in plants. In 1900, de Vries was one of several scientists who rediscovered Mendel's work.

Dobzhansky, Theodosius (1900-1975): Russian-born American geneticist who established evolutionary genetics as a viable discipline. His book *Genetics and the Origin of Species* (1937) represented the first application of Mendelian theory to Darwinian evolution.

Dulbecco, Renato (1914-   ): Among the first to study the genetics of tumor viruses. Dulbecco was awarded the 1975 Nobel Prize in Physiology or Medicine.

Ferguson, Margaret Clay (1863-1951): Plant geneticist whose use of *Petunia* as a model helped explain life cycles of various plants. Also noted for her description of the life cycle of pine trees.

Fink, Gerald R. (1940-   ): Isolation of specific mutants in yeast allowed the use of genetics in understanding biochemical mechanisms in that organism.

Fisher, Ronald Aylmer (1890-1962): British biologist whose application of statistics provided a means by which use of small sampling size could be applied to larger interpretations. Fisher's breeding of small animals led to an understanding of genetic dominance. He later applied his work to the study of inheritance of blood types in humans.

Franklin, Rosalind Elsie (1920-1958): British crystallographer whose X-ray diffraction studies helped confirm the double-helix nature of DNA. Franklin's work, along with that of Maurice Wilkins, was instrumental in confirming the structure of DNA as proposed by James Watson and Francis Crick. Franklin's early death precluded her receiving a Nobel Prize for her research.

Galton, Francis (1822-1911): British scientist who was an advocate of eugenics, the belief that human populations could be improved through "breeding" of desired traits. Galton was also the first to observe that fingerprints were unique to the individual.

Garnjobst, Laura Flora (1895-1977): Following her training under Nobel laureate Edward Tatum at both Stanford and Yale, Garnjobst spent her career in the study of genetics of the mold *Neurospora*.

Garrod, Archibald Edward (1857-1936): Applying his work on alkaptonuria, Garrod proposed that some human diseases result from a lack of specific enzymes. His theory of inborn errors of metabolism, published in 1908, established the genetic basis for certain hereditary diseases.

Gartner, Carl Friedrich von (1772-1850): German plant biologist and geneticist. Though Gartner did not generalize as to the significance of his work, his results provided the experimental basis for questions later developed by Gregor Mendel and Charles Darwin.

Giblett, Eloise (1921-   ): Discoverer of numerous genetic markers useful in defining blood groups and serum proteins. In the 1970's, Giblett discovered that certain immunodeficiency diseases result from the absence of certain enzymes necessary for immune cell development.

Gilbert, Walter (1932-   ): Developed method of sequencing DNA. With Paul Berg and Frederick Sanger, awarded the 1980 Nobel Prize in Chemistry.

Gilman, Alfred G. (1941-   ): Discovered the role of "G" proteins in regulating signal transduction in eukaryotic cells. With Martin Rodbell, won the Nobel Prize in Physiology or Medicine for 1994.

Goldschmidt, Richard B. (1878-1958): German-born geneticist who proposed that the chemical makeup of the chromosome determines heredity rather than the quantity of genes. He theorized that large mutations, or "genetic monsters," were important in generation of new species.

Goldstein, Joseph L. (1940-   ): Won the 1985 Nobel Prize in Physiology or Medicine, with Michael S. Brown. Brown and Goldstein conducted extensive research in the regulation of cholesterol in humans. They showed that in families with a history of high cholesterol, individuals who carry two copies of a mutant gene (homozygotes) have cholesterol levels several times higher than normal and those who have one mutant gene (heterozygotes) have levels about double normal. Their discoveries proved invaluable in managing heart disease and other cholesterol-related ailments.

Griffith, Frederick (1877-1941): British microbiologist who in 1928 reported the existence of a "transforming principle," an unknown substance that could change the genetic properties of bacteria. In 1944, Oswald Avery determined the substance to be DNA, three years after Griffith was killed during the German bombing of London.

Gruhn, Ruth (1907-1988): German geneticist who applied mathematical principles in the breeding of poultry and pigs.

Haeckel, Ernst Heinrich (1834-1919): German zoologist whose writings were instrumental in the dissemination of Charles Darwin's theories. Haeckel's "biogenetic law," since discarded, stated that "ontogeny repeats phylogeny," suggesting that embryonic development mirrors the evolutionary relationship of organisms.

Haldane, John Burdon Sanderson (1892-1964): British physiologist and geneticist who proposed that natural selection, and not mutation per se, was the driving force of evolution. Haldane was the first to determine an accurate rate of mutation for human genes, and he later demonstrated the genetic linkage of hemophilia and color blindness.

Hanafusa, Hidesaburo (1929-   ): Japanese-born scientist who played a key role in elucidating the role of oncogenes found among the RNA tumor viruses in transforming mammalian cells.

Hanawalt, Philip C. (1931-   ): In 1963, discovered the existence of a repair mechanism associated with DNA replication in bacteria. He later found a similar mechanism in eukaryotic cells. Hanawalt's later work included development of the technique of site mutagenesis in gene mapping.

Hardy, Godfrey Harold (1877-1947): British mathematician who, along with Wilhelm Weinberg, developed the Hardy-Weinberg law of population genetics. In a 1908 letter to the journal *Science*, Hardy used algebraic principles to confirm Mendel's theories as applied to populations, an issue then currently in dispute.

Hartwell, Leland H. (1939-   ): Discovered genes that regulate the movement of eukaryotic cells through the cell cycle. With Tim Hunt and Sir Paul Nurse, won the Nobel Prize in Physiology or Medicine in 2001.

Haynes, Robert Hall (1931-1998): Canadian molecular biologist who carried out much of the early work in the understanding of DNA repair mechanisms.

Hershey, Alfred Day (1908-1997): Molecular biologist who played a key role in understanding the replication and genetic structure of viruses. His experiments with Martha Chase confirmed that DNA carried the genetic information in some viruses. Hershey was awarded the 1969 Nobel Prize in Physiology or Medicine.

Herskowitz, Ira (1946-2003): His studies of gene conversion pathways in yeast led to an understanding of gene switching in control of mating types.

Hertwig, Paula (1889-1983): German embryologist who studied the effects of radiation on embryonic development of fish and animals.

Hippocrates (c. 460-377 B.C.E.): Greek physician who proposed the earliest theory of inheritance. Hippocrates believed that "seed material" was carried by body humors to the reproductive organs.

Hogness, David S. (1925-   ): One of the first to clone a gene from *Drosophila* (fruit flies). His technique of chromosomal "walking" allowed for the isolation of any known gene based on its ability to mutate. Also involved in identification of homeotic genes, genes which regulate development of body parts.

Holley, Robert William (1922-1993): Determined the sequence of nucleotide bases in transfer RNA (tRNA), the molecule that carries amino acids to ribosomes for protein synthesis. Holley's work provided a means for demonstrating the reading of the genetic code. He was awarded the Nobel Prize in Physiology or Medicine in 1968.

Horvitz, H. Robert (1947-   ): Harvard neurobiologist whose study of cell regulation in the nematode *Caenorhabditis* led to the discovery of genes that regulate cell death during embryonic development. With Sydney Brenner and John Sulston, he was awarded the Nobel Prize in Physiology and Medicine in 2002.

Hunt, R. Timothy (1943-   ): Discovered the existence and role of proteins called cyclins, which regulate the cell cycle in eukaryotic cells. With Leland Hartwell and Sir Paul Nurse, won the Nobel Prize in Physiology or Medicine in 2001.

Jacob, François (1920-   ): French geneticist and molecular biologist who, along with

Jacques Monod, elucidated a mechanism of gene and enzyme regulation in bacteria. The Jacob-Monod theory of gene regulation became the basis for understanding a wide range of genetic processes; they were awarded the 1965 Nobel Prize in Physiology or Medicine.

Jeffreys, Sir Alec (1950-   ): British biochemist who discovered the existence of introns in mammalian genes. His study of the pattern of repeat sequences in DNA was shown to be characteristic of individuals, and became the theoretical basis for DNA fingerprinting and DNA profiles.

Johannsen, Wilhelm L. (1857-1927): Danish botanist who introduced the term "genes," derived from "pangenes," factors suggested by Hugo de Vries to determine hereditary characteristics in plants. Johannsen also introduced the concepts of phenotype and genotype to distinguish between physical and hereditary traits.

Kenyon, Cynthia J. (1976-   ): Discovered the role of specific genes in regulation of cell migration and the aging process in the nematode *Caenorhabditis*, helping to clarify similar processes in more highly evolved eukaryotic organisms.

Khorana, Har Gobind (1922-   ): Developed methods for investigating the structure of DNA and deciphering the genetic code. Khorana synthesized the first artificial gene in the 1960's. He was awarded the Nobel Prize in Physiology or Medicine in 1968.

King, Helen Dean (1869-1955): By selective breeding of rodents, developed a method for production of inbred strains of animals for laboratory studies. The methodology was later applied to development of more desirable breeds of horses.

Klug, Aaron (1759-1853): Won the 1982 Nobel Prize in Chemistry. Klug used X-ray crystallography to investigate biochemical structures, especially that of viruses. He was able to link the assembly of viral protein subunits with specific sites on viral RNA, which helped in fighting viruses that cause disease in plants and, more basically, in understanding the mechanism of RNA transfer of genetic information. He also determined the

structure of transfer RNA (tRNA), which has a shape similar to that of a bent hair pin.

Knight, Thomas Andrew (1759-1853): Plant biologist who first recognized the usefulness of the garden pea for genetic studies because of its distinctive traits. Knight was the first to characterize dominant and recessive traits in the pea, though, unlike Gregor Mendel, he never determined the mathematical relationships among his crosses.

Kölreuter, Josef Gottlieb (1733-1806): A forerunner of Gregor Mendel, Kölreuter demonstrated the sexual nature of plant fertilization, in which characteristics were derived from each member of the parental generation in equivalent amounts.

Kornberg, Arthur (1918-   ): Carried out the first purification of DNA polymerase, the enzyme that replicates DNA. His work on the synthesis of biologically active DNA in a test tube culminated with his being awarded the 1959 Nobel Prize in Physiology or Medicine.

Kossel, Albrecht (1853-1927): Won the 1910 Nobel Prize in Physiology or Medicine. Isolated and described molecular constituents of the cell's nucleus, notably cytosine, thymine, and uracil. These molecules later proved to be constituents of the codons in DNA and RNA. Thus, Kossel's research prepared the way for understanding the biochemistry of genetics.

Lamarck, Jean-Baptiste (1744-1829): French botanist and evolutionist who introduced many of the earliest concepts of inheritance. Lamarck proposed that hereditary changes occur as a result of an organism's needs; his theory of inherited characteristics, since discredited, postulated that organisms transmit acquired characteristics to their offspring.

Leder, Philip (1934-   ): Along with Marshal Nirenberg, identified the genetic code words for amino acids. His later work has involved the transplantation of human oncogenes into mice, for the purpose of studying the effects of such genes in development of cancer.

Lederberg, Joshua (1925-   ): Established the occurrence of sexual reproduction in bacteria. Lederberg demonstrated that genetic manipulation of the DNA during bacterial conjugation could be used to map bacterial genes. He was awarded the 1958 Nobel Prize in Physiology or Medicine.

Levene, Phoebus Aaron (1869-1940): American biochemist who determined the components found in DNA and RNA. Levene described the presence of ribose sugar in RNA and of 2′-deoxyribose in DNA, thereby differentiating the two molecules. He also identified the nitrogen bases found in nucleic acid, though he was never able to determine the acid's molecular structure.

Lewis, Edward B. (1918-   ): Through the use of X-ray-induced mutations in *Drosophila* (fruit flies), Lewis was able to discover and map genes that regulate embryonic development. Among Lewis's discoveries was the existence of homeotic genes, genes that regulate development of body parts. Along with Christiane Nüsslein-Volhard and Eric Wieschaus, awarded the Nobel Prize in Physiology or Medicine in 1995.

Linnaeus, Carolus (1707-1778): Swedish naturalist and botanist most noted for establishing the modern method for classification of plants and animals. In his *Philosophia Botanica* (1751; *The Elements of Botany*, 1775), Linnaeus proposed that variations in plants or animals are induced by environments such as soil.

Luria, Salvador E. (1912-1991): A pioneer in understanding replication and genetic structure in viruses. The Luria-Delbrück fluctuation test, developed by Luria and Max Delbrück, demonstrated that genetic mutations precede environmental selection. Luria was awarded the 1969 Nobel Prize in Physiology or Medicine.

Lwoff, André (1902-1994): French biochemist and protozoologist. Lwoff's early work demonstrated that vitamins function as components of living organisms. He is best known for demonstrating that the genetic material of bacteriophage can become part of the host bacterium's DNA, a process known as lysogeny. Lwoff was awarded the 1965 Nobel Prize in Physiology or Medicine.

Lyon, Mary Frances (1925-   ): British cyto-

geneticist who proposed what became known as the Lyon hypothesis, that only a single X chromosome is active in a cell. Any other X chromosomes are observed as Barr bodies.

McClintock, Barbara (1902-1992): Demonstrated the existence in plants of transposable elements, or transposons, genes that "jump" from one place on a chromosome to another. The process was discovered to be widespread in nature. McClintock was awarded the 1983 Nobel Prize in Physiology or Medicine.

Macklin, Madge Thurlow (1893-1962): Developed a method to apply statistical analysis to understanding congenital diseases in human families. Her arguments were used to introduce genetics as a component of the curriculum in medical schools. Her support of eugenics for improvement of humans later made her views controversial.

McKusick, Victor A. (1921-   ): Cataloged and indexed many of the genes responsible for disorders that are passed in Mendelian fashion.

Margulis, Lynn (1938-   ): Developed the endosymbiont theory, which suggests that internal eukaryotic organelles, such as mitochondria and chloroplasts, originated as free-living prokaryotic ancestors. She proposed that free-living bacteria became incorporated in a larger, membrane-bound structure and developed a symbiotic relationship within the larger cell.

Mendel, Johann Gregor (1822-1884): The "father of genetics," Mendel was an Austrian monk whose studies on the transmission of traits in the garden pea established the mathematical basis of inheritance. Mendel's pioneering theories, including such fundamental genetic principles as the law of segregation and the law of independent assortment, were published in 1866 but received scant attention until the beginning of the twentieth century.

Meselson, Matthew Stanley (1930-   ): Demonstrated the nature of DNA replication, in which the two parental DNA strands are separated, each passing into one of the two daughter molecules. Also noted as a social activist.

Meyerowitz, Elliot M. (1951-): Discovered roles played by specific genes in differentiation of the plant *Arabidopsis*, as well as genes which regulate flowering in plants.

Miescher, Johann Friedrich (1844-1895): In 1869, Miescher discovered and purified DNA from cell-free nuclei obtained from white blood cells and gave the name "nuclein" to the extract. The substance was later known as nucleic acid.

Mintz, Beatrice (1921-): Noted for studies of the role of gene control in differentiation of cells and disease in humans. Developed a mouse model for the understanding of melanoma development in humans.

Monod, Jacques Lucien (1910-1976): French geneticist and molecular biologist who with François Jacob demonstrated a method of gene regulation in bacteria that came to be known as the Jacob-Monod model. Jacob and Monod were jointly awarded the 1965 Nobel Prize in Physiology or Medicine.

Moore, Stanford (1913-1982): Won the 1972 Nobel Prize in Chemistry, with William H. Stein. Moore and Stein supplemented Alfinsen's research by identifying the sequence of amino acids in ribonuclease, a clue to the structure of the gene responsible for it.

Morgan, Lilian Vaughan (1870-1952): Discovered the attached X and ring X chromosomes in *Drosophila* (fruit flies). Later contributed to studying the effects of polio vaccines in primates. Married to Thomas Hunt Morgan.

Morgan, Thomas Hunt (1866-1945): Considered the father of modern genetics, an embryologist whose studies of fruit flies (*Drosophila melanogaster*) established the existence of genes on chromosomes. Through his selective breeding of flies, Morgan also established concepts such as gene linkage, sex-linked characteristics, and genetic recombination. Won the 1933 Nobel Prize in Physiology or Medicine.

Müller, Hermann Joseph (1890-1967): Geneticist and colleague of Thomas Hunt Morgan. Muller's experimental work with fruit flies established the gene as the site of mutation. His work with X rays demonstrated a means of artificially introducing mutations into an organism. Won the 1946 Nobel Prize in Physiology or Medicine.

Mullis, Kary Banks (1944-): Devised the polymerase chain reaction (PCR), a method for duplicating small quantities of DNA. The PCR procedure became a major tool in research in the fields of genetics and molecular biology. Mullis was awarded the 1993 Nobel Prize in Chemistry.

Nathans, Daniel (1928-1999): Applied the use of restriction enzymes to the study of genetics. Nathans developed the first genetic map of SV40, among the first DNA viruses shown to transform normal cells into cancer. Nathans was awarded the 1978 Nobel Prize in Physiology or Medicine.

Neel, James Van Gundia (1915-2000): Considered to be the father of human genetics. Among his discoveries was the recognition of the genetic basis of sickle-cell disease. He was also noted for his study of the aftereffects of radiation on survivors of the atomic attack on Hiroshima and Nagasaki in World War II. He was the first to propose what was referred to as the thrifty-gene hypothesis, the idea that potentially lethal genes may have been beneficial to the human population earlier in evolution.

Nelson, Oliver Evans, Jr. (1920-2001): During the 1950's, carried out the first structural analysis of a gene in higher plants (corn), at the same time confirming the existence of transposable elements. His later work demonstrated the genetic significance of enzymatic defects in maize.

Neufeld, Elizabeth F. (1928-): French-born biochemist who found that many mucopolysaccharide storage diseases resulted from the absence of certain metabolic enzymes. Her work opened the way for prenatal diagnosis of such diseases.

Nirenberg, Marshall Warren (1927-): Molecular biologist who was among the first to decipher the genetic code. He later demonstrated the process of ribosome binding in protein synthesis and carried out the first cell-free synthesis of protein. Nirenberg was awarded the 1968 Nobel Prize in Physiology or Medicine.

Nurse, Sir Paul M. (1949-   ): British scientist who discovered the role of chemical modification (phosphorylation) in regulation of the cell cycle. With Tim Hunt and Leland Hartwell, he was awarded the Nobel Prize in Physiology or Medicine in 2001.

Nüsslein-Volhard, Christiane (1942-   ): German biologist whose genetic studies in *Drosophila* (fruit flies) led to the discovery of genes that regulate body segmentation in the embryo. Along with Edward Lewis and Eric Wieschaus, won the Nobel Prize in Physiology or Medicine in 1995.

Ochoa, Severo (1905-1993): Won the 1959 Nobel Prize in Physiology or Medicine, with Arthur Kornberg. Ochoa and Kornberg isolated enzymes involved in the synthesis of DNA and RNA, representing the first steps in decoding the biochemical instructions preserved in the structure of genes.

Olson, Maynard V. (1943-   ): Studied base-pair polymorphisms in the human genome and their significance to evolution. In 1987, with David Burke, Olson developed a new type of cloning vector, artificial chromosomes, that filled the need created by the Human Genome Project to clone very large insert DNAs (hundreds of thousands to millions of base pairs in length).

Pauling, Linus (1901-1994): American chemist who received the Nobel Prize in Chemistry in 1954 for his work on the nature of the chemical bond and the 1962 Nobel Peace Prize for his antinuclear activism. His 1950's investigations of protein structure contributed to the determination of the structure of DNA.

Punnett, Reginald C. (1875-1967): English biologist who collaborated with William Bateson in a series of important breeding experiments that confirmed the principles of Mendelian inheritance. Punnett also introduced the Punnett square, the standard graphical method of depicting hybrid crosses.

Rhabar, Shemooil (1929-   ): Iranian director at the University of Tehran, who became known as the most important immunologist in the Muslim world.

Roberts, Richard J. (1943-   ): Discovered that genes in eukaryotic cells and animal viruses are often discontinuous, with intervening sequences between segments of genetic material. With Philip Sharp, Roberts received the Nobel Prize in Physiology or Medicine in 1993.

Rodbell, Martin (1925-1998): Discovered the role of membrane-bound "G" proteins in regulation of signal transduction in eukaryotic cells. With Alfred Gilman, awarded the Nobel Prize in Physiology or Medicine in 1994.

Rowley, Janet (1925-   ): Cytogeneticist who developed the staining techniques for observation of cell structures. She demonstrated the role of chromosomal translocation as the basis for chronic myeloid leukemia, the first example of translocation as a cause of cancer.

Rubin, Gerald M. (1950-   ): Major figure in developing a structure/functional relationship of genes in *Drosophila* (fruit flies) through the use of insertion mutagenesis to inactivate specific genes.

Russell, Elizabeth Shull (1913-2001): Contributed to the understanding of the role played by specific genes in creating coat variations in animals. Her later work involved the identification of genetic defects in the aging process and in the development of diseases such as muscular dystrophy.

Russell, William (1910-2003): A pioneer in the genetic effects of radiation at Oak Ridge National Laboratory whose testing of mice led to standards for acceptable levels of human exposure to radiation. Winner of the 1976 Fermi Award.

Sager, Ruth (1918-1997): During the 1950's, demonstrated the existence of nonchromosomal heredity, also known as cytoplasmic inheritance, and hence the role of cytoplasmic genes in organelle development. Later involved in study of tumor suppressor and breast cancer genes.

Sageret, Augustin (1763-1851): French botanist who discovered the ability of different traits to segregate independently in plants.

Sanger, Frederick (1918-   ): Determined the method for sequencing DNA. His method separated the strands of DNA and then rebuilt them in stages that allowed the terminal nucleotides to be identified. This made it possible to sequence the entire genomes of organisms. With Paul Berg and Walter Gilbert, Sanger received the 1980 Nobel Prize in Chemistry.

Sharp, Phillip A. (1944-   ): Discovered that genes in eukaryotic cells or animal viruses are discontinuous, with segments divided by sections separated by intervening sequences of genetic material. With Richard Roberts, received the Nobel Prize in Physiology or Medicine in 1993.

Simpson, George Gaylord (1902-1984): American paleontologist who applied population genetics to the study of the evolution of animals. Simpson was instrumental in establishing a neo-Darwinian theory of evolution (the rejection of Lamarck's inheritance of acquired characteristics) during the early twentieth century.

Singer, Maxine (1931-   ): Applied the use of the newly discovered restriction enzymes in formation of recombinant DNA. Singer is most noted as a "voice of calm" in the debate over genetic research, emphasizing the application of such research, and the self-policing of scientists carrying out such work.

Smith, Hamilton Othanel (1931-   ): Pioneered the purification of restriction enzymes, winning the 1978 Nobel Prize in Physiology or Medicine, with Werner Arber and Daniel Nathans. Arber and the team of Nathans and Smith separately described the restriction-modification system by studying bacteria and bacteriophages; the system involves the action of site-specific endonuclease and other enzymes that cleave DNA into segments.

Smith, Michael (1932-   ): Won the 1995 Nobel Prize in Chemistry. Smith developed site-directed mutagenesis, a means for reconfiguring genes in order to create altered proteins with distinct properties. Smith's genetic engineering tool made it possible to treat genetic disease and cancer and to create novel plant strains.

Snell, George D. (1903-1996): Snell's discovery of the H-2 histocompatibility complex, which regulates the immune response in mice, led to the later discovery of the equivalent HLA complex in humans. Awarded the Nobel Prize in Physiology or Medicine in 1980.

Sonneborn, Tracy Morton (1905-1981): Discovered crossbreeding and mating types in paramecia, integrating the genetic principles as applied to multicellular organisms with single-celled organisms such as protozoa.

Spemann, Hans (1869-1941): Won the 1935 Nobel Prize in Physiology or Medicine. By transplanting bits of one embryo into a second, viable embryo, Spemann compiled evidence that an "organizer center" directs the development of an embryo and that different parts of the organizer governed distinct portions of the embryo. His experiments provided clues to the genetic control of growth from the earliest stages of an organism.

Spencer, Herbert (1820-1903): English philosopher influenced by the work of Charles Darwin. Spencer proposed the first general theory of inheritance, postulating the existence of self-replicating units within the individual which determine the traits. Spencer is more popularly known as the source of the notion of "survival of the fittest" as applied to natural selection.

Stanley, Wendell Meredith (1904-1971): American biochemist who was the first to crystallize a virus (tobacco mosaic virus), demonstrating its protein nature. Stanley was later a member of the team that determined the amino acid sequence of the TMV protein. Stanley spent the last years of his long career studying the relationship of viruses and cancer.

Stein, William H. (1911-1980): Won the 1972 Nobel Prize in Chemistry, with Stanford Moore. Stein and Moore supplemented Alfinsen's research by identifying the sequence of amino acids in ribonuclease, a clue to the structure of the gene responsible for it.

Stevens, Nettie Maria (1861-1912): Discovered the existence of the specific chromosomes that determine sex, now known as the X and Y chromosomes. Described the existence of chromosomes as paired structures within the cell.

Strobell, Ella Church (1862-1920): Developed the technique of photomicroscopy for analysis of chromosomal theory.

Sturtevant, Alfred Henry (1891-1970): Colleague of Thomas Hunt Morgan and among the pioneers in application of the fruit fly (*Drosophila*) in the study of genetics. In 1913, Sturtevant constructed the first genetic map of a fruit fly chromosome. His work became a major factor in chromosome theory. In the 1930's, his work with George Beadle led to important observations of meiosis.

Sulston, John E. (1942-   ): Developed first map of cell lineages in the model nematode *Caenorhabditis*, leading to the discovery of the first gene associated with programmed cell death. Sulston was also part of the team that sequenced the worm's genome. With Sydney Brenner and H. Robert Horvitz, he was awarded the Nobel Prize in Physiology or Medicine in 2002.

Sutton, Walter Stanborough (1877-1916): Biologist and geneticist who demonstrated the role of chromosomes during meiosis in gametes, and demonstrated their relationship to Mendel's laws. Sutton observed that chromosomes form homologous pairs during meiosis, with one member of each pair appearing in gametes. The particular member of each pair was subject to Mendel's law of independent assortment.

Tammes, Jantine (1871-1947): Dutch geneticist who demonstrated that the inheritance of continuous characters, traits that have a range of expression, could be explained in a Mendelian fashion. She developed a multiple allele hypothesis that helped explain some of the data.

Tan Jiazhen (C. C. Tan; 1909-   ): Considered the father of Chinese genetics. In a career spanning more than seven decades, Tan studied genetic structure and variation in a wide range of organisms. His most important work involved the study of evolution of genetic structures in *Drosophila* (fruit flies), as well as the concept of mosaic dominance in the beetle.

Tatum, Edward Lawrie (1909-1975): Along with George Beadle, Tatum demonstrated that the function of a gene is to encode an enzyme. Beadle and Tatum were awarded the 1958 Nobel Prize in Physiology or Medicine for their one gene-one enzyme hypothesis.

Temin, Howard Martin (1934-1994): Proposed that RNA tumor viruses replicate by means of a DNA intermediate. Temin's theory, initially discounted, became instrumental in revealing the process of infection and replication by such viruses. He later isolated the replicating enzyme, the RNA-directed DNA polymerase (reverse transcriptase). He was awarded the 1975 Nobel Prize in Physiology or Medicine, along with David Baltimore, for this work.

Todd, Alexander Robertus (1907-1997): Won the 1957 Nobel Prize in Chemistry. As part of wide-ranging research in organic chemistry, Todd revealed how ribose and deoxyribose bond to the nitrogenous bases on one side of a nucleotide unit and to the phosphate group on the other side. These discoveries provided necessary background for work by others that explained the structure of the DNA molecule.

Tonegawa, Susumu (1939-   ): Discovered the role of genetic rearrangement of DNA in lymphocytes, which plays a key role in generation of antibody diversity. In 1987, awarded the Nobel Prize in Physiology or Medicine.

Varmus, Harold Elliot (1939-   ): Elucidated the molecular mechanisms by which retroviruses (RNA tumor viruses) transform cells. Varmus was awarded the 1989 Nobel Prize in Physiology or Medicine.

Waelsch, Salome Gluecksohn (1907-   ): Studied the role genes play in abnormal cell differentiation and congenital abnormalities. Born in Germany, she fled to the United States from the Nazis in 1933. Her 1938 publication of the role of genes in the T (tailless) phenotype in mice is considered a genetic classic.

Waterston, Robert (1943-   ): Identified many of the genes that regulate muscle development in the nematode *Caenorhabditis*, as well as contributing to the sequence of the genome. His sequencing work was also applied in the Human Genome Project led by Francis Collins.

Watson, James Dewey (1928-   ): Along with Francis Crick, determined the double-helix structure of DNA. Together with Crick and Maurice Wilkins, Watson was awarded the 1962 Nobel Prize in Physiology or Medicine for their work in determining the structure of DNA.

Weinberg, Robert Allan (1942-   ): Molecular biologist who isolated the first human oncogene, the *ras* gene, associated with a variety of cancers, including those of the colon and brain. Weinberg later isolated the first tumor suppressor gene, the retinoblastoma gene. Weinberg is considered among the leading researchers in understanding the role played by oncogenes in development of cancer.

Weinberg, Wilhelm (1862-1937): German obstetrician who demonstrated that hereditary characteristics of humans such as multiple births and genetic diseases were subject to Mendel's laws of heredity. The mathematical application of such characteristics, published simultaneously (and independently) by Godfrey Hardy, became known as the Hardy-Weinberg equilibrium. The equation demonstrates that dominant genes do not replace recessive genes in a population; gene frequencies would not change from one generation to the next if certain criteria such as random mating and lack of natural selection were met.

Weismann, August (1834-1914): German zoologist noted for his chromosome theory of heredity. Weismann proposed that the source of heredity is in the nucleus only and that inheritance is based on transmission of a chemical or molecular substance from one generation to the next. Weismann's theory, which rejected the inheritance of acquired characteristics, came to be called neo-Darwinism. Though portions of Weismann's theory were later disproved, the nature of the chromosome was subsequently demonstrated by Thomas Hunt Morgan and his colleagues.

Wieschaus, Eric F. (1947-   ): Wieschaus's studies of genetic control in *Drosophila* (fruit flies) led to the discovery of genes that regulate cell patterns and shape in the embryo. Along with Edward Lewis and Christiane Nüsslein-Volhard, he was awarded the Nobel Prize in Physiology or Medicine in 1995.

Wilkins, Maurice Hugh Frederick (1916-   ): Studies on the X-ray diffraction patterns exhibited by DNA confirmed the double-helix structure of the molecule. Wilkins was a colleague of Rosalind Franklin, and it was their work that confirmed the nature of DNA as proposed by Watson and Crick. Wilkins was awarded the Nobel Prize for Physiology and Medicine in 1962, along with Watson and Crick.

Wilmut, Ian (1944-   ): Scottish embryologist and leader of a research team at the Roslin Institute near Edinburgh. In 1996, Wilmut and his colleagues succeeded in cloning an adult sheep, Dolly, the first adult mammal to be successfully produced by cloning.

Wilson, Edmund Beecher (1856-1939): His study of chromosomes in collaboration with Nettie Stevens led to the discovery of the X and Y chromosomes, playing a key role in the foundation of modern genetics. His later work involved the study of development and differentiation of the fertilized egg.

Witkin, Evelyn Maisel (1921-   ): Through her studies of induced or spontaneous mutations in bacterial DNA, discovered processes of enzymatic repair of DNA.

Woese, Carl R. (1928-   ): Based on his studies of ribosomal RNA differences in prokaryotes and eukaryotes, proposed that all life-forms exist in one of three domains: Bacteria, Archaea ("ancient" bacteria), and Eukarya (eukaryotic organisms, from microscopic plants to large animals). Woese expanded his theory in arguing that the Archaea represent the earliest form of life on Earth, and that they later formed a branch which became the eukaryotes.

Wright, Sewall (1889-1988): Discovered genetic drift of genetic traits. The "Sewall Wright" effect, the random drift of characters in small populations, was explained by the random loss of genes, even in the absence of natural selection.

Yanofsky, Charles (1925-   ): Confirmed that the genetic code involved groups of three bases by demonstrating colinearity of the bases and amino acid sequences. He applied this work in demonstrating similar colinearity of mutations in the tryptophan operon and changes in amino acid sequences.

Zinder, Norton (1928-   ): With Joshua Lederberg, discovered the role of bacteriophage in transduction, the movement of genes from one host to another by means of viruses. Zinder was also noted for his discovery of RNA bacteriophage and his work on the molecular genetics of such agents.

—Richard Adler

Nobel Prizes for Discoveries in Genetics

Physiology or Medicine

1910   **Albrecht Kossel (German)** isolated and described molecular constituents of the cell's nucleus, notably cytosine, thymine, and uracil. These molecules later proved to be constituents of the codons in deoxyribonucleic acid (DNA) and ribonucleic acid (RNA). Thus, Kossel's research prepared the way for understanding the biochemistry of genetics.

1933   **Thomas Hunt Morgan (American).** Experimenting with the fruit fly *Drosophila melanogaster,* Morgan discovered that the mechanism for the Mendelian laws of heredity lies in the chromosomes inside the nucleus of cells and that specific genes on the chromosomes govern specific somatic traits in the flies. Morgan confirmed the accuracy of Mendel's laws and ended a controversy over their physiological source.

1935   **Hans Spemann (German).** By transplanting bits of one embryo into a second, viable embryo, Spemann compiled evidence that an "organizer center" directs the development of an embryo and that different parts of the organizer governed distinct portions of the embryo. His experiments provided clues to the genetic control of growth from the earliest stages of an organism.

1946   **Hermann J. Muller (American).** Muller proved that X rays damage genes by altering their structure: radiation-induced mutation. Consequently, X rays also modify the structure of chromosomes. The mutations most often produce recessive and harmful traits in the irradiated organism.

1958   **George Beadle and Edward Tatum (both American).** In research on the fungus *Neurospora crassa,* Beadle and Tatum found that biotin was essential to cultivating certain mutant strains of the fungus; this fact demonstrated that genes regulate the synthesis of specific cellular chemicals, one or more of these genes being mutated in the biotin-dependent strain.

1958   **Joshua Lederberg (American).** Lederberg showed that the bacterium *Escherichia coli,* although not able to reproduce sexually, is capable of genetic recombination between chromosomes from different cells through a process called conjugation.

1959   **Severo Ochoa (Spanish) and Arthur Kornberg (American).** Ochoa and Kornberg isolated enzymes involved in the synthesis of deoxyribonucleic acid (DNA) and ribonucleic acid (RNA), representing the first steps in decoding the biochemical instructions preserved in the structure of genes.

1962   **Francis Crick (British), James Watson (American), and Maurice Wilkins (British).** Using X-ray diffraction analysis and molecular modeling, Wilkins, Crick, and Watson found that deoxyribonucleic acid (DNA) is structured in a double helix. They were able to identify the specific three-dimensional structure that is the basis for the ability of DNA to be replicated and transcribed.

1965   **François Jacob and Jacques Monod (both French).** Studying enzyme action, Jacob and Monod proved that messenger ribonucleic acid (mRNA) carries instructions from the nucleus to ribosomes, where molecules are assembled for use in the body, and they distinguished structural genes from regulatory genes.

1965   **André Lwoff (French).** Lwoff proposed that viral deoxyribonucleic acid (DNA) can become active after invading cells and cause the cells to divide out of control, producing cancerous tumors.

1968   **Robert W. Holley (American), Har Gobind Khorana (Indian), and Marshall W. Nirenberg (American).** Working separately, Holley, Khorana, and Nirenberg deciphered the genetic code in ribonucleic acid (RNA) and deoxyribonucleic acid (DNA). Their work anticipated DNA sequencing and genetic engineering.

1969 **Max Delbrück (German), Alfred D. Hershey (American), and Salvador E. Luria (Italian).** In joint studies of bacteriophages and their bacterial hosts, Delbrück, Hershey, and Luria described the conformation of bacteriophage deoxyribonucleic acid (DNA), showed that different strains exchange genetic information, and proved that bacterial DNA mutated to confer protection from attack, demonstrating that bacterial heredity is based on genetic exchange. The discovery explained why bacteria gradually become resistant to pharmaceuticals.

1975 **David Baltimore and Howard M. Temin (both American).** Working separately, Baltimore and Temin discovered reverse transcriptase, the enzyme that inserts viral deoxyribonucleic acid (DNA) into cellular DNA, which can cause cancer, They also identified retroviruses, a class of virus that includes the human immunodeficiency virus (HIV) that causes acquired immunodeficiency syndrome (AIDS).

1975 **Renato Dulbecco (Italian).** Dulbecco described how tumor viruses cause cellular transformation in somatic cells by suppressing the regulatory system that controls division; the cells then divide out of control.

1978 **Werner Arber (Swiss), Daniel Nathans (American), and Hamilton O. Smith (American).** Arber and the team of Nathans and Smith separately described the restriction-modification system by studying bacteria and bacteriophages; the system involves the action of site-specific endonuclease and other enzymes that cleave deoxyribonucleic acid (DNA) into segments.

1980 **Baruj Benacerraf (Venezuelan), Jean Dausset (French), and George D. Snell (American).** Benacerraf, Dausset, and Snell each explained the genetic components of the major histocompatibility complex (MHC), the key to a person's immune system, and how the system produces antibodies to such a wide variety of foreign molecules and pathogens, such as viruses, fungi, and bacteria.

1983 **Barbara McClintock (American)** investigated the genetics of maize (corn) and discovered a new mechanism of gene modification: Some "jumping genes" (now called transposable elements or transposons) move to new sites on chromosomes and either suppress nearby structural genes or inactivate suppressor genes. The discovery was a major breakthrough in understanding novel, non-Mendelian types of genetic variation.

1985 **Michael S. Brown and Joseph L. Goldstein (both American).** Brown and Goldstein conducted extensive research in the regulation of cholesterol in humans. They showed that in families with a history of high cholesterol, individuals who carry two copies of a mutant gene (homozygotes) have cholesterol levels several times higher than normal and those who have one mutant gene (heterozygotes) have levels about double normal. Their discoveries proved invaluable in managing heart disease and other cholesterol-related ailments.

1987 **Susumu Tonegawa (Japanese)** explained the diversity of antibodies by showing that the antigen-sensitive part of each antibody is created by segments of three genes; since the segments from each gene can vary in length, the possible combinations from three genes can produce billions of distinct antibodies.

1989 **J. Michael Bishop and Harold E. Varmus (both American).** Bishop and Varmus discovered that oncogenes (genes that play a role in initiating cancer) originate in normal cells and control cellular growth and are not solely derived from retroviruses, as previously thought. Their work greatly influenced subsequent studies of tumor development.

1993   **Richard J. Roberts (British) and Phillip A. Sharp (American).** Roberts and Sharp separately studied the relationship between deoxyribonucleic acid (DNA) and ribonucleic acid (RNA). They discovered that portions of a human gene can be divided among several DNA segments, called introns, separated by noncoding segments called exons. This discovery became important to genetic engineering and to understanding the mechanism for hereditary diseases.

1995   **Edward B. Lewis (American).** Lewis found that an array of master genes governs embryo development.

1995   **Christiane Nüsslein-Volhard (German) and Eric F. Wieschaus (American).** Nüsslein-Volhard and Wieschaus worked together to extend Lewis's investigations into the genetic control of embryo development through studies of fruit flies. They isolated more than five thousand participating genes and distinguished four types of "master" control genes: gap, pair-rule, segment polarity, and even-skipped.

2001   **Leland H. Hartwell (American), R. Timothy Hunt (British), and Paul M. Nurse (British).** Hartwell, Hunt, and Nurse conducted research on the regulation of cell cycles. Hartwell identified a class of genes that controls the cycle, including a gene that initiates it. Nurse cloned and described the genetic model of a key regulator, cyclin dependent kinase, while Hunt discovered cyclins, a class of regulatory proteins.

2002   **Sydney Brenner (American)** used the transparent nematode *Caenorhabditis elegans* to establish a simple model organism for studying how genes control the development of organs.

2002   **John E. Sulston (British)** studied cell division and cell lineages in *Caenorhabditis elegans* following Brenner's methods. He demonstrated that genetic control of specific lineages includes programmed cell death, called apoptosis, as part of the regulatory process, and he isolated the protein that degrades the deoxyribonucleic acid (DNA) of dead cells.

2002   **H. Robert Horvitz (American).** Using Brenner's *Caenorhabditis elegans* model, Horvitz discovered the first two "death genes" which instigate cell death. He further found that another gene helps protect cells from cell death.

Chemistry

1957   **Alexander Robertus Todd (British).** As part of wide-ranging research in organic chemistry, Todd revealed how ribose and deoxyribose bond to the nitrogenous bases on one side of a nucleotide unit and to the phosphate group on the other side. These discoveries provided necessary background for work by others that explained the structure of the deoxyribonucleic acid (DNA) molecule.

1972   **Christian B. Anfinsen (American).** Anfinsen, studying the three-dimensional structure of the enzyme ribonuclease, proved that its conformation was determined by the sequence of its amino acids and that to construct a complete enzyme molecule no separate structural information was passed on from the deoxyribonucleic acid (DNA) in the cell's nucleus.

1972   **Stanford Moore and William H. Stein (both American).** Moore and Stein supplemented Alfinsen's research by identifying the sequence of amino acids in ribonuclease, a clue to the structure of the gene responsible for it.

1980   **Paul Berg (American).** Berg invented procedures for removing a gene from a chromosome of one species and inserting it into the chromosome of an entirely different species, enabling him to study how the genetic information of the contributing organism interacts with host's deoxyribonucleic acid (DNA). The recombinant DNA technology, sometimes called gene splicing, became fundamental to the genetic engineering of transgenic species.

1980   **Walter Gilbert (American) and Frederick Sanger (British).** Gilbert and Sanger independently developed methods for determining the sequence of nucleic acids in DNA, thus decoding the genetic information. Gilbert's method cuts DNA into small units that reveals their structure when exposed to specific chemicals; Sanger's method separates the strands of DNA and then rebuilds them in stages that allow the terminal nucleotides to be identified. Their methods later made it possible to sequence the entire genomes of organisms.

1982   **Aaron Klug (British).** Klug used X-ray crystallography to investigate biochemical structures, especially that of viruses. He was able to link the assembly of viral protein subunits with specific sites on viral ribonucleic acid (RNA), which helped in fighting viruses that cause disease in plants and, more basically, in understanding the mechanism of RNA transfer of genetic information. He also determined the structure of transfer RNA (tRNA), which has a shape similar to that of a bent hair pin.

1989   **Sidney Altman (Canadian) and Thomas R. Cech (American).** Working independently Altman and Cech discovered that ribonucleic acid (RNA), like proteins, can act as a catalyst; moreover, Cech found that when ribosomal RNA participates in translation of mRNA and the synthesis of polypeptides, it acts as a catalyst in some steps.

1995   **Kary B. Mullis (American).** Mullis invented polymerase chain reaction (PCR), a method for swiftly making millions of copies of deoxyribonucleic acid (DNA). PCR soon became an important tool in genetic engineering, DNA fingerprinting, and medicine.

1995   **Michael Smith (British).** Smith developed site-directed mutagenesis, a means for reconfiguring genes in order to create altered proteins with distinct properties. Smith's genetic engineering tool made it possible to treat genetic disease and cancer and to create novel plant strains.

Peace

1970   **Norman Borlaug (American).** Borlaug was a key figure in the "green revolution" of agriculture. Working as a geneticist and plant physiologist in a joint Mexican-American program, he developed strains of high-yield, short-strawed, disease-resistant wheat. His goal was to increase crop production and alleviate world hunger.

—Roger Smith

Time Line of Major Developments in Genetics

12,000 B.C.E.   Humans begin domesticating plants and animals, the earliest form of artificial selection. Domestication involves selective breeding for certain traits. This form of "genetic engineering" allows for transition from hunter-gatherer societies to agrarian civilizations.

c. 323 B.C.E.   Aristotle theorizes about the nature of species, reproduction, and hybrids.

1651   William Harvey publishes *Exercitationes de generatione animalium* (*Anatomical Exercitations, Concerning the Generation of Living Creatures*, 1653), in which he suggests that all living things must originate in an egg.

1677   Antoni van Leeuwenhoek describes sperm and eggs and collects evidence that helps disprove the theory of spontaneous generation.

1691-1694   German botanist Rudolph Jacob Camerarius establishes the existence of sex in plants.

1759   Kaspar Friedrich Wolff publishes his epigenesis hypothesis, which states that the complex structures of chickens develop from initially homogeneous, structureless areas of the embryo. Many questions remain before this new hypothesis can be validated; other researchers focus their efforts on the sea squirt, a simpler organism with fewer differentiated tissues.

1760   Josef Gottlieb Kölreuter conducts studies on fertilization and hybridization, discovering the principle of incomplete dominance and laying the groundwork for later hybridizers.

1798   Thomas Robert Malthus publishes *An Essay on the Principle of Population*, in which he analyzes population growth and relates it to the struggle for existence, setting the stage for evolutionary theory.

1798   Edward Jenner develops vaccination. Jenner used the cowpox virus as a vaccine to induce immunity against the genetically and structurally similar, but lethal, virus that causes smallpox in humans.

1809   Jean-Baptiste Lamarck publishes *Philosophie zoologique* (*Zoological Philosophy*, 1914), in which he sets forth his laws of evolution, particularly his law of acquired characteristics. Although his notion that acquired traits are individually passed to the next generation was later disproved in favor of natural selection, Lamarck's book makes the link between evolution and inherited traits that lays a foundation for later evolutionary theory.

1838   G. J. Mulder precipitates a fibrous material from cells. He calls this material "protein" and believes it is the most important of the known components of living matter.

1850   Theodore Schwann, Matthias Jakob Schleiden, and Rudolph Virchow recognize that tissues are made up of cells. The cell theory contradicts the prevailing view of "vitalism," which states that no single part of an organism is alive (it was thought properties of living matter were somehow shared by the whole organism). The new theory considers the cell to be the basic and most fundamental unit of life.

1855   Alfred Russel Wallace publishes *On the Law Which Has Regulated the Introduction of New Species*; later, in 1858, he sends Charles Darwin a manuscript, "On the Tendency of Varieties to Depart Indefinitely from the Original Type." Today Wallace is recognized as having developed the theory of natural selection along with Darwin.

1857   Louis Pasteur begins research into fermentation. His "pasteurization" process is originally proposed as a means of preserving beer and wines. Through his work, Pasteur makes the important discovery that "life must be derived from life."

1859   Charles Darwin publishes *On the Origin of Species by Means of Natural Selection*, in which he sets forth his theory of natural selection. The actual mechanism of evolution is not understood at the time. Once genetics was studied as a discipline, it became clear that genetics and evolution are intimately associated. Genetic theories would later explain and prove the theory of evolution.

1862   The Organic Act establishes the U.S. Department of Agriculture (USDA). As one of its functions, the USDA is responsible for the collection of new and valuable seeds and plants and the distribution of them to agriculturists. The preservation and dissemination of agriculturally important plants was a necessity for maintaining and increasing the world's food supply.

1866   Ernst Haeckel develops the hypothesis that hereditary information is transmitted by the cell nucleus.

1866   Gregor Mendel, an Austrian monk, publishes a paper titled "Experiments in Plant Hybridization." Working with garden peas, Mendel used a systematic approach to study heredity, forming the theories of segregation and independent assortment. Although his work lies unnoticed for more than thirty years, it will eventually be rediscovered and become the foundation for the discipline of genetics.

1869   Francis Galton publishes *Hereditary Genius*, on the heredity of intelligence, which lays the foundation for the eugenics movement.

1869   Friedrich Miescher isolates "nuclein" from the nuclei of white blood cells. This substance is later found to be the nucleic acids DNA and RNA.

1875   Oskar Hertwig, a student of Ernst Haeckel, demonstrates the fertilization of an ovum in a sea urchin, thus establishing one of the basic principles of sexual reproduction: the union of egg and sperm cells.

1880   Walter Fleming first describes mitosis, one of the two major processes of cell division in higher organisms (the other being meiosis). This discovery is key to the understanding of inheritance, since microscopic observations of dividing cells helped early researchers connect Mendelian genetics with cellular biology.

1883   Galton founds the field of eugenics with the publication of *Inquiries into Human Faculty and Its Development*. The notion that the human species can be improved by selective breeding helps perpetuate racism and provides a scientific rationale for subsequent "ethnic cleansing" programs such as those of the Nazi Party fifty years later.

1883   Wilhelm Roux theorizes that mitosis must result in equal sharing of all chromosomal particles by the daughter cells and describes the process, but his work is generally ignored.

1883   E. van Beneden studies the processes of meiosis and fertilization in the parasitic worm *Ascaris*. Van Beneden was the first to observe that the chromosome number in somatic, or body, cells is twice the number that exist in gametes, or sex cells. He also realized that when fertilization occurs (the combination of two gametes, the egg from the female and the sperm from the male), the chromosome number of somatic cells is established.

1883   The first absolutely pure yeast culture (yeast propagated from a single cell) is introduced at Denmark's Carlsberg Brewery. The ability to propagate and maintain pure strains of organisms—genetically identical strains, or clones—will prove pivotal to future genetic research.

1886   August Weismann publishes *The Germ-Plasm: A Theory of Heredity*, in which he maintains that only the "germ cells" (eggs and sperm), not somatic cells, can transmit hereditary information and changes from one generation to the next; he disproved the Larmarckian notion of "acquired" characteristics.

1887-1890   Theodor Boveri investigates and describes chromosomes and their behavior, noting that they are preserved through the process of cell division and that sperm and egg contribute equal numbers of chromosomes.

1888-1889   Émile Maupas describes the relationship of conjugation (genetic recombination) and senescence.

1889   Richard Altmann renames "nuclein" (isolated by Miescher in 1869) "nucleic acid."

1896   Edmund B. Wilson publishes *The Cell in Development and Heredity*, in which he discusses the role of cells and chromosomes in inherited traits.

1897   Eduard Buchner shows that organic chemical transformations can be performed by cell extracts. He discovers that yeast extracts can convert glucose to ethyl alcohol. Buchner's was one of the first in vitro experiments. Performing such experiments outside the body allowed researchers to control conditions and to observe the effects of individual variables.

1899   The Royal Horticultural Society holds a meeting in Chiswick, London, in which William Bateson calls for research on discontinuous variations. The meeting later is renamed the First International Congress of Genetics, still held annually as of 2003.

1900   Hugo de Vries, Erich Tschermak von Seysenegg, and Carl Correns independently rediscover and reproduce Mendel's work. Mendel's theories provided a framework for other researchers. Studies in cytology, cellular biology, plant hybridization, and biochemistry support Mendel's assertions.

1900	Karl Landsteiner discovers human blood groups.

1901   Clarence McClung describes the role of the X chromosome in determining sex.

1902   Lucien Cuénot, William Bateson, and others begin to confirm Mendelian inheritance in animals.

1902   Austrian botanist Gottlieb Haberlandt completes the cell theory with his idea of totipotency: Cells must contain all of the genetic information necessary to create an entire, multicellular organism. Therefore, every plant cell is capable of developing into an entire plant.

1902   William Ernest Castle, director of the Bussey Institute at Harvard University, and his students begin research into mouse genetics. His laboratory produces some of the most influential mammalian geneticists of the twentieth century, including L. C. Dunn, Clarence Little, Sewall Wright, and George Snell.

1902   Theodor Boveri recognizes the correlation between Mendel's laws of inheritance and current studies of cellular biology; he deduces the haploid nature of sperm and egg cells (that each had equal amounts of hereditary information) and determines, by experimenting with sea urchin sperm and egg cells, that each must contribute half the total number of chromosomes to offspring for their normal development.

1903   Working independently of Boveri, Walter Sutton comes to similar conclusions using grasshoppers. Both Boveri and Sutton have formed the chromosomal theory of heredity. Mendel's notions of segregation and independent assortment coincided with Sutton's observations of how chromosomes segregated during cell division. This provided a cellular explanation for Mendel's observations.

1903   P. A. T. Levene establishes the distinction between DNA and RNA, showing that the thymine in DNA is replaced by uracil in RNA.

1905   William Bateson, E. R. Saunders, and R. C. Punnett discover the phenomenon of gene linkage when they observe a violation of the Mendelian rule of independent assortment, noting two traits that do not assort independently. Instead, these genes are carried, or linked, on the same chromosome. Bateson also coins the term "genetics" to describe the science of heredity.

1905   Nettie Stevens and Edmund Wilson independently describe the behavior of sex chromosomes. Their observations provide the first direct evidence to support the chromosomal theory of heredity.

1905-1933   The eugenics movement grows in popularity. It influences social policies and immigration and sterilization laws in the United States and other countries. The idea that human traits, notably behavior, are governed by simple genetic rules was used to discriminate against the "mentally deficient," immigrants from specific countries, and even the poor and homeless. The U.S. eugenics movement effectively ended after the theory became associated with the policies of Nazi Germany.

1908   Sir Archibald Garrod proposes that some human diseases are "inborn errors of metabolism." By studying the inheritance of human disorders, Garrod provides the first evidence of a specific relationship between genes and enzymes.

1908   George Shull self-pollinates plants for many generations to produce pure-breeding lines. Donald Jones performs similar experiments to increase productivity. These two researchers develop the scientific basis of modern agricultural genetics.

1908   Godfrey Hardy and Wilhelm Weinberg discover mathematical relationships between genotypic and phenotypic frequencies in populations. Known as the Hardy-Weinberg law, the rules governing these mathematical relationships help researchers understand the dynamics of population genetics and the evolution of species.

1909   Wilhelm Johannsen, working on the statistical analysis of continuous variation, expands the modern genetic vocabulary, coining the terms "gene," "genotype," and "phenotype."

1909   Hermann Nilsson-Ehle describes another violation of Mendelian inheritance. His studies with kernel color in wheat indicate this is a polygenic trait. This was one of the first demonstrations that many genes could influence a single trait. Depending on the alleles, each gene contributes to the trait in an additive fashion. Other examples of polygenic inheritance include skin color and height in humans.

1909   Carl Correns discovers another class of exceptions to Mendelian inheritance, one of the first examples of extranuclear inheritance. The notion that other cellular organelles besides the nucleus carry DNA was not recognized for decades. However, Correns's experiments in the plant *Mirabilis jalapa* showed inheritance of leaf color via the DNA in the chloroplasts.

1910-1928   Thomas Hunt Morgan clearly establishes the chromosomal theory of heredity after investigating a white-eyed fruit fly and finding that the trait does not segregate exactly according to Mendelian principles, but rather is influenced by the sex of the fly. This fly experiment becomes the cornerstone upon which theories of Mendelian, chromosomal, and sexual inheritance are built into a cohesive whole. Morgan also establishes the "Fly Room" at Columbia University, where he and his students will conduct groundbreaking experiments using *Drosophila* for the next quarter century. He will win the Nobel Prize in Physiology or Medicine in 1933.

1910   Albrecht Kossel wins the 1910 Nobel Prize in Physiology or Medicine for earlier work isolating and describing molecular constituents of the cell's nucleus, notably cytosine, thymine, and uracil.

1911   Peyton Rous produces cell-free extracts from chicken tumors that, when injected, can induce tumors in other chickens. The tumor-producing agent in the extract is later found to be a virus. Thus, Rous has discovered a link between cancer and viruses. He wins the Nobel Prize in Physiology or Medicine in 1966.

1913   Alfred H. Sturtevant, a student of Morgan, constructs the first gene maps of chromosomes. Maps indicate the order of genes as they exist physically on the chromosome. Knowledge of gene locations on chromosomes provided insights into inheritance, genetic diseases, and the function and regulation of DNA. In addition, isolation of specific genes often required knowledge of their chromosomal location.

1913   Eleanor Carothers reports her discovery of the chromosomal basis of independent assortment. By examining grasshopper chromosomes, Carothers observed the behavior of the X chromosome, responsible for sex determination, during cell division. These observations corresponded with Mendel's principle of independent assortment.

1914   Calvin Blackman Bridges uses the phenomenon of primary nondisjunction (a fault in cell division resulting in the failure of chromosomes to separate during metaphase I) to prove that genes are carried on chromomsomes.

1915   *The Mechanism of Mendelian Heredity*, by Morgan, Sturtevant, Muller, and Bridges, is published, establishing *Drosophila* as a model organism for genetics research and describing fundamentals of gene mapping.

1916   Research on the major histocompatibility complex begins with Clarence Little and E. E. Tyzzer's experiments transplanting tumors between mice.

1917   Félix d'Herelle discovers bacteriophages, viruses that infect bacteria. Bacteriophages played an important role in early genetics research, including confirmation that DNA is the hereditary material. Bacteriophages also became important in recombinant DNA applications.

1917   O. Winge publishes "The Chromosomes: Their Number and General Importance," which for the first time describes the relationship between chromosome doubling and allopolyploidy in plants.

1922-1932   In what would become known as the "modern synthesis," Ronald A. Fisher, J. B. S. Haldane, Sewall Wright, and S. S. Chetverikov independently publish papers on evolution, Mendelian inheritance, and natural selection, merging Darwin's theory of natural selection with Mendel's theory of genetic inheritance to create a field of population genetics that allows for genetic change through genetic drift. Haldane develops quantitative methods of studying the effects of selection, identifying the number of generations needed to alter gene frequencies for recessive and dominant traits, autosomal and sex-linked genes, and haploid and diploid organisms.

1925-1926   A. H. Sturtevant describes the position effect: An inversion may place a gene in another location in the chromosome, removing the gene from its regulatory elements and altering its expression. He also provides genetic proof of inversion.

1927   Hermann J. Muller, another student of Morgan, uses X rays to induce mutations in organisms. The ability to mutate DNA was a powerful tool to determine the function of specific genes. Muller receives the Nobel Prize in Physiology or Medicine in 1946.

1928   Frederick Griffith uses the bacterium that causes pneumonia to initiate his investigations into the "transforming principle," or transformation. The hereditary material has not yet been identified, but Griffith's experiments indicate that the transforming principle is DNA. Although not absolute proof, his experiment contributes significantly to the field and sparks ideas in other researchers.

1929   Clarence Little helps found the Jackson Laboratory in Bar Harbor, Maine, which will become one of the most influential genetics research institutions in North America, particularly in mouse (mammalian) genetics.

1931   Barbara McClintock and Harriet Creighton discover physical exchange between chromosomes in corn, a process known as "crossing over." Curt Stern uses a similar approach in the study of the X chromosome in *Drosophila*. Crossing over, or recombination, will be vital to mapping genes on chromosomes and to understanding inheritance involving linkage.

1932   Sewall Wright describes the relationship of genetic drift and evolution.

1933   Theophilus Painter discovers polytene chromosomes in *Drosophila* salivary glands. These special chromosomes, resulting from numerous rounds of DNA replications without separation, are large, with distinct banding patterns. They are used extensively in mapping genes to specific regions of the chromosome.

1933   Less than 1 percent of all the agricultural land in the Corn Belt has hybrid corn growing on it. However, by 1943, hybrids cover more than 78 percent of the same land. Techniques used to produce crops with desired properties rely heavily on an understanding of genetics. Through the process of producing hybrids, researchers attempt to breed the best traits of several varieties into one. This time-consuming and inexact process is to be superseded by the techniques of recombinant DNA technology.

1934   John Desmond Bernal examines protein structure by using X-ray crystallography.

1935   Ronald Aylmer publishes statistical analyses of Mendel's work. He finds errors in Mendel's interpretation of his data for a series of experiments. Aylmer does not dispute Mendel's theories but instead implies that an assistant was ultimately responsible for the error.

1935   Hans Spemann wins the 1935 Nobel Prize in Physiology or Medicine. By transplanting bits of one embryo into a second, viable embryo, Spemann compiled evidence that an "organizer center" directs the development of an embryo and that different parts of the organizer governed distinct portions of the embryo. His experiments provided clues to the genetic control of growth from the earliest stages of an organism.

1937   Theodosius Dobzhansky publishes *Genetics and the Origin of Species*. He shows that, in natural and experimental populations of *Drosophila* species, frequency changes and geographic patterns of variation in chromosome variants are consistent with the effects of natural selection.

1939   R. J. Gautheret demonstrates the first successful culture of isolated plant tissues as a continuously dividing callus tissue.

1940   Karl Landsteiner and A. S. Wiener describe the Rh blood groups.

1941   George Wells Beadle and Edward Tatum, working with a bread mold, *Neurospora*, publish results indicating that genes mediate cellular chemistry through the production of specific enzymes: the "one gene-one enzyme" experiment. This establishes the use of "simple" organisms as model systems to study genetics. Beadle and Tatum will receive the Nobel Prize in Physiology or Medicine in 1958.

1943   The Rockefeller Foundation, in collaboration with the Mexican government, initiates the Mexican Agricultural Program, the first use of plant breeding in foreign aid.

1944   Oswald T. Avery, Colin MacLeod, and Maclyn McCarty purify DNA and identify it as the "transforming principle" of Frederick Griffith's work. Although this experiment provides solid evidence that DNA is the hereditary material, most scientists still do not accept the notion.

1945   R. D. Owen conducts studies with two sets of cattle twins which demonstrate that their blood antigens could have come only from the opposite sires. These findings suggest the reciprocal passage of ancestral red blood cells. Owen's work has significant implications for immunology.

1945   Max Delbrück, Salvador Luria, and Alfred Hershey work on bacteriophage as a model system to study the mechanism of heredity. Delbrück organizes a course at Cold Spring Harbor, New York, to introduce researchers to the methods of working with bacteriophage. His course will be taught for twenty-six years, helping countless researchers to understand the use of model organisms in genetic investigations. Delbrück, Luria, and Hershey later share the 1969 Nobel Prize in Physiology or Medicine.

1946   Joshua Lederberg and Edward Tatum discover genetic recombination (conjugation) in bacteria, leading them to believe that bacteria, like eukaryotes, have a sexual reproductive cycle. This discovery forces researchers to realize that bacteria are genetic organisms, similar to the eukaryotes studied at the time. Lederberg wins the Nobel Prize in Physiology or Medicine in 1958; Tatum and George Beadle will also share in the 1958 prize, for their work with cellular chemistry, enzymes, and genetics.

1949   Linus Pauling proposes that sickle-cell disease is the result of a change in the normal amino acid sequence of hemoglobin that interferes with its binding properties. His later investigations into protein structure help determine the structure of DNA. He receives the Nobel Prize in Chemistry in 1954.

1950   Barbara McClintock first describes the theory that DNA is mobile and that certain of its elements can insert into different regions on the chromosome. The technical name for this phenomenon is transposition, and the genes affected are casually dubbed "jumping genes" and, more properly, transposable elements or transposons. McClintock's ideas were far ahead of her time. While most scientists were still trying to determine just how DNA works, McClintock is turning the field upside down. Her work will not be accepted until more evidence of transposons surfaces decades later. She will win the Nobel Prize in Physiology or Medicine in 1983.

1950 Erwin Chargaff discovers consistent one-to-one ratios of adenine to thymine and of guanine to cytosine in DNA. These four chemicals are the basic building blocks of DNA. Chargaff's observations become an important clue in determining the exact structure of DNA.

1951 Maurice Wilkins and Rosalind Franklin obtain X-ray diffraction photographs of DNA. These data indicate the exact shape of the DNA molecule; joined with Chargaff's data, these photographs begin to bring DNA into focus.

1952 Joshua and Esther Lederberg and Norton Zinder discover transduction, the transfer of genetic information by viruses. Using *Escherichia coli* and a bacteriophage called *P1*, the Lederbergs and Zinder are able to show that transduction can be used to map genes to the bacterial chromosome.

1952 Alfred Hershey and Martha Chase use bacteriophage and a blender to identify the transforming principle as DNA. They are able to show that DNA, and not protein, is responsible for transforming organisms. This experiment forms the conclusive piece of evidence confirming that DNA is the hereditary material.

1952 Investigations into bacteriophage by Salvador Luria and M. L. Human, and independently J. J. Weigle, lay the groundwork for the discovery of restriction endonucleases.

1953 The three-dimensional structure of DNA is outlined by James Watson and Francis Crick in a 900-word manuscript published in *Nature*, "Molecular Structure of Nucleic Acids: A Structure for Deoxyribose Nucleic Acid." This elegant and concise paper describes the structure of DNA and provides insight into its function. Watson and Crick, along with Maurice Wilkins, will win the Nobel Prize in Physiology or Medicine in 1962. Rosalind Franklin, who with Wilkins delineated the shape of DNA, did not share in the prize, having died several years earlier of cancer, almost certainly caused by her work with X rays.

1954 The first whole plant is regenerated, or cloned, from a single adult plant cell by W. H. Muir and colleagues.

1956 J. H. Tjio and A. Levan determine the chromosome number in humans to be forty-six. Until that time, the chromosome number was thought to be forty-eight. The advances that Tjio and Levan pioneered were instrumental in obtaining good chromosome preparations, allowing for significant advances in the field of cytogenetics.

1956-1958 Arthur Kornberg purifies the enzyme DNA polymerase from *Escherichia coli*. This is the enzyme responsible for DNA replication, making it possible to synthesize DNA. Kornberg, along with Severo Ochoa, wins the Nobel Prize in Physiology or Medicine in 1959.

1957	Heinz Fraenkel-Conrat and B. Singer show that tobacco mosaic virus contains RNA—the first concrete evidence that RNA, in addition to DNA, serves as the genetic material.
1957	In a landmark address to the British Society of Experimental Biology titled "On Protein Synthesis," Francis Crick articulates both the sequence hypothesis (the order of bases on a section of DNA codes for an amino acid sequence on a protein) and the "central dogma" of molecular genetics (genetic information moves from DNA to RNA to proteins, but not from proteins back to DNA).
1957	Alexander Robertus Todd wins the 1957 Nobel Prize in Chemistry. As part of wide-ranging research in organic chemistry, Todd revealed how ribose and deoxyribose bond to the nitrogenous bases on one side of a nucleotide unit and to the phosphate group on the other side. These discoveries provide a foundation for work by others that explains the structure of the DNA molecule.
1958	Matthew Meselson and Frank Stahl determine that DNA replicates in a semi-conservative manner: Each strand of the molecule serves as a template for the synthesis of a new, complementary strand.
1959	Jérôme Lejeune discovers that Down syndrome is caused by the presence of an extra chromosome. This was the first evidence that genetic disorders could be the result of changes in chromosome number, as opposed to changes in individual genes inherited in a Mendelian fashion.
1960's	Mitochondria—extranuclear organelles that are the site for ATP synthesis—are discovered to have their own DNA that is passed down maternally. In 1967, Lynn Margulis, resurrecting a theory proposed by Ivan Wallin in the 1920's, proposes that mitochondria in eukaryotic cells may have evolved from a symbiotic relationship between bacteria (prokaryotes) and ancient eukarotes.
1961	Sol Spiegelman and Benjamin Hall discover that single-stranded DNA will hydrogen bond to its complementary RNA. The discovery of the ability of DNA and RNA to form an association contributed greatly to the study of genes and their organization.
1961	Working initially with Johann Matthaei, biochemist Marshall Nirenberg discovers the first sequence of three bases of DNA that codes for an amino acid and "cracks" the genetic code. H. Gobind Khorana and Robert W. Holley extend the work and elucidate how the sequence of amino acids in a protein is encoded by the sequence of nucleic acids in a gene. Nirenberg, Khorana, and Holley receive the Nobel Prize in Physiology or Medicine in 1968.
1961	Jacques Monod, François Jacob, Sydney Brenner, and Francis Crick discover "messenger" RNA, reporting that it is the mechanism that carries the information from DNA to create proteins. This missing link between the genetic material of DNA and proteins was critical to the understanding of protein synthesis and hence gene expression. Monod and Jacob win the 1965 Nobel Prize in Physiology or Medicine for this work.

1962	Mary Lyon hypothesizes that during development, one of the two X chromosomes in normal mammalian females is inactivated at random. The inactivated X chromosome is called a Barr body, and her hypothesis is known as the Lyon hypothesis.
1962	Werner Arber finds bacteria that are resistant to infection by bacteriophage. It appears that some cellular enzymes destroy phage DNA, while others modify the bacterial DNA to prevent self-destruction. Several years later, Arber, Stuart Linn, Matthew Meselson, and Robert Yuan isolate the first restriction endonuclease and identify the modification of bacterial DNA as methylation. By this time, scientists are looking at how DNA regulates, and is regulated by, cellular activities in a new disciplne, molecular genetics. Arber wins the Nobel Prize in Physiology or Medicine in 1978.
1964	Robin Holliday proposes a model for the recombination of DNA. Although recombination, or crossing over, is not a new idea, the molecular mechanism behind the exchange of genetic information between DNA strands was not known. Holliday's model, widely accepted, explains the phenomenon.
1964	John Gurden transfers nuclei from adult toad cells into toad eggs. F. C. Steward grows single adult cells from a carrot into fully formed, normal plants. These experiments produced viable organisms, ushering in the era of cloning.
1964	The International Rice Research Institute introduces new strains of rice that double the yield of previous strains. This marks the beginning of the Green Revolution, which sought to enable all nations to grow sufficient quantities of food to sustain their own populations. The "father" of this movement, Norman Borlaug, will win the Nobel Peace Prize in 1970 for his role in developing high-yield grain varieties.
1965	Sydney Brenner and colleagues discover stop codons.
1965	André Lwoff shares the Nobel Prize in Physiology or Medicine with Monod and Jacob. Lwoff earlier demonstrated that the genetic material of bacteriophage can become part of the host bacterium's DNA, a process known as lysogeny.
1966	Victor McKusick publishes the first catalog of single genes responsible for traits, *Mendelian Inheritance in Man*, which will appear in many subsequent editions.
1967	Mary Weiss and Howard Green improve the process of gene mapping by using somatic cell hybridization.
1967	DNA ligase, the enzyme that joins DNA molecules, is discovered.
1968	Reiji Okazaki reports the discovery of short fragments of RNA later known as Okazaki fragments, showing the discontinuous synthesis of the lagging DNA strand.
1970	M. Mandel and A. Higa discover a method to increase the efficiency of bacterial transformation. They make the cells "competent" to take up DNA by treating bacteria with calcium chloride and then heat-shocking the cells. Introducing foreign DNA into cells was a key to the success of recombinant DNA methods.

1970   H. Gobind Khorana and twelve associates synthesize the first gene: the gene for an alanine transfer RNA in yeast.

1970   David Baltimore and Howard Temin independently discover reverse transcriptase, an enzyme used by viruses to convert their RNA into DNA. The reverse transcriptase enzyme becomes a key tool in genetic engineering, for which Baltimore and Temin will win the 1975 Nobel Prize in Physiology or Medicine.

1970   Hamilton O. Smith isolates the first restriction endonuclease that cuts at a specific DNA sequence—the first "site-specific restriction enzyme." Daniel Nathans uses this enzyme to create a restriction map of the virus SV40. The use of restriction enzymes, those that cut DNA, allowed for the detailed mapping and analysis of genes. It also was pivotal for recombinant DNA techniques, including the production of transgenic organisms. Nathans and Smith win the Nobel Prize in Physiology or Medicine in 1978 for their work on restriction enzymes.

1972   Paul Berg is the first to create a recombinant DNA molecule. He shows that restriction enzymes can be used to cut DNA in a predictable manner and that these DNA fragments can be joined together with fragments from different organisms. He is awarded the Nobel Prize in Physiology or Medicine in 1980.

1972   Stanford Moore and William H. Stein win the 1972 Nobel Prize in Chemistry for earlier work identifying the sequence of amino acids in ribonuclease, a clue to the structure of the gene responsible for it.

1973   Joseph Sambrook and other researchers at Cold Spring Harbor improve the method of separating DNA fragments based on size, a technique called agarose gel electrophoresis. This method makes it possible to acheive an accurate interpretation of the information in DNA.

1973   Stanley Cohen and Herbert Boyer develop recombinant DNA technology by producing the first recombinant plasmid in bacteria. Plasmids—small, circular pieces of DNA—occur naturally in bacteria. Using the newly discovered tools of molecular biology, Cohen and Boyer inserted a new, or "foreign," piece of DNA into an existing plasmid and had it propagate in a bacterial cell.

Feb., 1975   The Asilomar Conference is held in response to increasing concerns over safety and ethics of genetic engineering. Convening in Pacific Grove, California, under the auspices of the National Institutes of Health, 140 prominent international researchers and academicians, including Nobel laureate Phillip A. Sharp, air their opinions about recombinant DNA experimentation and advocate adoption of ethical guidelines. NIH later issues guidelines for recombinant DNA research to minimize potential hazards if genetically altered bacteria were released into the environment. The guidelines will be relaxed by 1981.

1975   Mary-Claire King and Allan Wilson report, based on results of a survey of protein and nucleic acid studies, that the average human protein is more than 99 percent identical to that of chimpanzees, which is confirmed by later research. The question of why two species that are so different can be as genetically similar as sibling species of other organisms remains open but is assumed to be a function of gene regulation as well as those relatively few mutations that make human DNA different.

1975   Edward Southern develops a method for transferring DNA from an agarose gel to a solid membrane. This technique, known as Southern blotting, becomes one of the most important methods used to identify cloned genes.

1975   Renato Dulbecco, David Baltimore, and Howard Temin receive the Nobel Prize in Physiology or Medicine for their work on the interaction between tumor viruses and the genetic material of the cell. Dulbecco applied phage genetic techniques to the study of animal viruses.

1976   Herbert Boyer and Robert Swanson form Genentech, a company devoted to the development and promotion of biotechnology and applications of genetical engineering.

1976   Susumu Tonegawa discovers the genetic principles for generation of antibody diversity. Tonegawa identified a novel mode of regulation of the genetic material. The genomic DNA of immune cells is actually cut and rejoined in different combinations. This explains how millions of different antibodies can be produced from a very small number of genes. Tonegawa wins the Nobel Prize in Physiology or Medicine in 1987.

1977   Allan Maxam and Walter Gilbert develop a method to determine the sequence of a piece of DNA. At the same time, Frederick Sanger develops a different method, the chain termination (dideoxy) sequencing method. It becomes possible, and relatively simple, to determine the exact sequence of adenine, guanine, thymine, and cytosine in any DNA molecule. Although both the Gilbert-Maxam and the Sanger methods are effective, the Sanger method becomes the dominant technique because it does not involve toxic chemicals. Gilbert and Sanger both receive the Nobel Prize in Physiology or Medicine in 1980.

1977   James Alwine develops the Northern blotting technique, which expands the basic blotting technique introduced by Southern to allow analysis of RNA and proteins.

1977   The U.S. Court of Customs and Patent Appeals rules that an inventor can patent new forms of microorganisms. The first patent granted for a recombinant organism, an oil-eating bacterium, is awarded in 1980. The legality and ethics of patenting recombinant organisms and other biological systems are highly controversial.

1977   Herbert Boyer synthesizes the human hormone somatostatin in *Escherichia coli*—the first successful use of recombinant DNA to produce a substance from the gene of a higher organism. Before, the first isolation of mammalian somatostatin required a half million sheep brains to produce 5 milligrams of the hormone. Now, with the use of recombinant DNA, only two gallons of bacterial culture are required to produce the same amount.

1977   Phillip A. Sharp and Richard Roberts discover that portions of a human gene can be divided among several DNA segments, called introns, separated by noncoding segments called exons. This discovery becomes important to genetic engineering and to understanding the mechanism for hereditary diseases. Sharp and Roberts win the 1993 Nobel Prize in Physiology or Medicine.

1978   Herbert Boyer discovers a synthetic version of the human insulin gene and inserts it into *Escherichia coli* bacteria. The bacteria serve as cloning vectors to maintain and replicate large amounts of human insulin. This application of recombinant DNA technology to produce human insulin for diabetics becomes the foundation for future industrial and medical applications of genetic engineering.

1978   P. C. Steptoe and R. G. Edwards successfully use in vitro fertilization and artificial implantation in humans. Louise Brown, the first "test-tube baby," is born July 25. The process gives hope to many childless couples who, prior to this development, have been unable to conceive. It also raises concerns from ethicists and others over the potential effects on both the individual child's long-term health and social implications.

1980   A team headed by David Botstein measures "restriction fragments" and finds that the length of such fragments often varies in individuals. Such variation, or "restriction fragment length polymorphism" (RFLP), is used to allow rapid discovery of the location of many human genes and genetic differences among individuals.

1980   The first transgenic mouse is created by J. W. Gordon.

June 16, 1980   The U.S. Supreme Court votes 5-4 that living organisms can be patented under federal law, and Ananda M. Chakrabarty receives the first patent for a genetically engineered organism, a form of bacteria, *Pseudomona originosa*, that can decompose crude oil for use in cleaning up oil spills.

1980   George Snell, Baruj Benacerraf, and Jean Dausset win the Nobel Prize in Physiology or Medicine for their discovery of and work on the major histocompatibility complex (MHC), the key to a person's immune system, and how the system produces antibodies to such a wide variety of foreign molecules and pathogens, such as viruses, fungi, and bacteria.

1981   J. Michael Bishop and Harold Varmus discover that oncogenes (genes that play a role in initiating cancer) originate in normal cells as genes that control cellular growth and are not solely derived from retroviruses, as previously thought. Their work greatly influences subsequent studies of tumor development. Varmus and Bishop win the Nobel Prize in Physiology or Medicine in 1989.

1982   The first genetically engineered product, human insulin, dubbed Humulin, is approved for sale by the U.S. government. The production of pharmaceuticals through recombinant DNA technology is becoming a driving force behind both the drug industry and agriculture.

1982   Aaron Klug wins a 1982 Nobel Prize in Chemistry. Klug used X-ray crystallography to investigate biochemical structures, especially that of viruses. He was able to link the assembly of viral protein subunits with specific sites on viral RNA, which helped in fighting viruses that cause disease in plants and, more basically, in understanding the mechanism of RNA transfer of genetic information. He also determined the structure of transfer RNA (tRNA), which has a shape similar to that of a bent hairpin.

1983   Nancy Wexler, Michael Conneally, and James Gusella determine the chromosomal location of the gene for Huntington's disease. Although close, they are unable to locate the gene itself; it will be discovered ten years later.

1983   Thomas Cech and Sidney Altman independently discover catalytic RNA. The idea that RNA can have an enzymatic function changes researchers' views on the role of this molecule, leading to important new theories about the evolution of life. Cech and Altman win the 1989 Nobel Prize in Chemistry.

1983   Bruce Cattanach provides evidence of genomic imprinting in mice. The phenomenon of imprinting is the modification of genes in male and female gametes. This leads to differential expression of these genes in the embryo after fertilization. Imprinting represents another exception to the rules of Mendelian inheritance.

1983   John Sulston, Sydney Brenner, and H. Robert Horvitz describe the cell lineage of the nematode *Caenorhabditis elegans*. The fixed developmental pattern of this small worm provides researchers with insights into how cells determine their own fates and how they influence the fates of neighboring cells. Sulston, Brenner, and Horvitz win the 2002 Nobel Prize in Physiology or Medicine.

1983-1984   William Bender's laboratory isolates and characterizes the molecular details of *Drosophila* homeotic genes. William McGinnis and J. Weiner discover that the base sequences of the homeotic genes they examined contain nearly the same sequence in the terminal 180 bases. They term the conserved 180-base sequence a "homeobox." These regulatory genes direct the development of body parts during gestation of most animals.

1983-1985   Kary B. Mullis invents the polymerase chain reaction (PCR). This revolutionary method of copying DNA from extremely small amounts of material changes the way molecular research is done in only a few short years. It also becomes important in medical diagnostics and forensic analysis. Mullis wins the Nobel Prize in Chemistry in 1995.

1984   The Plant Gene Expression Center, a collaborative effort between academia and the U.S. Department of Agriculture, is established to research plant molecular biology, sequence plant genomes, and develop genetically modified plants.

1984   Alec Jeffreys is the first to use DNA in identifying individuals. This technique, popularly known as "DNA fingerprinting," makes identification of individuals and construction of genetic relationships virtually indisputable.

1984   More than twenty-five scientists collaborate to isolate the gene that causes cystic fibrosis. As a result of technological advances, the identification, isolation, and sequencing of genes is becoming commonplace. Among the notable discoveries are genes implicated in Alzheimer's disease, diabetes, and even complex conditions such as cancer and heart disease.

1985   Michael S. Brown and Joseph L. Goldstein win the Nobel Prize in Physiology or Medicine for their work on the regulation of cholesterol in humans. They showed that in families with a history of high cholesterol, individuals who carry two copies of a mutant gene (homozygotes) have cholesterol levels several times higher than normal, and those who have one mutant gene (heterozygotes) have levels about double normal.

1985-1987   Robert Sinsheimer, Renato Dulbecco, and Charles DeLisi begin investigating the possibility of sequencing the entire human genome. DeLisi, head of the Department of Energy's Office of Health and Environmental Research, seeks federal funding. After the invention of automated sequencing (see below), the National Research Council and later the Office of Technology Assessment support the idea.

1986   Leroy Hood, a biologist at the California Institute of Technology, invents the automated sequencer, the most important advance in DNA sequencing technology since Gilbert and Sanger developed their sequencing methods in the 1970's. Automated sequencing replaces the use of dangerous radioactive labels for identifying the four DNA bases with colored fluorescent dyes. Each of the four DNA bases is coded with a different dye color to eliminate the need to run several reactions. Laser and computer technology are integrated at the end stage to gather data. The result is safer, more accurate, and much faster sequencing.

1986   The first release of a genetically modified crop, genetically engineered tobacco plants, is approved by the Environmental Protection Agency.

1987   Frostban, a genetically engineered bacterium designed to prevent freezing, is tested on strawberries in California. These bacteria are freely released outdoors, where it is hoped they will grow on the strawberries and prevent the fruit from being destroyed by frost late in the growing season. The environmental release of recombinant organisms is an important and controversial step in the application of genetic engineering.

1987   Calgene receives a patent for a DNA sequence that extends the shelf life of tomatoes.

1987   Carol Greider and Elizabeth Blackburn, using the model organism *Tetrahymena* (a protozoan), report evidence that telomeres are regenerated through an enzyme with an RNA component. Based on the action of DNA polymerase, telomeres (located at the tips of chromosomes) should become shorter during each round of cell division. Another enzyme, called telomerase, is found to be necessary to maintain the telomeres. Research in this field sparks interest in the possibility that declining levels of telomerase may contribute to aging and that the inappropriate expression of this enzyme in cells may be a factor in cancer.

1988   The Human Genome Organization (HUGO) is founded to coordinate and collect data from international efforts to sequence the human genome.

1988   The Food and Drug Administration (FDA) approves the sale of recombinant TPA (tissue plasminogen activator) as a treatment for blood clots. TPA shows promise in helping victims recovering from heart attack and stroke.

1989   Francis Collins, Lap-Chee Tsui, and researchers at Toronto's Hospital for Sick Children discover the *CF* gene, which codes the cystic fibrosis transmembrane conductance regulator (CFTR) protein.

1990   Gene therapy for severe combined immunodefiency disorder (SCID) is tested in clinical trials, with promising if not completely successful results.

1990   The Human Genome Project begins, initially headed by James Watson, under the auspices of the National Institues of Health, National Center for Human Genome Research. The project is to be completed by the year 2005. The ambitious project is designed to sequence the entire human genome in order to identify genes involved in biochemical processes such as disease pathology. Also included as part of the Human Genome Project is the sequencing of many model organisms.

1990   At the Plant Gene Expression Center, biologist Michael Fromm announces the use of a high-speed "gene gun" to transform corn. Gene guns are used to inject genetic material directly into cells via DNA-coated microparticles.

1990   The first human undergoes gene therapy. The patient is a four-year-old girl who was born without a functioning immune system as a result of a faulty gene that makes an enzyme called ADA (adenosine deaminase).

1991   J. Craig Venter of the National Institutes of Health demonstrates the use of automated sequencing and expressed sequence tags (ESTs)—cloned sequences of complementary DNA (cDNA) molecules stored in "libraries"—to identify genes and their functions rapidly and accurately.

1992   One of the first major accomplishments of the Human Genome Project is to publish a low-resolution linkage map of the entire human genome.

1993   The mutation that causes Huntington's disease is found, ten years after its chromosomal location was first identified. Fifty-eight scientists collaborated on the project.

1993   Gene therapy cures a mouse of cystic fibrosis.

1994   The Food and Drug Administration (FDA) approves the bovine hormone known as BST or BGH. The hormone is made from recombinant bacteria containing the bovine gene for BST. When injected into cows, the hormone increases milk production by up to 20 percent. Many supermarkets and manufacturers of dairy products refuse to carry or use milk from BST-injected cows, uncertain of what long-term effects this recombinant drug might have on the food chain.

1994   The Food and Drug Administration (FDA) gives approval for the marketing of the Flavr Savr tomato. This genetically altered tomato can be ripened on the vine before being picked and transported. Because the ripening process takes longer, the tomatoes do not rot on their way to the market.

1994   Alfred G. Gilman and Martin Rodbell receive the Nobel Prize in Physiology or Medicine for discovering the role of "G" proteins in regulating signal transduction in eukaryotic cells.

1995   A mutation in the gene *BRCA1*, found by Mark Skolnick and others, is implicated in breast cancer. More than any other gene previously identified, this discovery has wide potential for assessing cancer risk.

1995   J. Craig Venter of The Institute for Genome Research (TIGR) announces completion of the first DNA sequence of a nonviral, self-replicating, free-living organism, the bacterium *Haemophilus influenzae*, using "whole-genome random sequencing," nicknamed "shotgun" sequencing. This method, which precludes the need for a preliminary physical map of the genome, speeds the sequencing of other organisms significantly.

1995   Completion of the sequence of the smallest known bacterium, *Mycoplasma genitalium*, identifies the minimum number of genes required for independent life.

1995   Edward B. Lewis, Christiane Nüsslein-Volhard, and Eric Wieschaus win the Nobel Prize in Physiology or Medicine for their work on the genetic control of early development in *Drosophila*. These researchers took the fruit fly, a model organism from the age of classical genetics, into the age of molecular biology and discovered how genetics controls development. The same developmental mechanisms appear to be at work in other organisms, including humans.

1995   Nüsslein-Volhard completes a genetic mutation project involving zebra fish. Repeating her earlier work with *Drosophila*, Nüsslein-Volhard used similar techniques to begin an intensive study of development in a vertebrate system. This involved screening thousand of mutants to determine if any had developmental defects.

1995   Michael Smith wins the 1995 Nobel Prize in Chemistry for developing "site-directed mutagenesis," a means of reconfiguring genes in order to create altered proteins with distinct properties. Smith's technique makes it possible to treat genetic disease and cancer and to create novel plant strains.

1996   Kristen L. Kroll and Enrique Amaya create a technique to make stable transgenic *Xenopus* (frog) embryos.

1996   A group of more than six hundred researchers sequences the DNA of the yeast *Saccharomyces cereviseae*, the first eukaryotic organism to be sequenced.

1997   Ian Wilmut at the Roslin Institute in Scotland announces the successful cloning of a sheep. The clone is named Dolly, the first vertebrate cloned from the cell of an adult vertebrate. It is hoped that successful cloning of a mammal will allow for easier and cheaper development and propagation of transgenic animals.

1997   The United Nations Educational, Scientific, and Cultural Organization (UNESCO) adopts the Universal Declaration on the Human Genome and Human Rights.

1997   The genomic sequence of the bacterium *Escherichia coli* is reported by Frederick Blattner and colleagues. Although *E. coli* is not the first complete bacterial sequence reported, because of the importance of *E. coli*, the event represents a critical step forward.

1998   The genome of the bacterium *Mycobacterium tuberculosis* is sequenced.

1998   Celera Genomics is founded by former National Institutes of Health researcher J. Craig Venter. Its mission is to sequence the human genome in the private sector, using fast-working automated sequencers.

1998   The genome of the nematode *Caenorhabditis elegans* is the first genome of a multicellular organism to be completely sequenced.

1999   Laboratory tests suggest that the pollen of corn bioengineered to release the pesticide *Bacillus thuringiensis* (*Bt*) endangers monarch butterfly caterpillars. Although later evidence calls the finding into question, it prompts controversy over the safety of transgenic plants.

Sept., 1999   The first human death attributable to gene therapy during clinical trial is reported when an eighteen-year-old participant in a trial on gene therapy for hereditary ornithine transcarbamylase (OTC) deficiency dies of multiorgan failure caused by a severe immunological reaction to the disarmed adenovirus vector used in the trial.

Sept., 1999   Celera Genomics sequences the full genome of *Drosophila* and reports the results the following year in the May 24 issue of *Science*. Of the fly's 13,601 genes, many are shown to be closely related to human genes.

Dec. 1, 1999   The first human chromosome, chromosome 22, is completely sequenced.

Jan. 28, 2000   At a meeting in Montreal, Canada, the United Nations Convention on Biological Diversity approves the Cartegena Protocol on Biosafety, which sets the criterion internationally for patenting genetically modified organisms, including agricultural products.

2000   Chromosome 21, the smallest human chromosome, is completely sequenced; it is the second human chromosome to be completed.

2000   It is estimated that more than two-thirds of the processed foods in U.S. markets contain genetically modified ingredients, primarily soybeans or corn.

2000   The environmental organization Friends of the Earth reveals that StarLink, a genetically engineered corn variety meant only for animal fodder, has contaminated the human food supply. The news ignites public debate over the use of genetically modified food crops.

Dec. 13, 2000   At a press conference, a team of more than three hundred scientists from throughout the world announce that they have sequenced the genome of a plant for the first time. The plant is the model organism *Arabidopsis thaliana*.

2001   The third and fourth human chromosomes, chromosomes 20 and 14, are completely sequenced.

Feb., 2001   The first working drafts of the human genome sequence are published in *Science* (which reports the results from the private company Celera Genomics, headed by

J. Craig Venter) and in *Nature* (reporting the results from the publicly funded Human Genome Project). The relatively low number of human genes, estimated to be about 30,000, makes it necessary to revise the "one gene-one enzyme" hypothesis, since it appears that a single gene can encode more than one protein. The principle is therefore renamed the "one gene-one polypeptide" hypothesis. The paper published in *Science* notes that the DNA of all human beings is 99.9 percent the same, which redefines the notion of human "races" as primarily a social, rather than a biological, construct.

2001   Researchers complete the genomic sequence for rice, *Oryza sativa*.

Nov., 2001   Scientists report that genetic material from transgenic corn has mysteriously turned up in the genome of native corn species near Oaxaca, Mexico. Mexico banned transgenic crops three years earlier, and the closest known crop was located beyond the range of windborne pollen. The report raises concerns about the unintended ecological consequences of transgenic-wild hybrids, which could create problems such as "superweeds."

2001   Leland H. Hartwell, R. Timothy Hunt, and Paul M. Nurse win the 2001 Nobel Prize in Physiology or Medicine for their research on the regulation of cell cycles. Hartwell identified a class of genes that controls the cycle, including a gene that initiates it. Nurse cloned and described the genetic model of a key regulator, cyclin dependent kinase (cdk), and Hunt discovered cyclins, a class of regulatory proteins.

Dec., 2002   The mouse genome sequence is completed, using the shotgun method; it is compared with the draft of the human genome and found to be very similar; both organisms have about 30,000 genes and about 2,000 non-gene, or "junk DNA," regions.

Jan., 2003   A company called Clonaid announces the births of several babies they claim are the result of human cloning but later fails to produce any scientific evidence that the babies are clones. The apparent hoax, initially a media event, energizes the public debate over human cloning and its ramifications.

Feb., 2003   Dolly, the first vertebrate cloned from an adult cell, is euthanized after suffering advanced arthritis and lung disease. Researchers speculate about whether clones age prematurely as a result of the shortened telomere length of the chromosomes in the adult cells from which they are cloned.

Feb. 27, 2003   The U.S. House of Representatives passes the Human Prohibition Cloning Act of 2003, banning the cloning of human beings; the bill goes to the Senate.

April, 2003   The Human Genome Project completes its mission two years ahead of schedule: The entire human genome has now been sequenced.

—Nancy Morvillo, updated by Christina J. Moose

Glossary

A: the abbreviation for adenine, a purine nitrogenous base found in the structure of both DNA and RNA.

acentric chromosome: a chromosome that does not have a centromere and that is unable to participate properly in cell division; often the result of a chromosomal mutation during recombination.

acquired characteristic: a change in an individual organism brought about by its interaction with its environment.

acrocentric chromosome: a chromosome with its centromere near one end. *See also* metacentric chromosome and telocentric chromosome.

activator: a protein that binds to DNA, thus increasing the expression of a nearby gene.

active site: the region of an enzyme that interacts with a substrate molecule; any alteration in the three-dimensional shape of the active site usually has an adverse effect on the enzyme's activity.

adaptation: the evolution of a trait by natural selection, or a trait that has evolved as a result of natural selection.

adaptive advantage: increased reproductive potential in offspring as a result of passing on favorable genetic information.

adenine (A): a purine nitrogenous base found in the structure of both DNA and RNA.

adenosine triphosphate (ATP): the major energy molecule of cells, produced either through the process of cellular respiration or fermentation; it is also a component of DNA and RNA.

adult stem cell: an undifferentiated cell found among differentiated cells in a tissue or organ of an adult organism.

agarose: a chemical substance derived from algae and used to create gels for the electrophoresis of nucleic acids.

aggression: behavior directed toward causing harm to others.

Agrobacterium tumefaciens: a species of bacteria that causes disease in some plants and is able to transfer genetic information, in the form of Ti plasmids, into plant cells; modified Ti plasmids can be used to produce transgenic plants.

albinism: the absence of pigment such as melanin in eyes, skin, hair, scales, or feathers or of chlorophyll in plant leaves and stems.

albino: a genetic condition in which an individual does not produce the pigment melanin in the skin; other manifestations of the trait may be seen in the pigmentation of the hair or eyes; albino individuals occur in many animals and plants and are due to the absence of a variety of different pigments.

alcoholism: a medical diagnosis given when there is repeated use of alcohol over the course of at least a year, despite the presence of negative consequences; tolerance, withdrawal, uncontrolled use, unsuccessful efforts to quit, considerable time spent getting or using the drug, and a decrease in other important activities because of the use are part of this condition.

algorithm: a mathematical rule or procedure for solving a specific problem.

alkaptonuria: a genetic disorder, first characterized by geneticist Archibald Garrod, in which a compound called homogentisic acid accumulates in the cartilage and is excreted in the urine of affected individuals, turning both of these black (the name of the disorder literally means "black urine"); the specific genetic defect involves an inability to process by-products of phenylalanine and tyrosine metabolism.

allele: a form of a gene at a locus; each locus in an individual's chromosomes has two alleles, which may be the same or different.

allele frequency: the proportion of all the genetic variants at a locus within a population of organisms.

allergy: an abnormal immune response to a substance that does not normally provoke an immune response or that is not inherently dangerous to the body (such as plant pollens, dust, or animal dander).

allopatric speciation: a model of speciation in

which parts of a population may become geographically isolated, effectively preventing interbreeding, and over time may develop differences that lead to reproductive isolation and the development of a new species.

allopolyploid: a type of polyploid species that contains genomes from more than one ancestral species.

altruism: behavior that benefits others at the evolutionary (reproductive) cost of the altruist.

Alu sequence: a repetitive DNA sequence of unknown function, approximately three hundred nucleotides long, scattered throughout the genome of primates; the name comes from the presence of recognition sites for the restriction endonuclease Alu I in these sequences.

Alzheimer's disease: a degenerative brain disorder, usually found among the elderly, characterized by brain lesions leading to loss of memory, personality changes, and deterioration of higher mental functions.

amber codon: a stop codon (UAG) found in messenger RNA (mRNA) molecules that signals termination of translation.

ambiguous genitalia: external sexual organs that are not clearly male or female.

Ames test: a test devised by molecular biologist Bruce Ames for determining the mutagenic or carcinogenic properties of various compounds based on their ability to affect the nutritional characteristics of the bacterium *Salmonella typhimurium.*

amino acid: a nitrogen-containing compound used as the building block of proteins (polypeptides); in nature, there are twenty amino acids that can be used to build proteins.

aminoacyl tRNA: a transfer RNA (tRNA) molecule with an appropriate amino acid molecule attached; in this form, the tRNA molecule is ready to participate in translation.

amniocentesis: a procedure in which a small amount of amniotic fluid containing fetal cells is withdrawn from the amniotic sac surrounding a fetus; fetal cells, found in the fluid, are then tested for the presence of genetic abnormalities.

amniotic fluid: the fluid in which the fetus is immersed during pregnancy.

amyloid plaques: protein deposits in the brain formed by fragments from amyloid precursor proteins; amyloid plaques are characteristic of Alzheimer's disease.

anabolic steroids: drugs derived from androgens and inappropriately used to enhance performance in sports.

anabolism: the part of the cell's metabolism concerned with synthesis of complex molecules and cell structures.

anaphase: the third phase in the process of mitosis; in anaphase, sister chromatids separate at the centromere and migrate toward the poles of the cell.

anaphylaxis: a severe, sometimes fatal allergic reaction often characterized by swelling of the air passages, leading to inability to breathe.

ancient DNA: DNA isolated from archaeological artifacts or fossils; it is typically extensively degraded.

androgen receptors: molecules in the cytoplasm of cells that join with circulating male hormones.

androgens: steroid hormones that cause masculinization.

anencephalus: a neural tube defect characterized by the failure of the cerebral hemispheres of the brain and the cranium to develop normally.

aneuploid: a cell or individual with one or a few missing or extra chromosomes.

angstrom: a unit of measurement equal to one ten-millionth of a millimeter; a DNA molecule is 20 angstroms wide.

animal cloning: animal cloning is the process of generating a genetic duplicate of an animal starting with one of its differentiated cells.

annealing: the process by which two single-stranded nucleic acid molecules are converted into a double-stranded molecule through hydrogen bonding between complementary base pairs.

anthrax: an acute bacterial disease caused by *Bacillus anthracis* that affects animals and humans and that is especially deadly in its pulmonary form.

antibiotic: any substance produced naturally by a microorganism that inhibits the growth of other microorganisms; antibiotics are important in the treatment of bacterial infections.

antibody: an immune protein (immunoglobulin) that specifically recognizes an antigen; produced by B cells of the immune system.

anticodon: the portion of a transfer RNA (tRNA) molecule that is complementary in sequence to a codon in a messenger RNA (mRNA) molecule; because of this complementarity, the tRNA molecule can bind briefly to mRNA during translation and direct the placement of amino acids in a polypeptide chain.

antigen: any molecule that is capable of being recognized by an antibody molecule or of provoking an immune response.

antigenic drift or shift: minor changes in the H and N proteins of the influenza virus that enable the virus to evade the immune system of a potential host.

antioxidant: a molecule that preferentially reacts with free radicals, thus keeping them from reacting with other molecules and causing cellular damage.

antiparallel: a characteristic of the Watson-Crick double-helix model of DNA, in which the two strands of the molecule can be visualized as oriented in opposite directions; this characteristic is based on the orientation of the deoxyribose molecules in the sugar-phosphate backbone of the double helix.

antirejection medication: drugs developed to counteract the human body's natural immune system's reaction to transplanted organs.

antisense: a term referring to any strand of DNA or RNA that is complementary to a coding or regulatory sequence, for example, the strand opposite the coding strand (the sense strand) in DNA is called the antisense strand.

antisense RNA: an small RNA molecule that is complementary to the coding region of a messenger RNA (mRNA) and when bound to the mRNA prevents it from being translated.

antitoxin: a vaccine containing antibodies against a specific toxin.

Apo-B: a protein essential for cholesterol transport.

apoptosis: cell "suicide" occurring after a cell is too old to function properly, as a response to irreparable genetic damage, or as a function of genetic programming; apoptosis prevents cells from developing into a cancerous state and is a natural event during many parts of organismal development.

Archaea: the domain of life that includes diverse prokaryotic organisms distinct from the historically familiar Bacteria and which often require severe conditions for growth, such as high temperatures, high salinity, or lack of oxygen.

artificial selection: selective breeding of desirable traits, typically in domesticated organisms.

ascomycetes: organisms of the phylum *Ascomycota*, a group of fungi known as the sac fungi, which are characterized by a saclike structure, the ascus.

ascospore: a haploid spore produced by meiosis in ascomycete fungi.

ascus: a reproductive structure, found in ascomycete fungi, that contains ascospores.

asexual reproduction: reproduction of cells or organisms without the transfer or reassortment of genetic material; results in offspring that are genetically identical to the parent.

assortative mating: mating that occurs when individuals make specific mate choices based on the phenotype or appearance of others.

ATP: *See* adenosine triphosphate.

ATP synthase: the enzyme that synthesizes ATP.

autoimmune disorders: chronic diseases that arise from a breakdown of the immune system's ability to distinguish between the body's own cells (self) and foreign substances, leading to an individual's immune system attacking the body's own organs or tissues.

autoimmune response: an immune response of an organism against its own cells.

automated fluorescent sequencing: a modification of dideoxy termination sequencing which uses fluorescent markers to identify the terminal nucleotides, allowing the automation of sequencing.

autopolyploid: a type of polyploid species that contains more than two sets of chromosomes from the same species.

autosomal dominant allele: an allele of a gene (locus) on one of the nonsex chromosomes that is always expressed, regardless of the form of the other allele at the same locus.

autosomal recessive allele: an allele of a gene (locus) that will be expressed only if there are two identical copies at the same locus.

autosomal trait: a trait that typically appears just as frequently in either sex because an autosomal chromosome, rather than a sex chromosome, carries the gene.

autosomes: non-sex chromosomes; humans have forty-four autosomes.

auxotrophic strain: a mutant strain of an organism that cannot synthesize a substance required for growth and therefore must have the substance supplied in the growth medium.

azoospermia: the absence of spermatozoa in the semen.

B cells: a class of white blood cells (lymphocytes) derived from bone marrow and responsible for antibody-directed immunity.

B-DNA: the predominant form of DNA in solution and in the cell; a right-handed double helix most similar to the Watson-Crick model. *See also* Z-DNA.

B lymphocytes: *See* B cells.

B memory cells: descendants of activated B cells that are long-lived and that synthesize large amounts of antibodies in response to a subsequent exposure to the antigen, thus playing an important role in secondary immunity.

Bacillus thuringiensis (Bt): a species of bacteria that produces a toxin deadly to caterpillars, moths, beetles, and certain flies.

backcross: a cross involving offspring crossed with one of the parents. *See also* cross.

bacterial artificial chromosomes (BACs): cloning vectors used to clone large DNA fragments (up to 500 kb) that can be readily inserted in a bacterium, such as *Escherichia coli.*

bacteriophage: a virus that infects bacterial cells; often simply called a phage.

baculovirus: a type of virus that is capable of causing disease in a variety of insects.

Barr body: a darkly staining structure primarily present in female cells, believed to be an inactive X chromosome; used as a demonstration of the Lyon hypothesis.

base: a chemical subunit of DNA or RNA that encodes genetic information; in DNA, the bases are adenine (A), cytosine (C), guanine (G), and thymine (T); in RNA, thymine is replaced by uracil (U).

base pair (bp): often used as a measure of the size of a DNA fragment or the distance along a DNA molecule between markers; both the singular and plural are abbreviated bp.

base pairing: the process by which bases link up by hydrogen bonding to form double-stranded molecules of DNA or loops in RNA; in DNA, adenine (A) always pairs with thymine (T), and cytosine (C) pairs with guanine (G); in RNA, uracil (U) replaces thymine.

beta-amyloid peptide: the main constituent of the neuritic plaques in the brains of Alzheimer's patients.

bidirectional replication: a characteristic of DNA replication involving synthesis of DNA in both directions away from an origin of replication.

binary fission: cell division in prokaryotes in which the plasma membrane and cell wall grow inward and divide the cell in two.

biochemical pathway: the steps in the production or breakdown of biological chemicals in cells; each step usually requires a specific enzyme.

bioethics: the study of human actions and goals in a framework of moral standards relating to use and abuse of biological systems.

bioinformatics: the application of information technology to the management of biological information to organize data and extract meaning; a hybrid discipline that combines elements of computer science, information technology, mathematics, statistics, and molecular genetics.

biological clocks: genetically and biochemically based systems that regulate the timing and/or duration of biological events in an organism; examples of processes controlled by biological clocks include circadian rhythms, cell cycles, and migratory restlessness.

biological determinism: the concept that all

characteristics of organisms, including behavior, are determined by the genes the organism possesses; it is now generally accepted that the characteristics of organisms are determined both by genes and environment.

biological weapon (BW): a delivery system or "weaponization" of such pathological organisms as bacteria and viruses to cause disease and death in people, animals, or plants.

biometry: the measurement of biological and psychological variables.

biopesticides: chemicals or other agents derived from or involving living organisms that can be used to control the population of a pest species.

bioremediation: biologic treatment methods to clean up contaminated water and soils.

biotechnology: the use of biological molecules or organisms in industrial or commercial products and techniques.

bioterrorism: use of organisms as instruments or weapons of terror; for example, deliberate introduction of smallpox, anthrax, or other diseases in civilian populations.

blastocyst: a preimplantation embryo consisting of a hollow ball of two layers of cells.

blood type: one of the several groups into which blood can be classified based on the presence or absence of certain molecules called antigens on the red blood cells.

blotting: the transfer of nucleic acids or proteins separated by gel electrophoresis onto a filter paper, which allows access by molecules that will interact with only one specific sequence or molecule.

***BRCA1* and *BRCA2* genes:** the best known examples of genes associated with inherited breast cancers.

***Bt* toxin:** a toxic compound naturally synthesized by bacterium *Bacillus thuringiensis*, which kills insects.

C: the abbreviation for cytosine, a pyrimidine nitrogenous base found in the structure of both DNA and RNA.

C terminus: the end of a polypeptide with an amino acid that has a free carboxyl group.

C-value: the characteristic genome size for a species.

CAG expansion: a mutation-induced increase in the number of consecutive CAG nucleotide triplets in the coding region of a gene.

callus: a group of undifferentiated plant cells growing in a clump.

cAMP: *See* cyclic adenosine monophosphate.

cancer: a disease in which there is unrestrained growth and reproduction of cells, loss of contact inhibition, and, eventually, metastasis (the wandering of cancer cells from a primary tumor to other parts of the body); invasion of various tissues and organs by cancer cells typically leads to death.

capsid: the protective protein coating of a virus particle.

carcinogen: any physical or chemical cancer-causing agent.

carrier: a healthy individual who has one normal allele and one defective allele for a recessive genetic disease.

catabolism: the part of the cell's metabolism concerned with the breakdown of complex molecules, usually as an energy-generating mechanism.

catabolite repression: a mechanism of operon regulation involving an enzyme reaction's product used as a regulatory molecule for the operon that encodes the enzyme; a kind of feedback inhibition.

cDNA: *See* complementary DNA.

cDNA library: a collection of clones produced from all the RNA molecules in the cells of a particular organism, often from a single tissue. *See also* complementary DNA.

cell culture: growth and maintenance of cells or tissues in laboratory vessels containing a precise mixture of nutrients and hormones.

cell cycle: the various growth phases of a cell, which include (in order) G1 (gap phase 1), S (DNA synthesis), G2 (gap phase 2), and M (mitosis).

cell differentiation: a process during which a cell becomes specialized as a specific type of cell, such as a neuron, or undergoes programmed cell death (apoptosis).

cell line: a cell culture maintained for an indeterminate time.

cell signaling: communication between cells that occurs most commonly when one cell releases a specific "signaling" molecule that is received and recognized by another cell.

centiMorgan (cM): a unit of genetic distance between genes on the same chromosome, equal to a recombination frequency of 1 percent; also called a map unit, since these distances can be used to construct genetic maps of chromosomes.

central dogma: a foundational concept in modern genetics stating that genetic information present in the form of DNA can be converted to the form of messenger RNA (or other types of RNA) through transcription and that the information in the form of mRNA can be converted into the form of a protein through translation.

centriole: a eukaryotic cell structure involved in cell division, possibly with the assembly or disassembly of the spindle apparatus during mitosis and meiosis; another name for this organelle is the microtubule organizing center (MTOC).

centromere: a central region where a pair of chromatids are joined before being separated during anaphase of mitosis or meiosis; also, the region of the chromatids where the microtubules of the spindle apparatus attach.

checkpoint: the time in the cell cycle when molecular signals control entry to the next phase.

chemical mutagens: chemicals that can directly or indirectly create mutations in DNA.

chiasma (*pl.* chiasmata): the point at which two homologous chromosomes exchange genetic material during the process of recombination; the word literally means "crosses," which refers to the appearance of these structures when viewed with a microscope.

chi-square analysis: a nonparametric statistical analysis of data from an experiment to determine how well the observed data correlate with the expected data.

chloroplast: the cellular organelle in plants responsible for photosynthesis.

chloroplast DNA (cpDNA): circular DNA molecules found in multiple copies in chloroplasts; they contain some of the genes required for chloroplast functions.

cholera: an intestinal disease caused by the bacteria *Vibrio cholerae* which is often spread by water contaminated with human waste.

chorionic villus sampling: a procedure in which fetal cells are obtained from an embryonic structure called the chorion and analyzed for the presence of genetic abnormalities in the fetus.

chromatid: one half of a chromosome that has been duplicated in preparation for mitosis or meiosis; each chromatid is connected to its sister chromatid by a centromere.

chromatin: the form chromosomes take when not undergoing cell division; a complex of fibers composed of DNA, histone proteins, and nonhistone proteins.

chromatography: a separation technique involving a mobile solvent and a stationary, adsorbent phase.

chromosome: the form in which genetic material is found in the nucleus of a cell; composed of a single DNA molecule that is extremely tightly coiled, usually visible only during the processes of mitosis and meiosis.

chromosome jumping: similar to chromosome walking, but involving larger fragments of DNA and thus resulting in faster analysis of longer regions of DNA. *See also* chromosome walking.

chromosome map: a diagram showing the locations of genes on a particular chromosome; generated through analysis of linkage experiments involving those genes.

chromosome mutation: a change in chromosome structure caused by chromosome breakage followed by improper rejoining; examples include deletions, insertions, inversions, and translocations.

chromosome puff: an extremely unwound or uncoiled region of a chromosome indicative of a transcriptionally active region of the chromosome.

chromosome theory of inheritance: a concept, first proposed by geneticists Walter Sutton and Theodor Boveri, that genes are located on chromosomes and that the inheritance and movement of chromosomes during meiosis explain Mendelian principles on the cellular level.

chromosome walking: a molecular genetics technique used for analysis of long DNA fragments; the name comes from the technique of using previously cloned and characterized fragments of DNA to "walk" into uncharacterized regions of the chromosome that overlap with these fragments. *See also* chromosome jumping.

circadian rhythm: a cycle of behavior, approximately twenty-four hours long, that is expressed independent of environmental changes.

cirrhosis: a disease of the liver, marked by the development of scar tissue that interferes with organ function, that can result from chronic alcohol consumption.

cistron: a unit of DNA that is equivalent to a gene; it encodes a single polypeptide.

clinical trial: an experimental research study used to determine the safety and effectiveness of a medical treatment or drug.

clone: a molecule, cell, or organism that is a perfect genetic copy of another.

cloning: the technique of making a perfect genetic copy of an item such as a DNA molecule, a cell, or an entire organism.

cloning vector: a DNA molecule that can be used to transport genes of interest into cells, where these genes can then be copied.

codominance: a genetic condition involving two alleles at a locus in a heterozygous organism; each of these alleles is fully expressed in the phenotype of the organism.

codon: a group of three nucleotides in messenger RNA (mRNA) that represent a single amino acid in the genetic code; this is mediated through binding of a transfer RNA (tRNA) anticodon to the codon during translation.

color blindness: an inherited condition in people whose eyes lack one or more of the three color receptors.

complementary base pairing: hydrogen bond formation in DNA and RNA that occurs only between cytosine and guanine (in both DNA and RNA) or between adenine and thymine (in DNA) or adenine and uracil (in RNA).

complementary DNA (cDNA): a DNA molecule that is synthesized using messenger RNA (mRNA) as a template and which is catalyzed by the enzyme reverse transcriptase.

complementation testing: performing a cross between two individuals with the same phe-

notype to determine whether or not the mutations occur within the same gene.

composite transposon: a transposable element that contains genes other than those required for transposition.

concerted evolution: a process in which the members of a gene family evolve together.

concordance: the presence of a trait in both members of a pair of twins.

cones: the light-sensitive structures in the retina that are the basis for color vision.

congenital defect: a defect or disorder that occurs during prenatal development.

conjugation: a form of genetic transfer among bacterial cells involving the F pilus.

consanguine: of the same blood or origin; in genetics, the term implies the sharing of genetic traits or characteristics from the same ancestors (as cousins, for example).

consensus sequence: a sequence with no or only slight differences commonly found in DNA molecules from various sources, implying that the sequence has been actively conserved and plays an important role in some genetic process.

cosmid: a cloning vector partially derived from genetic sequences of lambda, a bacteriophage; cosmids are useful in cloning relatively large fragments of DNA.

cross: the mating of individuals to produce offspring by sexual reproduction.

crossing over: the exchange of genetic material between two homologous chromosomes during prophase I of meiosis, providing an important source of genetic variation; also called "recombination" or crossover.

cultivar: a variety of plant developed through controlled breeding techniques.

cyclic adenosine monophosphate (cAMP): an important cellular molecule involved in cell signaling and regulation pathways.

cyclins: a group of eukaryotic proteins with characteristic patterns of synthesis and degradation during the cell cycle; part of an elaborate mechanism of cell cycle regulation, and a key to the understanding of cancer.

cystic fibrosis: the most common recessive lethal inherited disease among Caucasians in the United States and the United Kingdom.

cytogenetics: the study of chromosome number and structure, including identification of abnormalities.

cytokines: soluble intercellular molecules produced by cells such as lymphocytes that can influence the immune response.

cytokinesis: the division of the cytoplasm, typically occurring in concert with nuclear division (mitosis or meiosis).

cytoplasmic inheritance: *See* extranuclear inheritance.

cytosine (C): a pyrimidine nitrogenous base found in the structure of both DNA and RNA.

cytoskeleton: the structure, composed of microtubules and microfilaments, that gives shape to a eukaryotic cell, enables some cells to move, and assists in such processes as cell division.

dalton: a unit of molecular weight equal to the mass of a hydrogen atom; cellular molecules such as proteins are often measured in terms of a kilodalton, equal to 1,000 daltons.

daughter cells: cells that result from cell division.

deamination: the removal of an amino group from an organic molecule.

degenerate: refers to a property of the genetic code via which two or more codons can code for the same amino acid.

deletion: a type of chromosomal mutation in which a genetic sequence is lost from a chromosome, usually through an error in recombination.

denaturation: changes in the physical shape of a molecule caused by changes in the immediate environment, such as temperature or pH level; denaturation usually involves the alteration or breaking of various bonds within the molecule and is important in DNA and protein molecules.

deoxyribonucleic acid (DNA): the genetic material found in all cells; DNA consists of nitrogenous bases (adenine, guanine, cytosine, and thymine), sugar (deoxyribose), and phosphate.

deoxyribose: a five-carbon sugar used in the structure of DNA.

diabetes: a syndrome in which the body cannot metabolize glucose appropriately.

diakinesis: a subphase of prophase I in meiosis in which chromosomes are completely condensed and position themselves in preparation for metaphase.

dicentric chromosome: a chromosome with two centromeres, usually resulting from an error of recombination.

dideoxy termination sequencing: *See* Sanger sequencing.

differentiation: the series of changes necessary to convert an embryonic cell into its final adult form, usually with highly specialized structures and functions.

dihybrid: an organism that is hybrid for each of two genes—for example, *AaBb;* when two dihybrid organisms are mated, the offspring will appear in a 9:3:3:1 ratio with respect to the traits controlled by the two genes.

diphtheria: an acute bacterial disease caused by *Corynebacterium diphtheriae;* symptoms are primarily the result of a toxin released by the bacteria.

diploid: a cell or organism with two complete sets of chromosomes, usually represented as $2N$, where N stands for one set of chromosomes; for example, humans have two sets of twenty-three chromosomes in their somatic cells, making them diploid.

diplotene: a subphase of prophase I in meiosis in which synapsed chromosomes begin to move apart and the chiasmata are clearly visible.

discontinuous replication: replication on the lagging strand of a DNA molecule, resulting in the formation of Okazaki fragments. *See also* Okazaki fragments.

discontinuous variation: refers to a set of related phenotypes that are distinct from one another, with no overlapping.

disjunction: the normal division of chromosomes that occurs during meiosis or mitosis; the related term "nondisjunction" refers to problems with this process.

disomy: a case in which both copies of a chromosome come from a single parent, rather than (as is usual) one being maternal and one being paternal.

dizygotic: developed from two separate zygotes; fraternal twins are dizygotic because they develop from two separate fertilized ova (eggs).

DNA: *See* deoxyribonucleic acid.

DNA fingerprinting: a DNA test used by forensic scientists to aid in the identification of criminals or to resolve paternity disputes which involves looking at known, highly variable DNA sequences; more correctly called DNA genotyping.

DNA footprinting: a molecular biology technique involving DNA-binding proteins that are allowed to bind to DNA; the DNA is then degraded by DNases, and the binding sites of the proteins are revealed by the nucleotide sequences protected from degradation.

DNA gyrase: a bacterial enzyme that reduces tension in DNA molecules that are being unwound during replication; a type of cellular enzyme called a topoisomerase.

DNA library: a collection of cloned DNA fragments from a single source, such as a genome, chromosome, or set of messenger RNA (mRNA) molecules; most common examples are genomic and cDNA libraries.

DNA ligase: a cellular enzyme used to connect pieces of DNA together; important in genetic engineering procedures.

DNA polymerase: the cellular enzyme responsible for making new copies of DNA molecules through replication of single-stranded DNA template molecules or, more rarely, using an RNA template molecule as in the case of RNA-dependent DNA polymerase or reverse transcriptase.

DNA replication: synthesis of new DNA strands complementary to template strands resulting in new double-stranded DNA molecules comprising the old template and the newly synthesized strand joined by hydrogen bonds; described as a semiconservative process in that half (one strand) of the original template is retained and passed on.

DNase: refers to a class of enzymes, deoxyribonucleases, which specifically degrade DNA molecules.

domain: the highest-level division of life, sometimes called a superkingdom.

dominant: an allele or a trait that will mask the presence of a recessive allele or trait.

dosage compensation: an equalization of gene products that can occur whenever there are more or fewer genes for specific traits than normal.

double helix: a model of DNA structure proposed by molecular biologists James Watson and Francis Crick; the major features of this model are two strands of DNA wound around each other and connected by hydrogen bonds between complementary base pairs.

down-regulation: generally used in reference to gene expression and refers to reducing the amount that a gene is transcribed and/or translated; up-regulation is the opposite.

Down syndrome: a genetic defect caused by possession of an extra copy of chromosome 21; symptoms include mental retardation, mongoloid facial features, and premature aging.

downstream: in relation to the left-to-right direction of DNA whose nucleotides are arranged in sequence with the 5′ carbon on the left and the 3′ on the right, downstream is to the right.

drug resistance: a phenomenon in which pathogens no longer respond to drug therapies that once controlled them; resistance can arise by recombination, by mutation, or by several methods of gene transfer, and is made worse by misuse of existing drugs.

duplication: a type of chromosomal mutation in which a chromosome region is duplicated because of an error in recombination during prophase I of meiosis; thought to play an important role in gene evolution.

dwarfism: the condition of adults of short stature who are less than 50 inches in height, which can be caused by genetic factors, endocrine malfunction, acquired conditions, or growth hormone deficiency; many dwarfs prefer to be called "little people."

E. coli: *See* Escherichia coli.

electron transport chain: a series of protein complexes that use high-energy electrons to do work such as pumping H^+ ions out of the mitochondrial matrix into the intermembrane space as a way of storing energy that is then used by ATP synthase to make ATP.

electrophoresis: *See* gel electrophoresis.

embryo: the term for a complex organism (particularly humans) during its earliest period of development, the stage of development that begins at fertilization and ends with the eighth week of development, after which the embryo is called a fetus.

embryology: the study of developing embryos.

embryonic stem cell: a cell derived from an early embryo that can replicate indefinitely in vitro and can differentiate into other cells of the developing embryo.

emerging disease: a disease whose incidence in humans or other target organisms has increased.

endemic: prevalent and recurring in a particular geographic region; for example, an organism that is specific to a particular region is characterized as endemic to that region.

endocrine gland: a gland that secretes hormones into the circulatory system.

endonuclease: an enzyme that degrades a nucleic acid molecule by breaking phosphodiester bonds within the molecule.

endosymbiotic hypothesis: a hypothesis stating that mitochondria and chloroplasts were once free-living bacteria that entered into a symbiotic relationship with early pre-eukaryotic cells; structural and genetic similarities between these organelles and bacteria provide support for this hypothesis.

enhancer: a region of a DNA molecule that facilitates the transcription of a gene, usually by stimulating the interaction of RNA polymerase with the gene's promoter.

enzyme: a protein that acts as a catalyst to speed up or facilitate a specific biochemical reaction in a cell.

epigenesis: the formation of differentiated cell types and specialized organs from a single, homogeneous fertilized egg cell without any preexisting structural elements.

epistasis: a genetic phenomenon in which a gene at one locus influences the expression of a second gene at another locus, usually by masking the effect of the second gene; however, only one trait is being controlled by these two genes, so epistasis is characterized by modified dihybrid ratios.

equational division: refers to meiosis II, in which the basic number of chromosome types remains the same although sister chromatids are separated from one another; after equational division occurs, functional haploid gametes are present.

Escherichia coli: a bacterium widely studied in genetics research and extensively used in biotechnological applications.

estrogens: steroid hormones or chemicals that stimulate the development of female sexual characteristics and control the female reproductive cycle.

ethidium bromide: a chemical substance that inserts itself (intercalates) into the DNA double helix; when exposed to ultraviolet light, ethidium bromide fluoresces, making it useful for the visualization of DNA molecules in molecular biology techniques.

etiology: the cause or causes of a disease or disorder.

euchromatin: chromatin that is loosely coiled during interphase; thought to contain transcriptionally active genes.

eugenics: a largely discredited field of genetics that seeks to improve humankind by selective breeding; can be positive eugenics, in which individuals with desirable traits are encouraged or forced to breed, or negative eugenics, in which individuals with undesirable traits are discouraged or prevented from breeding.

eukaryote: a cell with a nuclear membrane surrounding its genetic material (a characteristic of a true nucleus) and a variety of subcellular, membrane-bound organelles; eukaryotic organisms include all known organisms except bacteria, which are prokaryotic. *See also* prokaryote.

euploid: the normal number of chromosomes for a cell or organism.

eusociality: an extreme form of altruism and kin selection in which most members of the society do not reproduce but rather feed and protect their relatives; bees, for example, are eusocial.

euthanasia: the killing of suffering individuals; sometimes referred to as "mercy" killing.

exogenous gene: a gene produced or originating from outside an organism.

exon: a protein-coding sequence in eukaryotic genes, usually flanked by introns.

exonuclease: an enzyme that degrades a nucleic acid molecule by breaking phosphodiester bonds at either end of the molecule.

expressed sequence tags (EST): an STS (sequence tagged site) that has been derived from a cDNA library.

expression vector: a DNA cloning vector designed to allow genetic expression of inserted genes via promoters engineered into the vector sequence.

expressivity: the degree to which a genotype is expressed as a phenotype.

extranuclear inheritance: inheritance involving genetic material located in the mitochondria or chloroplasts of a eukaryotic cell; also known as maternal inheritance (because these organelles are generally inherited from the mother) and cytoplasmic inheritance (because the organelles are found in the cell's cytoplasm rather than its nucleus).

extreme halophiles: microorganisms that require extremely high salt concentrations for optimal growth.

F pilus: also called the fertility pilus; a reproductive structure found on the surface of some bacterial cells that allows the cells to exchange plasmids or other DNA during the process of conjugation.

F$_1$ generation: first filial generation; offspring produced from a mating of P (parental) generation individuals.

F$_2$ generation: second filial generation; offspring produced from a mating of F$_1$ generation individuals.

fate map: a description of the adult fate of embryonic cells.

fertilization: the fusion of two cells (egg and sperm) in sexual reproduction.

fitness: a measure of the ability of a genotype or individual to survive and reproduce; when fitness is compared to other genotypes or individuals it is called relative fitness.

fluorescent in situ hibridization (FISH): an extremely sensitive assay for determining the presence of deletions on chromosomes, which uses a fluorescence-tagged segment of DNA that binds to the DNA region being studied.

foreign DNA: DNA taken from a source other than the host cell that is joined to the DNA of the cloning vector; also known as "insert DNA."

forensic genetics: the use of genetic tests and principles to resolve legal questions.

formylmethionine (fMet): the amino acid used to start all bacterial proteins; it is attached to the initiator transfer RNA (tRNA) molecule.

frameshift mutation: a DNA mutation involving the insertion or deletion of one of several nucleotides that are not in multiples of three, resulting in a shift of the codon reading frame; usually produces nonfunctional proteins. *See also* open reading frame; reading frame.

fraternal twins: twins that develop and are born simultaneously but are genetically unique, being produced from the fertilization of two separate eggs; a synonymous term is "dizygotic twins."

free radical: *See* oxygen free radical.

G: the abbreviation for guanine, a purine nitrogenous base found in the structure of both DNA and RNA.

G$_0$: a point in the cell cycle at which a cell is no longer progressing toward cell division; can be considered a "resting" stage.

G$_1$ checkpoint: a point in the cell cycle at which a cell commits either to progressing toward cell division (by replicating its DNA and eventually engaging in mitosis) or to entering the G$_0$ phase, thereby withdrawing from the cell cycle either temporarily or permanently.

gamete: a sex cell, either sperm or egg, containing half the genetic material of a normal cell.

gel electrophoresis: a technique of molecular biology in which biological molecules are placed into a gel-like matrix (such as agarose or polyacrylamide) and then subjected to an electric current; using this technique, researchers can separate molecules of varying sizes and properties.

GenBank: a comprehensive, annotated collection of publicly available DNA sequences maintained by the National Center for

Biotechnology Information and available through its Web site.

gene: a portion of a DNA molecule containing the genetic information necessary to produce a molecule of messenger RNA (via the process of transcription) that can then be used to produce a protein (via the process of translation); also includes regions of DNA that are transcribed to RNA that does not get translated, but carries out other roles in the cell.

gene expression: the combined biochemical processes, called "transcription" and "translation," that convert the linearly encoded information in the bases of DNA into the three-dimensional structures of proteins.

gene families: multiple copies of the same or similar genes in the same genome; the copies can be identical and tandemly repeated, or they may differ slightly and be scattered on the same or different chromosomes.

gene flow: movement of alleles from one population to another by the movement of individuals or gametes.

gene frequency: the occurrence of a particular allele present in a population, expressed as a percentage of the total number of alleles present for the locus.

gene pool: the complete assortment of genes present in the gametes of the members of a population that are eligible to reproduce.

gene silencing: any form of genetic regulation in which the expression of a gene is completely repressed, either by preventing transcription (pre-transcriptional gene silencing) or after a messenger RNA (mRNA) has been transcribed (post-transcriptional gene silencing).

gene therapy: any procedure to alleviate or treat the symptoms of a disease or condition by genetically altering the cells of the patient.

gene transfer: the movement of fragments of genetic information, whole genes, or groups of genes between organisms.

genetic code: the correspondence between the sequence of nucleotides in DNA or messenger RNA (mRNA) molecules and the amino acids in the polypeptide a gene codes for.

genetic counseling: a discipline concerned with analyzing the inheritance patterns of a particular genetic defect within a given family, including the determination of the risk associated with the presence of the genetic defect in future generations and options for treatment of existing genetic defects.

genetic drift: chance fluctuations in allele frequencies within a population, resulting from random variation in the number and genotypes of offspring produced by different individuals.

genetic engineering: a term encompassing a wide variety of molecular biology techniques, all concerned with the modification of genetic characteristics of cells or organisms to accomplish a desired effect.

genetic load: the average number of the recessive deleterious (lethal or sublethal) alleles in individuals in a population.

genetic map: a "map" showing distances between genes in terms of recombination frequency; using DNA sequence data a physical map with distance in base pairs can also be produced.

genetic marker: a distinctive DNA sequence that shows variation in the population and can therefore potentially be used for identification of individuals and for discovery of disease genes.

genetic screening: the testing of individuals for disease-causing genes or genetic disease.

genetic testing: the use of the techniques of genetics research to determine a person's risk of developing, or status as a carrier of, a disease or other disorder.

genetically modified (GM) foods: foods produced through the application of recombinant DNA technology, whereby genes from the same or different species are transferred and expressed in crops that do not naturally harbor those genes.

genetically modified organism (GMO): an organism produced by using biotechnology to introduce a new gene or genes, or new regulatory sequences for genes, into it for the purpose of giving the organism a new trait, usually to adapt the organism to a new environment, provide resistance to pest species, or enable the production of new products from the organism. *See also* transgenic organism.

genetics: an area of biology involving the scientific study of heredity.

genome: all of the DNA in the nucleus or in one of the organelles, such as a chloroplast or mitochondrion.

genomic imprinting: a genetic phenomenon in which the phenotype associated with a particular allele depends on which parent donated the allele.

genomic library: a collection of clones that includes the entire genome of a single species as fragments ligated to vector DNA.

genomics: that branch of genetics dealing with the study of genetic sequences, including their structure and arrangement.

genotype: the genetic characteristics of a cell or organism, expressed as a set of symbols representing the alleles present at one or more loci.

germ cells: reproductive cells such as eggs and sperm.

germ-line gene therapy: a genetic modification in gametes or fertilized ova so all cells in the organism will have the change which potentially can be passed on to offspring.

germ-line mutation: a heritable change in the genes of an individual's reproductive cells, often linked to hereditary diseases.

gonad: an organ that produces reproductive cells and sex hormones; termed ovaries in females and testes in males.

Green Revolution: the introduction of scientifically bred or selected varieties of grain (such as rice, wheat, and maize) that, with high enough inputs of fertilizer and water, can greatly increase crop yields.

guanine (G): a purine nitrogenous base found in the structure of both DNA and RNA.

H substance: a carbohydrate molecule on the surface of red blood cells; when modified by certain monosaccharides, this molecule provides the basis of the ABO blood groups.

haplodiploidy: a system of sex determination in which males are haploid (developing from unfertilized eggs) and females are diploid.

haploid: refers to a cell or an organism with one set of chromosomes; usually represented as the N number of chromosomes, with $2N$ standing for the diploid number of chromosomes.

haplotype: a sequential set of genes on a single chromosome inherited together from one parent; the other parent provides a matching chromosome with a different set of genes.

Hardy-Weinberg law: a concept in population genetics stating that, given an infinitely large population that experiences random mating without mutation or any other such affecting factor, the frequency of particular alleles will reach a state of equilibrium, after which their frequency will not change from one generation to the next.

HeLa cells: the first human tumor cells shown to form a continuous cell line; they were derived from a cervical cancer tumor removed from a woman known as Henrietta Lacks.

helicase: a cellular enzyme that breaks hydrogen bonds between the strands of the DNA double helix, thus unwinding the helix and facilitating DNA replication.

helper T cells: a class of white blood cells (lymphocytes) derived from bone marrow that prompts the production of antibodies by B cells in the presence of an antigen.

hemizygous: characterized by having a gene present in a single copy, such as any gene on the X chromosome in a human male.

hemoglobin: a molecule made up of two alpha and two beta amino acid chains whose precise chemical and structural properties normally allow it to bind with oxygen in the lungs and transport it to other parts of the body.

hemophilia: an X-linked recessive disorder in which an individual's blood does not clot properly because of a lack of blood-clotting factors; as in all X-linked recessive traits, the disease is most common in males, the allele for the disease being passed from mother to son.

heredity: the overall mechanism by which characteristics or traits are passed from one generation of organisms to the next; genetics is the scientific study of heredity.

heritability: a proportional measure of the extent to which differences among organisms within a population for a particular character result from genetic rather than environmental causes (a measure of nature versus nurture).

hermaphrodite: an individual who has both male and female sex organs.

heterochromatin: a highly condensed form of chromatin, usually transcriptionally inactive.

heterochrony: a change in the timing or rate of development of characters in an organism relative to those same events in its evolutionary ancestors.

heteroduplex: a double-stranded molecule of nucleic acid with each strand from a different source, formed either through natural means such as recombination or through artificial means in the laboratory.

heterogametic sex: the particular sex of an organism that produces gametes containing two types of sex chromosome; in humans, males are the heterogametic sex, producing sperm that can carry either an X chromosome or a Y chromosome.

heterogeneous nuclear RNA (hnRNA): an assortment of RNA molecules of various types found in the nucleus of the cell and in various stages of processing prior to their export to the cytoplasm.

heterozygote: an individual with two different alleles at a gene locus.

heterozygous: composed of two alleles that are different, for example *Aa*; synonymous with "hybrid."

histones: specialized proteins in eukaryotic cells that bind to DNA molecules and cause them to become more compact; thought to be involved in regulation of gene expression as well.

HLA: *See* human leukocyte antigens.

hnRNA: *See* heterogeneous nuclear RNA.

holandric: refers to a trait passed from father to son via a sex chromosome such as the Y chromosome in human males.

homeobox: a DNA sequence encoding a highly basic protein known as a homeodomain; a homeodomain functions as a transcription factor and is thought to help regulate major events in the embryonic development of higher organisms.

homeotic gene: a gene that helps determine body plan early in development; the products of homeotic genes are transcription factors that control the expression of other genes.

homogametic sex: the particular sex of an organism that produces gametes containing only one type of sex chromosome; in humans, females are the homogametic sex, producing eggs with X chromosomes.

homologous: refers to chromosomes that are identical in terms of types of genes present and the location of the centromere; because of their high degree of similarity, homologous chromosomes can synapse and recombine during prophase I of meiosis.

homology: similarity resulting from descent from a common evolutionary ancestor.

homozygote: an individual with two identical alleles at a gene locus.

homozygous: characterized by a genotype composed of two alleles at the same locus that are the same, for example *AA* or *aa*; synonymous with "purebred."

Human Genome Project: a multi-year genetic research endeavor to sequence the entire human genome, as well as the genomes of related organisms; the human genome sequence was officially completed in 2003.

human leukocyte antigens (HLA): molecules found on the surface of cells that allow the immune system to differentiate between foreign, invading cells and the body's own cells.

hybrid: any cell or organism with genetic material from two different sources, through either natural processes such as sexual reproduction or more artificial processes such as genetic engineering.

hybridization: a process of base pairing involving two single-stranded nucleic acid molecules with complementary sequences; the extent to which two unrelated nucleic acid molecules will hybridize is often used as a way to determine the amount of similarity between the sequences of the two molecules.

hybridoma: a type of hybrid cancer cell created by artificially joining a cancer cell with an antibody-producing cell; hybridomas have useful applications in immunological research.

hydrogen bond: a bond formed between molecules containing hydrogen atoms with positive charges and molecules containing atoms such as nitrogen or oxygen that can possess a negative charge; a relatively weak but important bond in nature that, among other things, connects water molecules, allows DNA strands to base-pair, and contributes to the three-dimensional shape of proteins.

identical twins: a pair of genetically identical offspring that develop from a single fertilized egg; also known as monozygotic twins.

immune system: the system in the body that normally responds to foreign agents by producing antibodies and stimulating antigen-specific lymphocytes, leading to destruction of these agents.

in vitro: literally, "in glass"; an event occurring in an artificial setting such as in a test tube, as opposed to inside a living organism.

in vivo: literally, "in the living"; an event occurring in a living organism, as opposed to an artificial setting.

inborn error of metabolism: a genetic defect in one of a cell's metabolic pathways, usually at the level of an enzyme, that causes the pathway to malfunction; results in phenotypic alterations at the cellular or organismal level.

inbreeding: mating between genetically related individuals.

inbreeding depression: a reduction in the health and vigor of offspring from closely related individuals, a common and widespread phenomenon among nonhuman organisms.

inclusive fitness: an individual's total genetic contribution to future generations, comprising both direct fitness, which results from individual reproduction, and indirect fitness, which results from the reproduction of close relatives.

incomplete dominance: a phenomenon involving two alleles, neither of which masks the expression of the other; instead, the combination of the alleles in the heterozygous state produces a new phenotype that is usually intermediate to the phenotypes produced by either allele alone in the homozygous state.

independent assortment: a characteristic of standard Mendelian genetics referring to the random assortment or shuffling of alleles and chromosomes that occurs during meiosis I; independent assortment is responsible for the offspring ratios observed in Mendelian genetics.

inducer: a molecule that activates some bacterial operons, usually by interacting with regulatory proteins bound to the operator region.

induction: a process in which a cell or group of cells signals an adjacent cell or group of cells to pursue a different developmental pathway and so become differentiated from neighboring cells.

informed consent: the right of patients to know the risks of medical treatment and to determine what is done to their bodies, including the right to accept or refuse treatment based on this information.

initiation codon: also called the "start codon," a codon, composed of the nucleotides AUG, that signals the beginning of a protein-coding sequence in a messenger RNA (mRNA) molecule; in the genetic code, AUG always represents the amino acid methionine.

insert DNA: *See* foreign DNA.

insertion sequence: a small, independently transposable genetic element.

intelligence quotient (IQ): the most common measure of intelligence; it is based on the view that there is a single capacity for complex mental work and that this capacity can be measured by testing.

intercalary deletion: a type of chromosome deletion in which DNA has been lost from within the chromosome (as opposed to a terminal deletion involving a region of DNA lost from the end of the chromosome).

interference: in genetic linkage, a mathematical expression that represents the difference between the expected and the observed number of double recombinant offspring; this can be a clue to the physical location of linked genes on the chromosome.

interphase: the period of the cell cycle in which the cell is preparing to divide, consisting of two distinct growth phases (G_1 and G_2) separated by a period of DNA replication (S phase).

introgression: the transfer of genes from one species to another or the movement of genes between species (or other well-marked genetic populations) mediated by backcrossing.

intron: an intervening sequence within eukaryotic DNA, transcribed as part of a messenger RNA (mRNA) precursor but then removed by splicing before the mRNA molecule is translated; introns are thought to play an important role in the evolution of genes.

inversion: a chromosomal abnormality resulting in a region of the chromosome where the normal order of genes is reversed.

isotope: an alternative form of an element with a variant number of neutrons in its atomic nucleus; isotopes are frequently radioactive and are important tools for numerous molecular biology techniques.

jumping: *See* chromosome jumping

junk DNA: a disparaging (and now known to be inaccurate) characterization of the noncoding DNA content of a genome.

karyokinesis: division of a cell's nuclear contents, as opposed to cytokinesis (division of the cytoplasm). *See also* cytokinesis.

karyotype: the complete set of chromosomes possessed by an individual, usually isolated during metaphase and arranged by size and type as a method of detecting chromosomal abnormalities.

kilobase (kb): a unit of measurement for nucleic acid molecules, equal to 1,000 bases or nucleotides.

kinase: an enzyme that catalyzes phosphate addition to molecules.

kinetochore: a chromosome structure found in the region of the centromere and used as an attachment point for the microtubules of the spindle apparatus during cell division.

Klinefelter syndrome: a human genetic disorder in males who possess an extra X chromosome; Klinefelter males have forty-seven chromosomes instead of the normal forty-six and suffer from abnormalities such as sterility, body feminization, and mental retardation.

knockout: the inactivation of a specific gene within a cell (or whole organism, as in the case of knockout mice) to determine the effects of loss of function of that gene.

lactose: a disaccharide that is an important part of the metabolism of many bacterial species; lactose metabolism in these species is genetically regulated via the *lac* operon.

lagging strand: in DNA replication, the strand of DNA being synthesized in a direction opposite to that of replication fork movement; this strand is synthesized in a discontinuous fashion as a series of Okazaki fragments later joined together. *See also* Okazaki fragments.

Lamarckianism: the theory, originally proposed by Jean-Baptiste Lamarck, that traits acquired by an organism during its lifetime can be passed on to offspring.

lambda (λ) phage: a bacteriophage that infects bacteria and then makes multiple copies of itself by taking over the infected bacteria's cellular machinery.

lateral gene transfer: the movement of genes between organisms; also called horizontal gene transfer.

leading strand: in DNA replication, the strand of DNA being synthesized in the same direction as the movement of the replication fork; this strand is synthesized in a continuous fashion.

leptotene: a subphase of prophase I of meiosis in which chromosomes begin to condense and become visible.

lethal allele: an allele capable of causing the death of an organism; a lethal allele can be recessive (two copies of the allele are required before death results) or dominant (one copy of the allele produces death).

leucine zipper: an amino acid sequence, found in some DNA-binding proteins, characterized by leucine residues separated by sets of seven amino acids; two molecules of this amino acid sequence can combine via the leucine residues and "zip" together, creating a structure that can then bind to a specific DNA sequence.

linkage: a genetic phenomenon involving two or more genes inherited together because they are physically located on the same chromosome; Gregor Mendel's principle of independent assortment does not apply to linked genes, but genotypic and phenotypic variation is possible through crossing over.

linkage mapping: a form of genetic mapping that uses recombination frequencies to estimate the relative distances between linked genes.

locus (*pl.* loci): the specific location of a particular gene on a chromosome.

lymphocytes: sensitized cells of the immune system that recognize and destroy harmful agents via antibody and cell-mediated responses that include B lymphocytes from the bone marrow and T lymphocytes from the thymus.

Lyon hypothesis: a hypothesis stating that one X chromosome of the pair found in all female cells must be inactivated in order for those cells to be normal; the inactivated X chromosome is visible by light microscopy and stains as a Barr body.

lysis: the breaking open of a cell.

lysogeny: a viral process involving repression and integration of the viral genome into the genome of the host bacterial cell.

major histocompatibility complex (MHC): a group of molecules found on the surface of cells, allowing the immune system to differentiate between foreign, invading cells and the body's own cells; in humans, this group of molecules is called HLA (human leukocyte antigens). *See also* human leukocyte antigens.

map unit: *See* centiMorgan.

maternal inheritance: *See* extranuclear inheritance.

Maxam-Gilbert sequencing: a method of base-specific chemical degradation to determine DNA sequence; this method has largely been supplanted by the Sanger method. *See also* Sanger sequencing.

meiosis: a process of cell division in which the cell's genetic material is reduced by half and sex cells called gametes are produced; important as the basis of sexual reproduction.

melanism: the opposite of albinism, a condition that leads to the overproduction of melanin.

melting: a term sometimes used to describe the denaturation of a DNA molecule as it is heated in solution; as the temperature rises, hydrogen bonds between the DNA strands are broken until the double-strand molecule has been completely converted into two single-strand molecules.

Mendelian genetics: the genetics of traits that show simple inheritance patterns; based on the work of Gregor Mendel, a nineteenth century monk who studied the genetics of pea plants.

messenger RNA (mRNA): a type of RNA molecule containing the genetic information necessary to produce a protein through the process of translation; produced from the DNA sequence of a gene in the process of transcription.

metabolic pathway: a series of enzyme-catalyzed reactions leading to the breakdown or synthesis of a particular biological molecule.

metabolism: the collection of biochemical reactions occurring in an organism.

metacentric chromsome: a chromosome with the centromere located at or near the middle of the chromosome. *See also* acrocentric chromosome; telocentric chromosome.

metafemale: a term used to describe *Drosophila* (fruit fly) females that have more X chromosomes than sets of autosomes (for example, a female that has two sets of autosomes and three X chromosomes); also used in reference to human females with more than two X chromosomes.

metaphase: the second phase in the process of mitosis, involving chromosomes lined up in the middle of the cell on a line known as the equator.

methylation: the process of adding a methyl chemical group (one carbon atom and three hydrogen atoms) to a particular molecule, such as to the base portion of a nucleotide in a DNA nucleotide.

metric trait: *See* quantitative trait.

microarray: a flat surface on which 10,000 to 100,000 tiny spots of short DNA molecules (oligonucleotides) are fixed and are used to detect the presence of DNA or RNA molecules that are homologous to the oligonucleotides.

microsatellite DNA: a type of variable number tandem repeat (VNTR) in which the repeated motif is 1 to 6 base pairs; also called a simple sequence repeat (SSR) or a short tandem repeat (STR).

microtubule: a cell structure involved in the movement and division of chromosomes during mitosis and meiosis; part of the cell's cytoskeleton, microtubules can be rapidly assembled and disassembled.

microtubule organizing center (MTOC): *See* centriole.

minimal media: an environment that contains the simplest set of ingredients that a microorganism can use to produce all the substances required for reproduction and growth.

minisatellite DNA: a type of variable number tandem repeat (VNTR) in which the repeated motif is 12 to 500 base pairs in length.

miscegenation: sexual activity or marriage between members of two different human races.

mismatch repair: a cellular DNA repair process in which improperly base-paired nucleotides are enzymatically removed and replaced with the proper nucleotides.

missense mutation: a DNA mutation that changes an existing amino acid codon in a gene to some other amino acid codon; depending on the nature of the change, this can be a harmless or a serious mutation (for example, sickle-cell disease in humans is the result of a missense mutation).

mitochondrial genome (mtDNA): DNA found in mitochondria, which contains some of the genes that code for proteins involved in energy metabolism; it is a circular molecule similar in structure to the genome of bacteria.

mitochondrion: the organelle responsible for production of ATP through the process of cellular respiration in a eukaryotic cell; sometimes referred to as the "powerhouse of the cell."

mitosis: a process of cell division in which a cell's duplicated genetic material is evenly divided between two daughter cells, so that each daughter cell is genetically identical to the original parent cell.

model organism: an organism well suited for genetic research because it has a well-known genetic history, a short life cycle, and genetic variation between individuals in the population.

modern synthesis: the merging of the Darwinian mechanisms for evolution with Mendelian genetics to form the modern fields of population genetics and evolutionary biology; also called the neo-Darwinian synthesis.

molecular clock hypothesis: a hypothesis that predicts that amino acid changes in proteins and nucleotide changes in DNA are approximately constant over time.

molecular cloning: the process of splicing a piece of DNA into a plasmid, virus, or phage vector to obtain many identical copies of that DNA.

molecular genetics: the branch of genetics concerned with the central role that molecules, particularly the nucleic acids DNA and RNA, play in heredity.

monoclonal antibodies: identical antibodies (having specificity for the same antigen) produced by a single type of antibody-producing cell, either a B cell or a hybridoma cell line; important in various types of immunology research techniques.

monoculture: the agricultural practice of growing the same cultivar on large tracts of land.

monohybrid: an organism that is hybrid with respect to a single gene (for example, *Aa*); when two monohybrid organisms are mated, the offspring will generally appear in a 3:1 ratio involving the trait controlled by the gene in question.

monosomy: a genetic condition in which one chromosome from a homologous chromosome pair is missing, producing a $2n-1$ genotype; usually causes significant problems in the phenotype of the organism.

monozygotic: developed from a single zygote; identical twins are monozygotic because they develop from a single fertilized ovum that splits in two.

morphogen: a protein or other molecule made by cells in an egg that creates a concentration gradient affecting the developmental fate of surrounding cells by altering their gene expression or their ability to respond to other morphogens.

morphogenesis: the induction and formation of organized body parts or organs.

mosaicism: a condition in which an individual has two or more cell populations derived from the same fertilized ovum, or zygote, as in sex chromosome mosaics, in which some cells contain the usual XY chromosome pattern and others contain extra X chromosomes.

mRNA: *See* messenger RNA.

mtDNA: *See* mitochondrial genome.

MTOC: *See* centriole.

multifactorial: characterized by a complex interaction of genetic and environmental factors.

multiple alleles: a genetic phenomenon in which a particular gene locus is represented by more than two alleles in a population; the greater the number of alleles, the greater the genetic diversity.

mutagen: any chemical or physical substance capable of increasing mutations in a DNA sequence.

mutant: a trait or organism different from the normal, or wild-type, trait or organism seen commonly in nature; mutants can arise either through expression of particular alleles in the organism or through spontaneous or intentional mutations in the genome.

mutation: a change in the genetic sequence of an organism, usually leading to an altered phenotype.

N terminus: the end of a polypeptide with an amino acid that has a free amino group.

natural selection: a process involving genetic variation on the genotypic and phenotypic levels that contributes to the success or failure of various species in reproduction; thought to be the primary force behind evolution.

negative eugenics: improving human stocks through the restriction of reproduction by individuals with inferior traits or who are known to carry alleles for inferior traits.

neural tube: the embryonic precursor to the spinal cord and brain, which normally closes at small openings, or neuropores, by the twenty-eighth day of gestation in humans.

neurotransmitter: a chemical that carries messages between nerve cells.

neutral mutation: a mutation in a gene, or some other portion of the genome, that is considered to have no effect on the fitness of the organism.

neutral theory of evolution: Motoo Kimura's theory that nucleotide substitutions in the DNA often have no effect on fitness, and thus changes in allele frequencies in populations are caused primarily by genetic drift.

nondisjunction: refers to the improper division of chromosomes during anaphase of mitosis or meiosis, resulting in cells with abnormal numbers of chromosomes and sometimes seriously altered phenotypes.

nonhistone proteins: a heterogeneous group of acidic or neutral proteins found in chromatin that may be involved with chromosome structure, chromatin packaging, or the control of gene expression.

nonsense codon: another term for a termination or stop codon (UAA, UAG, or UGA).

nonsense mutation: a DNA mutation that changes an existing amino acid codon in a message to one of the three termination, or stop, codons; this results in an abnormally short protein that is usually nonfunctional.

Northern blot: a molecular biology procedure in which a labeled single-stranded DNA probe is exposed to cellular RNA immobilized on a filter; under the proper conditions, the DNA probe will seek out and bind to its complementary sequence in the RNA molecules if such a sequence is present.

nuclease: an enzyme that degrades nucleic acids by breaking the phosphodiester bond that connects nucleosides.

nucleic acid: the genetic material of cells, found in two forms: deoxyribonucleic acid (DNA) and ribonucleic acid (RNA); composed of repeating subunits called nucleotides.

nucleocapsid: a viral structure including the capsid, or outer protein coat, and the nucleic acid of the virus.

nucleoid: a region of a prokaryotic cell containing the cell's genetic material.

nucleolus: a eukaryotic organelle located in

the nucleus of the cell; the site of ribosomal RNA (rRNA) synthesis.

nucleoside: a building block of nucleic acids, composed of a sugar (deoxyribose or ribose) and one of the nitrogenous bases: adenine (A), cytosine (C), guanine (G), thymine (T), or uracil (U).

nucleosome: the basic unit molecule of chromatin, composed of a segment of a DNA molecule that is bound to and wound around histone molecules; DNA with nucleosomes appears as beads on a string when viewed by electron microscopy.

nucleotide: a building block of nucleic acids, composed of a sugar (deoxyribose or ribose), one of the nitrogenous bases (adenine, cytosine, guanine, thymine, or uracil) and one or more phosphate groups.

nucleus: the "control center" of eukaryotic cells, where the genetic material is separated from the rest of the cell by a membrane; site of DNA replication and transcription.

nullisomy: a genetic condition in which both members of a homologous chromosome pair are absent; usually, embryos with this type of genetic defect are not viable.

ochre codon: a stop codon (UAA) found in messenger RNA (mRNA) molecules that signals termination of translation.

Okazaki fragments: short DNA fragments, approximately two thousand or fewer bases in length, produced during discontinuous replication of the "lagging" strand of a DNA molecule.

oligonucleotide: a short molecule of DNA, generally fewer than twenty bases long and usually synthesized artificially; an important tool for numerous molecular biology procedures, including site-directed mutagenesis.

oncogene: any gene capable of stimulating cell division, thereby being a potential cause of cancer if unregulated; found in all cells and in many cancer-causing viruses.

oogenesis: the process of producing eggs in a sexually mature female organism; another term for meiosis in females.

opal codon: a stop codon (UGA) found in messenger RNA (mRNA) molecules; signals termination of translation.

open reading frame (ORF): a putative protein-coding DNA sequence, marked by a start codon at one end and a stop codon at the other end.

operator: a region of a bacterial operon serving as a control point for transcription of the operon; a regulatory protein of some type usually binds to the operator.

operon: a genetic structure found only in bacteria, whereby a set of genes are controlled together by the same control elements; usually these genes have a common function, such as the genes of the lactose operon in *Escherichia coli* for the metabolism of lactose.

oxygen free radical: a highly reactive form of oxygen in which a single oxygen atom has a free, unpaired electron; free radicals are common by-products of chemical reactions.

P generation: parental generation; the original individuals mated in a genetic cross.

pachytene: a subphase of prophase I in meiosis in which tetrads become visible.

palindrome: in general, a word that reads the same forwards and backwards (such as the words "noon" and "racecar"); in genetics, a DNA sequence that reads the same on each strand of the DNA molecule, although in opposite directions because of the antiparallel nature of the double helix; most DNA palindromes serve as recognition sites for restriction endonucleases.

pandemic: a worldwide outbreak of a disease.

paracentric inversion: an inversion of a chromosome's sequence that does not involve the centromere, taking place on a single arm of the chromosome.

parthenogenesis: production of an organism from an unfertilized egg.

paternal: coming from the father.

pedigree: a diagram of a particular family, showing the relationships between all members of the family and the inheritance pattern of a particular trait or genetic defect; especially useful for research into human traits that may otherwise be difficult to study.

penetrance: a quantitative term referring to the percentage of individuals with a certain genotype that also exhibit the associated phenotype.

peptide bond: a bond found in proteins; occurs between the carboxyl group of one amino acid and the amino group of the next, linking them together.

pericentric inversion: an inversion of a chromosome's sequence involving the centromere.

pharmacogenomics: the branch of human medical genetics that evaluates how an individual's genetic makeup influences his or her response to drugs.

phenotype: the physical appearance or biochemical and physiological characteristics of an individual, which is determined by both heredity and environment.

phenotypic plasticity: the ability of a genotype to produce different phenotypes when exposed to different environments.

phosphodiester bond: in DNA, the phosphate group connecting one nucleoside to the next in the polynucleotide chain.

photoreactivation repair: a cellular enzyme system responsible for repairing DNA damage caused by ultraviolet light; the system is activated by light.

phylogeny: often called an evolutionary tree, the branching patterns that show evolutionary relationships, with the taxa on the ends of the branches.

pilus: a hairlike reproductive structure possessed by some species of bacterial cells that allows them to engage in a transfer of genetic material known as conjugation.

plasmid: a small, circular DNA molecule commonly found in bacteria and responsible for carrying various genes, such as antibiotic resistance genes; important as a cloning vector for genetic engineering.

pleiotropy: a genetic phenomenon in which a single gene has an effect on two or more traits.

-ploid, -ploidy: a suffix that refers to a chromosome set; humans have two sets of chromosomes and are referred to as being diploid, whereas some plants may have four sets, called tetraploid. Other terms include "autoploidy" and "polyploidy." *See also* allopolyploid; aneuploid; autopolyploid; diploid; euploid; haplodiploidy; haploid; polyploid; triploid.

pluripotency: the ability of a cell to give rise to all the differentiated cell types in an embryo.

point mutation: a DNA mutation involving a single nucleotide.

polar body: a by-product of oogenesis used to dispose of extra, unnecessary chromosomes while preserving the cytoplasm of the developing ovum.

polycistronic: characterizing messenger RNA (mRNA) molecules that contain coding sequences for more than one protein, common in prokaryotic cells.

polygenic inheritance: expression of a trait depending on the cumulative effect of multiple genes; human traits such as skin color, obesity, and intelligence are thought to be examples of polygenic inheritance.

polymerase: a cellular enzyme capable of creating a phosphodiester bond between two nucleotides, producing a polynucleotide chain complementary to a single-stranded nucleic acid template; the enzyme DNA polymerase is important for DNA replication, and the enzyme RNA polymerase is involved in transcription.

polymerase chain reaction (PCR): a technique of molecular biology in which millions of copies of a single DNA sequence can be artificially produced in a relatively short period of time; important for a wide variety of applications when the source of DNA to be copied is either scarce or impure.

polymorphism: the presence of many different alleles for a particular locus in individuals of the same species.

polypeptide: a single chain of amino acids connected to one another by peptide bonds; all proteins are polypeptides, but a protein may comprise one or more polypeptide molecules.

polyploid: a cell or organism that possesses multiple sets of chromosomes, usually more than two.

polysome: a group of ribosomes attached to the same messenger RNA (mRNA) molecule and producing the same protein product in varying stages of completion.

population: a group of organisms of the same species in the same place at the same time and thus potentially able to mate; populations are the basic unit of speciation.

population genetics: the study of how genes behave in populations; often a highly mathematical branch of genetics in which evolutionary processes are modeled.

positive eugenics: selecting individuals to reproduce who have desirable genetic traits, as seen by those in control.

post-translational modification: chemical alterations to proteins after they have been produced at a ribosome that alters their properties.

prenatal testing: testing that is done during pregnancy to examine the chromosomes or genes of a fetus to detect the presence or absence of a genetic disorder.

primer: a short nucleic acid molecule used as a beginning point for the enzyme DNA polymerase as it replicates a single-stranded template.

prion: an infectious agent composed solely of protein; thought to be the cause of various human and animal diseases characterized by neurological degeneration, including scrapie in sheep, mad cow disease in cattle, and Creutzfeldt-Jakob disease in humans.

probe: in genetics research, typically a single-stranded nucleic acid molecule or antibody that has been labeled in some way, either with radioactive isotopes or fluorescent dyes; this molecule is then used to seek out its complementary nucleic acid molecule or protein target in a variety of molecular biology techniques such as Southern, Northern, or Western blotting.

product rule: a rule of probability stating that the probability associated with two simultaneous yet independent events is the product of the events' individual probabilities.

prokaryote: a cell that lacks a nuclear membrane (and therefore has no true nucleus) and membrane-bound organelles; bacteria are the only known prokaryotic organisms.

promoter: a region of a gene that controls transcription of that gene; a physical binding site for RNA polymerase.

prophase: the first phase in the process of mitosis or meiosis, in which the nuclear membrane disappears, the spindle apparatus begins to form, and chromatin takes on the form of chromosomes by becoming shorter and thicker.

propositus: the individual in a human pedigree who is the focus of the pedigree, usually by being the first person who came to the attention of the geneticist.

protein: a biological molecule composed of amino acids linked together by peptide bonds; used as structural components of the cell or as enzymes; the term "protein" can refer to a single chain of amino acids or to multiple chains of amino acids functioning in a concerted way, as in the molecule hemoglobin.

proteomics: the study of which proteins are expressed in different types of cells, tissues, and organs during normal and abnormal conditions.

proto-oncogene: a gene, found in eukaryotic cells, that stimulates cell division; ordinarily, expression of this type of gene is tightly controlled by the cell, but in cancer cells, proto-oncogenes have been converted into oncogenes through alteration or elimination of controlled gene expression.

pseudodominance: a genetic phenomenon involving a recessive allele on one chromosome that is automatically expressed because of the deletion of its corresponding dominant allele on the other chromosome of the homologous pair.

pseudogenes: DNA sequences derived from partial copies, mutated complete copies, or normal copies of functional genes that have lost their control sequences and therefore cannot be transcribed; may originate by gene duplication or retrotransposition and are apparently nonfunctional regions of the genome that may evolve at a maximum rate, free from the evolutionary constraints of natural selection.

pseudohermaphrodite: individual born with either ambiguous genitalia or external genitalia that are the opposite of the chromosomal sex.

punctuated equilibrium: a model of evolutionary change in which new species originate abruptly and then exist through a long period of stasis; important as an explanation of the stepwise pattern of species change seen in the fossil record.

purine: either of the nitrogenous bases adenine or guanine; used in the structure of nucleic acids.

pyrimidine: any of the nitrogenous bases cytosine, thymine, or uracil; used in the structure of nucleic acids.

quantitative trait: a trait, such as human height or weight, that shows continuous variation in a population and can be measured; also called a metric trait.

quantitative trait loci (QTLs): genomic regions that affect a quantitative trait, generally identified via DNA-based markers.

reaction norm: the relationship between environment and phenotype for a given genotype.

reading frame: refers to the manner in which a messenger RNA (mRNA) sequence is interpreted as a series of amino acid codons by the ribosome; because of the triplet nature of the genetic code, a typical messenger RNA (mRNA) molecule has three possible reading frames, although usually only one of these will actually code for a functional protein.

receptors: molecules to which signaling molecules bind in target cells.

recessive: a term referring to an allele or trait that will only be expressed if another, dominant, trait or allele is not also present.

reciprocal cross: a mating that is the reverse of another with respect to the sex of the organisms that possess certain traits; for example, if a particular cross were tall male × short female, then the reciprocal cross would be short male × tall female.

reciprocal translocation: a two-way exchange of genetic material between two nonhomologous chromosomes, resulting in a wide variety of genetic problems depending on which chromosomes are involved in the translocation.

recombinant DNA: DNA molecules that are the products of artificial recombination between DNA molecules from two different sources; important as a foundation of genetic engineering.

recombination: an exchange of genetic material, usually between two homologous chromosomes; provides one of the foundations for the genetic reassortment observed during sexual reproduction.

reductional division: refers to meiosis I, in which the amount of genetic material in the cell is reduced by half through nuclear division; it is at this stage that the diploid cell is converted to an essentially haploid state.

reductionism: the explanation of a complex system or phenomenon as merely the sum of its parts.

replication: the process by which a DNA or RNA molecule is enzymatically copied.

replicon: a region of a chromosome under control of a single origin of replication.

replisome: a multiprotein complex that functions at the replication fork during DNA replication; it contains all the enzymes and other proteins necessary for replication, including DNA polymerase.

repressor: a protein molecule capable of preventing transcription of a gene, usually by binding to a regulatory region close to the gene.

resistance plasmid (R plasmid): a small, circular DNA molecule that replicates independently of the bacterial host chromosome and encodes a gene for antibiotic resistance.

restriction endonuclease: a bacterial enzyme that cuts DNA molecules at specific sites; part of a bacterial cell's built-in protection against infection by viruses; an important tool of genetic engineering.

restriction enzyme: *See* restriction endonuclease.

restriction fragment length polymorphism (RFLP): a genetic marker, consisting of variations in the length of restriction fragments in DNA from individuals being tested, allowing researchers to compare genetic sequences from various sources; used in a variety of fields, including forensics and the Human Genome Project.

retrotransposon (retroposon): a DNA sequence that is transcribed to RNA and reverse transcribed to a DNA copy able to insert itself at another location in the genome.

retrovirus: a virus that carries reverse transcriptase that converts its RNA genome into a DNA copy that integrates into the host chromosome.

reverse transcriptase: a form of DNA polymerase, discovered in retroviruses, that uses an RNA template to produce a DNA molecule; the name indicates that this process is the reverse of the transcription process occurring naturally in the cell.

reverse-transcriptase polymerase chain reaction (RT-PCR): a technique, requiring isolated RNA, for quickly determining if a gene or a small set of genes are transcribed in a population of cells.

RFLP analysis: *See* restriction fragment length polymorphism.

Rh factor: a human red-blood-cell antigen, first characterized in rhesus monkeys, that contributes to blood typing; individuals can be either Rh positive (possessing the antigen on their red blood cells) or Rh negative (lacking the antigen).

ribonucleic acid (RNA): a form of nucleic acid in the cell used primarily for genetic expression through transcription and translation; in structure, it is virtually identical to DNA, except that ribose is used as the sugar in each nucleotide and the nitrogenous base thymine is replaced by uracil; present in three major forms in the cell: messenger RNA (mRNA), transfer RNA (tRNA), and ribosomal RNA (rRNA).

ribose: a five-carbon sugar used in the structure of ribonucleic acid (RNA).

ribosomal RNA (rRNA): a type of ribonucleic acid in the cell that constitutes some of the structure of the ribosome and participates in the process of translation.

ribosome: a cellular structure, composed of ribosomal RNA (rRNA) and proteins, that is the site of translation.

ribozyme: an RNA molecule that can function catalytically as an enzyme.

RNA: *See* ribonucleic acid.

RNA interference (RNAi): an artificial technique using small, interfering RNAs that cause gene silencing by binding to the part of a messenger RNA (mRNA) to which they are complementary, thus blocking translation.

RNA polymerase: the cellular enzyme required for making an RNA copy of genetic information contained in a gene; an integral part of transcription.

RNase: refers to a group of enzymes, ribonucleases, capable of specifically degrading RNA molecules.

rRNA: *See* ribosomal RNA.

Sanger sequencing: also known as dideoxy termination sequencing, a method using nucleotides that are missing the 3′ hydroxyl group in order to terminate the polymerization of new DNA at a specific nucleotide; the most common sequencing method, used almost exclusively.

segregation: a characteristic of Mendelian genetics, resulting in the division of homologous chromosomes into separate gametes during the process of meiosis.

semiconservative replication: a characteristic of DNA replication, in which every new DNA molecule is actually a hybrid molecule, being composed of a parental, preexisting strand and a newly synthesized strand.

sex chromosome: a chromosome carrying genes responsible for determination of an organism's sex; in humans, the sex chromosomes are designated *X* and *Y*.

sex-influenced inheritance: inheritance in which the expression of autosomal traits is influenced or altered relative to the sex of the individual possessing the trait; pattern baldness is an example of this type of inheritance in humans.

sex-limited inheritance: inheritance of traits expressed in only one sex, although these traits are usually produced by non-sex-linked genes (that is, they are genes located on autosomes instead of sex chromosomes).

sexual reproduction: reproduction of cells or organisms involving the transfer and reassortment of genetic information, resulting in offspring that can be phenotypically and genotypically distinct from either of the parents; mediated by the fusion of gametes produced during meiosis.

Shine-Dalgarno sequence: a short sequence in prokaryotic messenger RNA (mRNA) molecules complementary to a sequence in the prokaryotic ribosome; important for proper positioning of the start codon of the mRNA relative to the P site of the ribosome.

short interspersed sequences (SINES): short repeats of DNA sequences scattered throughout a genome.

shotgun cloning: a technique by which random DNA fragments from an organism's genome are inserted into a collection of vectors to produce a library of clones, which can then be used in a variety of molecular biology procedures.

sigma factor: a molecule that is part of RNA polymerase molecules in bacterial cells; allows RNA polymerase to select the genes that will be transcribed.

signal transduction: all of the molecular events that occur between the arrival of a signaling molecule at a target cell and its response; typically involves a cascading series of reactions that can eventually determine expression of many dozens of genes.

single nucleotide polymorphism (SNP): differences at the individual nucleotide level among individuals.

site-directed mutagenesis: a molecular genetics procedure in which synthetic oligonucleotide molecules are used to induce carefully planned mutations in a cloned DNA molecule.

small nuclear RNA (snRNA): small, numerous RNA molecules found in the nuclei of eukaryotic cells and involved in splicing of messenger RNA (mRNA) precursors to prepare them for translation.

snRNA: *See* small nuclear RNA.

snRNP: *See* small nuclear ribonucleoprotein.

sociobiology: the study of social structures, organizations, and actions in terms of underlying biological principles.

solenoid: a complex, highly compacted DNA structure consisting of many nucleosomes packed together in a bundle.

somatic mutation: a mutation occurring in a somatic, or nonsex, cell; because of this, somatic mutations cannot be passed to the next generation.

Southern blot: a molecular biology technique in which a labeled single-stranded DNA probe is exposed to denatured cellular DNA

immobilized on a filter; under the proper conditions, the DNA probe will seek out and bind to its complementary sequence among the cellular DNA molecules, if such a sequence is present.

speciation: the process of evolutionary change that leads to the formation of new species.

species: a group of organisms that can interbreed with one another but not with organisms outside the group; generally, members of a particular species share the same gene pool; defining a species is still controversial and remains a debated concept.

spermatogenesis: the process of producing sperm in a sexually mature male organism; another term for meiosis in males.

spindle apparatus: a structure, composed of microtubules and microfilaments, important for the proper orientation and movement of chromosomes during mitosis and meiosis; appears during prophase and begins to disappear during anaphase.

spliceosome: a complex of nuclear RNA and protein molecules responsible for the excision of introns from messenger RNA (mRNA) precursors before they are translated.

SRY: the sex-determining region of the Y chromosome; a gene encoding a protein product called testis determining factor (TDF), responsible for conversion of a female embryo to a male embryo through the development of the testes.

stem cell: an undifferentiated cell that retains the ability to give rise to other, more specialized cells.

sum rule: a rule of probability theory stating that the probability of either of two mutually exclusive events occurring is the sum of the events' individual probabilities.

supercoil: a complex DNA structure in which the DNA double helix is itself coiled into a helix; usually observed in circular DNA molecules such as bacterial plasmids.

sympatric speciation: the genetic divergence of populations, not separated geographically, that eventually results in formation of new species.

synapsis: the close association of homologous chromosomes occurring during early prophase I of meiosis; during synapsis, recombination between these chromosomes can occur.

T: the abbreviation for thymine, a pyrimidine nitrogenous base found in the structure of DNA; in RNA, thymine is replaced by uracil.

Taq polymerase: DNA polymerase from the bacterium *Thermus aquaticus*; an integral component of polymerase chain reaction.

tautomerization: a spontaneous internal rearrangement of atoms in a complex biological molecule that often causes the molecule to change its shape or its chemical properties.

taxon (*pl.* taxa): a general term used by evolutionists to refer to a type of organism at any level in a classification of organisms.

telocentric chromosome: a chromosome with a centromere at the end. *See also* acrocentric chromosome; metacentric chromosome.

telomere: the end of a eukaryotic chromosome, protected and replaced by the cellular enzyme telomerase.

telophase: the final phase in the process of mitosis or meiosis, in which division of the cell's nuclear contents has been completed and division of the cell itself occurs.

template: a single-stranded DNA molecule (or RNA molecule) used to create a complementary strand of nucleic acid through the activity of a polymerase.

teratogen: any chemical or physical substance, such as thalidomide, that creates birth defects in offspring.

testcross: a mating involving an organism with a recessive genotype for desired traits crossed with an organism that has an incompletely determined genotype; the types and ratio of offspring produced allow geneticists to determine the genotype of the second organism.

tetrad: a group of four chromosomes formed as a result of the synapsis of homologous chromosomes that takes place early in meiosis.

tetranucleotide hypothesis: a disproven hypothesis, formulated by geneticist P. A. Levene, stating that DNA is a structurally simple molecule composed of a repeating unit known as a tetranucleotide (composed, in turn, of equal amounts of the bases adenine, cytosine, guanine, and thymine).

thermal cycler: a machine that can rapidly heat and cool reaction tubes; used for performing PCR reactions.

theta structure: an intermediate structure in the bidirectional replication of a circular DNA molecule; the name comes from the resemblance of this structure to the Greek letter theta.

thymine (T): a pyrimidine nitrogenous base found in the structure of DNA; in RNA, thymine is replaced by uracil.

thymine dimer: a pair of thymine bases in a DNA molecule connected by an abnormal chemical bond induced by ultraviolet light; prevents DNA replication in the cell unless it is removed by specialized enzymes.

topoisomerases: cellular enzymes that relieve tension in replicating DNA molecules by introducing single- or double-stranded breaks into the DNA molecule; without these enzymes, replicating DNA becomes progressively more supercoiled until it can no longer unwind, and DNA replication is halted.

totipotent: the ability of a cell to produce an entire adult organism through successive cell divisions and development; as cells become progressively differentiated, they lose this characteristic.

trait: a phenotypic characteristic that is heritable.

transcription: the cellular process by which genetic information in the form of a gene in a DNA molecule is converted into the form of a messenger RNA (mRNA) molecule; dependent on the enzyme RNA polymerase.

transcription factor: a protein that is involved in initiation of transcription but is not part of the RNA polymerase.

transduction: DNA transfer between cells, with a virus serving as the genetic vector.

transfer RNA (tRNA): a type of RNA molecule necessary for translation to occur properly; provides the basis of the genetic code, in which codons in a messenger RNA (mRNA) molecule are used to direct the sequence of amino acids in a polypeptide; contains a binding site for a particular amino acid and a region complementary to a messenger RNA (mRNA) codon (an anticodon).

transformation: the process by which a normal cell is converted into a cancer cell; also refers to the change in phenotype accompanying entry of foreign DNA into a cell, such as in bacterial cells being used in recombinant DNA procedures.

transgenic organism: an organism possessing one or more genes from another organism, such as mice that possess human genes; important for the study of genes in a living organism, especially in the study of mutations within these genes. *See also* genetically modified organism (GMO).

transition mutation: a DNA mutation in which one pyrimidine (cytosine or thymine) takes the place of another, or a purine (adenine or guanine) takes the place of another.

translation: the cellular process by which genetic information in the form of a messenger RNA (mRNA) molecule is converted into the amino acid sequence of a protein, using ribosomes and RNA molecules as accessory molecules.

translocation: the movement of a chromosome segment to a nonhomologous chromosome as a result of an error in recombination; also refers to the movement of a messenger RNA (mRNA) codon from the A site of the ribosome to the P site during translation.

transposable element: *See* transposon.

transposon: a DNA sequence capable of moving to various places in a chromosome, discovered by geneticist Barbara McClintock; transposons are thought to be important as mediators of genetic variability in both prokaryotes and eukaryotes.

transversion: a DNA mutation in which a pyrimidine (cytosine or thymine) takes the place of a purine (adenine or guanine), or vice versa.

triploid: possessing three complete sets of chromosomes, or $3N$; important in the development of desirable characteristics in the flowers or fruit of some plants; triploids are often sterile.

trisomy: a genetic condition involving one chromosome of a homologous chromosome pair that has been duplicated in some way, giving rise to a $2N + 1$ genotype and causing serious phenotypic abnormalities; a well-known example is trisomy 21, or Down syndrome, in which the individual possesses three copies of chromosome 21 instead of the normal two copies.

tRNA: *See* transfer RNA.

tumor-suppressor genes: any of a number of genes that limit or halt cell division under certain circumstances, thereby preventing the formation of tumors in an organism; two well-studied examples are the retinoblastoma gene and the *p53* gene; mutations in tumor suppressor genes can lead to cancer.

Turner syndrome: a human genetic defect in which an individual has only forty-five chromosomes, lacking one sex chromosome; the sex chromosome present is an X chromosome, making these individuals phenotypically female, although with serious abnormalities such as sterility and anatomical defects.

uracil (U): a pyrimidine nitrogenous base found in the structure of RNA; in DNA, uracil is replaced by thymine.

variable number tandem repeat (VNTR): a repetitive DNA sequence of approximately fifty to one hundred nucleotides; important in the process of forensic identification known as DNA fingerprinting.

vector: a DNA molecule, such as a bacterial plasmid, into which foreign DNA can be inserted and then transported into a cell for further manipulation; important in a wide variety of recombinant DNA techniques.

virions: mature infectious virus particles.

viroids: naked strands of RNA, 270 to 380 nucleotides long, that are circular and do not code for any proteins that are able to cause disease in susceptible plants, many of them economically important. *See also* virusoid.

virus: a microscopic infectious particle composed primarily of protein and nucleic acid; bacterial viruses, or bacteriophages, have been important tools of study in the history of molecular genetics.

virusoids: similar to viroids, microscopic infec-

tious particles composed primarily of protein and nucleic acid; unlike viroids, virusoids are packaged in the protein coat of other plant viruses, referred to as helpers, and are therefore dependent
on the other virus. *See also* viroid.

VNTR: *See* variable number tandem repeats.

walking: *See* chromosome walking.

Western blot: a molecular biology technique involving labeled antibodies exposed to cellular proteins immobilized on a filter; under the proper conditions, the antibodies will seek out and bind to the proteins for which they are specific, if such proteins are present.

wild-type: a trait common in nature; usually contrasted with variants of the trait, which are known as mutants.

wobble hypothesis: a concept stating that the anticodon of a transfer RNA (tRNA) molecule is capable of interacting with more than one messenger RNA (mRNA) codon by virtue of the inherent flexibility present in the third base of the anticodon; first proposed by molecular biologist Francis Crick.

X linkage: a genetic phenomenon involving a gene located on the X chromosome; the typical pattern of X linkage involves recessive alleles, such as that for hemophilia, which exert their effects when passed from mother to son and are more likely to be exhibited by males than females.

X-ray diffraction: a method for determining the structure of molecules which infers structure by the way crystals of molecules scatter X rays as they pass through.

xenotransplants: transplants of organs or cellular tissue between different species of animals, such as between pigs and humans.

Y linkage: a genetic phenomenon involving a gene located on the Y chromosome; as a result, such a condition can be passed only from father to son.

yeast artificial chromosome (YAC): a cloning vector that has been engineered with all of the major genetic characteristics of a eukaryotic chromosome so that it will behave as such during cell division; YACs are used to clone extremely large DNA fragments from eukaryotic cells and are an integral part of the Human Genome Project.

Z-DNA: a zigzag form of DNA in which the strands form a left-handed helix instead of the normal right-handed helix of B-DNA; Z-DNA is known to be present in cells and is thought to be involved in genetic regulation. *See also* B-DNA.

zinc finger: an amino acid sequence, found in some DNA-binding proteins, that complexes with zinc ions to create polypeptide "fingers" that can then wrap around a specific portion of a DNA molecule.

zygote: a diploid cell produced by the union of a male gamete (sperm) with a female gamete (egg); through successive cell divisions, the zygote will eventually give rise to the adult form of the organism.

zygotene: a subphase of prophase I of meiosis involving synapsis between homologous chromosomes.

—Randall K. Harris, updated by Bryan Ness

Bibliography

GENERAL

Audesirk, Teresa, Gerald Audesirk, and Bruce E. Myers. *Biology: Life on Earth*. 6th ed. Upper Saddle River, N.J.: Prentice Hall, 2001.

Bailey, Philip S., Jr., Christina A. Bailey, and Robert J. Ouellette. *Organic Chemistry*. New York: Prentice Hall, 1997.

Banaszak, Leonard J. *Foundations of Structural Biology*. San Diego: Academic Press, 2000.

Berg, Paul, and Maxine Singer. *Dealing with Genes: The Language of Heredity*. Mill Valley, Calif.: University Science Books, 1992.

Brenner, Sydney, and Jeffrey K. Miller, eds. *Encyclopedia of Genetics*. New York: Elsevier, 2001.

Campbell, Neil A., and Jane B. Reece. *Biology*. 6th ed. San Francisco: Benjamin Cummings, 2002.

Clark, David P., and Lonnie D. Russell. *Molecular Biology Made Simple and Fun*. 2d ed. Vienna, Ill.: Cache River Press, 2000.

Diagram Group. *Genetics and Cell Biology on File*. Rev. ed. New York: Facts On File, 2003.

Fairbanks, Daniel J., and W. Ralph Anderson. *Genetics: The Continuity of Life*. New York: Brooks/Cole, 1999.

Finch, Caleb Ellicott. *Longevity, Senescence, and the Genome*. Reprint. Chicago: University of Chicago Press, 1994.

Hartl, Daniel L. *Genetics: Analysis of Genes and Genomes*. 5th ed. Boston: Jones and Bartlett, 2001.

Hartwell, L. H., L. Hood, M. L. Goldberg, A. E. Reynolds, L. M. Silber, and R. C. Veres. *Genetics: From Genes to Genomes*. 2d ed. New York: McGraw-Hill, 2003.

Hill, John, et al. *Chemistry and Life: An Introduction to General, Organic, and Biological Chemistry*. 6th ed. New York: Prentice Hall, 2000.

King, Robert C., and William D. Stansfield. *A Dictionary of Genetics*. 6th ed. New York: Oxford University Press, 2001.

Klug, William S. *Essentials of Genetics*. 3d ed. Upper Saddle River, N.J.: Prentice Hall, 1999.

Lerner, K. Lee. *World of Genetics*. Detroit, Mich.: Gale Research, 2001.

Lewin, Benjamin. *Genes VII*. New York: Oxford University Press, 2001.

Madigan, Michael T., John M. Martinko, and Jack Parker. *Brock Biology of Micro-organisms*. 10th ed. Upper Saddle River, N.J.: Prentice Hall, 2003.

Micklos, David A, Greg A. Freyer, and David A. Crotty. *DNA Science: A First Course*. Cold Spring Harbor, N.Y.: Cold Spring Harbor Laboratory Press, 2003.

Moore, David S. *Dependent Gene: The Fallacy of Nature vs. Nurture*. New York: W. H. Freeman, 2001.

Moore, John A. *Science as a Way of Knowing*. Reprint. Cambridge, Mass.: Harvard University Press, 1999.

Raven, Peter H., and George B. Johnson. *Biology*. 6th ed. New York: W. H. Freeman/Worth, 1999.

Reeve, Eric C. R. *Encyclopedia of Genetics*. Chicago: Fitzroy Dearborn, 2001.

Robinson, Richard, ed. *Genetics*. New York: Macmillan, 2002.

Russell, Peter J. *Genetics*. San Fransisco, Calif.: Benjamin Cummings, 2002.

Simon, Anne. *The Real Science Behind the X-Files: Microbes, Meteorites, and Mutants*. New York: Simon & Schuster, 1999.

Singer, Maxine, and Paul Berg, eds. *Exploring Genetic Mechanisms*. Sausalito, Calif.: University Science Books, 1997.

Starr, Cecie. *Biology: Concepts and Applications*. 5th ed. New York: Brooks/Cole, 2003.

Tamarin, Robert H. *Principles of Genetics*. Boston: McGraw-Hill, 2002.

Tortora, Gerard. *Microbiology: An Introduction*. 7th ed. San Francisco: Benjamin Cummings, 2001.

Weaver, Robert F., and Philip W. Hedrick. *Genetics*. 3d ed. New York: McGraw-Hill, 1997.

AGRICULTURE AND GENETICALLY MODIFIED FOODS

Altieri, Miguel A. *Genetic Engineering in Agriculture*. Chicago: LPC Group, 2001.

Anderson, Luke. *Genetic Engineering, Food, and*

Our Environment. White River Junction, Vt.: Chelsea Green, 1999.

Avery, Dennis T. *Saving the Planet with Pesticides and Plastic: The Environmental Triumph of High-Yield Farming.* 2d ed. Indianapolis, Ind.: Hudson Institute, 2000.

Carozzi, Nadine, and Michael Koziel, eds. *Advances in Insect Control: The Role of Transgenic Plants.* Bristol, Pa.: Taylor & Francis, 1997.

Chrispeels, Maarten J., and David E. Sadava. *Plants, Genes, and Crop Biotechnology.* Boston: Jones and Bartlett, 2003.

Engel, Karl-Heinz, et al. *Genetically Modified Foods: Safety Aspects.* Washington, D.C.: American Chemical Society, 1995.

Entwhistle, Philip F., Jenny S. Cory, Mark J. Bailey, and Steven R. Higgs. *Bacillus thuringiensis, an Environmental Biopesticide: Theory and Practice.* New York: Wiley, 1994.

Galun, Esra, and Adina Breiman. *Transgenic Plants.* London: Imperial College Press, 1997.

Goodman, David, et al. *From Farming to Biotechnology: A Theory of Agro-Industrial Development.* New York: Basil Blackwell, 1987.

Hall, Franklin R., and Julius J. Menn, eds. *Biopesticides: Use and Delivery.* Totowa, N.J.: Humana Press, 1999.

Henry, Robert J. *Practical Applications of Plant Molecular Biology.* New York: Chapman & Hall, 1997.

Koul, Opender, and G. S. Dhaliwal, eds. *Microbial Biopesticides.* New York: Taylor & Francis, 2002.

Krimsky, Sheldon, et al., eds. *Agricultural Biotechnology and the Environment: Science, Policy, and Social Issues.* Urbana: University of Illinois Press, 1996.

Lambrecht, Bill. *Dinner at the New Gene Cafe: How Genetic Engineering Is Changing What We Eat, How We Live, and the Global Politics of Food.* New York: Thomas Dunne Books, 2001.

Lurquin, Paul F. *The Green Phoenix: A History of Genetically Modified Plants.* New York: Columbia University Press, 2001.

Rissler, Jane, and Margaret Mellon. *The Ecological Risks of Engineered Crops.* Cambridge, Mass.: MIT Press, 1996.

Winston, Mark L. *Travels in the Genetically Modified Zone.* Cambridge, Mass.: Harvard University Press, 2002.

Zohary, Daniel, and Maria Hopf. *Domestication of Plants in the Old World: The Origin and Spread of Cultivated Plants in West Asia, Europe, and the Nile Valley.* 3d ed. New York: Oxford University Press, 2001.

BACTERIAL GENETICS

Adelberg, Edward A., ed. *Papers on Bacterial Genetics.* Boston: Little, Brown, 1966.

Birge, Edward R. *Bacterial and Bacteriophage Genetics.* 4th ed. New York: Springer, 2000.

Brock, Thomas D. *The Emergence of Bacterial Genetics.* Cold Spring Harbor, N.Y.: Cold Spring Harbor Laboratory Press, 1990.

Carlberg, David M. *Essentials of Bacterial and Viral Genetics.* Springfield, Ill.: Charles C Thomas, 1976.

Dale, Jeremy. *Molecular Genetics of Bacteria.* New York: John Wiley & Sons, 1994.

Day, Martin J. *Plasmids.* London: Edward Arnold, 1982.

Dean, Alastair Campbell Ross, and Sir Cyril Hinshelwood. *Growth, Function, and Regulation in Bacterial Cells.* Oxford: Clarendon Press, 1966.

De Bruijn, Frans J., et al., eds. *Bacterial Genomes: Physical Structure and Analysis.* New York: Chapman & Hall, 1998.

Dorman, Charles J. *Genetics of Bacterial Virulence.* Oxford: Blackwell Scientific, 1994.

Drlica, Karl, and Monica Riley, eds. *The Bacterial Chromosome.* Washington, D.C.: American Society for Microbiology, 1990.

Fry, John C., and Martin J. Day, eds. *Bacterial Genetics in Natural Environments.* London: Chapman & Hall, 1990.

Goldberg, Joanna B., ed. *Genetics of Bacterial Polysaccharides.* Boca Raton, Fla.: CRC Press, 1999.

Jacob, François, and Elie L. Wollman. *Sexuality and the Genetics of Bacteria.* New York: Academic Press, 1961.

Joset, Françoise, et al. *Prokaryotic Genetics: Genome Organization, Transfer, and Plasticity.* Boston: Blackwell Scientific Publications, 1993.

Miller, Jeffrey H. *A Short Course in Bacterial Ge-*

netics: A Laboratory Manual and Handbook for Escherichia coli and Related Bacteria. Cold Spring Harbor, N.Y.: Cold Spring Harbor Laboratory Press, 1999.

Murray, Patrick, ed. *Manual of Clinical Microbiology*. Washington, D.C.: ASM Press, 2003.

Schumann, Wolfgang, S. Dusko Ehrlich, and Naotake Ogasawara, eds. *Functional Analysis of Bacterial Genes: A Practical Manual*. New York: Wiley, 2001.

Siezen, Roland J., et al., eds. *Lactic Acid Bacteria: Genetics, Metabolism, and Applications*. 7th ed. Boston: Kluwer Academic, 2002.

Snyder, Larry, and Wendy Champness. *Molecular Genetics of Bacteria*. Washington, D.C.: ASM Press, 1997.

Summers, David K. *The Biology of Plasmids*. Malden, Mass.: Blackwell, 1996.

Thomas, Christopher M., ed. *The Horizontal Gene Pool: Bacterial Plasmids and Gene Spread*. Amsterdam: Harwood Academic, 2000.

U.S. Congress. Office of Technology Assessment. *New Developments in Biotechnology—Field-Testing Engineered Organisms: Genetic and Ecological Issues*. Washington, D.C.: National Technical Information Service, 1988.

Vaughan, Pat, ed. *DNA Repair Protocols: Prokaryotic Systems*. Totowa, N.J.: Humana Press, 2000.

BIOETHICS AND SOCIAL POLICY

Becker, Gerhold K., and James P. Buchanan, eds. *Changing Nature's Course: The Ethical Challenge of Biotechnology*. Hong Kong: Hong Kong University Press, 1996.

Bonnicksen, Andrea L. *Crafting a Cloning Policy: From Dolly to Stem Cells*. Washington, D.C.: Georgetown University Press, 2002.

Boylan, Michael, and Kevin E. Brown. *Genetic Engineering: Science and Ethics on the New Frontier*. Upper Saddle River, N.J.: Prentice Hall, 2001.

Brannigan, Michael C., ed. *Ethical Issues in Human Cloning: Cross-Disciplinary Perspectives*. New York: Seven Bridges Press, 2001.

Bulger, Ruth Ellen, Elizabeth Heitman, and Stanley Joel Reiser, eds. *The Ethical Dimensions of the Biological and Health Sciences*. 2d ed. New York: Cambridge University Press, 2002.

Caplan, Arthur. *Due Consideration: Controversy in the Age of Medical Miracles*. New York: Wiley, 1997.

Chadwick, Ruth, et al., eds. *The Ethics of Genetic Screening*. Boston: Kluwer Academic, 1999.

Chapman, Audrey R., ed. *Perspectives on Genetic Patenting: Religion, Science, and Industry in Dialogue*. Washington, D.C.: American Association for the Advancement of Science, 1999.

Charon, Rita, and Martha Montello, eds. *Stories Matter: The Role of Narrative in Medical Ethics*. New York: Routledge, 2002.

Comstock, Gary L., ed. *Life Science Ethics*. Ames: Iowa State Press, 2002.

Danis, Marion, Carolyn Clancy, and Larry R. Churchill, eds. *Ethical Dimensions of Health Policy*. New York: Oxford University Press, 2002.

Davis, Bernard D., ed. *The Genetic Revolution: Scientific Prospects and Public Perceptions*. Baltimore: The Johns Hopkins University Press, 1991.

De Waal, Franz. *Good Natured: The Origins of Right and Wrong in Humans and Other Animals*. Cambridge, Mass.: Harvard University Press, 1996.

Doherty, Peter, and Agneta Sutton, eds. *Man-Made Man: Ethical and Legal Issues in Genetics*. Dublin: Four Courts Press, 1997.

Espejo, Roman, ed. *Biomedical Ethics: Opposing Viewpoints*. San Diego: Greenhaven Press, 2003.

Evans, John Hyde. *Playing God? Human Genetic Engineering and the Rationalization of Public Bioethical Debate*. Chicago: University of Chicago Press, 2002.

Gonder, Janet C., Ernest D. Prentice, and Lilly-Marlene Russow, eds. *Genetic Engineering and Animal Welfare: Preparing for the Twenty-first Century*. Greenbelt, Md.: Scientists Center for Animal Welfare, 1999.

Grace, Eric S. *Biotechnology Unzipped: Promises and Reality*. Washington, D.C.: National Academy Press, 1997.

Harpignies, J. P. *Double Helix Hubris: Against Designer Genes*. Brooklyn, N.Y.: Cool Grove Press, 1996.

Holland, Suzanne, Karen Lebacqz, and Laurie Zoloth, eds. *The Human Embryonic Stem Cell Debate: Science, Ethics, and Public Policy (Basic*

Bioethics). Cambridge, Mass.: MIT Press, 2001.

Hubbard, Ruth, and Elijah Wald. *Exploding the Gene Myth: How Genetic Information Is Produced and Manipulated by Scientists, Physicians, Employers, Insurance Companies, Educators, and Law Enforcers.* Boston: Beacon Press, 1999.

Kass, Leon R. *Life, Liberty, and the Defense of Dignity: The Challenge for Bioethics.* San Francisco: Encounter Books, 2002.

Kevles, Daniel J. *In the Name of Eugenics: Genetics and the Uses of Human Heredity.* Cambridge, Mass.: Harvard University Press, 1995.

Kristol, William, and Eric Cohen, eds. *The Future Is Now: America Confronts the New Genetics.* Lanham, Md.: Rowman & Littlefield, 2002.

Leroy, Bonnie, Dianne M. Bartels, and Arthur L. Caplan, eds. *Prescribing Our Future: Ethical Challenges in Genetic Counseling.* New York: Aldine de Gruyter, 1993.

Long, Clarisa, ed. *Genetic Testing and the Use of Information.* Washington, D.C.: AEI Press, 1999.

MacKinnon, Barbara, ed. *Human Cloning: Science, Ethics, and Public Policy.* Urbana: University of Illinois Press, 2000.

May, Thomas. *Bioethics in a Liberal Society: The Political Framework of Bioethics Decision Making.* Baltimore: Johns Hopkins University Press, 2002.

O'Neill, Onora. *Autonomy and Trust in Bioethics.* New York: Cambridge University Press, 2002.

Rantala, M. L., and Arthur J. Milgram, eds. *Cloning: For and Against.* Chicago: Open Court, 1999.

Real, Leslie A., ed. *Ecological Genetics.* Princeton, N.J.: Princeton University Press, 1994.

Reilly, Philip R. *Abraham Lincoln's DNA and Other Adventures in Genetics.* Cold Spring Harbor, N.Y.: Cold Spring Harbor Laboratory Press, 2000.

Reiss, Michael J., and Roger Straughan, eds. *Improving Nature? The Science and Ethics of Genetic Engineering.* New York: Cambridge University Press, 2001.

Resnik, David B., Holly B. Steinkraus, and Pamela J. Langer. *Human Germline Gene Ther-* *apy: Scientific, Moral, and Political Issues.* Austin, Tex.: R. G. Landes, 1999.

Singer, Peter. *Unsanctifying Human Life: Essays on Ethics.* Edited by Helga Kuhse. Malden, Mass.: Blackwell, 2002.

Singer, Peter, and Deane Wells. *Making Babies: The New Science and Ethics of Conception.* New York: Charles Scribner's Sons, 1985.

U.S. Congress. Senate. Committee on Health, Education, Labor, and Pensions. *Fulfilling the Promise of Genetics Research: Ensuring Nondiscrimination in Health Insurance and Employment.* Washington, D.C.: Government Printing Office, 2001.

_____. *Protecting Against Genetic Discrimination: The Limits of Existing Laws.* Washington, D.C.: U.S. Government Printing Office, 2002.

Veatch, Robert M. *The Basics of Bioethics.* 2d ed. Upper Saddle River, N.J.: Prentice Hall, 2003.

Vogel, Fredrich, and Reinhard Grunwald, eds. *Patenting of Human Genes and Living Organisms.* New York: Springer, 1994.

Wailoo, Keith. *Dying in the City of the Blues: Sickle Cell Anemia and the Politics of Race and Health.* Chapel Hill: University of North Carolina Press, 2001.

Walters, LeRoy, and Julie Gage Palmer. *The Ethics of Human Gene Therapy.* Illustrated by Natalie C. Johnson. New York: Oxford University Press, 1997.

Yount, Lisa, ed. *The Ethics of Genetic Engineering.* San Diego: Greenhaven Press, 2002.

BIOINFORMATICS
See also Genomics and Proteomics

Barnes, Michael R. *Bioinformatics for Geneticists.* Hoboken, N.J.: Wiley, 2003.

Baxevanis, Andreas D., and B. F. Francis Ouellette. *Bioinformatics: A Practical Guide to the Analysis of Genes and Proteins.* 2d ed. Hoboken, N.J.: John Wiley & Sons, 2003.

Bergeron, Bryan P. *Bioinformatics Computing.* Upper Saddle River, N.J.: Prentice Hall, 2002.

Bird, R. Curtis, and Bruce F. Smith, eds. *Genetic Library Construction and Screening: Advanced Techniques and Applications.* New York: Springer, 2002.

Bishop, Martin J., ed. *Guide to Human Genome Computing.* 2d ed. San Diego: Academic Press, 1998.

Campbell, A. Malcolm. *Discovering Genomics, Proteomics, and Bioinformatics.* San Francisco: Benjamin Cummings, 2003.

Claverie, Jean-Michel, and Cedric Notredame. *Bioinformatics for Dummies.* Hoboken, N.J.: Wiley, 2003.

Clote, Peter. *Computational Molecular Biology: An Introduction.* New York: John Wiley, 2000.

Dwyer, Rex A. *Genomic Perl: From Bioinformatics Basics to Working Code.* Cambridge, England: Cambridge University Press, 2003.

Kohane, Isaac S. *Microarrays for an Integrative Genomics.* Cambridge, Mass.: MIT Press, 2003.

Krane, Dan E. *Fundamental Concepts of Bioinformatics.* San Francisco: Benjamin Cummings, 2003.

Krawetz, Stephen A., and David D. Womble. *Introduction to Bioinformatics: A Theoretical and Practical Approach.* Totowa, N.J.: Humana Press, 2003.

Lesk, Arthur M. *Introduction to Bioinformatics.* New York: Oxford University Press, 2002.

Lim, Hwa A. *Genetically Yours: Bioinforming, Biopharming, Biofarming.* River Edge, N.J.: World Scientific, 2002.

Mount, David W. *Bioinformatics: Sequence and Genome Analysis.* Cold Spring Harbor, N.Y.: Cold Spring Harbor Laboratory Press, 2001.

Rashidi, Hooman H. *Bioinformatics Basics: Applications in Biological Science and Medicine.* Boca Raton, Fla.: CRC Press, 2000.

Westhead, David R. *Bioinformatics.* Oxford, England: BIOS, 2002.

CELLULAR BIOLOGY

Attardi, Giuseppe M., and Anne Chomyn, eds. *Methods in Enzymology: Mitochondrial Biogenesis and Genetics.* Vols. 260, 264. San Diego: Academic Press, 1995.

Becker, Peter B. *Chromatin Protocols.* Totowa, N.J.: Humana Press, 1999.

Becker, W. M., L. J. Kleinsmith, and J. Hardin. *The World of the Cell.* 5th ed. San Francisco, Calif.: Benjamin Cummings, 2003.

Bell, John I., et al., eds. *T Cell Receptors.* New York: Oxford University Press, 1995.

Bickmore, Wendy A. *Chromosome Structural Analysis: A Practical Approach.* New York: Oxford University Press, 1999.

Blackburn, Elizabeth H., and Carol W. Greider, eds. *Telomeres.* Cold Spring Harbor, N.Y.: Cold Spring Harbor Laboratory Press, 1995.

Broach, J., J. Pringle, and E. Jones, eds. *The Molecular and Cellular Biology of the Yeast Saccharomyces.* 3 vols. Cold Spring Harbor, N.Y.: Cold Spring Harbor Laboratory Press, 1991-1997.

Carey, M., and S. T. Smale. *Transcriptional Regulation in Eukaryotes: Concepts, Strategies and Techniques.* Cold Spring Harbor, N.Y.: Cold Spring Harbor Laboratory Press, 2000.

Darnell, James, et al. *Molecular Cell Biology.* 4th ed. New York: W. H. Freedman, 2000.

Franklin, T. J., and G. A. Snow. *Biochemistry of Antimicrobial Action.* New York: Chapman and Hall, 2001.

Freshney, R. Ian. *Culture of Animal Cells.* New York: Wiley-Liss, 2000.

Gilchrest, Barbara A., and Vilhelm A. Bohr, eds. *The Role of DNA Damage and Repair in Cell Aging.* New York: Elsevier, 2001.

Gillham, Nicholas W. *Organelle Genes and Genomes.* London: Oxford University Press, 1997.

John, Bernard. *Meiosis.* New York: Cambridge University Press, 1990.

Kipling, David, ed. *The Telomere.* New York: Oxford University Press, 1995.

Lodish, Harvey, David Baltimore, and Arnold Berk. *Molecular Cell Biology.* 4th ed. New York: W. H. Freeman, 2000.

Marshak, Daniel R., Richard L. Gardner, and David Gottlieb, eds. *Stem Cell Biology.* Cold Spring Harbor, N.Y.: Cold Spring Harbor Laboratory Press, 2002.

Murray, A. W., and Tim Hunt. *The Cell Cycle: An Introduction.* New York: W. H. Freeman, 1993.

Pon, Liza, and Eric A. Schon, eds. *Mitochondria.* San Diego: Academic Press, 2001.

Scheffler, Immo E. *Mitochondria.* New York: John Wiley & Sons, 1999.

Sharma, Archana, and Sumitra Sen. *Chromosome Botany.* Enfield, N.H.: Science Publishers, 2002.

Vig, Baldev K., ed. *Chromosome Segregation and Aneuploidy.* New York: Springer-Verlag, 1993.

CLASSICAL TRANSMISSION GENETICS
See also History of Genetics

Berg, Paul, and Maxine Singer. *Dealing with Genes: The Language of Heredity.* Mill Valley, Calif.: University Science Books, 1992.

Cittadino, E. *Nature as the Laboratory.* New York: Columbia University Press, 1990.

Ford, E. B. *Understanding Genetics.* London: Faber & Faber, 1979.

Gardner, Eldon J. *Principles of Genetics.* 5th ed. New York: John Wiley & Sons, 1975.

Goodenough, Ursula. *Genetics.* 2d ed. New York: Holt, Rinehart and Winston, 1978.

Klug, William, and Michael Cummings. *Concepts of Genetics.* 4th ed. New York: Macmillan College, 1994.

Lewin, Benjamin. *Genes VII.* New York: Oxford University Press, 2001.

Mendel, Gregor. *Experiments in Plant-Hybridization.* Foreword by Paul C. Mangelsdorf. Cambridge, Mass.: Harvard University Press, 1965.

Rothwell, Norman V. *Understanding Genetics.* 4th ed. New York: Oxford University Press, 1988.

Russell, Peter. *Genetics.* 2d ed. Boston: Scott, Foresman, 1990.

Singer, Sam. *Human Genetics: An Introduction to the Principles of Heredity.* 2d ed. New York: W. H. Freeman, 1985.

Suzuki, David T., et al., eds. *An Introduction to Genetic Analysis.* 2d ed. San Francisco: W. H. Freeman, 1981.

Weaver, Robert F., and Philip W. Hendrick. 3d ed. *Genetics.* Dubuque, Iowa: Wm. C. Brown, 1997.

Wilson, Edward O. *The Diversity of Life.* New York: W. W. Norton, 1992.

CLONING
See also Genetic Engineering

Brannigan, Michael C., ed. *Ethical Issues in Human Cloning: Cross-Disciplinary Perspectives.* New York: Seven Bridges Press, 2001.

Chen, Bing-Yuan, and Harry W. Janes, eds. *PCR Cloning Protocols.* Rev. 2d ed. Totowa, N.J.: Humana Press, 2002.

DeSalle, Robert, and David Lindley. *The Science of Jurassic Park and the Lost World.* New York: BasicBooks, 1997.

Drlica, Karl. *Understanding DNA and Gene Cloning: A Guide for the Curious.* Rev. ed. New York: Wiley, 2003.

Jones, P., and D. Ramji. *Vectors: Cloning Applications and Essential Techniques.* New York: J. Wiley, 1998.

Klotzko, Arlene Judith, ed. *The Cloning Sourcebook.* New York: Oxford University Press, 2001.

Lauritzen, Paul, ed. *Cloning and the Future of Human Embryo Research.* New York: Oxford University Press, 2001.

Lu, Quinn, and Michael P. Weiner, eds. *Cloning and Expression Vectors for Gene Function Analysis.* Natick, Mass.: Eaton, 2001.

MacKinnon, Barbara, ed. *Human Cloning: Science, Ethics, and Public Policy.* Urbana: University of Illinois Press, 2000.

Prentice, David A. *Stem Cells and Cloning.* New York: Benjamin Cummings, 2003.

Rantala, M. L., and Arthur J. Milgram, eds. *Cloning: For and Against.* Chicago: Open Court, 1999.

Sambrook, Joseph, and David W. Russell, eds. *Molecular Cloning: A Laboratory Manual.* 3d ed. 3 vols. Cold Spring Harbor, N.Y.: Cold Spring Harbor Laboratory Press, 2001.

Shostak, Stanley. *Becoming Immortal: Combining Cloning and Stem-Cell Therapy.* Albany: State University of New York Press, 2002.

Wilmut, Ian, Keith Campbell, and Colin Tudge. *The Second Creation: The Age of Biological Control by the Scientists That Cloned Dolly.* London: Headline, 2000.

DEVELOPMENTAL GENETICS

Beurton, Peter, Raphael Falk, and Hans-Jorg Rheinberger, eds. *The Concept of the Gene in Development and Evolution: Historical and Epistemological Perspectives.* New York: Cambridge University Press, 2000.

Bier, Ethan. *The Coiled Spring: How Life Begins.* Cold Spring Harbor, N.Y.: Cold Spring Harbor Laboratory Press, 2000.

Bowman, John L. *Arabidopsis: An Atlas of Morphology and Development.* New York: Springer-Verlag, 1993.

Carroll, Sean B., Jennifer K. Grenier, and Scott D. Weatherbee. *From DNA to Diversity: Molecular Genetics and the Evolution of Animal*

Design. Malden, Mass.: Blackwell, 2001.

Cronk, Quentin C. B., Richard M. Bateman, and Julie A. Hawkins, eds. *Developmental Genetics and Plant Evolution.* New York: Taylor & Francis, 2002.

Davidson, Eric H. *Gene Activity in Early Development.* Orlando, Fla.: Academic Press, 1986.

DePamphilis, Melvin L., ed. *Gene Expression at the Beginning of Animal Development.* New York: Elsevier, 2002.

DePomerai, David. *From Gene to Animal: An Introduction to the Molecular Biology of Animal Development.* New York: Cambridge University Press, 1985.

Dyban, A. P., and V. S. Baranov. *Cytogenetics of Mammalian Embryonic Development.* Translated by V. S. Baranov. New York: Oxford University Press, 1987.

Gilbert, Scott F. *Developmental Biology.* Sunderland, Mass.: Sinauer Associates, 2003.

Gottlieb, Frederick J. *Developmental Genetics.* New York: Reinhold, 1966.

Gurdon, John B. *The Control of Gene Expression in Animal Development.* Cambridge, Mass.: Harvard University Press, 1974.

Hahn, Martin E., et al., eds. *Developmental Behavior Genetics: Neural, Biometrical, and Evolutionary Approaches.* New York: Oxford University Press, 1990.

Harvey, Richard P., and Nadia Rosenthal, eds. *Heart Development.* San Diego: Academic Press, 1999.

Hennig, W., ed. *Early Embryonic Development of Animals.* New York: Springer-Verlag, 1992.

Hsia, David Yi-Yung. *Human Developmental Genetics.* Chicago: Year Book Medical Publishers, 1968.

Hunter, R. H. F. *Sex Determination, Differentiation, and Intersexuality in Placental Mammals.* New York: Cambridge University Press, 1995.

Leighton, Terrance, and William F. Loomis, Jr. *The Molecular Genetics of Development.* New York: Academic Press, 1980.

McKinney, M. L., and K. J. McNamara. *Heterochrony: The Evolution of Ontogeny.* New York: Plenum Press, 1991.

Malacinski, George M., ed. *Developmental Genetics of Higher Organisms: A Primer in Developmental Biology.* New York: Macmillan, 1988.

Massaro, Edward J., and John M. Rogers, eds. *Folate and Human Development.* Totowa, N.J.: Humana Press, 2002.

Moore, Keith. *The Developing Human: Clinically Oriented Embryology.* 7th ed. Amsterdam: Elsevier Science, 2003.

Nüsslein-Volhard, Christiane, and J. Kratzschmar, eds. *Of Fish, Fly, Worm, and Man: Lessons from Developmental Biology for Human Gene Function and Disease.* New York: Springer, 2000.

Piontelli, Alessandra. *Twins: From Fetus to Child.* New York: Routledge, 2002.

Pritchard, Dorian J. *Foundations of Developmental Genetics.* Philadelphia: Taylor & Francis, 1986.

Raff, Rudolf. *The Shape of Life: Genes, Development, and the Evolution of Animal Form.* Chicago: University of Chicago Press, 1996.

Ranke, M., and G. Gilli. *Growth Standards, Bone Maturation, and Idiopathic Short Stature.* Farmington, Conn.: S. Karger, 1996.

Rao, Mahendra S., ed. *Stem Cells and CNS Development.* Totowa, N.J.: Humana Press, 2001.

Sang, James H. *Genetics and Development.* New York: Longman, 1984.

Saunders, John Warren, Jr. *Patterns and Principles of Animal Development.* New York: Macmillan, 1970.

Seidman, S., and H. Soreq. *Transgenic Xenopus: Microinjection Methods and Developmental Neurobiology.* Totowa, N.J.: Humana Press, 1997.

Stewart, Alistair D., and David M. Hunt. *The Genetic Basis of Development.* New York: John Wiley & Sons, 1982.

Tomanek, Robert J., and Raymond B. Runyan, eds. *Formation of the Heart and Its Regulation.* Foreword by Edward B. Clark. Boston: Birkhauser, 2001.

Ulijaszek, S. J., Francis E. Johnston, and Michael A. Preece. *Cambridge Encyclopedia of Human Growth and Development.* New York: Cambridge University Press, 1998.

Wilkins, Adam S. *Genetic Analysis of Animal Development.* New York: Wiley-Liss, 1993.

DISEASES AND SYNDROMES

General

American Psychiatric Association. *Diagnostic and Statistical Manual of Mental Disorders: DSM-IV-TR*. Rev. 4th ed. Washington, D.C.: Author, 2000.

Bennett, Robin L. *The Practical Guide to the Genetic Family History*. New York: Wiley-Liss, 1999.

Bianchi, Diana W., Timothy M. Crombleholme, and Mary E. D'Alton. *Fetology: Diagnosis and Management of the Fetal Patient*. New York: McGraw-Hill, 2000.

Brooks, G. F., J. S. Butel, and S. A. Morse. *Medical Microbiology*. 21st ed. Stamford, Conn.: Appleton and Lange, 1998.

Browne, M. J., and P. L. Thurlby, eds. *Genomes, Molecular Biology, and Drug Discovery*. San Diego: Academic Press, 1996.

Cotran, R. S., et al. *Robbins Pathologic Basis of Disease*. 6th ed. Philadelphia: Saunders, 1999.

Epstein, Richard J. *Human Molecular Biology; An Introduction to the Molecular Basis of Health and Disease*. Cambridge, England: Cambridge University Press, 2003.

Gallo, Robert C., and Flossie Wong-Staal, eds. *Retrovirus Biology and Human Disease*. New York: Marcel Dekker, 1990.

Gelehrter, Thomas D., Francis S. Collins, and David Ginsburg. *Principles of Medical Genetics*. 2d ed. Baltimore: Williams & Wilkins, 1998.

Gilbert, Patricia. *Dictionary of Syndromes and Inherited Disorders*. 3d ed. Chicago: Fitzroy Dearborn, 2000.

Hogenboom, Marga. *Living with Genetic Syndromes Associated with Intellectual Disability*. Philadelphia: Jessica Kingsley, 2001.

Jorde, L. B., J. C. Carey, M. J. Bamshad, and R. L. White. *Medical Genetics*. 2d ed. St. Louis, Mo.: Mosby, 2000.

Massimini, Kathy, ed. *Genetic Disorders Sourcebook: Basic Consumer Information About Hereditary Diseases and Disorders*. 2d ed. Detroit, Mich.: Omnigraphics, 2001.

Neumann, David, et al. *Human Variability in Response to Chemical Exposures: Measures, Modeling, and Risk Assessment*. Boca Raton, Fla.: CRC Press, 1998.

New, Maria I., ed. *Diagnosis and Treatment of the Unborn Child*. Reddick, Fla.: Idelson-Gnocchi, 1999.

O'Rahilly, S., and D. B. Dunger, eds. *Genetic Insights in Paediatric Endocrinology and Metabolism*. Bristol, England: BioScientifica, 1999.

Pai, G. Shashidhar, Raymond C. Lewandowski, Digamber S. Borgaonkar. *Handbook of Chromosomal Syndromes*. New York: John Wiley & Sons, 2002.

Pasternak, Jack J. *An Introduction to Human Molecular Genetics: Mechanisms of Inherited Diseases*. Bethesda, Md.: Fitzgerald Science Press, 1999.

Petrikovsky, Boris M., ed. *Fetal Disorders: Diagnosis and Management*. New York: Wiley-Liss, 1999.

Pilu, Gianluigi, and Kypros H. Nicolaides. *Diagnosis of Fetal Abnormalities: The 18-23-Week Scan*. New York: Parthenon Group, 1999.

Rakel, Robert E., et al., eds. *Conn's Current Therapy*. Philadelphia: W. B. Saunders, 2003.

Sasaki, Mutsuo, et al., eds. *New Directions for Cellular and Organ Transplantation*. New York: Elsevier Science, 2000.

Stephens, Trent D., and Rock Brynner. *Dark Remedy: The Impact of Thalidomide and Its Revival as a Vital Medicine*. Cambridge, Mass.: Perseus, 2001.

Twining, Peter, Josephine M. McHugo, and David W. Pilling, eds. *Textbook of Fetal Abnormalities*. New York: Churchill Livingstone, 2000.

Weaver, David D., with the assistance of Ira K. Brandt. *Catalog of Prenatally Diagnosed Conditions*. 3d ed. Baltimore: Johns Hopkins University Press, 1999.

Weiss, Kenneth M. *Genetic Variation and Human Disease: Principles and Evolutionary Approaches*. New York: Cambridge University Press, 1993.

Wynbrandt, James, and Mark D. Ludman. *The Encyclopedia of Genetic Disorders and Birth Defects*. 2d ed. New York: Facts On File, 2000.

Zallen, Doris Teichler. *Does It Run in the Family? A Consumer's Guide to DNA Testing for Genetic Disorders*. New Brunswick, N.J.: Rutgers University Press, 1997.

Alzheimer's Disease

Gauthier, S., ed. *Clinical Diagnosis and Management of Alzheimer's Disease*. 2d ed. London: Martin Dunitz, 2001.

Hamdy, Ronald, James Turnball, and Joellyn Edwards. *Alzheimer's Disease: A Handbook for Caregivers.* New York: Mosby, 1998.

Mace, M., and P. Rabins. *The Thirty-six Hour Day: A Family Guide to Caring for Persons with Alzheimer Disease, Related Dementing Illnesses, and Memory Loss in Later Life.* Baltimore: Johns Hopkins University Press, 1999.

Nelson, James Lindemann, and Hilde Lindemann Nelson. *Alzheimer's: Answers to Hard Questions for Families.* New York: Main Street Books, 1996.

Powell, L., and K. Courtice. *Alzheimer's Disease: A Guide for Families and Caregivers.* Cambridge, Mass.: Perseus, 2001.

Terry, R., R. Katzman, K. Bick, and S. Sisodia. *Alzheimer Disease.* 2d ed. Philadelphia: Lippincott Williams & Wilkins, 1999.

Cancer

Angier, Natalie. *Natural Obsessions: Striving to Unlock the Deepest Secrets of the Cancer Cell.* Boston: Mariner Books/Houghton Mifflin, 1999.

Bowcock, Anne M., ed. *Breast Cancer: Molecular Genetics, Pathogenesis, and Therapeutics.* Totowa, N.J.: Humana Press, 1999.

Coleman, William B., and Gregory J. Tsongalis, eds. *The Molecular Basis of Human Cancer.* Totowa, N.J.: Humana Press, 2002.

Cooper, Geoffrey M. *Oncogenes.* 2d ed. Boston: Jones and Bartlett, 1995.

Cowell, J. K., ed. *Molecular Genetics of Cancer.* 2d ed. San Diego: Academic Press, 2001.

Davies, Kevin, and Michael White. *Breakthrough: The Race to Find the Breast Cancer Gene.* New York: John Wiley, 1996.

Dickson, Robert B., and Marc E. Lipman, eds. *Genes, Oncogenes, and Hormones: Advances in Cellular and Molecular Biology of Breast Cancer.* Boston: Kluwer Academic, 1992.

Ehrlich, Melanie, ed. *DNA Alterations in Cancer: Genetic and Epigenetic Changes.* Natick, Mass.: Eaton, 2000.

Fisher, David E., ed. *Tumor Suppressor Genes in Human Cancer.* Totowa, N.J.: Humana Press, 2001.

Greaves, Mel F. *Cancer: The Evolutionary Legacy.* New York: Oxford University Press, 2000.

Habib, Nagy A., ed. *Cancer Gene Therapy: Past Achievements and Future Challenges.* New York: Kluwer Academic/Plenum, 2000.

Hanski, C., H. Scherübl, and B. Mann, eds. *Colorectal Cancer: New Aspects of Molecular Biology and Immunology and Their Clinical Applications.* New York: New York Academy of Sciences, 2000.

Heim, S., and Felix Mitelman. *Cancer Cytogenetics.* 2d ed. New York: J. Wiley, 1995.

Hodgson, Shirley V., and Eamonn R. Maher. *A Practical Guide to Human Cancer Genetics.* 2d ed. New York: Cambridge University Press, 1999.

Kemeny, Mary Margaret, and Paula Dranov. *Beating the Odds Against Breast and Ovarian Cancer: Reducing Your Hereditary Risk.* Reading, Mass.: Addison-Wesley, 1992.

Krupp, Guido, and Reza Parwaresch, eds. *Telomerases, Telomeres, and Cancer.* New York: Kluwer Academic/Plenum, 2003.

La Thangue, Nicholas B., and Lasantha R. Bandara, eds. *Targets for Cancer Chemotherapy: Transcription Factors and Other Nuclear Proteins.* Totowa, N.J.: Humana Press, 2002.

Lattime, Edmund C., and Stanton L. Gerson, eds. *Gene Therapy of Cancer: Translational Approaches from Preclinical Studies to Clinical Implementation.* 2d ed. San Diego: Academic Press, 2002.

Maruta, Hiroshi, ed. *Tumor-Suppressing Viruses, Genes, and Drugs: Innovative Cancer Therapy Approaches.* San Diego: Academic, 2002.

Mendelsohn, John, et al. *The Molecular Basis of Cancer.* 2d ed. Philadelphia: Saunders, 2001.

Mulvihill, John J. *Catalog of Human Cancer Genes: McKusick's Mendelian Inheritance in Man for Clinical and Research Oncologists.* Foreword by Victor A. McKusick. Baltimore: Johns Hopkins University Press, 1999.

National Cancer Institute. *Genetic Testing for Breast Cancer: It's Your Choice.* Bethesda, Md.: Author, 1997.

Ruddon, Raymond. *Cancer Biology.* 3d ed. New York: Oxford University Press, 1995.

Schneider, Katherine A. *Counseling About Cancer: Strategies for Genetic Counseling.* 2d ed. New York: Wiley-Liss, 2002.

Varmus, Harold, and Robert Weinberg. *Genes and the Biology of Cancer.* New York: W. H. Freeman, 1993.

Vogelstein, Bert, and Kenneth W. Kinzler, eds. *The Genetic Basis of Human Cancer.* 2d ed. New York: McGraw-Hill, 2002.

Wilson, Samuel, et al. *Cancer and the Environment: Gene-Environment Interaction.* Washington, D.C.: National Academy Press, 2002.

Cholera

Keusch, Gerald, and Masanobu Kawakami, eds. *Cytokines, Cholera, and the Gut.* Amsterdam: IOS Press, 1997.

Wachsmuth, Kate, et al., eds. *Vibrio Cholerae and Cholera: Molecular to Global Perspectives.* Washington, D.C.: ASM Press, 1994.

Color Blindness

Rosenthal, Odeda, and Robert H. Phillips. *Coping with Color Blindness.* Garden City Park, N.Y.: Avery, 1997.

Congenital Adrenal Hyperplasia

Speiser, Phyllis W., ed. *Congenital Adrenal Hyperplasia.* Philadelphia: W. B. Saunders, 2001.

Cystic Fibrosis

Hodson, Margaret E., and Duncan M. Geddes, eds. *Cystic Fibrosis.* 2d ed. New York: Oxford University Press, 2000.

Orenstein, David M., Beryl J. Rosenstein, and Robert C. Stern. *Cystic Fibrosis: Medical Care.* Philadelphia: Lippincott Williams & Wilkins, 2000.

Shale, Dennis. *Cystic Fibrosis.* London: British Medical Association, 2002.

Tsui, Lap-Chee, et al., eds. *The Identification of the CF (Cystic Fibrosis) Gene: Recent Progress and New Research Strategies.* New York: Plenum Press, 1991.

Yankaskas, James R., and Michael R. Knowles, eds. *Cystic Fibrosis in Adults.* Philadelphia: Lippincott-Raven, 1999.

Diabetes

American Diabetes Association. *American Diabetes Association Complete Guide to Diabetes: The Ultimate Home Reference from the Diabetes Experts.* New York: McGraw-Hill, 2002.

Becker, Gretchen. *The First Year: Type 2 Diabetes, An Essential Guide for the Newly Diagnosed.* New York: Marlowe, 2001.

Flyvbjerg, Allan, Hans Orskov, and George Alberti, eds. *Growth Hormone and Insulin-like Growth Factor I in Human and Experimental Diabetes.* New York: John Wiley & Sons, 1993.

Kahn, C. R., Gordon C. Weir, George L. King, Alan C. Moses, Robert J. Smith, and Alan M. Jacobson. *Joslin's Diabetes Mellitus.* 14th ed. Philadelphia: Lippincott Williams and Wilkins, 2003.

Lowe, William L., Jr., ed. *Genetics of Diabetes Mellitus.* Boston: Kluwer Academic, 2001.

Milchovich, Sue K., and Barbara Dunn-Long. *Diabetes Mellitus: A Practical Handbook.* 8th ed. Boulder, Colo.: Bull, 2003.

Down Syndrome and Mental Retardation

Broman, Sarah H., and Jordan Grafman, eds. *Atypical Cognitive Deficits in Developmental Disorders: Implications for Brain Function.* Hillsdale, N.J.: Lawrence Erlbaum, 1994.

Cohen, William I., Lynn Nadel, and Myra E. Madnick, eds. *Down Syndrome: Visions for the Twenty-first Century.* New York: Wiley-Liss, 2002.

Cunningham, Cliff. *Understanding Down Syndrome: An Introduction for Parents.* 1988. Reprint. Cambridge, Mass.: Brookline Books, 1999.

Dykens, Elisabeth M., Robert M. Hodapp, and Brenda M. Finucane. *Genetics and Mental Retardation Syndromes: A New Look at Behavior and Interventions.* Baltimore: Paul H. Brookes, 2000.

Faraone, Stephen V., Ming T. Tsuang, and Debby W. Tsuang. *Genetics of Mental Disorders: A Guide for Students, Clinicians, and Researchers.* New York: Guilford Press, 1999.

Lubec, G. *Protein Expression in Down Syndrome Brain.* New York: Springer, 2001.

Newton, Richard. *The Down's Syndrome Handbook: A Practical Guide for Parents and Caregivers.* New York: Arrow Books, 1997.

Selikowitz, Mark. *Down Syndrome: The Facts.* 2d ed. New York: Oxford University Press, 1997.

Shannon, Joyce Brennfleck, ed. *Mental Retardation Sourcebook: Basic Consumer Health Information About Mental Retardation and Its Causes, Including Down Syndrome, Fetal Alcohol Syndrome, Fragile X Syndrome, Genetic Conditions,*

Injury, and Environmental Sources. Detroit, Mich.: Omnigraphics, 2000.

Fragile X Syndrome

Hagerman, Paul J. *Fragile X Syndrome: Diagnosis, Treatment, and Research*. 3d ed. Baltimore: Johns Hopkins University Press, 2002.

Parker, James N., and Philip M. Parker, eds. *The 2002 Official Parent's Sourcebook on Fragile X Syndrome*. San Diego: ICON Health, 2002.

Gender Identity and Sex Errors

Berch, Daniel B., and Bruce G. Bender, eds. *Sex Chromosome Abnormalities and Human Behavior*. Boulder, Colo.: Westview Press, 1990.

Dreger, Alice Domurat. *Hermaphrodites and the Medical Invention of Sex*. Cambridge, Mass.: Harvard University Press, 1998.

Gilbert, Ruth. *Early Modern Hermaphrodites: Sex and Other Stories*. New York: Palgrave, 2002.

Money, John. *Sex Errors of the Body and Related Syndromes: A Guide to Counseling Children, Adolescents, and Their Families*. 2d ed. Baltimore: P. H. Brookes, 1994.

Heart Diseases and Defects

Braunwald, Eugene, Douglas P. Zipes, Peter Libby, and Douglas D. Zipes. *Heart Disease: A Textbook of Cardiovascular Medicine*. 6th ed. Philadelphia: W. B. Saunders, 2001.

Edwards, Jesse E. *Jesse E. Edwards' Synopsis of Congenital Heart Disease*. Edited by Brooks S. Edwards. Armonk, N.Y.: Futura, 2000.

Goldbourt, Uri, Kare Berg, and Ulf de Faire, eds. *Genetic Factors in Coronary Heart Disease*. Boston: Kluwer Academic, 1994.

Kramer, Gerri Freid, and Shari Maurer. *The Parent's Guide to Children's Congenital Heart Defects: What They Are, How to Treat Them, How to Cope with Them*. Foreword by Sylvester Stallone and Jennifer Flavin-Stallone. New York: Three Rivers Press, 2001.

Marian, Ali J. *Genetics for Cardiologists: The Molecular Genetic Basis of Cardiovascular Disorders*. London: ReMEDICA, 2000.

Hemophilia

Buzzard, Brenda, and Karen Beeton, eds. *Physiotherapy Management of Haemophilia*. Malden, Mass.: Blackwell, 2000.

Jones, Peter. *Living with Haemophilia*. 5th ed. New York: Oxford University Press, 2002.

Monroe, Dougald M., et al., eds. *Hemophilia Care in the New Millennium*. New York: Kluwer Academic/Plenum, 2001.

Resnik, Susan. *Blood Saga: Hemophilia, AIDS, and the Survival of a Community*. Berkeley: University of California Press, 1999.

Steinberg, Martin H., et al., eds. *Disorders of Hemoglobin: Genetics, Pathophysiology, and Clinical Management*. Foreword by H. Franklin Bunn. New York: Cambridge University Press, 2001.

Immune Disorders and Allergies

Abbas, Abul K., and Richard A. Flavell, eds. *Genetic Models of Immune and Inflammatory Diseases*. New York: Springer, 1996.

Bona, Constantin A., et al., eds. *The Molecular Pathology of Autoimmune Diseases*. 2d ed. New York: Taylor and Francis, 2002.

Clark, William R. *At War Within: The Double-Edged Sword of Immunity*. New York: Oxford University Press, 1995.

Cutler, Ellen W. *Winning the War Against Asthma and Allergies*. Albany, N.Y.: Delmar, 1998.

Hadley, Andrew G., and Peter Soothill, eds. *Alloimmune Disorders of Pregnancy: Anaemia, Thrombocytopenia, and Neutropenia in the Fetus and Newborn*. New York: Cambridge University Press, 2002.

Walsh, William. *The Food Allergy Book*. New York: J. Wiley, 2000.

Infectious and Emerging Diseases

Anderson, R. M., and R. M. May. *Infectious Diseases of Humans: Dynamics and Control*. Oxford, England: Oxford University Press, 1992.

DeSalle, Rob, ed. *Epidemic! The World of Infectious Disease*. New York: The New Press, 1999.

Drexler, Madeline. *Secret Agents: The Menace of Emerging Infections*. Washington, D.C.: Joseph Henry Press, 2002.

Garrett, Laurie. *The Coming Plague: Newly Emerging Diseases in a World Out of Balance*. New York: Penguin, 1995.

Hacker, J., and J. B. Kaper, eds. *Pathogenicity Islands and the Evolution of Pathogenic Microbes*. 2 vols. New York: Springer, 2002.

Kolata, Gina. *Flu: The Story of the Great Influenza Pandemic of 1918 and the Search for the Virus That Caused It.* New York: Simon and Schuster, 2001.

Lappe, Marc. *Breakout: The Evolving Threat of Drug-Resistant Disease.* San Francisco: Sierra Club Books, 1996.

Levy, Stuart B. *The Antibiotic Paradox: How Miracle Drugs Are Destroying the Miracle.* New York: Plenum Press, 1992.

McNeill, William H. *Plagues and Peoples.* New York: Anchor Books, 1998.

Parker, James N., and Philip M. Parker, eds. *The Official Patient's Sourcebook on E. Coli.* San Diego: ICON Health, 2002.

World Health Organization. *Future Research on Smallpox Virus Recommended.* Geneva, Switzerland: World Health Organization Press, 1999.

Klinefelter Syndrome

Bock, Robert. *Understanding Klinefelter Syndrome: A Guide for XXY Males and Their Families.* Bethesda, Md.: Department of Health and Human Services, Public Health Service, National Institutes of Health, National Institute of Child Health and Human Development, 1997.

Parker, James N., and Philip M. Parker, eds. *The Official Parent's Sourcebook on Klinefelter Syndrome: A Revised and Updated Directory for the Internet Age.* San Diego: ICON Health, 2002.

Probasco, Terri, and Gretchen A. Gibbs. *Klinefelter Syndrome: Personal and Professional Guide.* Richmond, Ind.: Prinit Press, 1999.

Lactose Intolerance

Auricchio, Salvatore, and G. Semenza, eds. *Common Food Intolerances 2: Milk in Human Nutrition and Adult-Type Hypolactasia.* New York: Karger, 1993.

Metabolic Disorders

Econs, Michael J., ed. *The Genetics of Osteoporosis and Metabolic Bone Disease.* Totowa, N.J.: Humana Press, 2000.

Evans, Mark I., ed. *Metabolic and Genetic Screening.* Philadelphia: W. B. Saunders, 2001.

Scriver, Charles, et al., eds. *The Metabolic and Molecular Bases of Inherited Disease.* 8th ed. 4 vols. New York: McGraw-Hill, 2001.

Mitochondrial Diseases

Lestienne, Patrick, ed. *Mitochondrial Diseases: Models and Methods.* New York: Springer, 1999.

Neural Defects

Bock, Gregory, and Joan Marsh, eds. *Neural Tube Defects.* New York: Wiley, 1994.

Goldstein, Sam, and Cecil R. Reynolds, eds. *Handbook of Neurodevelopmental and Genetic Disorders in Children.* New York: Guilford Press, 1999.

Phenylketonuria

Koch, Jean Holt. *Robert Guthrie, the PKU Story: A Crusade Against Mental Retardation.* Pasadena, Calif.: Hope, 1997.

Parker, James N., ed. *The Official Parent's Sourcebook on Phenylketonuria.* San Diego, Calif.: ICON Health Publications, 2002.

Prion Diseases

Baker, Harry F., ed. *Molecular Pathology of the Prions.* Totowa, N.J.: Humana Press, 2001.

Groschup, Martin H., and Hans A. Kretzschmar, eds. *Prion Diseases: Diagnosis and Pathogenesis.* New York: Springer, 2000.

Klitzman, Robert. *The Trembling Mountain: A Personal Account of Kuru, Cannibals, and Mad Cow Disease.* New York: Plenum Trade, 1998.

_____, ed. *Prion Biology and Diseases.* Cold Spring Harbor, N.Y.: Cold Spring Harbor Laboratory Press, 1999.

Rabenau, Holger F., Jindrich Cinatl, and Hans Wilhelm Doerr, eds. *Prions: A Challenge for Science, Medicine, and Public Health System.* New York: Karger, 2001.

Ratzan, Scott C., ed. *The Mad Cow Crisis: Health and the Public Good.* New York: New York University Press, 1998.

Prader-Willi Syndrome

Eiholzer, Urs. *Prader-Willi Syndrome: Effects of Human Growth Hormone Treatment.* New York: Karger, 2001.

Sickle Cell Disease

Anionwu, Elizabeth N., and Karl Atkin. *The Politics of Sickle Cell and Thalassaemia.* Philadelphia: Open University Press, 2001.

Serjeant, Graham R., and Beryl E. Serjeant. *Sickle Cell Disease*. 3d ed. New York: Oxford University Press, 2001.

Tapper, Melbourne. *In the Blood: Sickle Cell Anemia and the Politics of Race*. Philadelphia: University of Pennsylvania Press, 1999.

Tay-Sachs Disease

Desnick, Robert J., and Michael M. Kaback, eds. *Tay-Sachs Disease*. San Diego: Academic, 2001.

National Tay-Sachs and Allied Diseases Association. *A Genetics Primer for Understanding Tay-Sachs and the Allied Diseases*. Brookline, Mass.: Author, 1995.

Parker, James N., and Philip M. Parker, eds. *The Official Parent's Sourcebook on Tay-Sachs Disease: A Revised and Updated Directory for the Internet Age*. San Diego: Icon Press, 2002.

Turner Syndrome

Albertsson-Wikland, Kerstin, and Michael B. Ranke, eds. *Turner Syndrome in a Life Span Perspective: Research and Clinical Aspects*. New York: Elsevier, 1995.

Rieser, Patricia A., and Marsha Davenport. *Turner Syndrome: A Guide for Families*. Houston, Tex.: Turner Syndrome Society, 1992.

Rosenfeld, Ron G. *Turner Syndrome: A Guide for Physicians*. 2d ed. Houston, Tex.: Turner Syndrome Society, 1992.

EVOLUTION

Avital, Eytan, and Eva Jablonka. *Animal Traditions: Behavioural Inheritance in Evolution*. New York: Cambridge University Press, 2000.

Beurton, Peter, Raphael Falk, and Hans-Jorg Rheinberger, eds. *The Concept of the Gene in Development and Evolution: Historical and Epistemological Perspectives*. New York: Cambridge University Press, 2000.

Calos, Michele. *Molecular Evolution of Chromosomes*. New York: Oxford University Press, 2003.

Darwin, Charles. *The Descent of Man and Selection in Relation to Sex*. London: John Murray, 1871. Reprint. Princeton, N.J.: Princeton University Press, 1981.

_____. *On the Origin of Species by Means of Natural Selection: Or, The Preservation of Favored Races in the Struggle for Life*. London: John Murray, 1859. Reprint. New York: Random House, 1999.

Dawkins, Richard. *The Blind Watchmaker: Why the Evidence of Evolution Reveals a Universe Without Design*. New York: W. W. Norton, 1996.

_____. *Climbing Mount Improbable*. New York: W. W. Norton, 1997.

_____. *Extended Phenotype: The Long Reach of the Gene*. Rev. 2d ed. Afterword by Daniel Dennett. New York: Oxford University Press, 1999.

_____. *The Selfish Gene*. 2d ed. New York: Oxford University Press, 1990.

Depew, David, and Bruce Weber. *Darwinism Evolving: Systems Dynamics and the Genealogy of Natural Selection*. Boston: MIT Press, 1995.

Dobzhansky, Theodosius G. *Genetics and the Origin of Species*. New York: Columbia University Press, 1937.

Edey, Maitland A., and Donald C. Johnson. *Blueprints: Solving the Mystery of Evolution*. Reprint. New York: Viking, 1990.

Fisher, Ronald Aylmer. *The Genetical Theory of Natural Selection: A Complete Variorum Edition*. 1958. Edited with a foreword and notes by J. H. Bennett. New York: Oxford University Press, 1999.

Freeman, Scott, and Jon C. Herron. *Evolutionary Analysis*. 2d ed. Upper Saddle River, N.J.: Prentice Hall, 2000.

Gesteland, Raymond F., Thomas R. Cech, and John F. Atkins, eds. *The RNA World: The Nature of Modern RNA Suggests a Prebiotic RNA*. 2d ed. Cold Spring Harbor, N.Y.: Cold Spring Harbor Laboratory Press, 1999.

Gould, Stephen Jay. *Eight Little Piggies*. New York: W. W. Norton, 1994.

_____. *The Panda's Thumb*. New York: W. W. Norton, 1980.

_____. *The Structure of Evolutionary Theory*. Cambridge, Mass.: Harvard University Press, 2002.

Graur, Dan, and Wen-Hsiung Li. *Fundamentals of Molecular Evolution*. 2d ed. Sunderland, Mass.: Sinauer Associates, 1999.

Herrmann, Bernd, and Susanne Hummel, eds. *Ancient DNA: Recovery and Analysis of Genetic Material from Paleographic, Archaeological, Mu-*

seum, Medical, and Forensic Speciments. New York: Springer-Verlag, 1994.

Jones, Martin. *The Molecule Hunt: Archaeology and the Search for Ancient DNA.* New York: Arcade, 2002.

Keller, Laurent, ed. *Levels of Selection in Evolution.* Princeton, N.J.: Princeton University Press, 1999.

Kimura, Motoo, and Naoyuki Takahata, eds. *New Aspects of the Genetics of Molecular Evolution.* New York: Springer-Verlag, 1991.

Landweber, Laura F., and Andrew P. Dobson, eds. *Genetics and the Extinction of Species: DNA and the Conservation of Biodiversity.* Princeton, N.J.: Princeton University Press, 1999.

Langridge, John. *Molecular Genetics and Comparative Evolution.* New York: John Wiley & Sons, 1991.

Levy, Charles K. *Evolutionary Wars, a Three-Billion-Year Arms Race: The Battle of Species on Land, at Sea, and in the Air.* Illustrations by Trudy Nicholson. New York: W. H. Freeman, 1999.

Lewontin, R. *The Genetic Basis of Evolutionary Change.* New York: Columbia University Press, 1974.

Li, Wen-Hsiung. *Molecular Evolution.* Sunderland, Mass.: Sinauer Associates, 1997.

Lloyd, E. *The Structure of Evolutionary Theory.* Westport, Conn.: Greenwood Press, 1987.

Lynch, John M., ed. *Darwin's Theory of Natural Selection: British Responses, 1859-1871.* 4 vols. Bristol, England: Thoemmes Press, 2002.

Magurran, Anne E., and Robert M. May, eds. *Evolution of Biological Diversity: From Population Differentiation to Speciation.* New York: Oxford University Press, 1999.

Mayr, Ernst. *One Long Argument: Charles Darwin and the Genesis of Modern Evolutionary Thought.* Cambridge, Mass.: Harvard University Press, 1991.

Michod, Richard E. *Darwinian Dynamics: Evolutionary Transitions in Fitness and Individuality.* Princeton, N.J.: Princeton University Press, 1999.

Miller, Stanley L. *From the Primitive Atmosphere to the Prebiotic Soup to the Pre-RNA World.* NASA CR-2076334007116722. Washington, D.C.: National Aeronautics and Space Administration, 1996.

Nei, Masatoshi, and Sudhir Kumar. *Molecular Evolution and Phylogenetics.* New York: Oxford University Press, 2000.

Persell, Stuart Michael. *Neo-Lamarckism and the Evolution Controversy in France, 1870-1920.* Lewiston, N.Y.: Edwin Mellen Press, 1999.

Prothero, Donald R. *Bringing Fossils to Life: An Introduction to Paleobiology.* New York: McGraw-Hill, 2003.

Quammen, David. *Song of the Dodo.* New York: Simon & Schuster, 1997.

Raff, Rudolf. *The Shape of Life: Genes, Development, and the Evolution of Animal Form.* Chicago: University of Chicago Press, 1996.

Ryan, Frank. *Darwin's Blind Spot: Evolution Beyond Natural Selection.* Boston: Houghton Mifflin, 2002.

Selander, Robert K., et al., eds. *Evolution at the Molecular Level.* Sunderland, Mass.: Sinauer Associates, 1991.

Singh, Rama S., and Costas B. Krimbas, eds. *Evolutionary Genetics: From Molecules to Morphology.* New York: Cambridge University Press, 2000.

Somit, Albert, and Steven A. Peterson, eds. *The Dynamics of Evolution: The Punctuated Equilibrium Debate in the Natural and Social Sciences.* Ithaca, N.Y.: Cornell University Press, 1992.

Steele, Edward J., Robyn A. Lindley, and Robert V. Blanden. *Lamarck's Signature: How Retrogenes Are Changing Darwin's Natural Selection Paradigm.* Reading, Mass.: Perseus Books, 1998.

Williams, George C. *Adaptation and Natural Selection: A Critique of Some Current Evolutionary Thought.* 1966. Reprint. Princeton, N.J.: Princeton University Press, 1996.

Wolf, Jason B., Edmund D. Brodie III, and Michael J. Wade. *Epistasis and the Evolutionary Process.* New York: Oxford University Press, 2000.

FORENSIC GENETICS

Budowle, Bruce, et al. *DNA Typing Protocols: Molecular Biology and Forensic Analysis.* Natick, Mass.: Eaton, 2000.

Burke, Terry, R. Wolf, G. Dolf, and A. Jeffreys, eds. *DNA Fingerprinting: Approaches and Applications.* Boston: Birkhauser, 2001.

Coleman, Howard, and Eric Swenson. *DNA in the Courtroom: A Trial Watcher's Guide.* Seattle, Wash.: GeneLex Press, 1994.

Connors, Edward, et al. *Convicted by Juries, Exonerated by Science: Case Studies in the Use of DNA Evidence to Establish Innocence After Trial.* Washington, D.C.: U.S. Department of Justice, Office of Justice Programs, National Institute of Justice, 1996.

Fridell, Ron. *DNA Fingerprinting: The Ultimate Identity.* New York: Scholastic, 2001.

Goodman, Christi. *Paternity, Marriage, and DNA.* Denver, Colo.: National Conference of State Legislatures, 2001.

Hummel, Susanne. *Fingerprinting the Past: Research on Highly Degraded DNA and Its Applications.* New York: Springer-Verlag, 2002.

Jarman, Keith, and Norah Rudin. *An Introduction to Forensic DNA Analysis.* 2d ed. Boca Raton, Fla.: CRC Press, 2001.

Kirby, L. T. *DNA Fingerprinting: An Introduction.* New York: Stockton Press, 1990.

Lincoln, Patrick J., and Jim Thompson, eds. *Forensic DNA Profiling Protocols.* Vol. 98. Totowa, N.J.: Humana Press, 1998.

National Research Council. *DNA Technology in Forensic Science.* National Academy of Sciences Press, 1992.

Rudin, Norah, and Keith Inman. *An Introduction to Forensic DNA Analysis.* Boca Raton, Fla.: CRC Press, 2002.

Sonenstein, Freya L., Pamela A. Holcomb, and Kristin S. Seefeldt. *Promising Approaches to Improving Paternity Establishment Rates at the Local Level.* Washington, D.C.: Urban Institute, 1993.

United States National Research Council. *The Evaluation of Forensic DNA Evidence.* Rev. ed. Washington, D.C.: National Academy Press, 1996.

GENETIC ENGINEERING AND BIOTECHNOLOGY
See also Agriculture; Cloning

Aldridge, Susan. *The Thread of Life: The Story of Genes and Genetic Engineering.* New York: Cambridge University Press, 1996.

Bassett, Pamela. *Emerging Markets in Tissue Engineering: Angiogenesis, Soft and Hard Tissue Regeneration, Xenotransplant, Wound Healing, Biomaterials, and Cell Therapy.* Southborough, Mass.: D & MD Reports, 1999.

Boylan, Michael, and Kevin E. Brown. *Genetic Engineering: Science and Ethics on the New Frontier.* Upper Saddle River, N.J.: Prentice Hall, 2001.

Chrispeels, Maarten J., and David E. Sadava. *Plants, Genes, and Crop Biotechnology.* Boston: Jones and Bartlett, 2003.

Crocomo, O. J., ed. *Biotechnology of Plants and Microorganisms.* Columbus: Ohio State University Press, 1986.

Dale, Jeremy. *Molecular Genetics of Bacteria.* New York: John Wiley & Sons, 1994.

Davis, Bernard D., ed. *The Genetic Revolution: Scientific Prospects and Public Perceptions.* Baltimore: The Johns Hopkins University Press, 1991.

Evans, Gareth M. *Environmental Biotechnology: Theory and Application.* Hoboken, N.J.: Wiley, 2003.

Fincham, J. R. S. *Genetically Engineered Organisms: Benefits and Risks.* Toronto: University of Toronto Press, 1991.

Gaillardin, Claude, and Henri Heslot. *Molecular Biology and Genetic Engineering of Yeasts.* Boca Raton, Fla.: CRC Press, 1992.

Glick, Bernard R. *Molecular Biotechnology: Principles and Applications of Recombinant DNA.* Washington, D.C.: ASM Press, 2003.

Grange, J. M., et al., eds. *Genetic Manipulation: Techniques and Applications.* Boston: Blackwell Scientific, 1991.

Hill, Walter E. *Genetic Engineering: A Primer.* Newark, N.J.: Harwood Academic, 2000.

Jacobson, G. K., and S. O. Jolly. *Gene Technology.* New York: VCH Verlagsgesellschaft, 1989.

Joyner, Alexandra L., ed. *Gene Targeting: A Practical Approach.* New York: Oxford University Press, 1993.

Kiessling, Ann, and Scott C. Anderson. *Human Embryonic Stem Cells: An Introduction to the Science and Therapeutic Potential.* Boston: Jones and Bartlett, 2003.

Kontermann, Roland, and Stefan Dübel, eds. *Antibody Engineering.* New York: Springer, 2001.

Kreuzer, Helen, and Adrianne Massey. *Recombinant DNA and Biotechnology: A Guide for Teachers.* Washington, D.C.: ASM Press, 2001.

Krimsky, Sheldon. *Biotechnics and Society: The Rise of Industrial Genetics.* New York: Praeger, 1991.

Lappe, Marc. *Broken Code: The Exploitation of DNA.* San Francisco: Sierra Club Books, 1984.

Le Vine, Harry, III. *Genetic Engineering: A Reference Handbook.* Santa Barbara, Calif.: ABC-CLIO, 1999.

McGee, Glenn. *The Perfect Baby: A Pragmatic Approach to Genetics.* Lanham, Md.: Rowman & Littlefield, 1997.

McKelvey, Maureen D. *Evolutionary Innovations: The Business of Biotechnology.* New York: Oxford University Press, 1996.

Mak, Tak W., et al., eds. *The Gene Knockout Factsbook.* 2 vols. San Diego: Academic Press, 1998.

Mayforth, Ruth D. *Designing Antibodies.* San Diego: Academic Press, 1993.

Nicholl, Desmond S. T. *An Introduction to Genetic Engineering.* 2d ed. New York: Cambridge University Press, 2002.

Nossal, Gustav J. V., and Ross L. Coppel. *Reshaping Life: Key Issues in Genetic Engineering.* New York: Cambridge University Press, 1985.

Nottingham, Stephen. *Genescapes: The Ecology of Genetic Engineering.* New York: Zed Books, 2002.

Old, R. W., and S. B. Primrose. *Principles of Gene Manipulation: An Introduction to Genetic Engineering.* Boston: Blackwell Scientific, 1994.

Olson, Steve. *Biotechnology: An Industry Comes of Age.* Washington, D.C.: National Academy Press, 1986.

Oxender, Dale L., and C. Fred Fox, eds. *Protein Engineering.* New York: Liss, 1987.

Prokop, Ales, and Rakesh K. Bajpai. *Recombinant DNA Technology I.* New York: New York Academy of Sciences, 1991.

Russo, V. E. A., and David Cove. *Genetic Engineering: Dreams and Nightmares.* New York: W. H. Freeman, 1995.

Shannon, Thomas A., ed. *Genetic Engineering: A Documentary History.* Westport, Conn.: Greenwood Press, 1999.

Singer, Maxine, and Paul Berg, eds. *Exploring Genetic Mechanisms.* Sausalito, Calif.: University Science Books, 1997.

Sofer, William. *Introduction to Genetic Engineering.* Boston: Butterworth-Heinemann, 1991.

Spallone, Patricia. *Generation Games: Genetic Engineering and the Future for Our Lives.* Philadelphia: Temple University Press, 1992.

Steinberg, Mark, and Sharon D. Cosloy, eds. *The Facts On File Dictionary of Biotechnology and Genetic Engineering.* New ed. New York: Checkmark Books, 2001.

Thomas, John, et al., eds. *Biotechnology and Safety Assessment.* 2d ed. Philadelphia: Taylor & Francis, 1999.

Vega, Manuel A., ed. *Gene Targeting.* Boca Raton, Fla.: CRC Press, 1995.

Wade, Nicholas. *The Ultimate Experiment: Man-Made Evolution.* New York: Walker, 1977.

Walker, Mark, and David McKay. *Unravelling Genes: A Layperson's Guide to Genetic Engineering.* St. Leonards, N.S.W.: Allen & Unwin, 2000.

Walker, Matthew R., with Ralph Rapley. *Route Maps in Gene Technology.* Oxford, England: Blackwell Scientific, 1997.

Wang, Henry Y., and Tadayuki Imanaka, eds. *Antibody Expression and Engineering.* Washington, D.C.: American Chemical Society, 1995.

Warr, J. Roger. *Genetic Engineering in Higher Organisms.* Baltimore: E. Arnold, 1984.

Watson, James D., et al. *Recombinant DNA.* 2d ed. New York: W. H. Freeman, 1992.

Williams, J. G., A. Ceccarelli, and A. Wallace. *Genetic Engineering.* 2d ed. New York: Springer, 2001.

Wu-Pong, S., and Y. Rojanasakul. *Biopharmaceutical Drug Design and Development.* Totowa, N.J.: Humana Press, 1999.

GENOMICS AND PROTEOMICS
See also Bioinformatics

Bradbury, E. Morton, and Sandor Pongor, eds. *Structural Biology and Functional Genomics.* Boston: Kluwer Academic, 1999.

Dunn, Michael J., ed. *From Genome to Proteome: Advances in the Practice and Application of Proteomics.* New York: Wiley-VCH, 2000.

Gibson, Greg. *A Primer of Genome Science.* Sunderland, Mass.: Sinauer, 2002.

Innis, Michael A., David H. Gelfand, and John

J. Sninsky, eds. *PCR Applications: Protocols for Functional Genomics.* San Diego: Academic Press, 1999.

Jolles, P., and H. Jornvall, eds. *Proteomics in Functional Genomics: Protein Structure Analysis.* Boston: Birkhauser, 2000.

Kang, Manjit S. *Quantitative Genetics, Genomics, and Plant Breeding.* Wallingford, Oxon, England: CABI, 2002.

Liebler, David G. *Introduction to Proteomics: Tools for the New Biology.* Totowa, N.J.: Humana Press, 2001.

Link, Andrew J., ed. *2-D Proteome Analysis Protocols.* Totowa, N.J.: Humana Press, 1999.

Liu, Ben-Hui. *Statistical Genomics: Linkage, Mapping, and QTL Analysis.* Boca Raton, Fla.: CRC Press, 1998.

Pennington, S. R., and M. J. Dunn, eds. *Proteomics: From Protein Sequence to Function.* New York: Springer, 2001.

Rabilloud, Thierry, ed. *Proteome Research: Two-Dimensional Gel Electrophoresis and Identification Methods.* New York: Springer, 2000.

HISTORY OF GENETICS

Alibeck, Ken, with Stephen Handelman. *Biohazard: The Chilling True Story of the Largest Covert Biological Weapons Program in the World, Told from the Inside by the Man Who Ran It.* New York: Random House, 1999.

Ayala, Francisco J., and Walter M. Fitch, eds. *Genetics and the Origin of Species: From Darwin to Molecular Biology Sixty Years After Dobzhansky.* Washington, D.C.: National Academies Press, 1997.

Beighton, Peter. *The Person Behind the Syndrome.* Rev. ed. New York: Springer-Verlag, 1997.

Bowler, Peter J. *Evolution: The History of an Idea.* Berkeley: University of California Press, 1990.

Brookes, Martin. *Fly: The Unsung Hero of Twentieth-Century Science.* San Francisco: Harper-Collins, 2001.

Corcos, A., and F. Monaghan. *Mendel's Experiments on Plant Hybrids: A Guided Study.* New Brunswick, N.J.: Rutgers University Press, 1993.

Darwin, Charles. *Charles Darwin's Notebooks, 1836-1844.* Edited by P. H. Barrett et al. Ithaca, N.Y.: Cornell University Press, 1987.

_____. *The Correspondence of Charles Darwin.* Cambridge, England: Cambridge University Press, 1994.

Dover, Gabriel A. *Dear Mr. Darwin: Letters on the Evolution of Life and Human Nature.* Berkeley: University of California Press, 2000.

Dunn, L. C. *A Short History of Genetics: The Development of Some of the Main Lines of Thought, 1864-1939.* New York: McGraw-Hill, 1965.

Edelson, Edward. *Gregor Mendel and the Roots of Genetics.* New York: Oxford University Press, 1999.

Fast, Julius. *Blueprint for Life: The Story of Modern Genetics.* New York: St. Martin's Press, 1965.

Fitzgerald, Patrick J. *From Demons and Evil Spirits to Cancer Genes: The Development of Concepts Concerning the Causes of Cancer and Carcinogenesis.* Washington, D.C.: American Registry of Pathology, Armed Forces Institute of Pathology, 2000.

Fredrickson, Donald S. *The Recombinant DNA Controversy, a Memoir: Science, Politics, and the Public Interest, 1974-1981.* Washington, D.C.: ASM Press, 2001.

Fujimura, Joan H. *Crafting Science: A Sociohistory of the Quest for the Genetics of Cancer.* Cambridge, Mass.: Harvard University Press, 1996.

Gillham, Nicholas Wright. *A Life of Sir Francis Galton: From African Exploration to the Birth of Eugenics.* New York: Oxford University Press, 2001.

Haldane, J. B. S. *Selected Genetic Papers of J. B. S. Haldane.* Edited with an introduction by Krishna R. Dronamraju; foreword by James F. Crow. New York: Garland, 1990.

Henig, Robin Marantz. *The Monk in the Garden: The Lost and Found Genius of Gregor Mendel, the Father of Genetics.* Boston: Houghton Mifflin, 2000.

Hubbell, Sue. *Shrinking the Cat: Genetic Engineering Before We Knew About Genes.* Illustrations by Liddy Hubbell. Boston: Houghton Mifflin, 2001.

Iltis, Hugo. *Life of Mendel.* Translated by Eden Paul and Cedar Paul. London: Allen & Unwin, 1932.

Jacob, François. *The Logic of Life: A History of Heredity.* New York: Pantheon, 1973.

Johnson, George B. *How Scientists Think:*

Twenty-one Experiments That Have Shaped Our Understanding of Genetics and Molecular Biology. Dubuque, Iowa: Wm. C. Brown, 1996.

Judson, Horace Freeland. *The Eighth Day of Creation: Makers of the Revolution in Biology.* Rev. ed. New York: Cold Spring Harbor, N.Y.: Cold Spring Harbor Laboratory Press, 1997.

Kay, Lily E. *Who Wrote the Book of Life? A History of the Genetic Code.* Stanford, Calif.: Stanford University Press, 2000.

Keller, Evelyn Fox. *A Feeling for the Organism: The Life and Work of Barbara McClintock.* 10th anniversary ed. New York: W. H. Freeman, 1993.

Kornberg, Arthur. *For the Love of Enzymes: The Odyssey of a Biochemist.* Reprint. Cambridge, Mass.: Harvard University Press, 1991.

Kühl, Stefan. *The Nazi Connection: Eugenics, American Racism, and German National Socialism.* New York: Oxford University Press, 2002.

Laffin, John. *Hitler Warned Us: The Nazis' Master Plan for a Master Race.* Totowa, N.J.: Barnes and Noble, 1998.

McCarty, Maclyn. *The Transforming Principle: Discovering That Genes Are Made of DNA.* New York: W. W. Norton, 1994.

Maddox, Brenda. *Rosalind Franklin: The Dark Lady of DNA.* New York: HarperCollins, 2002.

Mayr, Ernst. *One Long Argument: Charles Darwin and the Genesis of Modern Evolutionary Thought.* Cambridge, Mass.: Harvard University Press, 1991.

Mendel, Gregor. *Experiments in Plant-Hybridization.* Foreword by Paul C. Mangelsdorf. Cambridge, Mass.: Harvard University Press, 1965.

Olby, Robert C. *Origins of Mendelism.* New York: Schocken Books, 1966.

Orel, Vítezslav. *Gregor Mendel: The First Geneticist.* Translated by Stephen Finn. New York: Oxford University Press, 1996.

Palladino, Paolo. *Plants, Patients, and the Historians: On (RE)Membering in the Age of Genetic Engineering.* New Brunswick, N.J.: Rutgers University Press, 2003.

Persell, Stuart Michael. *Neo-Lamarckism and the Evolution Controversy in France, 1870-1920.* Lewiston, N.Y.: Edwin Mellen Press, 1999.

Potts, D. M., and W. T. W. Potts. *Queen Victoria's Gene: Haemophilia and the Royal Family.* Stroud: Sutton, 1999.

Shannon, Thomas A., ed. *Genetic Engineering: A Documentary History.* Westport, Conn.: Greenwood Press, 1999.

Shermer, Michael. *In Darwin's Shadow: The Life and Science of Alfred Russel Wallace, a Biographical Study on the Psychology of History.* New York: Oxford University Press, 2002.

Stubbe, H. *A History of Genetics.* Cambridge, Mass.: MIT Press, 1968.

Sturtevant, A. H. *A History of Genetics.* 1965. Reprint. Introduction by Edward B. Lewis. Cold Spring Harbor, N.Y.: Cold Spring Harbor Laboratory Press, 2001.

Sulston, John, and Georgina Ferry. *The Common Thread: A Story of Science, Politics, Ethics, and the Human Genome.* Washington, D.C.: Joseph Henry Press, 2002.

Tudge, Colin. *In Mendel's Footnotes: An Introduction to the Science and Technologies of Genes and Genetics from the Nineteenth Century to the Twenty-Second.* London: Jonathan Cape, 2000.

Watson, James. *The Double Helix.* New York: Simon and Schuster, 2001.

Weiner, Jonathan. *The Beak of the Finch: A Story of Evolution in Our Time.* New York: Random House, 1995.

Wood, Roger J., and Vitezslav Orel. *Genetic Prehistory in Selective Breeding: A Prelude to Mendel.* New York: Oxford University Press, 2001.

HUMAN GENETICS

Adolph, Kenneth W., ed. *Human Molecular Genetics.* New York: Academic Press, 1996.

Alcock, John. *The Triumph of Sociobiology.* Reprint. New York: Oxford University Press, 2003.

Andreasen, Nancy C. *Brave New Brain: Conquering Mental Illness in the Era of the Genome.* New York: Oxford University Press, 2001.

Arking, Robert, ed. *Biology of Aging: Observations and Principles.* 2d ed. Sunderland, Mass.: Sinauer, 2001.

Austad, Steven N. *Why We Age: What Science Is Discovering About the Body's Journey Throughout Life.* New York: John Wiley & Sons, 1997.

Badcock, C. R. *Evolutionary Psychology: A Critical Introduction.* Malden, Mass.: Polity Press in association with Blackwell, 2000.

Baudrillard, Jean. *The Vital Illusion.* Edited by Julia Witwer. New York: Columbia University Press, 2000.

Blackmore, Susan J. *The Meme Machine.* Foreword by Richard Dawkins. New York: Oxford University Press, 1999.

Bock, Gregory R., and Jamie A. Goode. *Genetics of Criminal and Antisocial Behaviour.* New York: John Wiley & Sons, 1996.

Bock, Gregory R., Jamie A. Goode, and Kate Webb, eds. *The Nature of Intelligence.* New York: John Wiley & Sons, 2001.

Boyer, Samuel, ed. *Papers on Human Genetics.* Englewood Cliffs, N.J.: Prentice-Hall, 1963.

Brierley, John Keith. *The Thinking Machine: Genes, Brain, Endocrines, and Human Nature.* Rutherford, N.J.: Fairleigh Dickinson University Press, 1973.

Briley, Mike, and Fridolin Sulser, eds. *Molecular Genetics of Mental Disorders: The Place of Molecular Genetics in Basic Mechanisms and Clinical Applications in Mental Disorders.* Malden, Mass.: Blackwell, 2001.

British Medical Association. *Biotechnology, Weapons, and Humanity.* Amsterdam, Netherlands: Harwood Academic, 1999.

Burnet, Sir Frank Macfarlane. *Endurance of Life: The Implications of Genetics for Human Life.* New York: Cambridge University Press, 1978.

Burnham, Terry, and Jay Phelan. *Mean Genes: From Sex to Money to Food—Taming Our Primal Instincts.* Cambridge, Mass.: Perseus, 2000.

Carson, Ronald A., and Mark A. Rothstein. *Behavioral Genetics: The Clash of Culture and Biology.* Baltimore: Johns Hopkins University Press, 1999.

Carter, Cedric O. *Human Heredity.* Baltimore: Penguin Books, 1962.

Cartwright, John. *Evolution and Human Behavior: Darwinian Perspectives on Human Nature.* Cambridge, Mass.: MIT Press, 2000.

Cavalli-Sforza, L. L., and Francesco Cavalli-Sforza. *The Great Human Diasporas: The History of Diversity and Evolution.* Translated by Sarah Thorne. Reading, Mass.: Addison-Wesley, 1995.

Centers for Disease Control and Prevention. "Bioterrorism-Related Anthrax." *Emerging Infectious Diseases* 8 (October, 2002): 1013-1183.

Clark, William R., and Michael Grunstein. *Are We Hardwired? The Role of Genes in Human Behavior.* New York: Oxford University Press, 2000.

Clegg, Edward J. *The Study of Man: An Introduction to Human Biology.* London: English Universities Press, 1968.

Cole, Leonard A. *The Eleventh Plague: The Politics of Biological and Chemical Warfare.* New York: W. H. Freeman, 1996.

Cooper, Colin. *Intelligence and Abilities.* New York: Routledge, 1999.

Cooper, Necia Grant, ed. *The Human Genome Project: Deciphering the Blueprint of Heredity.* Foreword by Paul Berg. Mill Valley, Calif.: University Science Books, 1994.

Cronk, Lee. *That Complex Whole: Culture and the Evolution of Human Behavior.* Boulder, Colo.: Westview Press, 1999.

Crow, Tim J., ed. *The Speciation of Modern Homo Sapiens.* Oxford, England: Oxford University Press, 2002.

Cummings, Michael J. *Human Heredity: Principles and Issues.* 5th ed. Pacific Grove, Calif.: Brooks/Cole, 2000.

Curran, Charles E. *Politics, Medicine, and Christian Ethics: A Dialogue with Paul Ramsey.* Philadelphia: Fortress Press, 1973.

Cziko, Gary. *The Things We Do: Using the Lessons of Bernard and Darwin to Understand the What, How, and Why of Our Behavior.* Cambridge: MIT Press, 2000.

DeMoss, Robert T. *Brain Waves Through Time: Twelve Principles for Understanding the Evolution of the Human Brain and Man's Behavior.* New York: Plenum Trade, 1999.

Dennis, Carina, and Richard Gallagher. *The Human Genome.* London: Palgrave Macmillan, 2002.

Diamant, L., and R. McAnuity, eds. *The Psychology of Sexual Orientation, Behavior, and Identity: A Handbook.* Westport, Conn.: Greenwood Press, 1995.

Dobzhansky, Theodosius G. *Mankind Evolving: The Evolution of the Human Species.* New Haven: Yale University Press, 1967.

Edlin, Gordon. *Human Genetics: A Modern Synthesis.* Boston: Jones & Bartlett, 1990.

Fish, Jefferson M., ed. *Race and Intelligence: Separating Science from Myth.* Mahwah, N.J.: Lawrence Erlbaum, 2002.

Fishbein, Diana H., ed. *The Science, Treatment, and Prevention of Antisocial Behaviors: Application to the Criminal Justice System.* Kingston, N.J.: Civic Research Institute, 2000.

Fooden, Myra, et al., eds. *The Second X and Women's Health.* New York: Gordian Press, 1983.

Gallagher, Nancy L. *Breeding Better Vermonters: The Eugenics Program in the Green Mountain State.* Hanover, N.H.: University Press of New England, 2000.

Gardner, Howard. *Frames of Mind: The Theory of Multiple Intelligences.* 10th anniversary ed. New York: Basic Books, 1993.

Glassy, Mark C. *The Biology of Science Fiction Cinema.* Jefferson, N.C.: McFarland, 2001.

Goldhagen, Daniel J. *Hitler's Willing Executioners: Ordinary Germans and the Holocaust.* New York: Random House, 1996.

Gould, Stephen Jay. *The Mismeasure of Man.* Rev. ed. New York: W. W. Norton, 1996.

Graves, Joseph L., Jr. *The Emperor's New Clothes: Biological Theories of Race at the Millennium.* New Brunswick, N.J.: Rutgers University Press, 2001.

Hahn, Martin E., et al., eds. *Developmental Behavior Genetics: Neural, Biometrical, and Evolutionary Approaches.* New York: Oxford University Press, 1990.

Haldane, J. B. S. *Selected Genetic Papers of J. B. S. Haldane.* Edited with an introduction by Krishna R. Dronamraju; foreword by James F. Crow. New York: Garland, 1990.

Hamer, Dean, and Peter Copeland. *Living with Our Genes: Why They Matter More than You Think.* New York: Doubleday, 1998.

Harris, Harry. *The Principles of Human Biochemical Genetics.* New York: Elsevier/North-Holland Biomedical Press, 1980.

Herrnstein, Richard J., and Charles Murray. *The Bell Curve: Intelligence and Class Structure in America.* New York: Free Press, 1994.

Heschl, Adolf. *The Intelligent Genome: On the Origin of the Human Mind by Mutation and Selection.* Drawings by Herbert Loserl. New York: Springer, 2002.

Hsia, David Yi-Yung. *Human Developmental Genetics.* Chicago: Year Book Medical Publishers, 1968.

Jacquard, Albert. *In Praise of Difference: Genetics and Human Affairs.* Translated by Margaret M. Moriarty. New York: Columbia University Press, 1984.

Jensen, Arthur Robert. *Genetics and Education.* New York: Harper & Row, 1972.

Karlsson, Jon L. *Inheritance of Creative Intelligence.* Chicago: Nelson-Hall, 1978.

Kevles, Daniel J. *In the Name of Eugenics: Genetics and the Uses of Human Heredity.* New York: Alfred A. Knopf, 1985.

Korn, Noel, and Harry Reece Smith, eds. *Human Evolution: Readings in Physical Anthropology.* New York: Holt, 1959.

Krebs, J., and N. Davies. *An Introduction to Behavioral Ecology.* Malden, Mass.: Blackwell, 1991.

Lasker, Gabriel W., ed. *The Processes of Ongoing Human Evolution.* Detroit: Wayne State University Press, 1960.

Levitan, Max. *Textbook of Human Genetics.* New York: Oxford University Press, 1988.

Lewis, Ricki. *Human Genetics: Concepts and Applications.* 5th ed. New York: McGraw-Hill, 2003.

Lewontin, Richard C. *Human Diversity.* New York: Scientific American Library, 1995.

Ludmerer, Kenneth M. *Genetics and American Society: A Historical Appraisal.* Baltimore: The Johns Hopkins University Press, 1972.

Lynn, Richard. *Dysgenics: Genetic Deterioration in Modern Populations.* Westport, Conn.: Praeger, 1996.

McConkey, Edwin H. *Human Genetics: The Molecular Revolution.* Boston: Jones & Bartlett, 1993.

Macieira-Coelho, Alvaro. *Biology of Aging.* New York: Springer, 2002.

McKusick, Victor A., comp. *Mendelian Inheritance in Man: A Catalog of Human Genes and Genetic Disorders.* 12th ed. Baltimore: Johns Hopkins University Press, 1998.

McWhirter, David P., et al. *Homosexuality/Heterosexuality: Concepts of Sexual Orientation.* New York: Oxford University Press, 1990.

Mange, Elaine Johansen, and Arthur P. Mange. *Basic Human Genetics.* 2d ed. Sunderland, Mass.: Sinauer Associates, 1999.

Manuck, Stephen B., et al., eds. *Behavior, Health, and Aging.* Mahwah, N.J.: Lawrence Erlbaum, 2000.

Marks, Jonathan M. *Human Biodiversity: Genes, Race, and History.* New York: Aldine de Gruyter, 1995.

Marteau, Theresa, and Martin Richards, eds. *The Troubled Helix: Social and Psychological Implications of the New Human Genetics.* New York: Cambridge University Press, 1999.

Medina, John J. *The Clock of Ages: Why We Age, How We Age—Winding Back the Clock.* New York: Cambridge University Press, 1996.

Mielke, James H., and Michael H. Crawford, eds. *Current Developments in Anthropological Genetics.* New York: Plenum Press, 1980.

Miller, Judith, Stephen Engelberg, and William Broad. *Germs: Biological Weapons and America's Secret War.* New York: Simon & Schuster, 2001.

Miller, Orlando J., and Eeva Therman. *Human Chromosomes.* 4th ed. New York: Springer Verlag, 2001.

Moffitt, Terrie E., Avshalom Caspi, Michael Rutter, and Phil A. Silva. *Sex Differences in Antisocial Behaviour: Conduct Disorder, Delinquency, and Violence in the Dunedin Longitudinal Study.* New York: Cambridge University Press, 2001.

Moody, Paul Amos. *Genetics of Man.* New York: W. W. Norton, 1967.

Nelkin, Dorothy, and Laurence Tancredi. *Dangerous Diagnostics: The Social Power of Biological Information.* New York: Basic Books, 1989.

Ostrer, Harry. *Non-Mendelian Genetics in Humans.* New York: Oxford University Press, 1998.

Ott, Jurg. *Analysis of Human Genetic Linkage.* Baltimore: The Johns Hopkins University Press, 1991.

Pearson, Roger. *Eugenics and Race.* Los Angeles: Noontide Press, 1966.

Pierce, Benjamin A. *The Family Genetic Sourcebook.* New York: John Wiley & Sons, 1990.

Plomin, Robert, et al. *Behavioral Genetics.* 4th ed. New York: Worth, 2001.

Puterbaugh, Geoff. *Twins and Homosexuality: A Casebook.* New York: Garland, 1990.

Resta, Robert G., ed. *Psyche and Helix: Psychological Aspects of Genetic Counseling.* New York: Wiley-Liss, 2000.

Ricklefs, Robert E., and Caleb E. Finch. *Aging: A Natural History.* New York: W. H. Freeman, 1995.

Rifkin, Jeremy. *The Biotech Century: Harnessing the Gene and Remaking the World.* New York: Putnam, 1998.

Roderick, Gordon Wynne. *Man and Heredity.* New York: St. Martin's Press, 1968.

Rosenberg, Charles, ed. *The History of Hereditarian Thought: A Thirty-two Volume Reprint Series Presenting Some of the Classic Books in This Intellectual Tradition.* New York: Garland, 1984.

Ruse, M. *Sociobiology: Sense or Nonsense?* Boston: D. Riedel, 1979.

Santos, Miguel A. *Genetics and Man's Future: Legal, Social, and Moral Implications of Genetic Engineering.* Springfield, Ill.: Thomas, 1981.

Segerstråle, Ullica. *Defenders of the Truth: The Battle for Science in the Sociobiology Debate and Beyond.* New York: Oxford University Press, 2000.

Singh, Jai Rup, ed. *Current Concepts in Human Genetics.* Amritsar, India: Guru Nanak Dev University, 1996.

Steen, R. Grant. *DNA and Destiny: Nurture and Nature in Human Behavior.* New York: Plenum, 1996.

Strachan, Tom, and Andrew P. Read. *Human Molecular Genetics.* 2d ed. New York: Wiley-Liss, 1999.

Sussman, Robert, ed. *The Biological Basis of Human Behavior: A Critical Review.* 2d ed. New York: Simon and Schuster, 1998.

Suzuki, D., and P. Knudtson. *Genethics: The Clash Between the New Genetics and Human Value.* Cambridge, Mass.: Harvard University Press, 1989.

Terwilliger, Joseph Douglas, and Jurg Ott. *Handbook of Human Genetic Linkage.* Baltimore: Johns Hopkins University Press, 1994.

Thorner, M., and R. Smith. *Human Growth Hormone: Research and Clinical Practice.* Vol. 19. Totowa, N.J.: Humana Press, 1999.

Timiras, Paola S. *Physiological Basis of Aging and Geriatrics.* 3d ed. Boca Raton, Fla.: CRC Press, 2003.

Toussaint, Olivier, et al., eds. *Molecular and Cellular Gerontology.* New York: New York Academy of Sciences, 2000.

Turney, Jon. *Frankenstein's Footsteps: Science, Genetics, and Popular Culture*. New Haven, Conn.: Yale University Press, 1998.

Underwood, Jane H. *Human Variation and Human Microevolution*. Englewood Cliffs, N.J.: Prentice-Hall, 1979.

U.S. Congress. Office of Technology Assessment. *Mapping Our Genes: Genome Projects—How Big, How Fast?* Baltimore: The Johns Hopkins University Press, 1988.

Vandenberg, Steven G., ed. *Methods and Goals in Human Behavior Genetics*. New York: Academic Press, 1965.

Van der Dennen, Johan M. G., David Smillie, and Daniel R. Wilson, eds. *The Darwinian Heritage and Sociobiology*. Westport, Conn.: Praeger, 1999.

Varmus, Harold, and Robert Weinberg. *Genes and the Biology of Cancer*. New York: W. H. Freeman, 1993.

Vogel, Friedrich, and A. G. Motulsky. *Human Genetics: Problems and Approaches*. New York: Springer-Verlag, 1997.

Wasserman, David, and Robert Wachbroit, eds. *Genetics and Criminal Behavior*. New York: Cambridge University Press, 2001.

Weil, Jon. *Psychosocial Genetic Counseling*. New York: Oxford University Press, 2000.

Weir, Bruce S. *Human Identification: The Use of DNA Markers*. New York: Kluwer Academic, 1995.

Weiss, Kenneth M. *Genetic Variation and Human Disease: Principles and Evolutionary Approaches*. New York: Cambridge University Press, 1993.

Wexler, Alice. *Mapping Fate: A Memoir of Family, Risk, and Genetic Research*. Berkeley: University of California Press, 1996.

Whittinghill, Maurice. *Human Genetics and Its Foundations*. New York: Reinhold, 1965.

Wright, Lawrence. *Twins: And What They Tell Us About Who We Are*. New York: John Wiley & Sons, 1997.

Wright, William. *Born That Way: Genes, Behavior, Personality*. New York: Knopf, 1998.

Young, Ian D. *Introduction to Risk Calculation in Genetic Counseling*. 2d ed. New York: Oxford University Press, 1999.

Yu, Byung Pal, ed. *Free Radicals in Aging*. Boca Raton, Fla.: CRC Press, 1993.

IMMUNOGENETICS

Bell, John I., et al., eds. *T Cell Receptors*. New York: Oxford University Press, 1995.

Bibel, Debra Jan. *Milestones in Immunology*. New York: Springer-Verlag, 1988.

Bona, Constantin A., et al., eds. *The Molecular Pathology of Autoimmune Diseases*. 2d ed. New York: Taylor and Francis, 2002.

Clark, William R. *At War Within: The Double-Edged Sword of Immunity*. New York: Oxford University Press, 1995.

Coleman, Robert M., et al. *Fundamental Immunology*. Dubuque, Iowa: Wm. C. Brown, 1989.

Dwyer, John M. *The Body at War: The Miracle of the Immune System*. New York: New American Library, 1988.

Fudenberg, H. Hugh, et al. *Basic Immunogenetics*. New York: Oxford University Press, 1984.

Gallo, Robert C., and Flossie Wong-Staal, eds. *Retrovirus Biology and Human Disease*. New York: Marcel Dekker, 1990.

Holland, J. J., ed. *Current Topics in Microbiology and Immunology: Genetic Diversity of RNA Viruses*. New York: Springer-Verlag, 1992.

Janeway, Charles A., Paul Travers, et al. *Immunobiology: The Immune System in Health and Disease*. 5th rev. ed. Philadelphia: Taylor & Francis, 2001.

Joneja, Janice M. V., and Leonard Bielory. *Understanding Allergy, Sensitivity, and Immunity*. New Brunswick, N.J.: Rutgers University Press, 1990.

Kimball, John W. *Introduction to Immunology*. 3d ed. New York: Macmillan, 1990.

Kreier, Julius P., and Richard F. Mortensen. *Infection, Resistance, and Immunity*. New York: Harper & Row, 1990.

Kuby, Janis. *Immunology*. 4th ed. New York: W. H. Freeman, 2000.

Mak, Tak W., and John J. L. Simard. *Handbook of Immune Response Genes*. New York: Plenum Press, 1998.

Mizel, Steven B., and Peter Jaret. *In Self-Defense*. San Diego: Harcourt Brace Jovanovich, 1985.

Pines, Maya, ed. *Arousing the Fury of the Immune System*. Chevy Chase, Md.: Howard Hughes Medical Institute, 1998.

Roitt, Ivan, Jonathan Brostoff, and David Male, eds. *Immunology*. New York: Mosby, 2001.

Samter, Max. *Immunological Diseases.* 4th ed. Boston: Little, Brown, 1988.

Silverstein, Arthur M. *A History of Immunology.* San Diego: Academic Press, 1989.

Smith, George P. *The Variation and Adaptive Expression of Antibodies.* Cambridge, Mass.: Harvard University Press, 1973.

Stewart, John. *The Primordial VRM System and the Evolution of Vertebrate Immunity.* Austin, Tex.: R. G. Landes, 1994.

Theofilopoulos, A. N., ed. *Genes and Genetics of Autoimmunity.* New York: Karger, 1999.

Tizard, Ian R. *Immunology: An Introduction.* 2d ed. New York: W. B. Saunders, 1988.

MOLECULAR GENETICS

Adolph, Kenneth W., ed. *Gene and Chromosome Analysis.* San Diego: Academic Press, 1993.

_____. *Human Molecular Genetics.* San Diego: Academic Press, 1996.

Alberts, Bruce, Dennis Bray, Julian Lewis, Martin Raff, Keith Roberts, and James D. Watson. *Molecular Biology of the Cell.* 4th ed. New York: Garland, 2002.

Ausubel, Fredrick, Roger Brent, Robert Kingston, David Moore, J. Seidman, and K. Struhl. *Current Protocols in Molecular Biology.* Hoboken, N.J.: John Wiley & Sons, 1998.

Baltimore, David, ed. *Nobel Lectures in Molecular Biology, 1933-1975.* New York: Elsevier North-Holland, 1977.

Barry, John Michael, and E. M. Barry. *Molecular Biology: An Introduction to Chemical Genetics.* Englewood Cliffs, N.J.: Prentice-Hall, 1973.

Benjamin, Jonathan, Richard P. Ebstein, and Robert H. Belmaker, eds. *Molecular Genetics and the Human Personality.* Washington, D.C.: American Psychiatric Association, 2002.

Berul, Charles I., and Jeffrey A. Towbin, eds. *Molecular Genetics of Cardiac Electrophysiology.* Boston: Kluwer Academic, 2000.

Brändén, Carl-Ivar, and John Tooze. *Introduction to Protein Structure.* 2d ed. New York: Garland, 1999.

Broach, J., J. Pringle, and E. Jones, eds. *The Molecular and Cellular Biology of the Yeast Saccharomyces.* 3 vols. Cold Spring Harbor, N.Y.: Cold Spring Harbor Laboratory Press, 1991-1997.

Brown, Terence A. *Genetics: A Molecular Approach.* 3d ed. New York: Chapman & Hall, 1998.

Browning, Michael, and Andrew McMichael, eds. *HLA and MHC: Genes, Molecules, and Function.* New York: Academic Press, 1999.

Capy, Pierre, et al. *Dynamics and Evolution of Transposable Elements.* New York: Chapman & Hall, 1998.

Cotterill, Sue, ed. *Eukaryotic DNA Replication: A Practical Approach.* New York: Oxford University Press, 1999.

Crick, Francis. *Life Itself: Its Origin and Nature.* New York: Simon & Schuster, 1981.

_____. *What Mad Pursuit.* New York: Basic Books, 1988.

DePamphilis, Melvin L., ed. *Concepts in Eukaryotic DNA Replication.* Cold Spring Harbor, N.Y.: Cold Spring Harbor Laboratory Press, 1999.

De Pomerai, David. *From Gene to Animal: An Introduction to the Molecular Biology of Animal Development.* New York: Cambridge University Press, 1985.

Dillon, Lawrence S. *The Gene: Its Structure, Function, and Evolution.* New York: Plenum Press, 1987.

Dizdaroglu, Miral, and Ali Esat Karakaya, eds. *Advances in DNA Damage and Repair: Oxygen Radical Effects, Cellular Protection, and Biological Consequences.* New York: Plenum Press, 1999.

Eckstein, Fritz, and David M. J. Lilley, eds. *Catalytic RNA.* New York: Springer, 1996.

Elgin, Sarah C. R., and Jerry L. Workman, eds. *Chromatin Structure and Gene Expression.* 2d ed. New York: Oxford University Press, 2000.

Erickson, Robert P., and Jonathan G. Izant, eds. *Gene Regulation: Biology of Antisense RNA and DNA.* New York: Raven Press, 1992.

Frank-Kamenetskii, Maxim D. *Unraveling DNA.* Reading, Mass.: Addison-Wesley, 1997.

Freedman, Leonard P., and M. Karin, eds. *Molecular Biology of Steroid and Nuclear Hormone Receptors.* Boston: Birkauser, 1999.

Friedberg, Errol C., et al., eds. *DNA Repair and Mutagenesis.* Washington, D.C.: ASM Press, 1995.

Gomperts, Kramer, et al. *Signal Transduction.* San Diego, Calif.: Academic Press, 2002.

Gros, François. *The Gene Civilization*. New York: McGraw-Hill, 1992.

Gwatkin, Ralph B. L., ed. *Genes in Mammalian Reproduction*. New York: Wiley-Liss, 1993.

Hancock, John T. *Molecular Genetics*. Boston: Butterworth-Heinemann, 1999.

Harris, David A., ed. *Prions: Molecular and Cellular Biology*. Portland, Oreg.: Horizon Scientific Press, 1999.

Hatch, Randolph T., ed. *Expression Systems and Processes for rDNA Products.* Washington, D.C.: American Chemical Society, 1991.

Hatfull, Graham F., and William R. Jacobs, Jr., eds. *Molecular Genetics of Mycobacteria*. Washington, D.C.: ASM Press, 2000.

Hawkins, John D. *Gene Structure and Expression*. New York: Cambridge University Press, 1996.

Hekimi, Siegfried, ed. *The Molecular Genetics of Aging*. New York: Springer, 2000.

Henderson, Daryl S., ed. *DNA Repair Protocols: Eukaryotic Systems*. Totowa, N.J.: Humana Press, 1999.

Hoch, James A., and Thomas J. Silhavy, eds. *Two-Component Signal Transduction*. Washington, D.C.: ASM Press, 1995.

Holmes, Roger S., and Hwa A. Lim, eds. *Gene Families: Structure, Function, Genetics and Evolution*. River Edge, N.J.: World Scientific, 1996.

Joklik, Wolfgang K., ed. *Microbiology: A Centenary Perspective*. Washington, D.C.: ASM Press, 1999.

Kimura, Motoo, and Naoyuki Takahata, eds. *New Aspects of the Genetics of Molecular Evolution*. New York: Springer-Verlag, 1991.

Langridge, John. *Molecular Genetics and Comparative Evolution*. New York: John Wiley & Sons, 1991.

Lewin, Benjamin M. *Gene Expression*. New York: John Wiley & Sons, 1980.

Litvack, Simon. *Retroviral Reverse Transcriptases*. Austin, Tex.: R. G. Landes, 1996.

McDonald, John F., ed. *Transposable Elements and Genome Evolution*. London: Kluwer Academic, 2000.

MacIntyre, Ross J., ed. *Molecular Evolutionary Genetics*. New York: Plenum Press, 1985.

Maraia, Richard J., ed. *The Impact of Short Interspersed Elements (SINEs) on the Host Genome*. Austin, Tex.: R. G. Landes, 1995.

Miesfeld, Roger L. *Applied Molecular Genetics*. New York: John Wiley, 1999.

Müller-Hill, Benno. *The Lac Operon: A Short History of a Genetic Paradigm*. New York: Walter de Gruyter, 1996.

Murphy, Kenneth P. *Protein Structure, Stability, and Folding*. Totowa, N.J.: Humana Press, 2001.

Murray, James A. H., ed. *Antisense RNA and DNA*. New York: Wiley-Liss, 1992.

Nei, Masatoshi. *Molecular Evolutionary Genetics*. New York: Columbia University Press, 1987.

Pollack, Robert. *Signs of Life: The Language and Meanings of DNA*. New York: Houghton Mifflin, 1994.

Ptashne, Mark, and Alexander Gann. *Genes and Signals*. Cold Spring Harbor, N.Y.: Cold Spring Harbor Press, 2002.

Roe, Bruce A., Judy S. Crabtree, and Akbar S. Khan, eds. *DNA Isolation and Sequencing*. New York: John Wiley & Sons, 1996.

Sarma, Ramaswamy H., and M. H. Sarma. *DNA Double Helix and the Chemistry of Cancer*. Schenectady, N.Y.: Adenine Press, 1988.

Schleif, Robert F. *Genetics and Molecular Biology*. Reading, Mass.: Addison-Wesley, 1986.

Selander, Robert K., et al., eds. *Evolution at the Molecular Level*. Sunderland, Mass.: Sinauer Associates, 1991.

Simons, Robert W., and Marianne Grunberg-Manago, eds. *RNA Structure and Function*. Cold Spring Harbor, N.Y.: Cold Spring Harbor Laboratory Press, 1997.

Smith, Paul J., and Christopher J. Jones, eds. *DNA Recombination and Repair*. New York: Oxford University Press, 2000.

Smith, Thomas B., and Robert K. Wayne. *Molecular Genetic Approaches in Conservation*. New York: Oxford University Press, 1996.

Snyder, Larry, and Wendy Champness. *Molecular Genetics of Bacteria*. Washington, D.C.: ASM Press, 1997.

Stone, Edwin M., and Robert J. Schwartz. *Intervening Sequences in Evolution and Development*. New York: Oxford University Press, 1990.

Strachan, Tom, and Andrew P. Read. *Human Molecular Genetics*. 2d ed. New York: Wiley-Liss, 1999.

Trainor, Lynn E. H. *The Triplet Genetic Code: The*

Key to Molecular Biology. River Edge, N.J.: World Scientific, 2001.

Turner, Bryan. *Chromatin and Gene Regulation: Mechanisms in Epigenetics.* Malden, Mass.: Blackwell, 2001.

Watson, James D. *The Double Helix.* New York: Atheneum, 1968.

_____., et al. *Molecular Biology of the Gene.* 5th ed. 2 vols. Menlo Park, Calif.: Benjamin Cummings, 2003.

Weaver, Robert F. *Molecular Biology.* 2d ed. New York: McGraw-Hill, 2002.

White, Robert J. *Gene Transcription: Mechanisms and Control.* Malden, Mass.: Blackwell, 2001.

POPULATION GENETICS

Avise, John, and James Hamrick, eds. *Conservation Genetics: Case Histories from Nature.* New York: Chapman and Hall, 1996.

Ayala, Francisco J. *Population and Evolutionary Genetics: A Primer.* Menlo Park, Calif.: Benjamin/Cummings, 1982.

Boorman, Scott A., and Paul R. Levitt. *The Genetics of Altruism.* New York: Academic Press, 1980.

Bushman, Frederick. *Lateral Gene Transfer: Mechanisms and Consequences.* Cold Spring Harbor, N.Y.: Cold Spring Harbor Laboratory Press, 2001.

Charlesworth, Brian. *Evolution in Age-Structured Populations.* New York: Cambridge University Press, 1980.

Christiansen, Freddy B. *Population Genetics of Multiple Loci.* New York: Wiley, 2000.

Costantino, Robert F., and Robert A. Desharnais. *Population Dynamics and the Tribolium Model: Genetics and Demography.* New York: Springer-Verlag, 1991.

Crow, James F. *Basic Concepts in Population, Quantitative, and Evolutionary Genetics.* New York: W. H. Freeman, 1986.

Dawson, Peter S., and Charles E. King, eds. *Readings in Population Biology.* Englewood Cliffs, N.J.: Prentice-Hall, 1971.

De Waal, Franz. *Good Natured: The Origins of Right and Wrong in Humans and Other Animals.* Cambridge, Mass.: Harvard University Press, 1996.

Falconer, D. S., and Trudy F. MacKay. *Introduc-* *tion to Quantitative Genetics.* 4th ed. Reading, Mass.: Addison-Wesley, 1996.

Gale, J. S. *Population Genetics.* New York: John Wiley & Sons, 1980.

Gillespie, John H. *Population Genetics: A Concise Guide.* Baltimore: Johns Hopkins University Press, 1997.

Gould, Steven Jay. *Ontogeny and Phylogeny.* Cambridge, Mass.: Belknap Press, 1977.

Harper, J. L. *Population Biology of Plants.* New York: Academic Press, 1977.

Hartl, Daniel L. *A Primer of Population Genetics.* Rev. 3d ed. Sunderland, Mass.: Sinauer Associates, 2000.

Hedrick, Philip W. *Genetics of Populations.* 2d ed. Boston: Jones and Bartlett, 2000.

Hoelzel, A. R., ed. *Molecular Genetic Analysis of Populations: A Practical Approach.* New York: IRL Press at Oxford University Press, 1992.

Kang, Manjit S., and Hugh G. Gauch, Jr. *Genotype-by-Environment Interaction.* Boca Raton, Fla.: CRC Press, 1996.

Kingsland, S. E. *Modeling Nature: Episodes in the History of Population Ecology.* Chicago: University of Chicago Press, 1985.

Lack, D. *The Natural Regulation of Animal Numbers.* Oxford, England: Clarendon Press, 1954.

Laikre, Linda. *Genetic Processes in Small Populations: Conservation and Management Considerations with Particular Focus on Inbreeding and Its Effects.* Stockholm: Division of Population Genetics, Stockholm University, 1996.

Lynn, Richard. *Dysgenics: Genetic Deterioration in Modern Populations.* Westport, Conn.: Praeger, 1996.

McKinney, M. L., and K. J. McNamara. *Heterochrony: The Evolution of Ontogeny.* New York: Plenum Press, 1991.

Magurran, Anne E., and Robert M. May, eds. *Evolution of Biological Diversity: From Population Differentiation to Speciation.* New York: Oxford University Press, 1999.

Papiha, Surinder S., Ranjan Deka, and Ranajit Chakraborty, eds. *Genomic Diversity: Applications in Human Population Genetics.* New York: Kluwer Academic/Plenum, 1999.

Provine, William B. *The Origins of Theoretical Population Genetics.* 2d ed. Chicago: University of Chicago Press, 2001.

Real, Leslie A., ed. *Ecological Genetics.* Princeton, N.J.: Princeton University Press, 1994.

Ruse, M. *Sociobiology: Sense or Nonsense?* Boston: D. Riedel, 1979.

Schonewald-Cox, Christine M., et al., eds. *Genetics and Conservation: A Reference for Managing Wild Animal and Plant Populations.* Menlo Park, Calif.: Benjamin/Cummings, 1983.

Slatkin, Montgomery, and Michel Veuille, eds. *Modern Developments in Theoretical Population Genetics: The Legacy of Gustave Malécot.* New York: Oxford University Press, 2002.

Syvanen, Michael, and Clarence Kado. *Horizontal Gene Transfer.* 2d ed. Burlington, Mass.: Academic Press, 2002.

Thornhill, Nancy Wilmsen, ed. *The Natural History of Inbreeding and Outbreeding: Theoretical and Empirical Perspectives.* Chicago: University of Chicago Press, 1993.

Wilson, E. O. *Sociobiology: The New Synthesis.* Cambridge, Mass.: Belknap Press, 1975.

REPRODUCTIVE TECHNOLOGY

Andrews, Lori B. *The Clone Age: Adventures in the New World of Reproductive Technology.* New York: Henry Holt, 1999.

Bentley, Gillian R., and C. G. Nicholas Mascie-Taylor. *Infertility in the Modern World: Present and Future Prospects.* New York: Cambridge University Press, 2000.

Bonnicksen, Andrea L. *In Vitro Fertilization: Building Policy from Laboratories to Legislature.* Reprint. New York: Columbia University Press, 1991.

Brinsden, Peter R., ed. *A Textbook of In Vitro Fertilization and Assisted Reproduction: The Bourn Hall Guide to Clinical and Laboratory Practice.* 2d ed. New York: Parthenon, 1999.

Campbell, Annily. *Childfree and Sterilized: Women's Decisions and Medical Responses.* New York: Cassell, 1999.

Elder, Kay, and Brian Dale. *In Vitro Fertilization.* 2d ed. New York: Cambridge University Press, 2000.

Ettore, Elizabeth. *Reproductive Genetics, Gender, and the Body.* New York: Routledge, 2002.

Heyman, Bob, and Mette Henriksen. *Risk, Age, and Pregnancy: Case Study of Prenatal Genetic Screening and Testing.* New York: Palgrave, 2001.

Jansen, Robert, and D. Mortimer, eds. *Towards Reproductive Certainty: Fertility and Genetics Beyond 1999.* Boca Raton, Fla.: CRC Press, 1999.

McElreavey, Ken, ed. *The Genetic Basis of Male Infertility.* New York: Springer, 2000.

McGee, Glenn. *The Perfect Baby: A Pragmatic Approach to Genetics.* Lanham, Md.: Rowman & Littlefield, 1997.

Mader, Sylvia S. *Human Reproductive Biology.* 3d ed. New York: McGraw-Hill, 2000.

Marrs, Richard, et al. *Dr. Richard Marrs' Fertility Book.* New York: Dell, 1997.

Parry, Vivienne. *Antenatal Testing Handbook: The Complete Guide to Testing in Pregnancy.* Collingdale, Pa.: DIANE, 1998.

Rapp, Rayna. *Testing Women, Testing the Fetus: The Social Impact of Amniocentesis in America.* New York: Routledge, 1999.

Rosenthal, M. Sara. *The Fertility Sourcebook: Everything You Need to Know.* 2d ed. Los Angeles: Lowell House, 1998.

Rothman, Barbara Katz. *The Tentative Pregnancy: How Amniocentesis Changes the Experience of Motherhood.* Rev. ed. New York: Norton, 1993.

Seibel, Machelle M., and Susan L. Crockin, eds. *Family Building Through Egg and Sperm Donation.* Boston: Jones and Bartlett, 1996.

Trounson, Alan O., and David K. Gardner, eds. *Handbook of In Vitro Fertilization.* 2d ed. Boca Raton, Fla.: CRC Press, 2000.

TECHNIQUES AND METHODOLOGIES
See also Bioinformatics

Adolph, Kenneth W., ed. *Gene and Chromosome Analysis.* San Diego: Academic Press, 1993.

Braman, Jeff, ed. *In Vitro Mutagenesis Protocols.* 2d ed. Totowa, N.J.: Humana Press, 2002.

Budowle, Bruce, et al. *DNA Typing Protocols: Molecular Biology and Forensic Analysis.* Natick, Mass.: Eaton, 2000.

Chen, Bing-Yuan, and Harry W. Janes, eds. *PCR Cloning Protocols.* Rev. 2d ed. Totowa, N.J.: Humana Press, 2002.

Crawley, Jacqueline N. *What's Wrong with My Mouse? Behavioral Phenotyping of Transgenic and Knockout Mice.* New York: Wiley-Liss, 2000.

Davis, Rowland H. *Neurospora: Contributions of a Model Organism.* New York: Oxford University Press, 2000.

Double, John A., and Michael J. Thompson, eds. *Telomeres and Telomerase: Methods and Protocols.* Totowa, N.J.: Humana Press, 2002.

Farrell, Robert. *RNA Methodologies.* 2d ed. San Diego, Calif.: Academic Press, 1998.

Gjerde, Douglas T., Christopher P. Hanna, and David Hornby. *DNA Chromatography.* Weinheim, Germany: Wiley-VCH, 2002.

Grange, J. M., et al., eds. *Genetic Manipulation: Techniques and Applications.* Boston: Blackwell Scientific, 1991.

Guilfoile, P. *A Photographic Atlas for the Molecular Biology Laboratory.* Englewood, Colo.: Morton, 2000.

Hames, B. D., and D. Rickwood, eds. *Gel Electrophoresis of Nucleic Acids: A Practical Approach.* 2d ed. New York: Oxford University Press, 1990.

Harlow, Ed, and David Lane, eds. *Using Antibodies: A Laboratory Manual.* Rev. ed. Cold Spring Harbor, N.Y.: Cold Spring Harbor Laboratory Press, 1999.

Jackson, J. F., H. F. Linskens, and R. B. Inman, eds. *Testing for Genetic Manipulation in Plants.* New York: Springer, 2002.

Kochanowski, Bernd, and Udo Reischl, eds. *Quantitative PCR Protocols.* Methods in Molecular Medicine 26. Totowa, N.J.: Humana Press, 1999.

Lai, Eric, and Bruce W. Birren, eds. *Electrophoresis of Large DNA Molecules: Theory and Applications.* Cold Spring Harbor, N.Y.: Cold Spring Harbor Laboratory, 1990.

Lloyd, Ricardo V., ed. *Morphology Methods: Cell and Molecular Biology Techniques.* Totowa, N.J.: Humana Press, 2001.

McPherson, M. J., and S. G. Møller. *PCR Basics.* Oxford, England: BIOS Scientific, 2000.

McRee, Duncan Everett. *Practical Protein Crystallography.* 2d ed. San Diego: Academic Press, 1999.

O'Connell, Joe, ed. *RT-PCR Protocols.* Totowa, N.J.: Humana Press, 2002.

Pacifici, O. G. M., Julio Collado-Vides, and Ralf Hofestadt, eds. *Gene Regulation and Metabolism: Postgenomic Computational Approaches.* Cambridge: MIT Press, 2002.

Sandor, Suhai, ed. *Theoretical and Computational Methods in Genome Research.* New York: Plenum Press, 1997.

Silver, Lee. *Mouse Genetics: Concepts and Applications.* New York: Oxford University Press, 1995.

Suzuki, D. T., et al. *An Introduction to Genetic Analysis.* 7th ed. New York: W. H. Freeman, 2000.

Wilson, Zoe A. *Arabidopsis: A Practical Approach.* New York: Oxford University Press, 2000.

VIRAL GENETICS

Becker, Yechiel, and Gholamreza Darai, eds. *Molecular Evolution of Viruses: Past and Present.* Boston: Kluwer Academic, 2000.

Cann, Alan J. *DNA Virus Replication.* New York: Oxford University Press, 2000.

Carlberg, David M. *Essentials of Bacterial and Viral Genetics.* Springfield, Ill.: Charles C Thomas, 1976.

Domingo, Esteban, Robert Webster, and John Holland, eds. *Origin and Evolution of Viruses.* New York: Academic Press, 1999.

Gallo, Robert C., and Flossie Wong-Staal, eds. *Retrovirus Biology and Human Disease.* New York: Marcel Dekker, 1990.

Holland, J. J., ed. *Current Topics in Microbiology and Immunology: Genetic Diversity of RNA Viruses.* New York: Springer-Verlag, 1992.

Kolata, Gina. *The Story of the Great Influenza Pandemic of 1918 and the Search for the Virus That Caused It.* New York: Simon and Schuster, 2001.

World Health Organization. *Future Research on Smallpox Virus Recommended.* Geneva, Switzerland: World Health Organization Press, 1999.

Web Sites

The sites listed below were visited by the editors of Salem Press in May of 2003. Because URLs frequently change or are moved, their accuracy cannot be guaranteed; however, long-standing sites—such as those of university departments, national organizations, and government agencies—generally maintain links when sites move or otherwise may upgrade their offerings and hence remain useful. Sites with an "N/A" affiliation are, to our knowledge, unallied, mounted by an individual, or "uncredited." –*Roger Smith*

General Genetics

BioMedNet
Genetics Gateway
http://reviews.bmn.com/?subject=Genetics

A repository of research articles for scientists, links to databases, and news and feature articles written for general readers on all aspects of genetics. Users must register, for which there is no charge, although access to the full text of articles and databases requires a fee in some cases.

Dolan DNA Learning Center, Cold Spring
    Harbor Laboratory
Gene Almanac
http://www.dnalc.org

An online science center devoted to public education in genetics at high school and college levels, this is the best entry point to the Web for newcomers to genetics. It provides information on DNA science, genetics and medicine, and biotechnology through interactive features, animated tutorials, and downloads. With an extensive list of links.

Genetics Society of America
Home Page
http://www.genetics-gsa.org

Although dedicated to subscribers who are professional geneticists, the society's Web site supports genetics education for all ages and offers a history of the organization and short position statements on evolution and genetically modified organisms. There are also links to databases and related Web sites.

N/A
Kimball's Biology Pages
http://biology-pages.info

A collection of clear, well-illustrated explanations of topics in biology by university professor John W. Kimball, including all aspects of genetics and biotechnology, with news updates. A reliable place to start for those new to genetics.

MedBioWorld
Genetics, Genomics & Biotechnology
http://www.medbioworld.com

A list of links to the major journals in genetics, many of which allow free access to abstracts, some articles, and news postings. A fast search vehicle for the most recent information about a topic in genetics.

National Public Radio
The DNA Files
http://www.dnafiles.org/home.html

The text from an award-winning public radio series, including programs about the human genome, genetics and ecology, genetics and medicine, biotechnology, and the genetics of identity. Lucid, in-depth treatments for the nonscientist.

Nature Publishing Group
genetics@nature.com
http://www.nature.com/genetics

Part of the Web site for *Nature*, Britain's premier science journal. Written for educated general readers, the sections offer news and recently published articles, commentary, and an encyclopedia of life sciences. There are links to specialty journals concerning topics in genetics and biotechnology. Some articles are accessible by the general public, but full use of the site requires a subscription.

Netspace
MendelWeb
http://www.mendelweb.org

An educational site for teachers and students concerning the origins of classical genetics and elementary plant science. It reproduces early publications by such pioneers as Gregor Mendel and William Bateson, accompanied by commentaries and reference resources.

Rutgers University
Morgan
http://morgan.rutgers.edu/
   MorganWebFrames/How_to_use/
   HTU_intro.html

A multimedia tutorial for advanced high school students or beginning college students. Its six levels review basic principles in genetics with particular attention to molecular interactions.

U.S. Department of Energy Office of Science
Virtual Library on Genetics
http://www.ornl.gov/TechResources/
   Human_Genome/genetics.html

A comprehensive catalog of Web site links, arranged by subject, pertaining to the Human Genome Project. The links lead to gene and chromosome databases and specific information on genetics, bioinformatics, and genetic disorders.

University of Massachusetts
DNA Structure
http://molvis.sdsc.edu/dna/index.htm

An interactive, animated tutorial on the molecular composition and structure of DNA for high school students and college freshmen. It can be downloaded and is available in Spanish, German, and Portuguese.

University of Utah
Genetic Science Learning Center
http://gslc.genetics.utah.edu

Designed for students, this site posts essays on the basics of DNA, genetic disorders, cloning, stem cells, and genetic testing. It also describes simple experiments, such as how to extract DNA material.

Bioinformatics

Bioinformatics.org
Home Page
http://bioinformatics.org

The site belongs to an international organization dedicated to the exchange of genetic information and includes online databases and analysis tools, software, explanations of frequently asked questions about bioinformatics, and news postings.

N/A
Earl's Forensic Page
http://members.aol.com/EarlNMeyer/
   DNA.html

With a variety of illustrations, Earl Meyer summarizes how genetic fingerprinting works and its use in crime investigations and in determining paternity.

European Bioinformatics Institute
The Path to Knowledge
http://www.ebi.ac.uk

A research center, this institute maintains databases concerning nucleic acids, protein sequences, and macromolecular structures. Also posts news, events, and descriptions of ongoing scientific projects.

Technical University of Denmark
Center for Biological Sequencing Analysis
http://www.cbs.dtu.dk

The Center conducts basic research in bioinformatics and here offers its sequencing analysis databases, genome atlases, analysis tools, and news. Primarily meant for researchers and university students.

Biotechnology

Bio-Link
Educating the Bio-Link Workforce
http://Bio-Link.org

Intended for technicians, this site offers information and instruction covering recent advances in biotechnology, as well as a virtual laboratory and library, news postings, and details about regional education centers.

Carolina
Biotechnology and Genetics
http://www.carolina.com/biotech

A rich resource designed for teachers and students. It offers a newsletter, workshops, articles, classroom activities, and videos, all focusing on the use of biotechnology and laboratory techniques.

Dow AgroSciences
Plant Genetics and Biotechnology
http://www.dowagro.com/homepage/
   index.htm

This corporate Web site concerns the marketing of its agricultural biotechnology. Given that bias, it contains news releases, descriptions of products, data sheets, articles and position statements, and a media kit of interest to general readers.

The Hastings Center
Home Page
http://www.thehastingscenter.org

The Hastings Center, an independent nonprofit organization, specializes in bioethics, particularly in health care and biotechnology. Its site contains news postings, articles on bioethics and different aspects of genetics science, and announcements of events and publications.

National Institutes of Health
National Center for Biotechnology
   Information
http://www.ncbi.nlm.nih.gov

This Web site for the main health agency of the United States contains links to the various specialized institutes under its umbrella as well as public databases in genomics and sequencing, articles and handbooks on a wide range of biotechnology topics, and more than a dozen types of free software for analyzing genetic data. Primarily intended for researchers, but with the public in mind—the site includes a science primer and other resources for educators, students, and other nonspecialists.

Genomics
Göteborg University, Sweden
RatMap, the Rat Genome Database
http://ratmap.gen.gu.se

A professional database of information on approximately six thousand rat genes, their positions on chromosomes, pertinent nomenclature, and gene functions.

Johns Hopkins University
The Genome Database
http://gdbwww.gdb.org

The official central storage center for gene-mapping data compiled in the Human Genome Initiative, an international effort to decode and analyze human DNA. Intended for scientists, the site presents information in three categories: regions, maps, and variations of the human genome.

Lawrence Berkeley National Laboratory
Human Genome Sequencing Department
http://www-hgc.lbl.gov

The department is part of the Joint Genome Institute, which includes genome laboratories at Lawrence Livermore and Los Alamos. The site describes its directed sequencing method, explains its work on sequencing the human genome and the genome of the fruit fly *Drosophila*, and provides access to sequencing archives.

Massachusetts Institute of Technology
Whitehead Institute for Biomedical Research
http://www-genome.wi.mit.edu

The institute's home page affords access to news of genomics research, software, and sequencing databases, all intended for scientists and university students. However, its information about the Human Genome Project has general background articles, photos, and a video that will be of interest to nonspecialists.

National Center for Biotechnology
   Information
GeneMap '99
http://www.ncbi.nlm.nih.gov/genemap99

Starting with a general introduction to the human genome and the process of gene mapping, this site provides charts of the known genes on each chromosome, articles about the Human Genome Project and gene-related medical research, and links to other genome sites and databases.

National Human Genome Research Institute
Home Page
http://www.genome.gov

In addition to information for researchers, this site contains a comprehensive introduction to the Human Genome Project, a glossary of genetic terms, fact sheets, multimedia education kits, and links to online education resources, all for the general public.

New York University/Bell Atlantic/Center for
    Advanced Technology
The Student Genome Project
http://www.cat.nyu.edu/sgp/parent.html

Uses interactive multimedia and three-dimensional technology to present tutorials and games related to the human genome and genetics for middle school and high school students. Also provides news about New York-based science events and links.

Sanger Institute, Wellcome Trust
Home Page
http://www.sanger.ac.uk

This research institute is dedicated to genomics. Accordingly, the site offers news updates, a searchable database, explanations of gene sequencing and computer software aids, and descriptions of genomics research projects. All information is intended for scientists.

U.S. Department of Energy Biological and
    Environmental Research Program
Human Genome Research
http://www.er.doe.gov/production/ober/
    hug_top.html

This site describes the Department of Energy's contribution to the Human Genome Project with pages containing a history, a time line, project information, an essay on the science of genetics, and abstracts of recent research. There is also a section of links designed for young students and science teachers, as well as information about fellowships and research opportunities.

Medicine and Genetics
American College of Medical Genetics
Home Page
http://www.acmg.net

Designed for physicians, this site also contains information for patients about treatments, research, standards for gene therapy, and health, along with news articles and links to related sites.

Genethon
Gene Therapies Research and Applications
    Center
http://www.genethon.fr/php/index_us.php

Supported by the French Muscular Dystrophy Association, Genethon sponsors research in genetic and cellular therapies for rare diseases. This site discusses research methods, the organization's services, and, in a section accompanied by computer graphics, the theory of gene therapy.

National Center for Biotechnology
    Information
Online Mendelian Inheritance in Man
http://www.ncbi.nlm.nih.gov/Omim

The Online Mendelian Inheritance in Man (OMIM) is a catalog of human genes and genetic disorders for scientists. The site also offers maps of genes and diseases, statistical summaries, and links to similar sites devoted to medical literature and biotechnology.

National Fragile X Foundation
Xtraordinary Accomplishments
http://www.nfxf.org

Provides extensive general information about fragile X syndrome, a cause of inherited mental impairments, and advises care-givers on testing, medical treatment, education, and life-planning.

National Organization for Rare Disorders
    (NORD).
Home Page
http://www.rarediseases.org.

Offers a very useful and long list of genetically related disorders and diseases, each of which links to a fact sheet that ends with a list of organizations for additional information. Also posts articles about rare genetic conditions and diseases, accessible through searchable databases.

National Society of Genetic Counselors
Society Home Page
http://www.nsgc.org

Although much of this site is devoted to society members, it has a search engine for locating genetic counselors in the United States and a newsroom with press releases and fact sheets about the counseling services.

The University of Nottingham
OMNI
http://omni.ac.uk/subject-listing/QH426.html

This gateway to information about biomedicine has links to sites devoted to the science, applications, and ethics of medical genetics.

Transgenics

Center for Life Sciences, Colorado State
    University
Transgenic Crops: An Introduction and
    Resource Guide
http://www.colostate.edu/programs/
    lifesciences/TransgenicCrops/

The purpose of this site is to provide balanced information about the technology and safety issues involved in genetically modified food crops. Compiled by genetics researchers. With links, a history of plant breeding, and glossary. Also in Spanish.

National Academy of Sciences
Transgenic Plants and World Agriculture

http://www.nap.edu/html/transgenic/

An online, downloadable pamphlet published in July, 2000, by a consortium of leading research societies around the world. It assesses the need to modify crops genetically in order to feed the increasing world population and then discusses examples of the technology, its safety, effects on the environment, funding sources, and intellectual property issues.

Oak Ridge National Laboratory
Transgenic and Targeted Mutant Animal
    Database
http://www.ornl.gov/TechResources/Trans/
    hmepg.html

A searchable professional database about lines of genetically modified animals, methods used to create them, and descriptions of the modified DNA, the expression of transgenes, and how transgenes are named.

University of Michigan
Transgenic Animal Model Core
http://www.med.umich.edu/tamc

A professional Web site for researchers seeking a host animal to test transgenes. However, it contains much useful general information about transgenics (especially transgenic rats), vectors, and laboratory procedures. With links and a photo gallery.

—*Roger Smith*

Encyclopedia of
Genetics
Revised Edition

Category Index

Articles are listed under all of the following categories that apply:

Bacterial Genetics
Bioethics
Bioinformatics
Cellular Biology
Classical Transmission Genetics
Developmental Genetics
Diseases and Syndromes
Evolutionary Biology
Genetic Engineering and Biotechnology
History of Genetics
Human Genetics and Social Issues
Immunogenetics
Molecular Genetics
Population Genetics
Techniques and Methodologies
Viral Genetics

Bacterial Genetics
Anthrax, 35
Bacterial Genetics and Cell Structure, 54
Bacterial Resistance and Super Bacteria, 61
Cholera, 137
Diphtheria, 214
Gene Regulation: Bacteria, 291
Gene Regulation: *Lac* Operon, 298
Model Organism: *Escherichia coli*, 527
Transposable Elements, 742

Bioethics
Bioethics, 73
Cloning: Ethical Issues, 170
Gene Therapy: Ethical and Economic Issues, 309
Genetic Engineering: Risks, 347
Genetic Engineering: Social and Ethical Issues, 351
Genetic Testing: Ethical and Economic Issues, 364
Insurance, 471
Miscegenation and Antimiscegenation Laws, 501
Patents on Life-Forms, 594

Bioinformatics
Bioinformatics, 79
cDNA Libraries, 115
Genomic Libraries, 380
Icelandic Genetic Database, 447

Cellular Biology
Archaea, 45
Bacterial Genetics and Cell Structure, 54
Cell Culture: Animal Cells, 117
Cell Culture: Plant Cells, 120
Cell Cycle, The, 122
Cell Division, 125
Chromosome Mutation, 144
Chromosome Structure, 147
Cytokinesis, 198
Extrachromosomal Inheritance, 274
Gene Regulation: Bacteria, 291
Gene Regulation: Eukaryotes, 295
Gene Regulation: *Lac* Operon, 298
Gene Regulation: Viruses, 301
Mitosis and Meiosis, 509
Nondisjunction and Aneuploidy, 579
Stem Cells, 710
Telomeres, 728
Totipotency, 736

Classical Transmission Genetics
Chromosome Structure, 147
Chromosome Theory of Heredity, 152
Classical Transmission Genetics, 160
Complete Dominance, 184
Dihybrid Inheritance, 210
Epistasis, 255
Incomplete Dominance, 465
Mendelian Genetics, 494
Monohybrid Inheritance, 555
Multiple Alleles, 559
Polygenic Inheritance, 609

Developmental Genetics
Developmental Genetics, 201
Homeotic Genes, 416
Steroid Hormones, 717
X Chromosome Inactivation, 759

Diseases and Syndromes
Albinism, 9
Alcoholism, 11
Alzheimer's Disease, 19
Autoimmune Disorders, 51
Biopharmaceuticals, 96
Breast Cancer, 101
Burkitt's Lymphoma, 106
Cancer, 109
Cholera, 137
Color Blindness, 179
Congenital Defects, 187
Consanguinity and Genetic Disease, 191
Cystic Fibrosis, 195
Diabetes Mellitus, 207
Diphtheria, 214
Down Syndrome, 244
Dwarfism, 248
Emerging Diseases, 252
Fragile X Syndrome, 282
Heart Disease, 392
Hemophilia, 396
Hereditary Diseases, 399
Hermaphrodites, 411
Huntington's Disease, 434
Hypercholesterolemia, 445
Inborn Errors of Metabolism, 458
Infertility, 468
Klinefelter Syndrome, 479
Lactose Intolerance, 484

Metafemales, 499
Mitochondrial Diseases, 503
Neural Tube Defects, 572
Nondisjunction and Aneuploidy, 579
Phenylketonuria (PKU), 604
Prader-Willi and Angelman Syndromes, 623
Prion Diseases: Kuru and Creutzfeldt-Jakob Syndrome, 631
Pseudohermaphrodites, 648
Sickle-Cell Disease, 692
Smallpox, 700
Swine Flu, 720
Tay-Sachs Disease, 727
Testicular Feminization Syndrome, 731
Thalidomide and Other Teratogens, 733
Turner Syndrome, 748
XYY Syndrome, 764

Evolutionary Biology
Ancient DNA, 27
Artificial Selection, 48
Evolutionary Biology, 267
Genetics, Historical Development of, 370
Lamarckianism, 485
Molecular Clock Hypothesis, 547
Natural Selection, 568
Punctuated Equilibrium, 650
RNA World, 686

Genetic Engineering and Biotechnology
Animal Cloning, 31
Biofertilizers, 77
Biological Weapons, 88
Biopesticides, 92
Biopharmaceuticals, 96
Chromosome Walking and Jumping, 158
Cloning, 166
Cloning Vectors, 174
Cloning: Ethical Issues, 170
DNA Isolation, 220
DNA Replication, 227
DNA Sequencing Technology, 233
Gene Therapy, 304
Gene Therapy: Ethical and Economic Issues, 309
Genetic Engineering, 326
Genetic Engineering: Agricultural Applications, 332
Genetic Engineering: Historical Development, 335

Genetic Engineering: Industrial Applications, 339
Genetic Engineering: Medical Applications, 343
Genetic Engineering: Risks, 347
Genetic Engineering: Social and Ethical Issues, 351
Genetically Modified (GM) Foods, 366
Genetics, Historical Development of, 370
High-Yield Crops, 413
Knockout Genetics and Knockout Mice, 481
Parthenogenesis, 592
Polymerase Chain Reaction, 611
Restriction Enzymes, 667
Reverse Transcriptase, 670
Shotgun Cloning, 691
Synthetic Genes, 725
Transgenic Organisms, 739
Xenotransplants, 761

History of Genetics
Chromosome Theory of Heredity, 152
Classical Transmission Genetics, 160
Eugenics: Nazi Germany, 264
Genetic Code, Cracking of, 319
Genetic Engineering: Historical Development, 335
Genetics, Historical Development of, 370
Genetics in Television and Films, 376
Human Genome Project, 428
Lamarckianism, 485
Mendelian Genetics, 494
Miscegenation and Antimiscegenation Laws, 501
One Gene-One Enzyme Hypothesis, 586
Sociobiology, 704

Human Genetics and Social Issues
Aggression, 1
Aging, 3
Biochemical Mutations, 70
Bioethics, 73
Biological Clocks, 83
Biological Determinism, 86
Biological Weapons, 88
Cloning: Ethical Issues, 170
Criminality, 193
DNA Fingerprinting, 216
Eugenics, 259
Eugenics: Nazi Germany, 264

Forensic Genetics, 279
Gender Identity, 287
Gene Therapy, 304
Gene Therapy: Ethical and Economic Issues, 309
Genetic Counseling, 321
Genetic Engineering: Social and Ethical Issues, 351
Genetic Screening, 357
Genetic Testing, 360
Genetic Testing: Ethical and Economic Issues, 364
Genetics in Television and Films, 376
Heredity and Environment, 406
Homosexuality, 419
Human Genetics, 421
Human Genome Project, 428
Human Growth Hormone, 432
In Vitro Fertilization and Embryo Transfer, 454
Insurance, 471
Intelligence, 474
Miscegenation and Antimiscegenation Laws, 501
Patents on Life-Forms, 594
Paternity Tests, 596
Prenatal Diagnosis, 626
Race, 658
Sociobiology, 704
Stem Cells, 710
Sterilization Laws, 715

Immunogenetics
Allergies, 13
Antibodies, 38
Autoimmune Disorders, 51
Hybridomas and Monoclonal Antibodies, 441
Immunogenetics, 449
Organ Transplants and HLA Genes, 588
Synthetic Antibodies, 723

Molecular Genetics
Ancient DNA, 27
Antisense RNA, 42
Biochemical Mutations, 70
Bioinformatics, 79
Central Dogma of Molecular Biology, 128
Chemical Mutagens, 131
Chloroplast Genes, 133
Chromatin Packaging, 140
Chromosome Mutation, 144

DNA Isolation, 220
DNA Repair, 223
DNA Structure and Function, 237
Gene Families, 289
Gene Regulation: Bacteria, 291
Gene Regulation: Eukaryotes, 295
Gene Regulation: *Lac* Operon, 298
Gene Regulation: Viruses, 301
Genetic Code, 313
Genetic Code, Cracking of, 319
Genome Size, 378
Genomics, 384
Human Growth Hormone, 432
Mitochondrial Genes, 505
Molecular Clock Hypothesis, 547
Molecular Genetics, 549
Mutation and Mutagenesis, 561
Noncoding RNA Molecules, 575
Oncogenes, 583
One Gene-One Enzyme Hypothesis, 586
Plasmids, 606
Polymerase Chain Reaction, 611
Protein Structure, 634
Protein Synthesis, 638
Proteomics, 643
Pseudogenes, 646
Repetitive DNA, 664
Restriction Enzymes, 667
Reverse Transcriptase, 670
RNA Isolation, 674
RNA Structure and Function, 676
RNA Transcription and mRNA Processing, 681
RNA World, 686
Signal Transduction, 696
Steroid Hormones, 717
Transposable Elements, 742
Tumor-Suppressor Genes, 746

Population Genetics
Altruism, 16
Artificial Selection, 48
Behavior, 65
Consanguinity and Genetic Disease, 191
Evolutionary Biology, 267
Genetic Load, 354
Hardy-Weinberg Law, 389
Hybridization and Introgression, 437
Inbreeding and Assortative Mating, 461
Lateral Gene Transfer, 489

Natural Selection, 568
Pedigree Analysis, 599
Penetrance, 602
Polyploidy, 613
Population Genetics, 617
Punctuated Equilibrium, 650
Quantitative Inheritance, 654
Sociobiology, 704
Speciation, 708

Techniques and Methodologies
Amniocentesis and Chorionic Villus Sampling, 23
Bioinformatics, 79
Blotting: Southern, Northern, and Western, 98
cDNA Libraries, 115
Cell Culture: Animal Cells, 117
Cell Culture: Plant Cells, 120
Chromosome Walking and Jumping, 158
Complementation Testing, 181
Gel Electrophoresis, 285
Genomic Libraries, 380
Human Genome Project, 428
Icelandic Genetic Database, 447
Linkage Maps, 491
Model Organism: *Arabidopsis thaliana*, 513
Model Organism: *Caenorhabditis elegans*, 516
Model Organism: *Chlamydomonas reinhardtii*, 520
Model Organism: *Drosophila melanogaster*, 522
Model Organism: *Escherichia coli*, 527
Model Organism: *Mus musculus*, 533
Model Organism: *Neurospora crassa*, 536
Model Organism: *Saccharomyces cerevisiae*, 539
Model Organism: *Xenopus laevis*, 542
Model Organisms, 545
Pedigree Analysis, 599
Polymerase Chain Reaction, 611
Proteomics, 643
RFLP Analysis, 672
Twin Studies, 750

Viral Genetics
Emerging Diseases, 252
Gene Regulation: Viruses, 301
Oncogenes, 583
Smallpox, 700
Swine Flu, 720
Viral Genetics, 754
Viroids and Virusoids, 756

Personages Index

Allen, Garland, 261
Allinen, Minna, 103
Allison, A. C., 392
Altman, Sidney, 130, 679, 687, 767
Alwine, James, 100
Alzheimer, Alois, 20
Amaya, Enrique, 543
Andrews, Tommy Lee, 280
Anfinsen, Christian B., 635, 767
Angelman, Harry, 623
Arber, Werner, 338, 767
Aristotle, 767
Athma, Prasanna, 103
Auerbach, Charlotte, 131, 767
Avery, Oswald, 57, 227, 238, 372, 725, 767

Bailey, Catherine, 767
Bakewell, Robert, 50
Baltimore, David, 103, 116, 130, 670, 767
Barr, Murray, 759, 767
Basson, Craig, 206
Bateson, William, 154, 255, 260, 372, 767
Baur, Erwin, 274
Beadle, George, 183, 372, 537, 587, 767
Beckwith, Jonathan R., 767
Bell, Julia, 767
Benacerraf, Baruj, 768
Bender, William, 417
Berg, Paul, 336, 768
Bickel, Horst, 606
Binet, Alfred, 86, 409
Bishop, J. Michael, 103, 768
Bluhm, Agnes, 768
Bonnet, Charles, 268
Boring, Alice Middleton, 768
Borlaug, Norman, 370, 656, 768
Botstein, David, 157, 768
Bouchard, T. J., Jr., 475
Boveri, Theodor, 147, 201, 523, 809
Boyer, Herbert, 174, 340, 351, 768
Bradley, Allan, 104

Brakefield, Paul, 408
Braun, A., 514
Breggin, Peter, 194
Brennan, William J., Jr., 595
Brenner, Sydney, 205, 319, 768
Bridges, Calvin, 144, 147, 155, 372, 499
Brigham, C. C., 476
Broca, Paul, 86
Brody, William R., 473
Bronstein, M. H., 103
Brown, Louise, 456
Brown, Michael S., 768
Brown, Nicole, 281
Buffon, comte de, 268, 487
Burger, Warren, 594
Burke, David, 178
Burkitt, Denis, 106
Burnet, Frank Macfarlane, 450, 768
Burnette, W. N., 100
Bush, George W., 714

Cairns, John, 530, 768
Cann, Rebecca, 508, 661
Cano, Raúl, 30
Capecchi, Mario, 329, 481
Carrel, Alexis, 117
Carroll, Christiane Mendrez, 768
Cech, Thomas R., 130, 679, 687, 768
Centerwall, Willard, 606
Chakrabarty, Ananda M., 340, 594
Chamberlain, Houston Stewart, 265
Chargaff, Erwin, 240, 373, 768
Chase, Martha, 57, 227, 239, 725
Claus, Elizabeth, 103
Cohen, Stanley, 174, 340, 351, 769
Collins, Francis S., 158, 429, 769
Connor, J. Michael, 400
Correns, Carl Erich, 134, 153, 274, 372, 465, 769

Coryell, Charles, 693
Creighton, Harriet, 156, 743
Crick, Francis, 129, 147, 228, 234, 239, 319, 322, 336, 351, 374, 639, 678, 686, 725, 745, 769
Cuénot, Lucien, 533

D'Andrea, Alan, 103
Darlington, Cyril, 769
Darre, R. Walther, 265
Darwin, Charles, 17, 61, 259, 268, 371, 488, 568-572, 650, 704, 708, 769
Darwin, Erasmus, 769
Darwin, Horace, 260
Darwin, Leonard, 260
Dausset, Jean, 769
Davenport, Charles, 260, 715
Dawkins, Richard, 270, 570
De Gobineau, Joseph-Arthur, 265
Delbrück, Max, 769
Delihas, Nicholas, 680
Demerec, Milislav, 769
Depew, David J., 260
De Vries, Hugo, 372, 769
Dight, Charles F., 322
Dobzhansky, Theodosius, 263, 622, 769
Doolittle, W. Ford, 745
Driesch, Hans, 201
Dulbecco, Renato, 770

Einstein, Albert, 594
Eldredge, Niles, 271-272, 651
Emery, Alan, 400
Enders, John, 119
Escherich, Theodor, 528

Ferguson, Margaret Clay, 770
Fink, Gerald R., 770
Fisher, Ronald A., 260, 408, 569, 610, 618, 654, 770
Fleming, Alexander, 61
Fleming, Walter, 372
Følling, Asbjørn, 605
Franklin, Rosalind, 228, 239, 322, 374, 428, 770

Gallie, Daniel, 642
Galton, Francis, 259, 265, 502, 654, 715, 750, 770
Gardner, Howard, 476
Garnjobst, Laura Flora, 770
Garrod, Archibald, 183, 372, 458, 586, 770
Gartner, Carl Friedrich von, 770
Gebhard, Paul, 288
Gey, George, 118
Giblett, Eloise, 770
Gilbert, Walter, 428, 687, 770
Gilman, Alfred G., 770
Goddard, Henry, 261
Goldman, Ronald, 281
Goldschmidt, Richard B., 770
Goldstein, Joseph L., 770
Golenberg, Edward, 30
Goodfellow, Peter, 143
Göring, Hermann, 266
Gould, Stephen Jay, 86, 263, 272, 651
Gram, Hans Christian, 55
Greenblatt, Charles L., 30
Griffith, Frederick, 56, 238, 372, 489, 725, 770
Gruhn, Ruth, 771
Grunberg-Manago, Marianne, 725
Guthrie, Robert, 606

Haeckel, Ernst Heinrich, 771
Haldane, J. B. S., 569, 610, 618, 771
Hamilton, W. D., 17
Hanafusa, Hidesaburo, 771
Hanawalt, Philip C., 771
Hardy, Godfrey, 372, 389, 462, 617, 771
Harrison, Ross, 117
Hartwell, Leland H., 771
Harvey, William, 734
Hasty, Paul, 104
Hayes, William, 56
Hayflick, Leonard, 3, 119
Haynes, Robert Hall, 771
Henson, Jim, 254
Herrick, James B., 693
Herrnstein, Richard, 86, 477
Hershberger, Scott L., 753
Hershey, Alfred D., 57, 227, 239, 725, 771
Herskowitz, Ira, 771

Hertwig, Paula, 771
Heydrich, Reinhard, 266
Himmler, Heinrich, 265
Hippocrates, 771
Hitler, Adolf, 263, 265, 392
Hogness, David S., 771
Holley, Robert W., 725, 771
Horvitz, H. Robert, 205, 771
Hunt, R. Timothy, 771
Huxley, Aldous, 262
Huxley, Julian, 269
Huxley, Thomas Henry, 260

Ingram, Vernon, 694
Inouye, Masayori, 680
Itano, Harvey, 693

Jacob, François, 56, 59, 291, 299, 374, 771
Jeffreys, Sir Alec, 218, 597, 772
Jensen, Arthur R., 86, 476
Johannsen, Wilhelm, 164, 372, 407, 772

Kainu, Tommi, 103
Kellogg, John Harvey, 262
Kelsey, Frances O., 735
Kenyon, Cynthia, 772
Khorana, Har Gobind, 314, 320, 374, 726, 772
Kimura, Motoo, 270, 548, 622
King, Helen Dean, 772
King, Mary Claire, 103
Klebs, Edwin, 214
Klee, Harry, 516
Klinefelter, Harry, Jr., 479
Klug, Aaron, 772
Knight, Thomas Andrew, 772
Köhler, Georges, 441, 723
Kölreuter, Joseph Gottlieb, 772
Kornberg, Arthur, 374, 725, 772
Kossel, Albrecht, 725, 772
Kozak, Marilyn, 640
Kroll, Kristen L., 543
Krontiris, Theodore, 103
Kunkel, Louis, 160

Labhart, Alexis, 624
Laibach, Freidrich, 514
Lamarck, Jean-Baptiste, 268, 485, 772
Landsteiner, Karl, 372
Langridge, Peter, 515
Lanza, Robert, 730

Laughlin, Harry, 261
Leder, Philip, 320, 772
Lederberg, Esther, 57
Lederberg, Joshua, 56, 772
Lee, Nancy, 293
Lehrach, Hans, 158
Lejeune, Jérôme, 244, 323
Lenz, Widukind, 735
Levene, Phoebus Aaron, 773
Levine, Arnold, 103
Lewis, Edward B., 203, 417, 773
Li, Quan Yi, 206
Liaw, Danny, 103
Lindegren, Carl, 537
Linnaeus, Carolus, 268, 501, 658, 773
Little, Clarence, 533
Livingston, David, 104
Loehlin, John C., 752
Löffler, Friedrich, 214
Lombroso, Cesare, 193
Lovell-Badge, Robin, 143
Luria, Salvador E., 773
Lwoff, André, 773
Lyell, Charles, 488
Lynch, Henry, 103
Lyon, Mary Frances, 759, 773

McCarty, Maclyn, 57, 227, 372, 725
McClintock, Barbara, 156, 490, 564, 742, 743, 773
McClung, Clarence, 372
McGue, Matt, 475
Macklin, Madge Thurlow, 773
McKusick, Victor A., 773
MacLeod, Colin, 57, 227, 372, 725
Malkin, David, 103
Margulis, Lynn, 278, 506, 773
Marks, Joan, 322
Martienssen, Robert, 516
Mather, Kenneth, 610
Matthaei, Heinrich, 314, 319, 725
Maxson, Robert, 206
Mayr, Ernst, 269, 652, 708
Mendel, Gregor, 50, 88, 134, 147, 153, 161, 184, 210, 260, 336, 371, 400, 407, 465, 491, 494-499, 523, 546, 555, 559, 568, 654, 773
Meselson, Matthew, 234, 374, 529, 550, 773

Meyerowitz, Elliot M., 774
Michurin, Ivan V., 487
Miescher, Johann Friedrich, 725, 774
Miki, Yoshio, 103
Milstein, Cesar, 441, 723
Mintz, Beatrice, 774
Mitchell-Olds, Thomas, 409
Miyashita, Norikazu, 730
Monod, Jacques, 59, 291, 299, 374, 774
Moore, Stanford, 774
Morgan, Lilian Vaughan, 774
Morgan, Thomas Hunt, 154, 164, 238, 372, 523, 774
Müller, Hermann J., 262, 372, 564, 774
Mullis, Kary B., 27, 130, 428, 611, 774
Murray, Charles, 86, 477

Nathans, Daniel, 338, 774
Neel, James, 391, 774
Nefertiti, 29
Nelson, Oliver Evans, Jr., 774
Neufeld, Elizabeth F., 774
Newton, Sir Isaac, 389, 594
Nichols, Robert C., 752
Nieuwkoop, Pieter, 543
Nilsson-Ehle, Herman, 609, 654
Nirenberg, Marshall, 314, 319, 374, 725, 774
Noller, Harry, 688
Nurse, Paul M., 775
Nüsslein-Volhard, Christiane, 203, 775

Ochoa, Severo, 374, 725, 775
Olson, Maynard V., 178, 775
Oparin, Aleksandr, 687
Orgel, Leslie, 745
Osborn, Frederick, 263

Painter, Thomas S., 157
Pauling, Linus, 547, 644, 693, 775
Pearson, Karl, 260
Penrose, Lionel, 244
Pigliucci, Massimo, 409
Pizarro, Francisco, 88
Pollender, Aloys-Antoine, 35
Pott, Percivall, 132
Potter, Van Rensselaer, 170

Prader, Andrea, 624
Prusiner, Stanley B., 632
Punnett, Reginald C., 154, 256, 372, 775
Pyeritz, Reed, 400

Redei, George, 515
Reinholz, Erna, 515
Rhabar, Shemooil, 775
Richter, Melissa, 322
Rifkin, Jeremy, 351
Rimoin, David, 400
Robbelen, G., 515
Roberts, Richard J., 775
Rodbell, Martin, 775
Rous, Peyton, 110
Rowley, Janet, 775
Rubin, Gerald M., 775
Russell, Bertrand, 262
Russell, Elizabeth Shull, 775
Russell, William, 775
Ryan, Clarence, 95

Sabin, Albert, 119
Sachs, Bernard, 727
Sager, Ruth, 277, 775
Sageret, Augustin, 775
Saiki, R. K., 27
Saint-Hilaire, Étienne and Isidore, 734
Salk, Jonas, 119
Sanger, Frederick, 428, 775
Sapienza, Carmen, 745
Schleif, Robert, 293
Schlichting, Carl, 409
Schmalhausen, Ivan, 408
Scott, Ronald B., 695
Sharp, Phillip A., 776
Shaw, George Bernard, 262
Sheldon, Peter, 651
Shelley, Mary, 351
Simpson, George Gaylord, 269, 776
Sinclair, Andrew, 160
Singer, Maxine, 776
Slamon, Dennis, 103
Smith, Hamilton O., 338, 351, 667, 776
Smith, Harry, 409
Smith, Michael, 776
Smithies, Oliver, 481
Snell, George D., 776
Sonenberg, Nahum, 640
Sonneborn, Tracy Morton, 776

Southern, Ed, 99
Spemann, Hans, 543, 776
Spencer, Herbert, 259, 776
Sperling, Karl, 581
Stadler, Lewis, 743
Stahl, Franklin, 234, 374, 529, 550
Stanley, Wendell Meredith, 776
Stebbins, G. Ledyard, 269
Steffánsson, Kári, 447
Stein, William H., 776
Steitz, Joan, 680
Stern, Curt, 157
Stevens, Nettie Maria, 776
Strobell, Ella Church, 777
Sturtevant, Alfred H., 144, 154, 164, 372, 525, 777
Sullenger, Bruce, 689
Sullivan, Louis, 194
Sulston, John E., 205, 777
Sutton, Walter, 147, 152-158, 322, 372, 523, 777, 809
Swanson, Robert, 340
Swift, Michael, 103

Tammes, Jantine, 777
Tan Jiazhen (C. C. Tan), 777
Tatum, Edward, 56, 183, 372, 537, 587, 777
Taussig, Helen Brooke, 735
Tay, Warren, 727
Temin, Howard, 116, 119, 130, 670, 777
Terman, Lewis M., 86, 261
Theologis, Athanasios, 515
Todd, Alexander Robertus, 777
Tomizawa, Jun-ichi, 680
Tonegawa, Susumu, 450, 777
Tschermak von Seysenegg, Erich, 153, 372
Tsui, Lap-Chee, 159
Turner, Henry H., 748

Ugolini, François, 103

Van der Veen, J. H., 515
Varmus, Harold E., 103, 777
Veleminsky, J., 515
Venter, Craig, 430
Vogelstein, Bert, 225
Vogt, Peter, 103
Von Nägeli, Carl, 497
Vrba, Elisabeth, 271

Waelsch, Salome Glueksohn, 777
Wakayama, Teruhiko, 730
Wallace, Alfred Russel, 269, 371, 568
Wallace, Douglas C., 508
Wasserman, David, 194
Waterston, Robert, 777
Watson, James, 129, 147, 228, 234, 239, 322, 336, 351, 374, 429, 639, 686, 725, 778
Weber, Bruce H., 260
Weinberg, Robert Allan, 778
Weinberg, Wilhelm, 372, 389, 462, 617, 778

Weiner, J., 417
Weismann, August, 147, 201, 778
Weissman, Sherman, 158
Weldon, Walter Frank, 260
Wieschaus, Eric, 203, 417, 778
Wilkins, Maurice, 228, 239, 322, 374, 778
Willi, Heinrich, 624
Williamson, Peter, 652
Wilmut, Ian, 171, 778
Wilson, Allan C., 508
Wilson, Edmund Beecher, 778
Wilson, Edward O., 409, 704-708
Witkin, Evelyn Maisel, 778

Woese, Carl R., 45, 778
Wolff, Kaspar Friedrich, 201
Wollman, Elie, 56
Woodcock, Chris, 141
Wooster, Richard, 103
Wright, Sewall, 263, 463, 610, 778

Yanofsky, Charles, 319, 779
Yanofsky, Martin, 516
Yaswen, Paul, 103
Yerkes, Robert M., 86, 261

Zinder, Norton, 57, 779
Zuckerkandl, Émile, 547

Subject Index

A page range in **boldface** indicates a full article devoted to the topic.

A1PI deficiency. *See* Alpha-1 proteinase inhibitor (A1PI) deficiency

AAT gene, 329

ABO blood types, 466. *See also* Blood groups

Abortion. *See* Spontaneous abortion

ACE gene, 395

Achondroplasia, 249, 601

Acquired characteristics, 486

Acquired immunodeficiency syndrome (AIDS), 41, 106, 243, 252, 590, 672, 680

Acromegaly, 433

Activators, 298

Acute leukemia, 106, 245

ADA. *See* Adenosine deaminase

Adaptation, 267, 568-569

Adaptive advantage, 592

Addison's disease, 52

Adenine, 549, 640

Adenosine deaminase (ADA), 329, 338

Adenosine triphosphate (ATP), 300, 504-506, 680

Adenoviruses, 197, 306, 330

ADHD. *See* Attention deficit hyperactivity disorder

Adleman, Leonard, 243

Adoption studies, 475

Adult-onset diabetes, 209

Adult stem cells, 710. *See also* Stem cells

AEC. *See* Atomic Energy Commission

AFP. *See* Alpha fetoprotein

African Americans, 661

African clawed frog. *See Xenopus laevis*

Africans, 659, 661

Agenesis, 734

Aggression, **1-2**, 765

Aging, **3-9**, 728; crop plants, 333

Agonistic behavior. *See* Aggression

Agriculture; and chloroplast genes, 136; genetic engineering, 332-335

Agrobacterium rhizogenes, 177

Agrobacterium tumefaciens, 92-93, 122, 177, 330, 332, 514, 553, 608, 740

AIDS. *See* Acquired immunodeficiency syndrome

Aire gene, 209

Albinism, **9-11**, 71, 185, 460, 556

Alcoholism, **11-13**; and aggression, 1; and criminality, 193; genetic tests, 365

Aldosteronism, 720

Algae, 77. *See also Chlamydomonas reinhardtii*

Algorithms, 80

Alkaptonuria, 183, 586

Alkylation, 132

Allele frequencies, 389-390

Alleles; biochemical mutations, 70; complementation testing, 181; complete dominance, 184; consanguinity and genetic disease, 191; dihybrid inheritance, 210; dwarfism, 248; epistasis, 255; forensic genetics, 279; gene regulation, 291; genetic screening, 357; inbreeding and assortative mating, 461; incomplete dominance, 465; linkage maps, 491; monohybrid inheritance, 555; multiple, **559-561**; organ transplants and HLA genes, 589; pedigree analysis, 599; population genetics, 617

Allelic variation, 618

Allen, Garland, 261

Allergens, 13

Allergies, **13-16**, 454

Allinen, Minna, 103

Allison, A. C., 392

Allopatric isolation, 709

Allopatric speciation, 650, 652, 708

Allopolyploidy, 614

Alpha fetoprotein (AFP), 628

Alpha helix, 636

Alpha-1 proteinase inhibitor (A1PI) deficiency, 98

Alpha-globin gene family, 290

Alternation of generations, 512

Alternative splicing, 297

Altman, Sidney, 130, 679, 687, 767

Altruism, **16-19**; in gay men, 420; and kin selection, 571; and sociobiology, 704

Alu sequences, 665

Alwine, James, 100

Alzheimer, Alois, 20

Alzheimer's disease, **19-23**, 67, 83, 243, 246, 324, 338, 471-472, 519; and prion diseases, 633

Amaya, Enrique, 543

American Eugenics Society, 261

Americans with Disabilities Act, 363

Ames, Bruce, 132

Ames test, 132, 565

Amino acids, 319, 550, 634, 638

Amish people, 192

Amniocentesis, **23-27**, 323, 360

Amniotic fluid, 24, 626

Amphibians as model organisms, 543

Amphidiploid organisms, 615

Amyloid plaques, 19

Anabolic steroids, 718

Analysis of variance, 408

Anaphase (mitosis), 123, 127, 199, 511

Anaphylaxis, 14, 214

Andrews, Tommy Lee, 280

Androgen-insensitivity syndrome, 732

Androgen receptors, 731

Androgenesis, 736

Androgens, 718, 731

Anemia; pernicious, 52; sickle-cell, 692-696. *See also* Sickle-cell disease
Anencephalus, 573-574
Anencephaly, 360, 735
Aneuploidy, 244, 402, **579-582**, 614, 627, 764-765
Anfinsen, Christian B., 635, 767
Angelman, Harry, 623
Angelman syndrome (AS), **623-626**
Angina pectoris, 392
Angiogenesis, 110
Angiogenesis inhibitors, 113
Angiotensin-cleaving enzyme (ACE), 392
Angiotensin II, 395
Animal cloning, **31-35**, 738. *See also* Cloning
Animal experimentation, 352
Animal organs, 761
Animal rights, 763
Ankylosing spondylitis, 52, 453
Annotation (genomic libraries), 386
Antagonistic pleiotropy theory of aging, 6
Antennapedia, 417
Anthrax, **35-37**, 88, 608
Antibacterial resistance, **61-65**, 608
Antibacterial soaps, 63
Antibiotic resistance, 175, 252, 348, 745
Antibiotic resistance genes, 691
Antibiotics, 347; cholera, 137; overuse, 61-65
Antibodies, **38-42**; allergies, 13; and autoimmunity, 51; blood groups, 560; cloning, 166; diphtheria, 214; genetic engineering, 338; HLA genes, 590; humanized, 96; immunogenetics, 450; molecular structure, 553; monoclonal, **441-445**; synthetic, 723-725
Anticodons, 316, 319, 551, 679
Antigenic drift, 720-721
Antigenic shift, 720
Antigens, 441; allergies, 13; autoimmune disorders, 51; blood groups, 466, 560; clonal selection theory, 450;

genetically engineered, 340; immune response, 39; neonatal, 52; organ transplants, 589; and synthetic antibodies, 723
Anti-inflammatory drugs, 719
Antimicrobial drugs, 61
Antimiscegenation laws, **501-503**
Antioxidants, 3
Antiparallel structure of DNA, 240
Antirejection drugs, 591, 724, 761
Antisense, 42
Antisense RNA, **42-45**, 577, 680
Antisense technology, 295, 297
Antisera, 441
Antisocial behaviors (ASBs), 1-2, 68
Antiterminators, 302
Antitoxin, 214
Anxiety, 67
APC gene, 112, 747
Ape-human genetics, 146
APO A1. *See* Apolipoproteins
APOA1 gene, 393
Apolipoproteins, 393, 445
Apoptosis, 6, 151, 450, 747
Arabidopsis Genome Initiative, 515
Arabidopsis thaliana, 513-516, 546
Arabinose operon, 293
Aralast (biopharmaceutical), 98
Arber, Werner, 338, 767
Archaea, **45-48**, 54; relationship with bacteria, 489
Aristotle, 767
Arms of chromosomes, 148
Aromatic amines, 132
Arthritis, 52
Artificial chromosomes, 178
Artificial selection, **48-51**, 568; breeding programs, 271; and eugenics, 260; gene therapy, 305; plants, 710
Aryan "race," 264
AS. *See* Angelman syndrome
ASBs. *See* Antisocial behaviors
Ascomycetes, 537
Ascus, 539
Asexual reproduction, 31, 125; bacteria, 56
Asilomar Conference (1975), 168, 337

Assortative mating, 461-465, 618-619
AT gene, 102
Ataxia telangiectasia, 4, 102
Atherosclerosis, 393
Athma, Prasanna, 103
Atomic Energy Commission (AEC), 428
Atopy, 454
ATP. *See* Adenosine triphosphate
ATP synthase, 136, 505-506
ATT gene, 344
Attention deficit hyperactivity disorder (ADHD), 68
Auerbach, Charlotte, 131, 767
Autoimmune disorders, 15, **51-53**, 453
Autoimmune hemolytic anemia, 52
Autoimmune polyendocrinopathy-candidiasis-ectodermal dystrophy (APECED), 209
Autoimmune response, 207
Automated sequencing, 233, 236, 430
Autopolyploidy, 614
Autosomal dominant alleles, 248, 558, 599
Autosomal dominant disorders; albinism, 10
Autosomal recessive alleles, 248, 558, 599
Autosomal recessive disorders, 4, 460; albinism, 10; Tay-Sachs disease, 727
Autosomes, 479, 499
Auxins, 121
Auxotrophic organisms, 537
Avery, Oswald, 57, 227, 238, 372, 725, 767
Azathioprine, 591
Azidothymidine (AZT), 130
Azoospermia, 479
AZT. *See* Azidothymidine

B cells, 38, 52, 106-107, 450, 482, 723
B memory cells, 38
B vitamins, 689
Baby Fay case (1984), 762
Bacilli, 54
Bacillus anthracis, 608
Bacillus cereuss, 342

Bacillus thuringiensis (*Bt*), 92, 93, 332, 366, 741
Back-mutation, 565
Backcrossing. *See* Introgression
BACs. *See* Bacterial artificial chromosomes
Bacteria, **54-60**, 527-533; relationship with archaea, 489; gene regulation, 291-295
Bacterial artificial chromosomes (BACs), 384, 520
Bacterial gene swapping, 254
Bacterial transformation, 489
Bacteriophage, 301; antibodies, 444; archaea, 47; cloning vectors, 174; DNA, 227; lambda, 327, 383; restriction, 667; RNA, 130; transduction, 57, 489. *See also* Lambda (λ) phage
Baculoviruses, 92, 94, 118
Bailey, Catherine, 767
Baker's yeast. *See Saccharomyces cerevisiae* (yeast)
Bakewell, Robert, 50
Baltimore, David, 103, 116, 130, 670, 767
Barr, Murray, 759, 767
Barr bodies, 499, 759
Basal transcription factor, 295
Base pairing, 129, 224, 228, 233, 240, 326, 641, 677
Bases, 129, 223, 228, 319, 549
Basson, Craig, 206
Bateson, William, 154, 255, 260, 372, 767
Baur, Erwin, 274
Beadle, George, 183, 372, 537, 587, 767
Beagle, HMS, 371
Beckwith, Jonathan R., 767
Bees; circadian rhythms, 83; haplodiploidy and diploidy, 17; parthenogenesis, 592; social status gene, 67; sociobiology, 705
Behavior, **65-70**; biological determinism, 86-88; criminality, 193-195; homosexuality, 419; human, 409; XYY syndrome, 764-766
Behavioral ecology, 705
Behavioral genomics, 68
Behaviorism, 263

Bell, Julia, 767
Benacerraf, Baruj, 768
Bender, William, 417
Berg, Paul, 336, 768
Beta sheets, 636
Beta-amyloid peptide, 19
Beta-galactosidase, 175, 292, 299, 484, 529
Beta-globin gene family, 290
Beta-thalassemia, 142, 684
Bickel, Horst, 606
Bicoid gene, 203
Binary fission, 56, 126, 199, 509
Binet, Alfred, 86, 409
Biochemical pathways, 70
Biodiversity, 50, 351, 708-710
Bioethics, **73-77**, 309-313; cloning, 170; stem cell research, 713
Biofertilizers, **77-79**
Biofuels, 341
Bioinformatics, **79-83**, 421
Biological and Toxin Weapons Convention (1972), 89
Biological clocks, **83-85**
Biological determinism, **86-88**
Biological species concept, 708
Biological weapons, **88-92**
Biomass, 339
Biometry, 259
Biopesticide resistance, 95
Biopesticides, **92-96**
Biopharmaceuticals, **96-98**
Bioremediation, 339
Biotechnology, 326, 351. *See also* Genetic engineering
Bioterrorism, 88, 90, 700
Birth defects. *See* Congenital defects; Diseases (hereditary)
Birth weight, 655
Bishop, J. Michael, 103, 768
Bithorax complex, 417
Black Death, 88
Blastocysts, 710
Blastulas, 202
Bleeding and hemophilia, 396-399
Blood clotting factors. *See* Clotting factors
Blood groups; ABO, 71, 185; forensic analysis, 279; multiple alleles, 559-561; paternity testing, 162, 596
Blood pressure, 395

Blotting techniques, **98-101**
Bluhm, Agnes, 768
Blumenbach, Johann Friedrich, 658
Bonnet, Charles, 268
Boring, Alice Middleton, 768
Borlaug, Norman, 370, 656, 768
Botstein, David, 157, 768
Bouchard, T. J., Jr., 475
Boveri, Theodor, 147, 201, 523, 809
Boyer, Herbert, 174, 340, 351, 768
Brachydactyly, 734
Brachyury gene, 206
Bradley, Allan, 104
Brain; and aggression, 1; Alzheimer's disease, 20; androgens, 419; anencephaly, 360, 735; atherosclerosis, 445; and behavior, 65; cells, 711; and criminality, 193; DES, 419; diseases, 4; and Down syndrome, 246; of gay men, 420; Hex A, 727; Huntington's disease, 435; hypothalamus, 83; and intelligence, 477; mouse models, 536; neural tube, 573; phenylketonuria, 357, 361, 460; pituitary gland, 432; Prader-Willi and Angelman syndromes, 624; prion diseases, 631; stem cells, 738
Brakefield, Paul, 408
Brassicaceae, 513
Braun, A., 514
Brave New World (Huxley), 262
BRCA1 and *BRCA2* genes, 101, 112, 226, 746
BRCA3 gene, 103
Bread mold. *See Neurospora crassa*
Breast cancer, **101-106**; genes, 103, 112, 226, 338, 746; genetic testing, 365; insurance, 472
Breggin, Peter, 194
Brennan, William J., Jr., 595
Brenner, Sydney, 205, 319, 768
Brewer's yeast. *See Saccharomyces cerevisiae*
Bridges, Calvin, 144, 147, 155, 372, 499

Brigham, C. C., 476
Broca, Paul, 86
Brody, William R., 473
Bromouracil, 131
Bronstein, M. H., 103
Brown, Louise, 456
Brown, Michael S., 768
Brown, Nicole, 281
Bt. See Bacillus thuringiensis
BTWC. *See* Biological and Toxin
    Weapons Convention
Buck v. Bell (1927), 716
Budding; megakarocytes, 616;
    viral, 756; yeast, 539
Buffon, comte de, 268, 487
Bunyaviruses, 756
Burger, Warren, 594
Burke, David, 178
Burkitt, Denis, 106
Burkitt's lymphoma, **106-108**,
    583
Burnet, Frank Macfarlane, 450,
    768
Burnette, W. N., 100
Bush, George W., 714
Bussey Institute, Harvard
    University, 533
Butterflies, 93

C-value paradox, 379
Caenorhabditis elegans
    (nematode), 43, 205, 516-
    520, 546
CAG expansion, 434
CAH. *See* Congenital adrenal
    hyperplasia
Cairns, John, 530, 768
Calgene (Davis, California), 333
Callus cells, 120, 737
cAMP. *See* Cyclic adenosine
    monophosphate
Campath (biopharmaceutical),
    98
Cancer, **109-115**; and aging, 4;
    antisense RNA, 45; breast,
    101-106, 338; and the cell
    cycle, 124; cell division, 127;
    chromsomal modifications,
    151; compared to embryonic
    cells, 206; gene therapy, 330;
    genetic testing, 365, 423;
    homeobox genes, 418;
    immortal cells, 731;
    insurance, 472; mouse

studies, 534; oncogenes, 583-
    586; proto-oncogenes, 685;
    skin, 9; somatic mutations,
    562; stem cell research, 712;
    synthetic antibodies, 725;
    tumor-suppressor genes, 746-
    748
Cancer Genetics Network
    (National Cancer Institute),
    105
Cann, Rebecca, 508, 661
Cannibalism, 632
Cano, Raúl, 30
Canola, 368
Capecchi, Mario, 329, 481
Capsid, 754
Capsules (bacteria), 55
Carcinogens, 109
Carcinoma, 112
Carrel, Alexis, 117
Carrier screening, 358
Carrier testing, 361
Carroll, Christiane Mendrez,
    768
Catalytic receptors, 698
Caucasians, 662
CDC. *See* Centers for Disease
    Control and Prevention
cdk's. *See* Cyclin-dependent
    protein kinases
cDNA. *See* Complementary DNA
cDNA libraries, **115-117**, 522,
    578, 675, 692. *See also*
    Complementary DNA;
    Genomic libraries
Cech, Thomas R., 130, 679, 687,
    768
Celera Genomics, 430
Cell culture; animal cells,
    117-120; plant cells, **120-122**
Cell cycle, **122-125**; breast
    cancer, 101; cell division, 127;
    cytokinesis, 199; DNA
    replication, 229; mitosis and
    meiosis, 510; proto-
    oncogenes, 111; signal
    transduction, 699; tumor-
    suppressor genes, 746
Cell differentiation, 517, 710,
    736
Cell division, **125-128**, 579; and
    aging, 3; binary fission, 56;
    and cancer, 4, 125, 424;
    chromosomes, 551;

cytokinesis, 198-200; DNA in,
    242; errors, 564; eukaryotes,
    229; gametes, 500; meiosis,
    455; and plasmids, 607;
    polyploidy, 614; proto-
    oncogenes, 107, 111;
    Saccharomyces cerevisiae, 540;
    telomeres, 729; and tumor-
    suppressor genes, 746. *See
    also* Binary fission;
    Cytokinesis; Mitosis; Meiosis
Cell-division-cycle-2 (*cdc2*) gene,
    124
Cell lines, 117-118. *See also* Germ
    cells; Stem cells
Cell plates, 512
Cell signaling, 696
Cell strains, 118
Centers for Disease Control
    (CDC), 254, 703
Centerwall, Willard, 606
CentiMorgan (cM), 492
Central dogma of molecular
    biology, **128-131**, 670, 678,
    686
Centromeres, 148, 199, 509-510,
    551
Ceramics, bioactive, 342
CF. *See* Cystic fibrosis
Chakrabarty, Ananda M., 340,
    594
Chamberlain, Houston Stewart,
    265
Chargaff, Erwin, 240, 373, 768
Chase, Martha, 57, 227, 239, 725
CHD. *See* Coronary heart
    disease
Checkpoint (cell cycle), 122
Chediak-Higashi syndrome, 10
CHEK2 gene, 103
Chemical mutagens, **131-133**,
    527, 565
Chemolithotrophic bacteria,
    342
Chemotherapy, 112
Chernobyl nuclear accident
    (1986), 356, 581
Chiasmas, 553
Chiasmata, 512
Chimerism, 412
Chimpanzee-human genetics,
    146
Chinchilla coat mutations
    (rabbits), 71, 186, 465

Chlamydomonas reinhardtii, 277, 520-522

Chloroform/isoamyl alcohol (CIA), 221

Chlorophyll, 134

Chloroplast DNA (cpDNA), **133-137**, 276, 385; isolation, 222

Chloroplasts, 133, 275; *Chlamydomonas reinhardtii*, 520

Cholera, **137-139**

Cholesterol, 393

Chondrodystrophic dwarfism, 249

Chorionic gonadotropin, 542

Chorionic villi, 24

Chorionic villus sampling (CVS), **23-27**, 324, 360, 628

Choroideremia gene, 160

Christmas disease, 397

Chromatids, 126-127, 509-510

Chromatin, 126, 140, 149, 511, 550

Chromatin packaging, **140-144**, 296

Chromatin remodeling, 295

Chromatography, 643

Chromosomal defects, 399

Chromosome mutations, **144-147**, 563-564

Chromosome painting, 149

Chromosome theory of heredity, 147, **152-158**, 370, 523

Chromosomes, 126, 160, 378, 550; in cell division, 199; replication, 509; structure, **147-152**

"Chromosomes in Heredity, The" (Sutton), 154

Chronic illness, 471

Chronic lymphocytic leukemia, 98

Chronic myelogenous leukemia (CML), 151, 583

Chronic progressive external ophthalmoplegia (CPEO), 504

Chymosin, 167

CIA. *See* Chloroform/isoamyl alcohol

Circadian rhythm, 83

Circular DNA. *See* Plasmids

Cirrhosis, 12, 52

Cis-platinum, 132

Cistrons, 182

Claiciviruses, 756

Classification, Linnaean, 659

Claus, Elizabeth, 103

Cleft lip, 324

Cleft palate, 189, 400, 735

Climate change; plankton proliferation, 138

Clinical genetics, 425

Clinical trials; biopharmaceuticals, 96; gene therapy, 306, 311

CLOCK mutation, 84

Clonaid, 169

Clonal deletion theory, 52

Clonal selection theory, 450

Cloning, 54, **166-170**, 326, 332, 336; animals, **31-35**; biopharmaceuticals, 343; *Escherichia coli*, 575; ethical issues, **170-174**; vs. gene therapy, 305; genes, 58, 166; human, 74; organisms, 166; parthenogenesis, 592-594; shotgun, **691-692**

Cloning vectors, 59, 116, 166, **174-179**, 339, 691. *See also* Vectors; Viral vectors

Clotting factor VIII, 59, 167, 330, 398

Clotting factor IX, 167, 397

Clotting factor XII, 396

Clotting factors, 167; heart disease risks, 394; hemophilia, 396-399

Clubfoot, 734

cM. *See* CentiMorgan

CML. *See* Chronic myelogenous leukemia

Coat color, 467; cats, 566; mice, 533; rabbits, 71, 186, 465, 560

Cocci, 54

Cockayne syndrome, 4

Coding regions, 640, 679. *See also* Exons

CODIS. *See* Combined DNA Index System

Codominance, 71, 186, 465, 557, 559

Codons, 128, 313, 319, 547, 575, 640, 678; mitochondrial, 276

Coenzymes, 689

Cohen, Stanley, 174, 340, 351, 769

Colinearity, 128

Collins, Francis S., 158, 429, 769

Colonies of bacteria, 56

Color blindness, **179-181**, 212, 400, 557

Colorectal cancer, 111, 747

Combined DNA Index System (CODIS), 219

Commensalism, 606

Comparative genomics, 387, 515

Complementary base pairing. *See* Base pairing

Complementary bases, 129

Complementary DNA (cDNA), 115, 520, 675. *See also* cDNA libraries

Complementary gene action, 257

Complementation testing, **181-184**

Complete dominance, **184-187**, 465

Composite transposons, 742, 745

Concerted evolution, 289

Concordance, 419

Condensation, 149

Conditional knockout mice, 483

Cones, 180

Congenital adrenal hyperplasia (CAH), 419, 649, 719

Congenital defects, **187-191**, 399-400; chromosomal abnormalities, 374; inborn errors of metabolism, 458-461; lactose intolerance, 484; neural tube defects, 572-575; prenatal testing, 23-27, 324, 360, 626-628, 630-631; thalidomide, 733-736

Conjugal plasmids, 608

Conjugation, 45, 489, 531; in archaea, 47; in bacteria, 56

Connor, J. Michael, 400

Consanguinity, **191-193**

Constrictions (centromeres), 148

Contact dermatitis, 15

Controlling site, 291

Cooley's anemia, 685

Copepods, 138

Copy-choice recombination, 254

Copy DNA. *See* cDNA libraries; Complementary DNA

Core promoters, 295

Corn. *See* Maize; *Zea mays* (maize)

Coronary heart disease (CHD), 392-396

Correns, Carl Erich, 134, 153, 274, 372, 465, 769

Cortisol, 719

Coryell, Charles, 693

Corynebacterium diphtheriae, 214

Cosmids, 520

Cosuppression, 553

Cotton, genetically modified, 341

Council for Responsible Genetics (Boston), 353

cpDNA. *See* Chloroplast DNA

CPEO. *See* Chronic progressive external ophthalmoplegia

Craniosynostosis, 206

Creighton, Harriet, 156, 743

Creutzfeldt-Jakob syndrome, **631-634**

CRG. *See* Council for Responsible Genetics

Cri du chat syndrome, 146, 148, 403; detection, 25

Crick, Francis, 129, 147, 228, 234, 239, 319, 322, 336, 351, 374, 639, 678, 686, 725, 745, 769

Crime; and aggression, 2

Criminality, **193-195**

Crop plants; polyploidy, 616

Crop yields, 367. *See also* High-yield crops; Transgenic organisms

Crossbreeding, 161

Crossed eyes. *See* Strabismus

Crossing over, 152, 491, 512, 537

Crossover frequency, 491

Crossover maps. *See* Linkage maps

Cryopreservation, 457

Cuénot, Lucien, 533

CVS. *See* Chorionic villus sampling

Cyclic adenosine monophosphate (cAMP), 138, 300

Cyclin-dependent protein kinases (cdk's), 126

Cyclin-dependent protein kinases (cdk's), 124, 128

Cyclin-dependent kinases (cdk's), 699

Cyclins, 122, 126

Cyclosporine, 591

Cystic fibrosis, **195-198**, 328, 358; amniocentesis, 25; carriers, 460; gene therapy, 197; genetic testing, 365; incidence, 459; multiple alleles, 561; single-gene defect, 189, 400; testing, 324

Cytochromes, 505

Cytogenetics, 479, 537

Cytokines, 40, 450

Cytokinesis, 123, 127, **198-200**, 512; polyploidy, 614

Cytokinins, 121

Cytoplasm, 274

Cytosine, 549, 640

Cytosol, 319

Cytotoxic hypersensitivity reaction, 15

D'Andrea, Alan, 103

Danio rerio, 546

Darlington, Cyril, 769

Darre, R. Walther, 265

Darwin, Charles, 17, 61, 259, 268, 371, 488, 568-572, 650, 704, 708, 769

Darwin, Erasmus, 769

Darwin, Horace, 260

Darwin, Leonard, 260

Darwinism, 498

Databases; genomics, 80; Icelandic, 447-449; proteomics, 645

Daughter cells, 199, 509

Dausset, Jean, 769

Davenport, Charles, 260, 715

Dawkins, Richard, 270, 570

DBA mice, 533

DCC gene, 112

DDAVP. *See* Desmopressin acetate

Deamination, 131

Death genes, 205

deCODE Genetics, 447

Defensin (antibiotic), 196

De Gobineau, Joseph-Arthur, 265

Delbrück, Max, 769

Deleterious alleles, 354-357, 391

Deletions, 144, 564, 623

Delihas, Nicholas, 680

Dementia, 631

Demerec, Milislav, 769

Denaturing, 285, 675

Dentatorubral-pallidoluysian atrophy, 143

Deoxyribonucleic acid. *See* DNA

Deoxyribonucleoside triphosphate (dNTP), 670

Department of Energy (DOE), 428

Depew, David J., 260

Depression, 67

DES. *See* Diethylstilbestrol

Desmopressin acetate (DDAVP), 398

Determinism, 86; biological, 86-88

Deterministic mutations, 19

Deuteranopes, 180

Development, 566; *Drosophila melanogaster* studies, 526; pseudohermaphrodism, 648

Developmental biology; amphibian studies, 543

Developmental genetics, **201-207**; homeotic genes, 416-418

Developmental rate, 379

De Vries, Hugo, 153, 372, 769

DHT. *See* Dihydrotestosterone

Diabetes mellitus, 52, 59, 97, **207-210**, 328, 343, 479, 504

Diagnosis, medical, 344

Diamond, Sidney A., 594

Diamond v. Chakrabarty, 594-596

Diastropic dysplasia, 249

Dichromats, 180

Diethylstilbestrol (DES), 419, 734

Differentiated cell, 31

Differentiation, 201, 732

Dight, Charles F., 322

Dihybrid inheritance, **210-214**

Dihybrids, 72, **210-214**, 255, 491

Dihydrotestosterone (DHT), 649

Diphtheria, 138, **214-216**

Diploid cells, 153, 455, 509, 537, 539, 592

Diploid organisms, 613

Directional selection, 621

Diseases, emerging, 61, 252-255

Diseases, hereditary, 311, 328, 360, **399-406**, 457, 692-696, 727-728, 731-733, 748-750; albinism, 9-11; autoimmune disorders, 51-53; breast cancer, 101-106; Burkitt's lymphoma, 106-108; consanguinity, **191-193**; cystic fibrosis, 195-198; Down syndrome, 244-248; Fragile X syndrome, 282-284; genetic counseling and, 321-326; genetic testing and, 360-366; hemophilia, 396-399; inborn errors of metabolism, 458-461; Klinefelter syndrome, 479-480; lactose intolerance, 484-485; neural tube defects, 572-575; nondisjunction, 579-582; prenatal diagnosis, 324, 626-628, 630-631; protein markers, 645; XYY syndrome, 764-766

Diseases, mitochondrial, 503-505

Diseases, prion, 631-634

Disomy, 623

Disposable soma theory of aging, 6

Dizygotic organism, 474, 750

DNA, **237-244**; ancient, **27-31**; damage to, 223-227, 552, 564; double helical structure, 129; early studies, 57; isolation, **220-223**; repair, **223-227**, 552; repetitive, **664-667**; replication, 131, 140-144, 174, 224, 227-233, 509; structure and function, 140-144, 223, 227, 373; synthesis, 229. See also Double helix

DNA cloning, 166

DNA fingerprinting, **216-220**, 279, 363, 426, 596, 666

DNA hybridization, 166

DNA immunization, 179

DNA isolation, **220-223**, 673

DNA libraries, 115-117

DNA ligase, 327, 336

DNA microarrays, 345, 520

DNA polymerase, 27, 224, 230, 611, 670

DNA repair, 48, **223-227**; and cancer, 424; and life span, 7

DNA replication, 126, **227-233**, 242; Escherichia coli, 529

DNA sequence analysis, 326-327

DNA sequencers, 429

DNA sequencing technology, 80, **233-236**

DNA structure, 140, **237-244**

dNTP. See Deoxyribonucleoside triphosphate

Dobzhansky, Theodosius, 263, 622, 769

Doctrine of similitudes, 96

DOE. See Department of Energy

Dolly the sheep, 32-33, 168, 171, 232, 456, 730, 738

Dominance; absence of, 213; complete, 184-187; incomplete, 465-468

Dominant alleles, 71, 364, 555, 559, 599; pea plants, 495

Doolittle, W. Ford, 745

Dosage compensation, 759

Double helix (DNA), 140, 237

Double knockout mice, 482

Double-minutes (DMs), 583

Down-regulation, 42

Down syndrome, 146, 189, **244-248**, 323, 360, 374, 400, 403, 563, 579; and aging, 4; detection, 25; familial, 247

Driesch, Hans, 201

Drosophila melanogaster, 84, 131, 144, 154, 203, 372, 417, 463, 522-527, 546, 564-565, 569, 618; crossover, 492; lateral gene transfer, 490

Drug resistance, 62, 252

Druse, 192

Duchenne muscular dystrophy. See Muscular dystrophy

Dulbecco, Renato, 770

Dunkers, 192

Duplicate gene interaction, 258

Duplications, 144, 157, 290, 564

Dwarfism, **248-251**, 330, 432; and consanguinity, 192

Dysmorphology, 421

E. coli Genome Project, 530

Ebola virus, 252

EBV. See Epstein-Barr virus

EcoRI enzyme, 668

Edwards' syndrome, 360, 374, 403, 580

Ehrlich, Paul, 61

Eight-cell stage, 171, 457, 627

Einstein, Albert, 594

El Niño, 138

Eldredge, Niles, 271-272, 651

Electron transport chain, 505

Electrophoresis, 99, 285-287, 672, 693. See also Gel electrophoresis

Electroporation, 329, 332

Eli Lilly, 340

Ellis-van Creveld syndrome, 192

Elongation phase, 642

Embryo transfer, **454-458**

Embryology, 542

Embryonic development, 201-207, 418

Embryonic stem cells, 481, 533, 710. See also Stem cells

Emerging diseases, 61, **252-255**. See also Viruses

Emery, Alan, 400

Encephalocele, 574

Endemic, 137

Enders, John, 119

Endocrine glands, 432

Endomitosis, 616

Endosymbiont theory, 277

Englesberg, Ellis, 293

Enhancers, 295, 301

Enterococcus, 63

Entrez system, 80

Environment, heredity and, 406-411. See also Nature vs. nurture

Enzymes, 466, 550, 634, 667. See also specific names

EPCs. See Epiphysial plate cells

Epidermal growth factor, 59

Epigenesis, 201

Epilepsy, 504

Epiphysial plate cells (EPCs), 433

Epistasis, 72, **255-259**

Epithelial cells, 195

Epitopes, 39

Epstein-Barr virus (EBV), 106, 724

Error catastrophe of aging, 7

Erythropoietin, 167

Escherich, Theodor, 528

Escherichia coli, 527-533; beta-galactosidase, 299; biopharmaceuticals, 374;

cloning vectors, 174, 340; DNA repair, 545; genetic code research, 316; genetic mapping, 155; genome sequencing, 530; lateral gene transfer, 489; monoclonal antibodies, 41; mutualisms, 54; pathogenic, 62; plasmids, 336, 608; rRNA, 290; synthetic genes, 726. *See also* *E. coli* Genome Project

Estradiol, 718

Estrogens, 718, 748; DES, 419

ESTs. *See* Expressed sequence tags

Ethanol, 78

Ethnic weapons, 88

Etiology, 573

Euchromatin, 149

"Eugenical Sterilization" (Laughlin), 263

Eugenics, 65, **259-264**, 322, 353, 376, 392, 501-502, 658, 715; Nazi Germany, **264-267**, 392, 716; progeny, 312

Eugenics Education Society of London, 260

Eukaryotes, 54, 682, 728; DNA, 729; gene regulation, 295-298; lateral gene transfer, 490

Euphenics, 263

Eusociality, 704

Euthanasia, 265-266

Evolution, 267, 568-572; and altruism, 16; ape-human genetics, 146; and biological determinism, 87; chloroplasts, 136; concerted, 289; and gene families, 290; human, 426; plasmids' role, 608

Evolutionary biology, **267-274**, 420, 486, 568, 622; ancient DNA, 27-31; artificial selection, 48-51; and genetics, 370-376; Lamarckianism, 485-489; molecular clock hypothesis, 547-549; natural selection, 568-572; punctuated equilibrium, 650-653; RNA world, 686-690

Evolutionary theory, 371

Exogenous genes, 332

Exons, 101, 129-130, 297, 551, 679, 684

Explants, 120

Expressed sequence tags (ESTs), 522; library, 384

Expression cassettes, 304, 306

Expression vectors, 177

Expressivity, 602, 604

Extra X aneuploidy. *See* Metafemales

Extrachromosomal inheritance, **274-278**

Extreme halophiles, 45

Eye color; fruit flies, 72, 183, 467, 523; humans, 659

F plasmid, 608

F_1 generation, 255

F_2 generation, 255

Fabry disease, 727

Factors (genes), 162, 184

Factors, clotting. *See* Clotting factors

FAD. *See* Familial Down syndrome

Familial Alzheimer's disease, 19

Familial Down syndrome, 247, 403

Familial glucocorticoid deficiency (FGD), 719

Familial hypercholesterolemia, 446

Fate maps, 202, 542-543

FDA. *See* Food and Drug Administration

Ferguson, Margaret Clay, 770

Fermentation (biofertilization), 78

Fertilization, 455, 592

Fertilizers, 77-79

Fetal alcohol syndrome, 12, 189

Fetal blood sampling, 324

FGD. *See* Familial glucocorticoid deficiency

FGFR1 gene, 103

Fibroblast growth factor receptor gene 1. *See* *FGFR1* gene

Film depictions of genetics, **376-378**

Filoviruses, 756

Fink, Gerald R., 770

First International Peace Conference (1899), 88

FISH. *See* Fluorescent in situ hybridization

Fishel, Richard, 225

Fisher, Ronald A., 260, 408, 569, 610, 618, 654, 770

Fission, 144

Fitness, 267, 347, 568-569, 617. *See also* Natural selection

Flagellin operons, 294

Flaviviruses, 756

Flavr Savr tomato, 168, 351, 741

Fleming, Alexander, 61

Fleming, Walter, 372

Flu. *See* Influenza

Fluorescent in situ hybridization (FISH), 116, 623, 628

FMR1 gene, 282

Følling, Asbjørn, 605

Food and Drug Administration (FDA), 169, 311

Food safety, 369

For gene (bees), 67

Ford, C. E., 748

Foreign DNA, 174

Forensic genetics, **279-282**, 421, 426, 596, 674; DNA fingerprinting, 219; PCR, 613

Fossil record, 272, 651

Foundations of the Nineteenth Century, The (Chamberlain), 265

Four-o'clock plants, 134, 274, 465

Fragile X syndrome, 67, 143, 232, **282-284**, 666

Frame shifts, 563. *See also* Open reading frames; Reading frames

Frankenstein (Shelley), 351

Franklin, Rosalind, 228, 239, 322, 374, 428, 770

Fraternal twins, 475. *See also* Twin studies

Free radical theory of aging, 7

Free radicals, 3

Free-running cycle, 83

Frogs. *See* *Xenopus laevis*

Fruit fly. *See* *Drosophila melanogaster*

Functional genomics, 387

Fungal Genome Initiative, 538

Fungi; biopesticides, 94; sequencing projects, 385. *See also* *Neurospora crassa*

Fusion, 144

G₁. *See* Gap 1

G₂. *See* Gap 2

GAL1 gene, 541

Galactosemia, 25, 71, 358, 484

Galápagos Islands, 371

Gallie, Daniel, 642

Galton, Francis, 259, 265, 502, 654, 715, 750, 770

Gametes, 455, 494

Gametophytes, 512

Gap 1 (G₁), 123, 127

Gap 2 (G₂), 123, 127

Gap genes, 204

Gardner, Howard, 476

Garnjobst, Laura Flora, 770

Garrod, Archibald, 183, 372, 458, 586, 770

Gartner, Carl Friedrich von, 770

Gastrulation, 202

Gaucher disease, 727

Gay orientation. *See* Homosexuality

Gebhard, Paul, 288

Gel electrophoresis, 99, **285-287**, 672. *See also* Electrophoresis

GenBank, 79, 385, 423

Gender identity, **287-289**, 411-413, 648, 731-733, 759-761

Gene expression, 201, 237, 242, 299; masking, 256

Gene families, **289-291**

Gene flow, 389; GM to wild outcrossing, 349-350, 369, 438, 490; isolating mechanisms, 652

Gene guns, 122. *See also* Shotgun cloning

Gene mutations, 563; *APP*, 21; *CF* gene, 160; vs. chromosome mutations, 144; circadian rhythms, 84; sexual phenotypes, 287. *See also* Chromosome mutations; Mutations

Gene pool, 389, 561, 617

Gene regulation; bacteria and, 59, **291-295**; eukaryotes, **295-298**; *lac* operon and, **298-300**; viruses, **301-304**

Gene replacement therapy, 197, 424

Gene silencing, 42, 553

Gene splicing. *See* Genetic engineering

Gene swapping, 254

Gene targeting, 343

Gene therapy, 178, 243, **304-309**, 338, 343, 345, 421, 554, 689; and antisense RNA, 45; cancer, 113; cystic fibrosis, 197; dominant disorders, 187; ethical and economic issues, **309-313**, 352; pre-implantation, 457; stem cell research, 713

Gene transfer, 56, 329, 421, 489; lateral, 489-490

Genentech, 337, 340, 351

General Electric, 594

Genes, defined, 161, 291, 606

Genetic anticipation, 434

Genetic code, **313-318**; cracking of, **319-321**; second, 317

Genetic counseling, **321-326**, 425, 601

Genetic database, 447

Genetic determinism, 376

Genetic discrimination, 325

Genetic disorders, 357. *See also* Congenital defects; Diseases

Genetic diversity; artifical selection, 48; ethnic and racial, 658; genetic engineering, 350; genetic load, 355; mutagenesis, 561; patents on life-forms, 595; speciation, 709. *See also* Biodiversity

Genetic drift; evolutionary biology, 267; Hardy-Weinberg law, 389; population genetics, 620; speciation, 709

Genetic engineering, **326-331**, 366, 554; agriculture, **332-335**, 414; biological weapons, 88; historical development, **335-339**; industrial applications, **339-343**; medicine, **343-347**; risks, **347-350**; social and ethical issues, **351-354**

Genetic load, **354-357**

Genetic mapping, 153, 513. *See also* Fate maps; Genome maps; Linkage maps

Genetic merit, 48

Genetic profile, 447

Genetic recombination, 744

Genetic screening, 323, **357-360**, 421; insurance, 471; prenatal diagnosis, 626-628, 630-631; Tay-Sachs disease, 727; XYY syndrome, 764

Genetic testing, **360-364**, 421; amniocentesis and chorionic villus sampling, 23-27; bioethics, 73; cystic fibrosis, 198; ethical and economic issues, **364-366**; prenatal diagnosis, 626-628, 630-631

Genetic toxicology, 131

Genetic transformation, 339

Genetic variation, 48, 656

Genetical Theory of Natural Selection, The (Fisher), 260

Genetically modified (GM) foods, 332, 351, **366-370**; allergenicity, 347

Genetically modified organisms (GMOs), 168, 306, 366, 437, 489-490; patents, 594-596; plants, 122. *See also* Transgenic organisms

Geneva Protocol (1925), 89

Genitalia, ambiguous, 648

Genome maps, 50

Genomes, 133, 274, 332, 347; *Arabidopsis thaliana*, 514; bacteria, 56; bacteriophage, 130; bioinformatics, 79; *Caenorhabditis elegans*, 519; cDNA library probes, 116; *Chlamydomonas reinhardtii*, 277, 521; chloroplast, 135; *Drosophila melanogaster*, 527; duplications, 144; *Escherichia coli*, 532; eukaryotes, 729; gene therapy, 305; human, 140; Human Genome Project, 428-432; influenza, 254; knockouts, 481; mechanisms of change, 552; mitochondrial, 356, 507; model organisms, 545; *Mus musculus*, 535; *Neurospora crassa*, 538; nuclear vs. organelle, 550; patenting, 595; plasmons, 275; prokaryotes, 728; pseudogenes, 646; repetitive DNA, 548, 664; retroviruses, 116; RNA, 670; *Saccharomyces*

cerevisiae, 539; sequencing, 234; size, **378-380**; smallpox, 700; trangenics, 739; transposition, 744; variation among individuals, 473; viral, 178, 301, 754; *Xenopus laevis,* 543. *See also* Cloning vectors; Genetic engineering; Human Genome Project
Genomic libraries, 58-59, 158, 326, **380-384**, 691. *See also* cDNA libraries
Genomic mutations, 563-564
Genomic walking, 383
Genomics, 79, 97, **384-388**, 421, 428, 431, 533; medical, 473
Genotypes, 70, 184, 267, 347, 407, 411, 617, 648, 654
Genotypic variation, 618
Germ cells, 309, 730. *See also* Cell lines; Stem cells
Germ-line gene therapy, 343
Germ-line mutation, 101
Germ-plasm theory of development, 201
Germany, Nazi, 264; eugenics, **264-267**
Germinal mutation, 562
Germ-line therapy, 309
Gey, George, 118
Gibberellins, 121
Giblett, Eloise, 770
Giemsa staining, 149
Gigantism, 433
Gilbert, Walter, 428, 687, 770
Gilman, Alfred G., 770
Glanzmann thrombasthenia, 666
Global Influenza Surveillance Network, 722
Glomerulonephritis, 52
Glucocorticoids, 718
Glycomics, 645
Glyphosate, 333, 368, 440
GM foods. *See* Genetically modified (GM) foods
GMOs. *See* Genetically modified organisms
Goddard, Henry, 261
Golden rice, 333, 368, 741
Goldman, Ronald, 281
Goldschmidt, Richard B., 770
Goldstein, Joseph L., 770

Golenberg, Edward, 30
Gonads, 411, 648
Gonorrhea, 61
Goodfellow, Peter, 143
Goodwin, Frederick, 194
Göring, Hermann, 266
Gould, Stephen Jay, 86, 263, 272, 651
Gradualism, 272, 650
Grain yield, 656
Gram, Hans Christian, 55
Gram stain technique, 55
Grammar of Science (Pearson), 260
Graves' disease, 52
Green manure, 78
Green Revolution, 413, 656
Greenblatt, Charles L., 30
Greenpeace, 353
Griffith, Frederick, 56, 238, 372, 489, 725, 770
gRNAs. *See* Guide RNAs
Group selection, 568
Growth factors, 641. *See also* Epidermal growth factor
Growth hormone. *See* Human growth hormone
Gruhn, Ruth, 771
Grunberg-Manago, Marianne, 725
GTP-binding proteins, 697
Guanine, 549, 640
Guide RNAs (gRNAs), 577
Guthrie, Robert, 606
Guthrie test, 357, 606
Gynecomastia, 479

H5N1 influenza virus, 252
Habitat destruction, 93
Haeckel, Ernst Heinrich, 771
Haemophilus influenzae, 430
Hageman factor, 396
Haldane, J. B. S., 569, 610, 618, 771
Halobacteria, 55
Halobacterium salinarum, 47
Haloferax volcanii, 47
Hamilton, W. D., 17
Hanafusa, Hidesaburo, 771
Hanawalt, Philip C., 771
Haplodiploidy, 16-17
Haploid cells, 153, 455, 509, 537, 539, 592
Haplotype, 450

Happy puppet syndrome. *See* Angelman syndrome
Hardy, Godfrey, 372, 389, 462, 617, 771
Hardy-Weinberg law, **389-392**, 462, 569, 617
Harrison, Ross, 117
Hartwell, Leland H., 771
Harvey, William, 734
Harvey rat sarcoma oncogene 1. *See HRAS1* gene
Hasty, Paul, 104
Hayes, William, 56
Hayflick, Leonard, 3, 119
Hayflick limit, 3, 119
Haynes, Robert Hall, 771
Head-to-tail arrays, 664
Health insurance, 471-474
Heart disease, **392-396**, 400
Heart transplants, 762
Height genes, 465
HeLa cells, 117-118
Hemagluttinin, 720
Hemizygotes, 599
Hemizygous, 399
Hemoglobin, 693
Hemoglobin gene, 466
Hemophilia, 25, 146, 330, **396-399**, 666; hemophilia A, 396; hemophilia B, 396
Hemostasis, 396
Henson, Jim, 254
Hepadnaviruses, 301, 756
Hepatitis; vaccine, 338, 344
HER-2/neu growth factor gene, 103
Herbicide resistance, 366, 439
Herceptin, 103
Hereditary Genius (Galton), 260, 502
Heredity and environment, 66, **406-411**. *See also* Environment; Nature vs. nurture
Heritability, 48, 65, 407, 419, 609, 654
Hermansky-Pudlak syndrome, 10
Hermaphrodism, 287-288, **411-413**, 648-650, 735; *Caenorhabditis elegans,* 517
Herpesviruses, 306, 756
Herrick, James B., 693
Herrnstein, Richard, 86, 477

Hershberger, Scott L., 753
Hershey, Alfred D., 57, 227, 239, 725, 771
Herskowitz, Ira, 771
Hertwig, Paula, 771
Heterochromatin, 149
Heterochrony, 650, 652
Heteroplasmy, 503-504
Heterozygote advantage, 693
Heterozygotes, 70-71, 184, 210, 400, 462, 465, 599, 621, 693
Hex A. *See* Hexosaminidase A
Hexosaminidase A (Hex A), 727
Heydrich, Reinhard, 266
High-density lipoprotein (HDL), 393, 445
High risk (insurance term), 471
High-yield crops, 333, **413-416**
Himmler, Heinrich, 265
*Hin*dII nuclease, 667
Hippocampus, 19
Hippocrates, 161, 771
Histocompatibility antigens, 589
Histone proteins, 140, 147, 149, 290, 550
History of genetics, **370-376**
Hitler, Adolf, 263, 265, 392
HIV. *See* Human immunodeficiency virus
HLA. *See* Human leukocyte antigen
Hogness, David S., 771
Holley, Robert W., 725, 771
Holocaust, 266
Homeoboxes, 526
Homeotic genes, 203, 272, **416-418**, 526
Homo sapiens, 659. *See also* Human Genome Project
Homogeneously staining regions (HSRs), 583
Homologous chromosomes, 147, 491, 614
Homologous organisms, 545
Homologous pairs, 509
Homologs, 509
Homosexuality, 68, 288, **419-421**; twin studies, 67
Homozygotes, 70, 184, 210, 400, 462, 465, 599, 621, 693
Homunculus, 201

Hong Kong flu, 252. *See also* Influenza
Horizontal gene transfer. *See* Lateral gene transfer
Hormonal therapy, 250, 480
Hormones, 718; and autoimmune disorders, 52; and homosexuality, 419; human growth, 432-434; plant, 121, 177, 737; signal transduction, 696-699; steroid, 717-720. *See also* specific hormones
Horvitz, H. Robert, 205, 771
Hostility, 67
HRAS1 gene, 102-103
HSRs. *See* Homogeneously staining regions
Human Cloning Prohibition Act, 74, 169
Human genetics, **421-428**; nature vs. nurture, 409
Human Genome Diversity Project, 658, 662
Human Genome Project, **428-432**; bioinformatics, 80; cloning vectors, 178; comparative genomics, 490; gene therapy, 305; genetic counseling, 323, 601; hereditary diseases, 328, 405; immune system, 91; QTLs, 67; RFLP analysis, 157
Human growth hormone, 59, 167, 250, 330, 338, **432-434**, 748; genetically engineered, 374
Human immunodeficiency virus (HIV), 41, 252, 330, 680
Human leukocyte antigen (HLA), 452, 561, **588-591**, 596
Human Proteome Organization (HUPO), 645
Humulin, 343
Hunchback gene, 204
Hunt, R. Timothy, 771
Hunter-Thompson chondrodysplasia, 250
Hunter's syndrome, 727
Huntington's disease, 67, 83-84, 143, 160, 232, 324, 362, 364, **434-437**, 472, 602, 666; single-gene defect, 400

HUPO. *See* Human Proteome Organization
Hurler's syndrome, 25, 727
Hutchinson-Gilford progeria, 4
Hutterites, 192
Huxley, Aldous, 262
Huxley, Julian, 269
Huxley, Thomas Henry, 260
Hybrid vigor, 655
Hybridization, 99, **437-440**, 501, 709; in situ, 115; probe, 326
Hybridomas, 40, **441-445**, 723
Hybrids, 494
Hydrocephaly, 735
Hydrogen bonding, 129, 240, 634, 636, 641
Hypercholesterolemia, 393, **445-446**, 666
Hypersensitivity, 13
Hypertension, 395, 719
Hypogonadism, 479
Hypophysectomy, 432
Hypophysis. *See* Pituitary gland
Hypopituitarism, 250
Hypostatic genes, 256
Hypothalamus, 83

Icelandic Genetic Database, **447-449**
Identical twins, 475. *See also* Twin studies
Immigration, and eugenics, 259
Immortal cells, 731
Immune complex hypersensitivity reaction, 15
Immune response, 39, 453, 589; cloning vectors, 179; monoclonal antibodies, 443; organ transplants, 589; signal transduction, 698; swine flu, 721; vaccines, 344
Immune system, 14, 51, 88, 179; and aging, 4; and cancer, 113; monoclonal antibodies, 443; mouse studies, 482; organ transplants, 761
Immunization, 215; passive, 723. *See also* Vaccination
Immunogenetics, **449-454**
Immunogens, 13
Immunoglobulins, 38, 441, 450; IgE, 52. *See also* Antibodies
Imprinting, 623

Impulsiveness, 67; and aggression, 1
In situ hybridization, 115
In vitro, 481
In vitro fertilization (IVF), **454-458**, 468
In vivo, 481
Inborn errors of metabolism, 71, 182, 357, 361, 372, 400, **458-461**. *See also* Mendelian defects; Single-gene defects
Inbreeding, 462, 533, 618; and assortative mating, **461-465**
Inbreeding depression, 355-356, 462, 655
Inclusive fitness, 16-17
Incomplete dominance, 71, 184, 187, **465-468**
Independent assortment, 153, 214, 275, 491, 494, 497
Induced mutant strains, 482
Induction, 201-202
Infertility, **468-470**, 479, 719; nuclear transplantation, 456; and Turner syndrome, 749. *See also* Klinefelter syndrome
Influenza, 252, 590; 1918 pandemic, 253; swine flu, 720-723; viruses, 304
Informed consent, 73, 447
Ingram, Vernon, 694
Inhibiting gene action, 257
Initiation codons, 640
Inouye, Masayori, 680
Insects; pests, 92, 348; social, 592
Insert DNA, 174
Insertion, 744
Insertion elements, 531
Insertion sequence, 45
Insertional mutagenesis, 520
Institute for Genome Research. *See* The Institute for Genome Research
Insulators, 295
Insulin, 309; and diabetes, 207-210; genetically engineered, 59, 97, 167, 329, 337, 340, 374
Insulin (human), 294, 328
Insulin, genetically engineered, 343
Insulin-dependent diabetes, 208
Insulin-like growth factors, 433

Insurance, 325, 363, 365, **471-474**, 630
Intelligence, 68, 194, 283, 409, **474-478**, 655, 765; metafemales, 500
Intelligence quotient (IQ), 68, 86, 474, 477, 655
Intelligence testing, 86, 261, 409, 476
Intercalating agents, 132
Interference RNA. *See* RNA interference
Interferon, 167, 338
Interphase, 199, 510
Interphase (mitosis), 123, 127, 199
Interphase chromosomes, 142
Interracial marriage, 501-503
Interrupted mating, 56
Interspecific transmission, 252
Introgression, **437-440**
Introns, 115-116, 129-130, 297, 547, 551, 575-576, 646, 679, 684
Inversions, 144, 149, 157, 553, 564
IQ. *See* Intelligence quotient
Iraq and biological weapons, 90
Irritability and aggression, 1
Isolating mechanisms, reproductive, 192, 438, 652, 708-709
Isolation, RNA, 674-676
Itano, Harvey, 693
ITGB3 gene, 394
IVF. *See* In vitro fertilization

Jackson Laboratory (Bar Harbor, Maine), 482
Jacob, François, 56, 59, 291, 299, 374, 771
Jeffreys, Sir Alec, 218, 597, 772
Jensen, Arthur R., 86, 476
Johannsen, Wilhelm, 164, 372, 407, 772
Johnson Act (1924), 261
Jumping genes, **158-160**. *See also* Transposable elements; Transposons
Junk DNA, 378, 425, 578, 666. *See also* Repetitive DNA; Selfish DNA
Juvenile-onset diabetes, 208

Kainu, Tommi, 103
Karyotyping, 147, 149, 411, 479, 579, 628, 648, 748
Katskee v. Blue Cross Blue Shield of Nebraska (1994), 471
kb. *See* Kilobase pairs
Kearn-Sayre syndrome, 278, 504
Kellogg, John Harvey, 262
Kelsey, Frances O., 735
Kennedy's disease, 143
Kenyon, Cynthia, 772
Khorana, Har Gobind, 314, 320, 374, 726, 772
Kilobase pairs (kb), 158
Kimura, Motoo, 270, 548, 622
Kin recognition, 18
Kin selection, 16-17, 571, 704
Kinases, 122, 229
Kinetochores, 511, 551
King, Helen Dean, 772
King, Mary Claire, 103
Klebs, Edwin, 214
Klee, Harry, 516
Klinefelter, Harry, Jr., 479
Klinefelter syndrome, 400, 470, **479-480**, 564, 580, 760
Klug, Aaron, 772
Knight, Thomas Andrew, 772
Knirps gene, 204
Knockout genetics, 343, 345, **481-483**
Knockout mice, 329, **481-483**
Knudsen, Alfred, 746
Köhler, Georges, 441, 723
Kölreuter, Josef Gottlieb, 437, 772
Kornberg, Arthur, 374, 725, 772
Kossel, Albrecht, 725, 772
Kozak, Marilyn, 640
Kroll, Kristen L., 543
Krontiris, Theodore, 103
Krüppel gene, 204
Kunkel, Louis, 160
Kuru disease, **631-634**

Labhart, Alexis, 624
Laboratory for Experimental Evolution, 260
lac operon, 59, 292, 298-300, 529, 682
Lactase, 484
Lactose, 291, 484, 682
Lactose intolerance, **484-485**
Lagging-strand synthesis, 230

Laibach, Freidrich, 514

Lamarck, Jean-Baptiste, 268, 485, 772

Lamarckianism, **485-489**

Lambda (λ) phage, 291, 301-304, 327, 380, 383. *See also* Bacteriophage

Landsteiner, Karl, 372

Langridge, Peter, 515

Lanza, Robert, 730

Lateral gene transfer, 47, **489-490**, 553

Laughlin, Harry, 261

Lazy eye. *See* Strabismus

LDLR gene, 393

Leber's hereditary optic neuropathy (LHON), 504

Leclerc, Georges-Louis. *See* Buffon, comte de

Leder, Philip, 320, 772

Lederberg, Esther, 57

Lederberg, Joshua, 56, 772

Lee, Nancy, 293

Left ventricular hypertrophy (LVH), 394

Lehrach, Hans, 158

Leigh disease, 278

Lejeune, Jérôme, 244, 323

Lentiviruses, 306

Lenz, Widukind, 735

Lesbian orientation. *See* Homosexuality

Lesch-Nyhan syndrome, 25, 481, 483

Lethal alleles, 354-357, 391, 601

Leukemia; acute, 245; chronic lymphocytic, 98; chronic myelogenous, 151

Levene, Phoebus Aaron, 773

Levine, Arnold, 103

Lewis, Edward B., 203, 417, 773

LHON. *See* Leber's hereditary optic neuropathy

Li, Quan Yi, 206

Li-Fraumeni syndrome, 102, 746

Liability, 1

Liaw, Danny, 103

Libraries, genomic, 380-384

Liddle syndrome, 720

Life span, 6

Ligase, 166

Ligation, 380

Lindegren, Carl, 537

LINES. *See* Long interspersed sequences

Linkage, 65, 67, 153, 161, 214

Linkage analysis, 361

Linkage groups, 151

Linkage maps, 153, 155, **491-493**, 524, 537

Linked traits, 523

Linnaeus, Carolus, 268, 501, 658, 773

Linnean Society, 371

Liposomes, 329, 740

Little, Clarence, 533

Livestock pedigrees, 50

Livingston, David, 104

Locus concept, 182, 184, 208, 255, 357, 491

Locus control regions, 142

Loehlin, John C., 752

Löffler, Friedrich, 214

Lombroso, Cesare, 193

Long interspersed sequences (LINES), 646-647, 665

Lovell-Badge, Robin, 143

Low-density lipoprotein (LDL), 393, 445

Low-density lipoprotein receptor (LDLR), 393

Lung cancer, 112

Lupus erythematosus, 52

Luria, Salvador E., 773

LVH. *See* Left ventricular hypertrophy

Lwoff, André, 773

Lyell, Charles, 488

Lymphocytes, 38, 51, 482. *See also* B cells; T cells

Lymphoma; Burkitt's, 106-108; non-Hodgkins, 725

Lynch, Henry, 103

Lyon, Mary Frances, 759, 773

Lyon hypothesis, 499, 759

Lyonization, 759

Lysenkoism, 486-487

Lysis, 221, 675; organelle preservation, 222

Lysogeny, 301

Lysosomes, 727

McCarty, Maclyn, 57, 227, 372, 725

McClintock, Barbara, 156, 490, 564, 742, 743, 773

McClung, Clarence, 372

McGinnis, William, 417

McGue, Matt, 475

Macklin, Madge Thurlow, 773

McKusick, Victor A., 558, 773

MacLeod, Colin, 57, 227, 372, 725

Macroevolution, 272, 391

Mad cow disease, 631

Maize, 742-743; *Bt* maize, 368; cpDNA, 278. *See also Zea mays*

Major histocompatibility complex (MHC), 18, 51, 452, 590

Malaria, 392

MALDI-TOF mass spectrometry, 645

Male infertility, 469

Malkin, David, 103

Manx allele, 187

MAOA. *See* Monoamine oxidase A

Maps. *See* Fate maps; Genome maps; Linkage maps; Physical mapping; Restriction maps

Margulis, Lynn, 278, 506, 773

Markers, 158, 360, 691

Marks, Joan, 322

Marriage, interracial, 501-503

Martienssen, Robert, 516

Masking gene action, 258

Mass spectroscopy, 81, 643

Maternal altruism, 16

Maternal effect genes, 204

Maternal inheritance, 503

Mather, Kenneth, 610

Mating types (yeast), 539

Matthaei, Heinrich, 314, 319, 725

Maturation promoting factor (MPF), 124, 128

Maxam-Gilbert sequencing, 233, 235

Maxson, Robert, 206

Mayr, Ernst, 269, 652, 708

MCH. *See* Molecular clock hypothesis

Medical genetics, 343-347

Medical genomics, 473

Megakaryocytes, 616

Meiosis, 126, **509-513**; chromosome disjoining, 155; chromosome structure, 151; classical transmission genetics, 161; crossing over,

537; cytokinesis, 199; Down syndrome, 244; *Drosophila melanogaster*, 525; in vitro fertilization, 455; linkage maps, 491; metafemales, 499; nondisjunction and aneuploidy, 579; parthenogenesis, 592; XYY syndrome, 764. *See also* Cell division; Mitosis

Melanin, 185, 658

Melanism, 9

Melanoma, 747

Mendel, Gregor, 50, 88, 134, 147, 153, 161, 184, 210, 260, 336, 371, 400, 407, 465, 491, 494-499, 523, 546, 555, 559, 568, 654, 773

Mendelian defects. *See also* Inborn errors of metabolism; Single-gene defects

Mendelian genetics, 134, 153, 162, 184, 210, 268, 370, 389, **494-499**, 555, 568

Mendelian Inheritance in Man (McKusick), 558

"Mendelian Proportions in a Mixed Population" (Hardy), 389

Mendelian traits, 399

Mendelism vs. Darwinism, 568

Meningocele, 574

Meningomyelocele, 574

Mental retardation, 470. *See* Chromosomal disorders; Cri du chat syndrome; Down syndrome; Fragile X syndrome; Galactosemia; Klinefelter syndrome; Nondisjunction; Phenylketonuria; X chromosome inactivation

Mercury II, 340

Meristic traits, 609

MERRF. *See* Myoclonic epilepsy with ragged-red fiber disease

Meselson, Matthew, 234, 374, 529, 550, 773

Messenger RNA (mRNA), 551; central dogma of molecular biology, 129; DNA structure and function, 242; *Escherichia coli*, 531; gene regulation,

298; genetic code, 316, 319; mutation and mutagenesis, 562; Northern blotting, 100; processing, **681-686**; protein synthesis, 639; RNA structure and function, 676; synthetic genes, 725

Metabolic pathways, 193, 458, 586

Metabolic rate and longevity, 6

Metabolism, 458

Metafemales, 403, **499-501**, 760. *See also* Multiple-X syndrome

Metals, toxic, 340

Metaphase (mitosis), 123, 127, 199, 511; chromosome structure, 142

Metaphase plate, 511

Metarhizium anisopliae, 94

Metastasis, 109

Methane, 341

Methanococcus voltae, 47

Methanogens, 45, 55

Methionine, 320

Methylation, 667, 760

Metric traits, 654. *See also* Quantitative traits

Meyerowitz, Elliot M., 774

MHC. *See* Major histocompatibility complex

Mice. *See* Mouse; *Mus musculus*

Michurin, Ivan V., 487

Microarray analysis, 675

Microarrays, 79, 345, 520

Microbes, genetically modified, 341

Microcephaly, 735

Microevolution, 272, 389-392

Microinjection, 329

Micropropagation, 117

microRNA (miRNA), 43

Microsatellites, 216, 665

Miescher, Johann Friedrich, 221, 725, 774

Migration. *See* Gene flow

Miki, Yoshio, 103

Milk, 484-485

Milstein, Cesar, 441, 723

Mineralocorticoid excess syndrome, 720

Mineralocorticoids, 718

Minimal media, 537

Minisatellites, 216, 665

Minnesota Twin Registry, 752

Mintz, Beatrice, 774

Mirabilis jalapa, 134, 274

Miracle rice, 333

miRNA. *See* microRNA

Miscarriage. *See* Spontaneous abortion

Miscegenation, **501-503**

Mismatch repair, 225

Mitchell, Peter, 506

Mitchell-Olds, Thomas, 409

Mitochondria, 274, 503, 550

Mitochondrial chromosomes, 506

Mitochondrial diseases, 399-400, **503-505**; extrachromosomal inheritance, 278

Mitochondrial DNA (mtDNA), 3, 7, 276, 356, 456, 504, **505-509**, 661; clones, 33; evolution, 426; human origins, 29

Mitochondrial genome, 31

Mitochondrial ribosomes, 505

Mitogens, 108

Mitosis, 123, 126, **509-513**; chromosome theory of heredity, 153; cytokinesis, 199; DNA structure and function, 242; nondisjunction and aneuploidy, 579. *See also* Cell division; Meiosis

Miyashita, Norikazu, 730

Mode of inheritance, 399

Model organisms, **545-547**; and aging, 3; *Arabidopsis thaliana*, **513-516**; *Caenorhabditis elegans*, 205, **516-520**; *Chlamydomonas reinhardtii*, **520-522**; *Drosophila melanogaster*, **522-527**; *Escherichia coli*, **527-533**; *Mus musculus*, **533-536**; *Neurospora crassa*, **536-539**; *Saccharomyces cerevisiae*, **539-542**; *Xenopus laevis*, **542-544**

Modern synthesis, 270, 498, 568, 617

Modifying gene action, 257

Mold. *See Neurospora crassa*

Molecular biology, 242

Molecular clock hypothesis (MCH), **547-549**, 730

Molecular cloning, 611

Molecular genetics, 59, 336, 428, **549-555**

Mongolism, 244

Monoamine oxidase A (MAOA), 68

Monoclonal antibodies, 40, 441-445, 454, 723

Monoculture, 413

Monod, Jacques, 59, 291, 299, 374, 774

Monohybrid inheritance, 210, 400, **555-559**

Monosomy, 403, 579

Monozygotic organisms, 474, 750

Moore, Stanford, 774

Morgan, Lilian Vaughan, 774

Morgan, Thomas Hunt, 154, 164, 238, 372, 523, 774

Morphogenesis, 120

Morphogens, 201

Morphology (embryonic development), 202

Morquio syndrome, 249

Morton, Samuel George, 86

Mosaicism; development, 201; hermaphrodites, 288, 412; Klinefelter syndrome, 479; mutation and mutagenesis, 562; trisomy 21, 247; X chromosome inactivation, 759

Moth, peppered, 570

Mouse; knockout mice, 481-483; model organisms, 533-536. *See also Mus musculus*

Mouse antibodies, 724. *See also* Hybridomas; Monoclonal antibodies

Mouse-ear cress. *See Arabidopsis thaliana*

MPF. *See* Maturation promoting factor

MPS. *See* Mucopolysaccharidosis

mRNA. *See* Messenger RNA

MSX2 gene, 206

mtDNA. *See* Mitochondrial DNA

Mucopolysaccharidosis (MPS), 249

Müller, Hermann J., 262, 372, 564, 774

Müller's ratchet, 356

Mullis, Kary B., 27, 130, 428, 611, 774

Multi-drug resistant (MDR) bacterial strains, 608

Multifactorial traits, 189, 328, 399, 573, 601, 609-611. *See also* Polygenic traits

Multinational Coordinated *Arabidopsis* Genome Research Project, 515

Multiple alleles, **559-561**

Multiple cloning site, 175

Multiple intelligence, 476-477

Multiple-resistant bacteria, 63

Multiple sclerosis, 453

Multiple X syndrome, 499-501. *See also* Metafemales

Multipotent cells, 710, 736

Murray, Charles, 86, 477

Mus musculus (mouse), 18, 533-536, 546

Muscular dystrophy, 25, 160, 328; Duchenne, 232, 328, 666

Mustard gas, 131

Mutagenesis, **561-567**; insertional, 347, 520

Mutagens, 562; chemical, 131-133

Mutation and mutagenesis, **561-567**

Mutation rates, 562

Mutations, 54, 131-133, 374, 552; and aging, 3, 7; biochemical, **70-73**; cancer, 424; disease-causing, 338; dominant alleles, 186; genetic diversity, 355; homeobox genes, 418; mtDNA, 278; neutral, 547; petite, 275; random, 659

Myasthenia gravis, 52, 453

Mycobacterium tuberculosis, 62, 252

Myelomas; multiple, 40

Myoclonic epilepsy with ragged-red fiber disease (MERRF), 504

Myotonic dystrophy, 142

Nanotechnology, 340, 342

Nathans, Daniel, 338, 351, 774

National Bioethics Advisory Commission, 169

National Cancer Institute, 105

National Center for Biotechnology Information (NCBI), 79, 385, 423

National Institutes of Health (NIH), 429

National Society of Genetic Counselors, 324

Natural selection, 16, 146, 267, 269, 347, 355, **568-572**, 621, 658; vs. artificial selection, 49; and molecular clock hypothesis, 548; and social Darwinism, 259; and speciation, 708

Nature vs. nurture, 67, 86-88, 161, 406, 474, 603, 750; heart disease, 395

Nazi Germany, 264; eugenics, 262, **264-267**

NCBI. *See* National Center for Biotechnology Information

ncRNA. *See* Noncoding RNA molecules

Neanderthals, 426

Neel, James, 391, 774

Nefertiti, 29

Negative eugenics, 259, 501, 715

Nelson, Oliver Evans, Jr., 774

Nematodes. *See Caenorhabditis elegans*

Neo-Darwinian synthesis. *See* Modern synthesis

Neonatal screening, 357. *See also* Prenatal diagnosis

Neonatal testing, 361

Neufeld, Elizabeth F., 774

Neural tube defects, **572-575**, 628

Neural tubes, 573, 626

Neuraminidase, 720

Neuroendocrine theory, 419

Neurofibrillary tangles, 19

Neurofibromatosis, 160, 666, 746

Neuropores, 573

Neurospora crassa, 183, 276, 536-539, 587, 743

Neurotic behavior, 67

Neurotransmitters, 65, 193; and aggression, 1; and behavior, 68

Neutral mutations, 547

Neutral theory of evolution, 548, 617, 622

Neutral theory of molecular evolution, 646
Newton, Sir Isaac, 389, 594
NF1 gene, 746
Nichols, Robert C., 752
Nieuwkoop, Pieter, 543
NIH. *See* National Institutes of Health
Nilsson-Ehle, Herman, 609, 654
Nirenberg, Marshall, 314, 319, 374, 725, 774
Nitrogen fixation, 78, 531
Nitrous acid, 131
N-myc gene, 583
Nobel Prize winners, 780
Noller, Harry, 688
Noncoding DNA molecules, 425. *See also* Repetitive DNA
Noncoding RNA molecules, 43, **575-579**; *table*, 576
Nondirective counseling, 322
Nondisjunction, 155, 244, 499, **579-582**, 764; Down syndrome, 247
Nonhistone proteins, 140
Nonprocessed pseudogenes, 647
Nonrandom mating, 391, 619
Nonsense codons, 642. *See also* Stop codons
Nontarget species; and biopesticies, 93
Nordic, 265
Norms of reaction, 602
Norris Cancer Center, 206
Northern blotting, 98-101
Nppa gene, 395
Nuclear division, 512
Nuclear genome, 31
Nuclear transplantation, 456
Nucleases, 667
Nucleic acids, 221, 237, 549. *See also* DNA; RNA
Nucleoli, 511
Nucleosomes, 140, 141, 149, 550
Nucleotides, 140, 223, 237, 313, 319, 549, 640, 664, 667; and mutation, 562
Nurse, Paul M., 775
Nüsslein-Volhard, Christiane, 203, 417, 775
Nystagmus, 9

Ochoa, Severo, 374, 725, 775
Ocular albinism, 10
Oil recovery, 342
Oil spills, 340
Okazaki, Reiji, 230
Okazaki fragments, 230
Oligonucleotides, 344
Olins, Ada, 141
Olins, Donald, 141
Olson, Maynard V., 178, 775
On the Natural Variety of Mankind (Blumenbach), 658
On the Origin of Species by Means of Natural Selection (Darwin), 61, 259, 268, 371, 568
Oncogenes, 101-106, 109-110, 122, 125, 308, 424, **583-586**, 746; discovery, 119
Oncoretroviruses, 304, 306
One gene-one enzyme hypothesis, 183, 373, 537, **586-588**
One gene-one polypeptide hypothesis, 183, 588
Oocyte maturation, 123
Oparin, Aleksandr, 687
Open reading frames (ORFs), 134, 136, 316, 665
Operators, 298, 301
Operons, 59, 291, 298, 528; *Escherichia coli*, 529
Ophthalmoplegia, 504
Optic neuropathy, 504
ORC. *See* Origin recognition complex
ORFs. *See* Open reading frames
Organ transplants, **588-591**; xenotransplants, 761-764
Organelle DNA, 274-278; isolation, 222. *See also* Chloroplast DNA; Mitochondrial DNA
Organismal cloning, 168
Organogenesis, 202, 418
Orgel, Leslie, 745
Origin of replication (plasmid sequence), 174
Origin recognition complex (ORC), 229
Osborn, Frederick, 263
Osmotic shock, 221

Out of Africa theory, 426
Ovarian cancer, 103
Oxidizing agents, 132

P elements, 490
p16 gene, 747
p53 gene, 101, 103, 112, 125, 746-747
Packaging DNA, 176
PAH gene, 605
Painter, Thomas S., 157
Pair-rule genes, 204
Pandemics, 137, 253, 720
Parasites and lateral gene transfer, 490
Parathyroid hormone, 167
Parthenogenesis, **592-594**, 736
Partial dominance. *See* Incomplete dominance
Particle bombardment, 329
Parvoviruses, 756
Patau syndrome, 360, 374, 403, 580
Patents on life-forms, **594-596**; human genes, 352; mice, 535
Paternity exclusion, 596
Paternity testing, 220, 280, **596-599**, 674; blood type, 561
Pauling, Linus, 547, 644, 693, 775
PB2 gene, 253
PCR. *See* Polymerase chain reacion
PDGF. *See* Platelet derived growth factor
Pea-plant experiments, 134, 162, 210, 371, 495, 545, 559
Pearson, Karl, 260
Pearson's syndrome, 278
Pedigree analysis, 50, 322, 560, **599-602**; familial Down syndrome, 247; homosexuality, 419
Pelargonium geraniums, 274
Pemphigus vulgaris, 52
Penetrance, **602-604**
Penrose, Lionel, 244
Peppered moth, 570
Peptide bonding, 635, 638
Peripheral proteins, 643
Pernicious anemia, 52
Peromelia, 734-735
Pesticides, 92-96

Petite mutations, 275

PGD. *See* Preimplantation genetic diagnosis

PGRs. *See* Plant growth regulators

Phage. *See* Bacteriophage; Lambda (λ) phage

Pharm animals, 740

Pharmaceuticals; cloning vectors, 177; genetically engineered, 343-347; proteins, 368; transgenic, 740

Pharmacogenomics, 96, 98, 421, 447-448

Phases of cell division, 126

Phases of the cell cycle, 126

Phenol, 221

Phenotypes; biochemical mutations, 70; complete dominance, 184; heredity and environment, 407; hermaphrodites, 411; incomplete dominance, 465; *Mus musculus*, 533; mutation and mutagenesis, 562; penetrance, 602; pseudohermaphrodites, 648; quantitative inheritance, 654; testicular feminization syndrome, 732

Phenotypic plasticity, 407, 602

Phenylalanine, 604

Phenylalanine hydroxylase, 605

Phenylketonuria (PKU), 71, 357, 361, 400, 459-460, **604-606**

Phenylpyruvic acid, 605

Phi X174 virus, 428

Philadelphia chromosome, 146, 151

Phocomelia, 735

Phosphorylation, 124, 126, 747

Photodynamic therapy, 113

Photophobia, 9

Photoreactivation, 224

Photosynthesis, 134

Phragmoplasts, 200

Phyletic gradualism, 650

Phylogenetic systematics, 270

Phylogenomics, 388

Phylogeny, 267, 426, 547

Physical mapping, 153, 157, 386, 493, 515, 526

Phytochromes, 409

Piebaldism, 9

Pigliucci, Massimo, 409

Pigmentation; cats, 71, 566; flowers, 257; fruit flies, 72; humans, 467, 658; maize, 742; skin, 185. *See also* Coat color; Eye color; Skin color

Pituitary gland, 432

Pizarro, Francisco, 88

PKU. *See* Phenylketonuria

Placenta, 626

Plague, 28, 88, 608. *See also Yersinia pestis*

Plankton, and cholera, 138

Plant growth regulators (PGRs), 120, 121

Plant hormones, 177

Plant protoplasts, 122

Plants, 133-137; hormones, 737; poloyploidy, 615. *See also* Agriculture; Crop plants; Genetically modified (GM) foods; High-yield crops

Plasma cells, 38

Plasmacytoma, 441

Plasmagenes, 274

Plasmids, **606-609**; anthrax, 35; antisense RNA, 43; bacterial genetics, 54; bacterial resistance, 61; biopharmaceuticals, 97; cloning vectors, 59, 166, 174, 327, 332, 336, 340; extrachromosomal inheritance, 275; prokaryotes, 380; *Saccharomyces cerevisiae*, 540; shotgun cloning, 691; transgenic organisms, 739; transposable elements, 745

Plasmons, 274

Plastics, genetically engineered, 342

Plastids, 274, 550; inheritance, 274; isolation, 222

Plastomes, 274

Platelet derived growth factor (PDGF), 584

Pleiotropy, 3, 601

Pluripotent cells, 710, 736

Point mutations, 563

Poison gas, 131

Poliovirus, 119

Pollender, Aloys-Antoine, 35

Pollination; lateral gene transfer, 490

Poly-A-binding protein (PABP), 642

Poly-A tails, 551, 640, 679, 684

Polyarteritis nodosa, 52

Polychlorinated biphenyls (PCBs), 341

Polyclonal antibodies, 39-40

Polydactyly, 186

Polygenic traits, **609-611**. *See also* Multifactorial traits

Polyglutamine tract, 434

Polyhydroxybutyrate, 342

Polymerase chain reaction (PCR), 27, 35, 79, 216, 338, 428, 457, **611-613**, 665, 674; automated sequencing, 236; and blotting, 101; synthetic antibodies, 724

Polymorphism, 157, 589, 664

Polypeptides, 586, 634, 638

Polyploidy, 380, 402, **613-617**, 709

Polyproteins, 552

Population databases, 447

Population genetics, 270, 498, 569, **617-623**; race, 658; speciation, 708

Positive eugenics, 259, 501, 715

Post-transcriptional control, 297

Post-transcriptional gene silencing, 577

Post-translational modification, 586

Potato spindle tuber viroid (PSTVd), 757

Pott, Percivall, 132

Potter, Van Rensselaer, 170

Poxviruses, 302, 700, 755

Prader, Andrea, 624

Prader-Willi syndrome (PWS), 148, **623-626**

Prairie dogs, 705

Pre-diabetes, 208

Preembryonic stage, 188

Preexisting conditions, 471

Preformationism hypothesis of embryonic development, 201

Pregnancy, 188; and age, 470

Preimplantation genetic diagnosis (PGD), 457, 627

Pre-messenger RNA, 297, 679, 683

Prenatal diagnosis, 322, 324, 360, **626-631**, 727; sickle-cell disease, 673
Prenatal testing, 24
Presenilin, 21, 519
President's Council on Bioethics (est. 2001), 74
Primary cells, 117
Primary protein structure, 635
Primers, 27, 234, 670
Prions; discovery of, 632; diseases, **631-634**
Privacy issues, 425; Icelandic Genetic Database, 448
Probe, 99
Probe hybridization, 326
Processed pseudogenes, 647
Progeroid syndromes, 4, 730
Progesterone, 718
Progestins, 718
Programmed senescence theory of aging, 6
Progressive Era eugenics, 262
Prokaryotes, 45, 54, 682, 728; DNA, 729; *Escherichia coli*, 527-533; gene regulation, 291-295; gene transfer, 489; and mitochondria, 277
Prokaryotic cells, 546. *See also* Prokaryotes
Prolastin, 167
Prometaphase (mitosis), 511
Promoters, 296, 298, 301, 416, 551, 678
Proofreading, 232, 424, 670
Prophase (mitosis), 123, 127, 199, 511
Prophylaxis, 137
Prostate cancer, 112, 338
Protanopes, 180
Protein folding structure, 643
Protein genetic code, 314
Protein markers, 643
Protein synthesis, 319, 551, 563, **638-643**; mitochondrial, 507; RNA world, 688. *See also* Translation
Proteins; composition, 319; functions, 637; structure, 550, **634-638**
Proteomics, 79, 81, 387, 421, 428, 431, **643-646**
Proto-oncogenes, 107, 125, 583-584, 685; and aging, 4

Protoplasts, 332
Proviruses, 680
Prozac, 193
Prusiner, Stanley B., 632
Pseudogenes, 289-290, **646-648**
Pseudohermaphrodites, **648-650**, 685
Pseudomona originosa, 594
Pseudomonas aeruginosa, 62
Psoralens, 132
PSTVd. *See* Potato spindle tuber viroid
Psychometricians, 474
PTEN gene, 103
Punctuated equilibrium, 272, 532, **650-653**
Punnett, Reginald C., 154, 256, 372, 775
Punnett squares, 390
PWS. *See* Prader-Willi syndrome
Pyeritz, Reed, 400
Pyrenoids, 520

QTLs. *See* Quantitative trait loci
Quantitative genetics, 408
Quantitative inheritance, **654-657**
Quantitative trait loci (QTLs), 67, 393, 407, 609
Quantitative traits, 567, 609
Quaternary protein structure, 637

R factors. *See* Resistance factors (R factors)
R groups, 634
R plasmids. *See* Resistance factors (R factors)
Rabbits; brachydactyly, 734; coat color, 71, 186, 465, 560
Race, 86, 501, **658-664**; and eugenics, 259
Race Betterment Foundation, 262
Racial discrimination, 358
Racial Hygiene Society, 265
Radiation therapy, 113
Random mating, 462
ras gene, 112, 124, 585
Rb gene, 125, 746
rbcL gene, 136
rDNA. *See* Recombinant DNA
Reaction norms, 407

Reading frames, 313, 316, 563, 640
Reassociation kinetics, 378
Reassortment; influenza viruses, 254
Receptor tyrosine kinases (RTKs), 698
Receptors, 696
Recessive alleles; biochemical mutations, 71; consanguinity, 192; cystic fibrosis, 195; genetic load, 355; genetic testing, 364; incomplete dominance, 465; monohybrid inheritance, 555; multiple alleles, 559; pea plants, 495; pedigree analysis, 599
Reciprocal altruism, 17, 704
Reciprocal translocation, 106
Recombinant DNA; bacterial genetics, 54; biological weapons, 88; cloning vectors, 174; gene therapy, 305, 309; GM foods, 332, 366; historical development, 336; industrial applications, 340; restriction enzymes, 669; shotgun cloning, 691; social and ethical issues, 351; technology, 166
Recombinant DNA technology. *See* Genetic engineering
Recombinant factor VIII. *See* Clotting factor VIII
Recombination, 512, 553
Recombinational repair process, 226
Redei, George, 515
Reductional division, 163
Reductionism, 86
Reed, Sheldon, 322
Regulative development, 201
Regulators, 643
Reification, 86
Reinholz, Erna, 515
Rejection, 761
Rennin, 167
Reoviruses, 756
Repeat-induced point (RIP) mutations, 538
Repetitive DNA, 151, 289, 379, 548, **664-667**, 729; LINES and SINES, 647; telomeres, 3

Replication, 129, 227, 528, 549; plasmids, 607; semiconservative, 550. *See also* DNA replication
Replication slippage, 290
Replicative segregation, 504
Replicons, 229
Repressor protein, 292
Repressors, 298
Reproduction technologies; parthenogenesis, 592-594
Reproductive cloning, 166
Reproductive isolating mechanisms. *See* Isolating mechanisms, reproductive
Reproductive technology; in vitro fertilization, 454-458; infertility, 468-470; prenatal diagnosis, 324
Resistance. *See* Antibacterial resistance; Biopesticide resistance; Resistance factors
Resistance factors (R factors), 61-62, 608, 742, 745
Resistance plasmids. *See* Resistance factors (R factors)
Restriction endonucleases. *See* Restriction enzymes
Restriction enzymes, **667-670**; bacterial genetics, 58, 552; biopharmaceuticals, 96; chromosome theory of heredity, 157; cloning, 166, 174, 691; genetic engineering, 326; genetic testing, 362; genomic libraries, 382; historical development, 336, 374; RFLP analysis, 672; social and ethical issues, 351; synthetic genes, 725
Restriction fragment length polymorphism analysis. *See* RFLP analysis
Restriction maps, 669
Retinoblastoma, 111, 125, 746
Retroposons. *See* Retrotransposons
Retropseudogenes, 665
Retrotransposition, 664
Retrotransposons, 553, 646-647, 744
Retroviruses; central dogma of molecular biology, 129; gene

therapy, 306; genetically engineered, 329; HIV, 41; oncogenes, 583; pseudogenes, 647; reverse transcriptase, 670; RNA structure and function, 676; RNA world, 689; transposable elements, 744; vectors, 176; viral genetics, 756
Reverse transcriptase (RT), 115, 646, **670-672**, 680. *See also* RT inhibitors
Reverse transcriptase polymerase chain reaction (RT-PCR), 675
Reverse transcription, 129, 725
Reverse transcription polymerase chain reaction (RT-PCR), 130
RFLP analysis, 100, 157, 287, 327, 362, 611, 669, **672-674**
Rhabar, Shemooil, 775
Rheumatic fever, 52
Rheumatoid arthritis, 52, 453
Ri plasmid, 177
Ribonuclease. *See* RNase
Ribonucleic acid. *See* RNA
Ribonucleoprotein complexes, 680
Ribosomal RNA (rRNA), 319, 639, 677, 686; repetitive, 290
Ribosomes, 319, 505, 551, 562, 639, 677, 686, 754
Ribozymes, 129-130, 346, 576, 677, 686-687
Ribulose biphosphate carboxylase (RuBP carboxylase). *See* Rubisco
Rice; genetically engineered, 333
Rich, Alexander, 242
Richter, Melissa, 322
Rifkin, Jeremy, 351
Rimoin, David, 400
RISC. *See* RNA-induced silencing complex
RNA, 3, 237, 313-314, 638, 675; antisense, 42-45; chloroplast, 276; early studies, 57; mitochondrial, 276, 504; noncoding, 575-579; structure and function, **676-681**

RNA-induced silencing complex (RISC), 44
RNA interference (RNAi), 43-44, 295, 297, 346, 519, 577, 680
RNA isolation, **674-676**
RNA polymerase, 142, 295, 298, 301, 678, 682, 756
RNA replicase, 687
RNA transcription, **681-686**
RNA world, **686-690**
RNAi. *See* RNA interference
RNase, 57, 222, 635, 675, 678, 756
RNase H, 116, 671
RNase P, 576, 687
RNomics, 578
Robbelen, G., 515
Roberts, Richard J., 775
Robertsonian translocation, 145
Rodbell, Martin, 775
Rotaviruses, 254
Roundworms. *See* Caenorhabditis elegans
Rous, Peyton, 110, 583
Rous sarcoma virus, 110, 583, 670
Rowley, Janet, 775
rRNA. *See* Ribosomal RNA
RS enzymes, 316-317
RT. *See* Reverse transcriptase
RT inhibitors, 672
RTKs. *See* Receptor tyrosine kinases
RT-PCR. *See* Reverse transcriptase polymerase chain reaction
Rubin, Gerald M., 775
Rubisco, 136, 276
Russell, Bertrand, 262
Russell, Elizabeth Shull, 775
Russell, William, 775
Ryan, Clarence, 95

Sabin, Albert, 119
Saccharomyces cerevisiae (yeast), 178, 275, 493, 539-542, 546, 553, 576. *See also* Yeast
Sachs, Bernard, 727
Sager, Ruth, 277, 775
Sageret, Augustin, 775
Saiki, R. K., 27
Saint-Hilaire, Étienne and Isidore, 734

Salk, Jonas, 119
Salmonella, 90
Sandhoff disease, 666
Sanger, Frederick, 428, 775
Sanger sequencing, 234-235
Sapienza, Carmen, 745
Sarcomas, 106, 110, 583, 670
SARS. *See* Severe acute
    respiratory syndrome
Satellite DNA, 665, 729
Satellite RNA, 757
Schizophrenia, 68, 338
Schizosaccharomyces pombe, 546
Schleif, Robert, 293
Schlichting, Carl, 409
Schmalhausen, Ivan, 408
Schutzstaffel (SS), 265
SCID. *See* Severe combined
    immunodeficiency
    disorder
SCID-X. *See* Severe combined
    immunodeficiency
    syndrome X
Scleroderma, 52
SCN. *See* Suprachiasmatic
    nucleus
SCNT. *See* Somatic cell nuclear
    transfer
Scott, Ronald B., 695
Scrapie, 631
Secondary protein structure,
    636
Segment polarity genes, 204
Segmentation genes, 204
Segregation, 211, 371, 494, 496,
    556
Selection. *See* Artificial selection;
    Natural selection
Selective markers, 175
Self-fertilization, 463
Self-incompatibility, 655
Selfish DNA, 271, 425, 666, 705,
    742. *See also* Junk DNA
Selfish Gene, The (Dawkins), 270,
    570
Semidominance. *See* Incomplete
    dominance
Senescence. *See* Aging
Senile plaques, 643
Sensitive period, 187
Sequenced organisms, 385. *See
    also* Genomes; Genomics
Serotonin, 67, 193; and
    aggression, 1

Severe acute respiratory
    syndrome (SARS), 252
Severe combined
    immunodeficiency disorder
    (SCID), 306, 329, 338, 359,
    453
Severe combined
    immunodeficiency syndrome
    (SCID-X), 307
Sewer sludge (biofertilizer),
    78
Sex chromosomes; *Drosophila
    melanogaster*, 523; fragile X
    syndrome, 282; hereditary
    diseases, 402; infertility, 468;
    Klinefelter syndrome, 479;
    metafemales, 499; mutation
    and mutagenesis, 564;
    Turner syndrome, 748; X
    chromosome inactivation,
    759; XYY syndrome, 764. *See
    also* Gender identity; Sex-
    linked traits; X-linked traits
Sex determination, 287, 732
Sex differences and aggression,
    2
Sex linkage, 557
Sex-linked traits, 211, 460, 467,
    524, 558, 599; eye color, 155;
    fragile X syndrome, 282, 419;
    gender identity, 287-289;
    hemophilia, 396
Sexual identity, 411-413, 648,
    731-733, 759-761. *See also*
    Gender identity
Sexual orientation, 287
Sexual reproduction, 161; cell
    division, 200. *See also* Cell
    division; Meiosis; Sex
    chromosomes
Sharp, Phillip A., 776
Shaw, George Bernard, 262
Sheared DNA, 382
Sheldon, Peter, 651
Shelley, Mary, 351
Short interspersed sequences
    (SINES), 646-647, 664
Short-limb dwarfism, 249
Shotgun cloning, **691-692**
Shotgun sequencing, 386
Sibling mating, 463
Sickle-cell disease, 25, 96, 328,
    358, 363, 391, 467, 563, 673,
    692-696

Signal transduction, **696-699**
Silicon chips, 342
Simpson, George Gaylord, 269,
    776
Simpson, O. J., trial, 281
Sinclair, Andrew, 160
SINES. *See* Short interspersed
    sequences
Singer, Maxine, 776
Single-gene defects, 305, 400.
    See also Inborn errors of
    metabolism; Mendelian
    defects
Single-gene traits, 555-559
Single nucleotide
    polymorphisms (SNPs), 35,
    82, 428
siRNAs. *See* Small interfering
    RNAs
sis oncogene, 584
Sister chromatids, 199, 510
Site-directed mutagenesis,
    726
Skeletal dysplasias, 249
Skin cancer, 9, 112
Skin color (human), 658
Slamon, Dennis, 103
Slippage, 290
Small interfering RNAs
    (siRNAs), 44, 577
Small non-messenger RNAs
    (snmRNAs), 576
Small nuclear
    ribonucleoproteins
    (snRNPs), 576, 684
Small nuclear RNAs (snRNAs),
    43, 576, 680, 684
Small nucleolar RNAs
    (snoRNAs), 43
Small RNAs (sRNAs), 576
Small subunit ribosomal RNAs
    (ssu rRNAs), 45
Smallpox, 88, **700-704**
SMaRT gene therapy, 197
Smith, Hamilton O., 338, 351,
    667, 776
Smith, Harry, 409
Smith, Michael, 776
Smithies, Oliver, 481
Snell, George D., 776
snmRNAs. *See* Small non-
    messenger RNAs
snoRNAs. *See* Small nucleolar
    RNAs

SNPs. *See* Single nucleotide polymorphisms
snRNAs. *See* Small nuclear RNAs
snRNPs. *See* Small nuclear ribonucleoproteins
SNRPN gene, 624
Social Darwinism, 259
Society, 704
Sociobiology, 409, **704-708**
Sociobiology: The New Synthesis (Wilson), 704
Sociopathic behavior, 480
Solenoids, 141
Somatic cell nuclear transfer (SCNT), 714
Somatic cell therapy, 309, 343
Somatic cells, 738
Somatic embryos, 120
Somatic mutation/DNA damage theory of aging, 7
Somatic mutations, 562
Somatomedin, 433
Somatostatin, 337, 532
Sonenberg, Nahum, 640
Sonneborn, Tracy Morton, 776
SOS response, 226
Southern, Ed, 99
Southern blotting, 98-101, 665, 672
Soviet Union; biological weapons, 90; evolutionary theory, 487
Soybeans, 368
Spacers, 505
Spanish flu pandemic (1918), 253
Speciation, 272, 652, **708-710**
Specific-locus test, 565
Spemann, Hans, 543, 776
Spencer, Herbert, 259, 776
Sperling, Karl, 581
Sperm cells, 469
Spin columns, 222
Spina bifida, 324, 360, 573, 735
Spindle apparatus, 511
Spindle fibers, 127, 147
Spinocerebellar ataxia, 143
Spirilla, 54
Spirochetes, 54
Spliceosomes, 575, 679, 684
Splicing, 679
Split genes, 679. *See also* Introns
Spondyloepiphyseal dysplasia, 249

Spontaneous abortion, 24, 189, 402, 470, 479, 581
Spontaneous mutations, 565
Sporophytes, 512
src gene, 583
sRNAs. *See* Small RNAs
SRY gene, 143, 648
SS. *See* Schutzstaffel
ssu rRNAs. *See* Small subunit ribosomal RNAs
Stadler, Lewis, 743
Stahl, Franklin, 234, 374, 529, 550
Stanley, Wendell Meredith, 776
Staphylococcus bacteria, 62, 64
Start codons, 320
Stebbins, G. Ledyard, 269
Steffánsson, Kári, 447
Stein, William H., 776
Steitz, Joan, 680
Stem cells, 166, 168, 343, 346, 436, **710-715**; totipotency, 738
Sterilization laws, 262, 265, **715-717**
Stern, Curt, 157
Steroid hormones, 697, **717-720**
Stevens, Nettie Maria, 776
Sticky ends, 96, 116, 225, 668
Stop codons, 320
Strabismus, 10
Streptococcus bacteria, 41, 56, 238, 254
Stress hormones, 420
Strobell, Ella Church, 777
Structural genomics, 387
Stt7 gene, 522
Sturtevant, Alfred H., 144, 154, 164, 372, 525, 777
Sulfolobus, 47
Sullenger, Bruce, 689
Sullivan, Louis, 194
Sulston, John E., 205, 777
Super bacteria, **61-65**
Superfemales. *See* Metafemales; Multiple X syndrome
Superweeds, 349, 369
Suprachiasmatic nucleus (SCN), 83
Supreme Court, U.S., 716
Surrogates, 455
Survival of the fittest, 568
Sutton, Walter, 147, 152-158, 322, 372, 523, 777, 809
Svedberg units, 640

Swanson, Robert, 340
Swift, Michael, 103
Swine flu, 253, **720-723**
sxl gene, 588
Symbiosis, 77
Sympatric speciation, 708
Syndrome, 748
Syntenic genes, 491
Synthesis (S), 123
Synthetic antibodies, **723-725**
Synthetic genes, 532, **725-726**
Syphilis, 62

T-cell receptors, 452
T cells, 38, 52, 452, 482
T2 phage, 57, 239
Tammes, Jantine, 777
Tan Jiazhen (C. C. Tan), 777
Tandem repetitive DNA (TR-DNA), 664
Taq polymerase, 27
Target cells, 696
Target species; biopesticides, 93
Targeted gene inactivation, 481
Tatum, Edward, 56, 183, 372, 537, 587, 777
Tau protein, 19
Taussig, Helen Brooke, 735
Tautomerization, 131
Taxon, 547
Tay, Warren, 727
Tay-Sachs disease, 324, 360, 460, 628, 666, **727-728**; genetic testing, 365; screening, 358
TBX5 gene, 206
T-DNA. *See* Transferred DNA
Television depictions of genetics, **376-378**
Telocentric chromosomes, 148
Telomerase, 232, 729
Telomeres, 550, **728-731**; aging, 3; animal cloning, 31; chromosome structure, 151, 232; discovery, 743; length in clones, 32, 730; reverse transcriptase, 672
Telophase (mitosis), 123, 127, 199, 511
Temin, Howard, 116, 119, 130, 670, 777
Teratogens, 187, **733-736**
Teratology, 187
Terman, Lewis M., 86, 261
Termination, 642

Terminator, 678
Tertiary period, 652
Tertiary protein structure, 636
Testicular feminization
    syndrome, 288, **731-733**
Testosterone, 648, 718
Tetrads (chromatids), 512
Thalassemia, 142, 560. *See also*
    Beta-thalassemia
Thalidomide, **733-736**
The Institute for Genome
    Research (TIGR), 430
Theologis, Athanasios, 515
Theraputic cloning, 166
Thermal cyclers, 27
Thermophilic bacteria, 55, 342,
    489
Threshold traits, 609
Thrombate III, 167
Thrombospondin (TSP), 394
Thymine, 550
Thymocytes, 453
Thymus, 452
Ti plasmid, 93, 177, 330, 332,
    608
TIGR. *See* The Institute for
    Genome Research
Tissue culture. *See* Cell culture
Tissue plasminogen activator
    (tPA), 97, 167, 295, 329, 338,
    344, 740
Tobacco; genetically
    engineered, 95; heart
    disease, 395
Todd, Alexander Robertus, 777
Togaviruses, 756
Toluene, 340
Tomatoes, 95
Tomizawa, Jun-ichi, 680
Tonegawa, Susumu, 450, 777
Totipotency, 120, 710, **736-739**
Toxicogenomics, 421
tPA. *See* Tissue plasminogen
    activator
Transcription, 129, 227, 237,
    298, 314, 530, 549, 643, 677,
    681-686; bacteria, 291; and
    DNA repair, 224
Transcription factors, 295, 301,
    416, 456
Transduction, 57, 489, 531
Transfer RNA (tRNA), 129, 313,
    316, 320, 531, 551, 588, 639,
    677, 679, 725

Transformation, 56, 92, 116-117,
    327, 531, 739; *Chlamydomonas
    reinhardtii*, 520, 522; plants,
    122
Transformist theory of
    evolution, 486
Transgenes, 329, 351, 437, 739
Transgenic human growth
    hormone, 433
Transgenic organisms, 326, 328,
    739-742; animals, 343, 351,
    739; biopesticides, 92; crop
    plants, 332; crops, 348; frogs,
    542; genetic code, 318; GM
    foods, 367; mice, 481, 535;
    plants, 41, 94, 740; risks, 347.
    See also Genetically modified
    organisms
Transgenic pharming, 344
Transgenic proteins, 432
Transgenics, 533
Transition, 131
Translation, 129, 227, 237, 242,
    291, 313, 531, 549, 551, 638,
    640, 643, 677, 682;
    prevention of, 297. *See also*
    Protein synthesis
Translocation, 144, 151, 157,
    564, 623, 642
Transmissibility, 700
Transmission genetics,
    160-165
Trans-NIH *Xenopus* Initiative,
    544
Transplant rejection, 589
Transplantation, 589
Transplants, 588-591
Transposable elements, 553,
    564; SINES, 665. *See also*
    Transposons
Transposase, 742
Transposons, 61, 450, 489-490,
    531, 606, **742-746**. *See also*
    Jumping genes; Transposable
    elements
Transsexualism, 288
TR-DNA. *See* Tandem repetitive
    DNA
Trichoderma reese, 41
Trichromats, 180
Triplet repeat diseases, 232
Triploid organisms, 614
Trisomy, 244, 403, 579, 626
Trisomy 13. *See* Patau syndrome

Trisomy 18. *See* Edwards'
    syndrome
Trisomy 21. *See* Down syndrome
Tritanopes, 180
Triticale, 616
Triticum aestivum, 616
tRNA. *See* Transfer RNA
trp operon, 293, 529
Tryptophan, 293
Tschermak von Seysenegg,
    Erich, 153, 372
TSP. *See* Thrombospondin
Tsui, Lap-Chee, 159
Tuberculosis, 55, 62, 252
Tuberous sclerosis, 10
Tumor-suppressor genes, 102,
    109, 122, 424, 585, **746-748**;
    mouse studies, 535;
    retinoblastoma, 111
Tumors, 109
Turner, Henry H., 748
Turner syndrome, 250, 400, 403,
    469, 564, 579-580, **748-750**,
    759; detection, 25
Twin studies, **750-753**; diabetes,
    209; heart disease, 395;
    intelligence, 475; QTLs, 68;
    sexual orientation, 67, 288,
    419
Two-hybrid system, 541
Type I diabetes, 208
Type II diabetes, 209
Typological species concept,
    708
Tyrosine, 605

UBE3A gene, 624
Ugolini, François, 103
Ultrasound for prenatal testing,
    324, 360, 628
Ultraviolet (UV) radiation, 563
Unequal crossing over, 290
Uniparental disomy (UPD), 624
Unipotent cells, 736
United States v. Yee (1990-1991),
    280
Up-regulation, 43
UPD. *See* Uniparental disomy
Uracil, 132, 550, 640

Vaccination, 179. *See also*
    Immunization
Vaccines, 340; edible, 344
Vaccinia virus, 330, 703

Van der Veen, J. H., 515
Variable number tandem repeats (VNTRs), 35, 216, 664
Varmus, Harold E., 103, 777
Vectors. *See also* Cloning vectors; Viral vectors
Vectors, cloning, 116, 174-179, 304, 326-327, 332, 739
Vegetative petites, 275
Veleminsky, J., 515
Venter, Craig, 430
Vernalization of wheat, 487
Verticillium lecanii, 94
Very highly repetitive segments (VRS's), 729
Vibrio cholerae, 137-138
Violence Initiative, 194
Vir genes, 177
Viral genetics, **754-756**; RNA, 550
Viral vectors, 306, 327; for cystic fibrosis, 197. *See also* Cloning vectors; Vectors
Virginia Twin Registry, 752
Virions, 754; smallpox, 701
Viroids, **756-758**
Viruses, 57, 590; bacterial, 754; and cancer, 746; cancer-causing, 110; cell culture techniques, 119; emerging, 252-255; gene regulation, 301-304; influenza, 720-723; retroviruses, 329, 583, 680; smallpox, 700-704; *T2*, 239; as transposons, 744; vaccinia, 703. *See also* Bacteriophages; Retroviruses
Virusoids, **756-758**
VNTRs. *See* Variable number tandem repeats
Vogelstein, Bert, 225
Vogt, Peter, 103
Von Gierke's disease, 459
Von Behring, Emil Adolf, 214
Von Nägeli, Carl, 497
Vrba, Elisabeth, 271
VRS's. *See* Very highly repetitive segments

Waardenberg syndrome, 9
Waelsch, Salome Glueksohn, 777

Wakayama, Teruhiko, 730
Walking (chromosomes), **158-160**
Wallace, Alfred Russel, 269, 371, 568
Wallace, Douglas C., 508
Wasserman, David, 194
Waste management, 341
Waterston, Robert, 777
Watson, James, 129, 147, 228, 234, 239, 322, 336, 351, 374, 429, 639, 686, 725, 778
Weapons; biological, 88-92
Weber, Bruce H., 260
Weeds, 348
Weinberg, Robert Allan, 778
Weinberg, Wilhelm, 372, 389, 462, 617, 778
Weiner, J., 417
Weismann, August, 147, 201, 778
Weissman, Sherman, 158
Weldon, Walter Frank, 260
Werner's syndrome, 4
Wernicke-Korsakoff's syndrome, 12
Western blotting, 98-101
Wheat, 616
Whitehead Institute Center for Genome Research, 538
WHO. *See* World Health Organization
Wieschaus, Eric, 203, 417, 778
Wild-type genes, 562
Wilkins, Maurice, 228, 239, 322, 374, 778
Willi, Heinrich, 624
Williamson, Peter, 652
Wilmut, Ian, 171, 778
Wilson, Allan C., 508
Wilson, Edmund Beecher, 778
Wilson, Edward O., 409, 704-708
Witkin, Evelyn Maisel, 778
Woese, Carl R., 45, 778
Wolff, Kaspar Friedrich, 201
Wollman, Elie, 56
Woodcock, Chris, 141
Wooster, Richard, 103
World Health Organization (WHO), 254, 703, 722
Wright, Sewall, 263, 463, 610, 778

X chromosome. *See* Sex chromosomes; Sex-linked traits; X-linked traits
X chromosome inactivation, **759-761**
X inactivation center (XIC), 760
X-linked traits, 557, 599; albinism, 10
X radiation, 564
X-ray diffraction, 227
Xenopus laevis (frog), 542-544, 546, 576, 738
Xenotransplants, **761-764**
Xeroderma pigmentosa, 112
XIC. *See* X inactivation center
XIST gene, 760
XXX syndrome. *See* Metafemales; Multiple-X syndrome
XXY syndrome, 479-480. *See also* Klinefelter syndrome.
XYY syndrome, 155, 194, **764-766**

Y chromosome. *See* Sex chromosomes; Sex-linked traits; X-linked traits
YAC. *See* Yeast artificial chromosome
Yanofsky, Charles, 319, 779
Yanofsky, Martin, 516
Yaswen, Paul, 103
Yeast, 124, 677; genetically modified, 342. *See also* Saccharomyces cerevisiae
Yeast artificial chromosome (YAC), 178, 327
Yerkes, Robert M., 86, 261
Yersinia pestis (plague), 28, 608

Z-DNA. *See* Zigzag DNA
Zea mays, 493, 579. *See also* Maize
ZFN127 gene, 624
Zigzag DNA (Z-DNA), 242
Zinder, Norton, 57, 779
ZNF217 gene, 103
Zoological Philosophy (Lamarck), 488
Zuckerkandl, Émile, 547
Zygosity, 750
Zygotes, 455, 592, 750